Second Edition

Restoration of Contaminated Aquifers

Petroleum Hydrocarbons and Organic Compounds

Second Edition

Restoration of Contaminated Aquifers

Petroleum Hydrocarbons and Organic Compounds

Stephen M. Testa • Duane L. Winegardner

LEWIS PUBLISHERS
Boca Raton London New York Washington, D.C.

Library of Congress Cataloging-in-Publication Data

Testa, Stephen M.

Restoration of contaminated aquifers: petroleum hydrocarbons and organic compounds / by Stephen M. Testa, Duane L. Winegardner.—2nd ed.

p. cm.

Rev. ed. of: Restoration of petroleum-contaminated aquifers. c1991.

Includes bibliographical references and index.

ISBN 1-56670-320-4 (alk. paper)

1. Oil pollution of water. 2. Groundwater—Purification. I. Winegardner, Duane L. II. Testa, Stephen M. Restoration of petroleum-contaminated aquifers. III. Title.

TD427.P4 T47 2000
628.1′6833—dc21 00-025884
CIP

Direct all inquiries to CRC Press LLC, 2000 N.W. Corporate Blvd., Boca Raton, Florida 33431.

International Standard Book Number 1-56670-320-4
Library of Congress Card Number 00-025884
Printed in the United States of America 1 2 3 4 5 6 7 8 9 0
Printed on acid-free paper

Preface

This book acquaints the beginning and seasoned practicing environmental professional with the fundamentals of the more important aspects of restoring aquifers impacted by petroleum hydrocarbons and organic compounds. Throughout this book, an effort has been made to limit the theoretical discussion to that required for basic understanding and to emphasize practical applications. Some of the material will undoubtedly be familiar to some; however, readers will find material new to them or that will serve to supplement their current knowledge. Numerous case histories and examples have been added that veer from the idealized situation, and are intended to demonstrate technology and approaches that were deemed appropriate for a particular site or situation under existing regulations.

Although no two sites are exactly alike, the organization of this book is intended to guide the reader toward solutions to many of the situations typically encountered at a variety of sites, from small-scale retail service stations to larger-scale industrial complexes such as refineries. It may prove helpful to read the book in its entirety quickly, then to concentrate on those sections of greater concern or interest. The authors clearly recognize that other procedures or interpretative opinions may be well suited for solution of a particular problem or site situation. The discussion presented is representative of widely recognized and proven practices that reflect over 50 years of combined experience.

The objective of this book is to present the state-of-the-art knowledge on restoration of aquifers impacted by petroleum hydrocarbons and other organic compounds and their derivatives. This book is intended for use by managers, regulators, consultants, and, notably, the office and field remediation specialist and practitioner. For those readers interested in advanced discussions, there is a continually increasing and burgeoning amount of information in the literature, and much effort has gone into presenting and updating the reference sections following each chapter, which include listings of pertinent papers and books of a developmental, overview, and theoretical nature.

The book is organized into 13 chapters and four appendixes, and includes 174 figures and 59 tables. An introduction to the subject is provided in Chapter 1, which presents a brief overview of the use of petroleum hydrocarbons and organic compounds in society, the role of the environmental professional, and environmental challenges facing us as we enter the new millennium. The regulatory environment and framework, which provide the mechanism and incentive for which environmental issues are addressed, are discussed in Chapter 2. Thus, it is difficult to discuss subsurface restoration without some understanding of the subsurface environment. Discussion of fundamental geologic and hydrogeologic principles was added to this second edition as presented in Chapter 3. A general discussion of the geochemistry of hydrocarbons and forensic geochemical techniques is presented in Chapter 4. The subsurface behavior of nonaqueous-phase liquids (NAPLs) and fate-and-transport principles are provided in Chapter 5. Chapter 6 presents a detailed discussion of techniques used to characterize the occurrence, lateral and vertical distribution, and quantity of NAPLs in the subsurface. Particular attention is directed to solving problems associated with apparent vs. actual thickness measurements, effects of rising and falling water tables, and NAPL volume determination, and how these factors affect recoverability. Remedial strategies for the recovery of NAPLs from the subsurface and appropriate pumping technologies and strategies under certain subsurface conditions are presented in Chapter 7. With all subsurface groundwater remediation and restoration programs, large quantities of coproduced water may be generated. This water is in most cases contaminated and requires special handling. Coproduced water handling and management is discussed in Chapter 8. Chapter 9 is new to this second edition, and addresses dissolved constituents in groundwater and conventional remedial strategies for aquifer restoration, with an emphasis on pump-and-treat, air sparging, and bioremediation. Soil issues and *in situ* remedial strategies for dealing with impacted soil are discussed in Chapter 10. Economic considerations are discussed in Chapter 11. Actual case histories of various impacted sites where aquifer restoration in association with NAPL presence and recovery has been performed by the authors and close associates are presented in Chapter 12. It is hoped that these studies will serve as conceptual models and will aid the reader in selection of the most appropriate technology and equipment available. This chapter and the last have also been added to this second edition. Site closure, including crossing the bridge from impact to natural attenuation to "no further action," is discussed in Chapter 13, with the intent to guide the practitioner to successful project completion in an efficient and cost-effective manner.

The authors would like to thank the numerous clients over the years who placed their trust in us. Appreciation also goes out to Lydia Testa for editorial and word-processing assistance. We, however, take responsibility for any errors that may exist.

Stephen M. Testa
President
Testa Environmental Corporation
Mokelumne Hill, California

Duane L. Winegardner
Senior Hydrogeologist
Cardinal Engineering, Inc.
Oklahoma City, Oklahoma

The Authors

Stephen M. Testa is president and founder of Testa Environmental Corporation located in Mokelumne Hill, California. Mr. Testa received his B.S. and M.S. in geology from California State University at Northridge. For the past 20 years, Mr. Testa has worked as a consultant specializing in the areas of geology, hydrogeology, engineering and environmental geology, and hazardous waste management.

Mr. Testa has participated in hundreds of projects pertaining to subsurface hydrogeologic site characterization associated with nuclear hydroelectric power plants, hazardous waste disposal facilities, numerous refineries, tank farms, industrial and commercial complexes, and aboveground and underground storage tank sites. Maintaining overall management and technical responsibilities in engineering geology, hydrogeology, and hazardous waste–related projects, he has participated in hundreds of projects involving geologic and hydrogeologic characterization, environmental studies, water quality assessment, mine reclamation, remediation of soil and groundwater, hydrocarbon recovery, development of remedial strategies, and design and development of groundwater monitoring and aquifer remediation programs. In addition, Mr. Testa has provided technical assistance and litigation support and testimony for a number of nationwide law firms.

Mr. Testa is the author of over 90 technical papers and 11 books, and is professionally active. He is a member of numerous organizations including the American Institute of Professional Geologists, where he served as president in 1998. He is also a member of the American Association for the Advancement of Science, Association of Engineering Geologists, American Association of Petroleum Geologists (where he currently serves as editor-in-chief of its Division of Environmental Geosciences peer-review journal *Environmental Geosciences*), National Ground Water Association, Geological Society of America, California Groundwater Association, and Sigma Xi, among others. Mr. Testa also conducts numerous workshops on various environmental aspects of the subsurface presence of petroleum hydrocarbons, and taught hazardous waste management, geology, and mineralogy at California State University at Fullerton, and petroleum environmental engineering at the University of Southern California.

Duane L. Winegardner is a senior hydrogeologist for Cardinal Engineering, Inc., located in Oklahoma City, Oklahoma, and former president of American Environmental Consultants, Inc., located in Norman, Oklahoma. In this capacity, most of Mr. Winegardner's work involves investigation, evaluations, and engineering designs for remediation of contaminated soil and groundwater. During the past 3 years, Mr. Winegardner has also served as an adjunct professor at the Oklahoma City campus of Oklahoma State University, teaching classes and seminars related to soil and groundwater characterization and remediation.

Mr. Winegardner received his B.S. in geology and M.S. in geology and hydrology from the University of Toledo, Toledo, Ohio. Subsequently, he has achieved registration as a Professional Engineer (Civil), and is currently licensed in several states. For the past 28 years, his work has focused on applied technology in the construction and environmental industries. Mr. Winegardner's past employers have included Toledo Testing Laboratory, St. Johns River Water Management District (Florida), Environmental Science and Engineering, O. H. Materials Corporation, and Engineering Enterprises, Inc. For the past 8 years, his emphasis has focused on the cleanup of petroleum contamination in soil and groundwater. Many of his remediation designs have been based on new applications of existing technology as well as development of unique processes for specific geologic and chemical settings.

Mr. Winegardner is the author of numerous papers, and the author of *An Introduction to Soils for Environmental Professionals* published by CRC/Lewis Publishers, and coauthor of *Principles of Technical Consulting and Project Management* published by Lewis Publishers. He is a member of the National Ground Water Association and the Oklahoma Society of Environmental Professionals.

Dedication

To my wife Lydia
Stephen M. Testa

To my wife Jane
Duane L. Winegardner

Table of Contents

Chapter 12 LNAPL Recovery Case Histories

Chapter 13 Site Closure

Appendixes

1 Introduction

"Water is truly a mineral resource, dependent not merely upon the degree of usefulness, but upon the scarcity."

1.1 PETROLEUM HYDROCARBON USE IN SOCIETY

Petroleum hydrocarbons have been known and used by civilizations for thousands of years. Several oil-producing regions throughout the world, such as in the Middle East, have been known for centuries. The industry of providing petroleum as fuel for lamps was described by Marco Polo in 1291. Early seafarers used the asphalt associated with the world's natural oil seeps as a means to caulk and waterproof their sailing vessels. Since the development of the internal combustion engine, petroleum has become the energy of choice for transportation and many other applications, including use as an ingredient for many materials commonly used today.

It becomes quite ironic when one views petroleum from an evolutionary perspective. With the advent of the industrial revolution at the turn of the 20th century, a demand developed for an inexpensive fuel for lighting, which further evolved with the introduction of the car into a demand for cheap fuel for transportation. The use of petroleum was viewed as an effective solution to air pollution caused by the use and burning of coal (Figure 1.1). Oil has since been the key resource throughout the 20th century, and there is every sign that it will continue to be so as the new millennium begins.

The petroleum industry today is the largest single industry both in the United States and throughout the world. This industry employs the largest number of people in a variety of ways, and contributes the largest number of dollars to taxes at various governmental levels. It is thus understandable that over the past several decades, much attention has been focused upon the environmental impacts of petroleum in the environment. This attention is due in part to the increasing number of media reports on the release and impact of oil and petroleum hydrocarbon products into the environment from aboveground storage tanks (Figure 1.2) and underground storage tanks (USTs; Figure 1.3), reservoirs, pipelines, land-based and off-shore oil spills, air pollution, environmental terrorism, and an overall increasing concern regarding the quality of the nation's soil, water, and air.

In the United States, contamination of water wells and streams by petroleum hydrocarbons extends back to the turn of the century. In Marion, Indiana, local streams and the single source of groundwater involving 200 to 300 surface and rock wells were found to be contaminated by adjacent petroleum production activities.

FIGURE 1.1 In Glasgow, Scotland, and other cities throughout the British Isles, evidence of former air pollution from the burning of coal for heat in the late 19th century can still be observed on many stone buildings.

FIGURE 1.2 Breached valve from an aboveground storage tank resulted in the release of hydrocarbon product into a bermed but unlined enclosure (A), and leakage through a corroded unlined tank bottom (B).

FIGURE 1.3 Excavation of a corroded UST.

Significant releases of petroleum hydrocarbons from unlined surface impoundments in oil fields have also been reported as far back as the early 1900s. One unlined surface oil reservoir located in the Kern River field, southern California, had a reported fluid loss on the order of 500,000 barrels. Excavated pits showed oil penetration to depths exceeding 20 ft. Another loss of 1 million barrels over a period of 6 years occurred from another unlined reservoir in the same field, although some of this loss was through evaporation.

Over the past two decades, the recognition of petroleum hydrocarbon–impacted groundwater has become widespread throughout the United States and internationally. Much subsurface attention has been focused upon uncontrolled releases of petroleum product stored in USTs. These types of releases in numerous cases have resulted in adverse impacts on groundwater quality and overall groundwater resources. Uncontrolled releases from USTs, although ubiquitous, were generally localized in their vertical and lateral extent. In the late 1980s, large-scale regional impact of petroleum products and their derivatives on groundwater resources became widely recognized in densely urbanized and industrialized areas. This reflected the uncontrolled and virtually unnoticed loss of hundreds of thousands of barrels of product being released into the subsurface, notably from refineries, bulk-liquid aboveground storage terminals and pipeline corridors, and densely concentrated industrial areas (Figure 1.4). As a result, a large level of effort has been invested in the restoration of aquifers impacted by spilled or leaked petroleum products and their derivatives.

The impact from petroleum hydrocarbons and their derivatives in the environment can take many forms. Petroleum hydrocarbons in the form of fuels (i.e., gasoline, diesel, jet fuel, etc.) are very common subsurface contaminants. Their release into the environment is not necessarily well understood by the public at large.

FIGURE 1.4 Areal photographs showing a refinery in southern California in the 1920s (a), and the same area with urban encroachment in the 1960s (b).

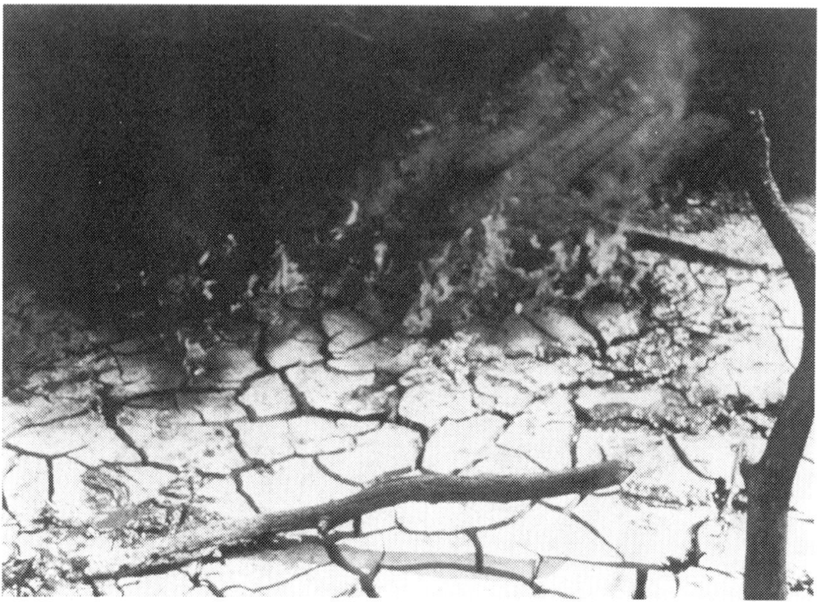

FIGURE 1.5 Hydrocarbon vapors emitting from an LNAPL pool overlying a shallow water table at a refinery site.

Environmental issues associated with the subsurface release of petroleum hydrocarbons and other organics fall into four areas: (1) vapors (Figure 1.5), (2) impacted soils, (3) the presence of nonaqueous phase liquids (NAPLs), and (4) dissolved constituents (i.e., benzene, toluene, ethylbenzene, and xylenes (BTEX), and other components) in groundwater.

The subsurface presence of hydrocarbons, notably, the potential for NAPLs and dissolved constituents, is detected by drilling borings and subsequently installing monitoring wells. Evaluation of the magnitude of the problem and the ultimate remediation strategy depends initially upon the accurate detection of NAPLs and volume evaluation thereof. Remediation strategies are then developed for the delineation and recovery of recoverable NAPLs, which can serve as a continued source for groundwater contamination until removed. In addition to NAPLs, dissolved hydrocarbon constituents in groundwater and impacted soils must also be addressed.

The environmental concern regarding petroleum hydrocarbons has recently taken a new turn, and not for the better. The addition of oxygenated compounds in gasoline was mandated by the Clean Air Act Amendments in 1990. Methyl tertiary butyl ether (MTBE), an oxygenated compound, was added to gasoline to enhance gasoline combustion by increasing the octane ratio of gasoline, thus combustion efficiency, and to improve air quality by lowering atmospheric ozone and carbon monoxide. In 1992, the EPA mandated that gasoline was to comprise at least 2.7% oxygen by weight, without any significant toxicity testing performed prior to its use. MTBE quickly became a frequently encountered groundwater contaminant as a result of leakage of gasoline from USTs.

MTBE behavior in the subsurface environment is a reflection of its chemical properties. MTBE is characterized by high solubility and mobility in the subsurface environment, with a high vapor pressure in its pure phase. In gasoline its vapor pressure is relatively high, and its Henry's law constant, the ratio of a partial pressure of a compound in air to the concentration of that water at a given temperature, is low. A low Henry's law constant indicates that the concentration of MTBE in the soil vapor phase would be low. The result is low extractability from water, which explains to some degree why conventional remedial technologies such as vapor extraction have been ineffective. Recent regional studies have also shown that its presence may depend more on high solubility and mobility, with less influence from hydraulic gradients, hydrogeology, and land use. With a low affinity for organic carbon, and because its primary attenuation mechanism is dispersion, retardation is minimal coincident with increasing BTEX concentrations.

MTBE presence in groundwater has been reported in thousands of wells nationwide. MTBE has been shown to easily migrate over 1000 ft hydraulically downgradient from its source and has been reported to travel as far as 3 miles from its source, with reported widths up to 250 ft. This is in contrast to dissolved BTEX plumes whose maximum downgradient length is less than 300 ft, with vertical migration through the upper portion of the water column on the order of 20 to 30 ft, at best. What this clearly implies is that in many, if not most, cases the dissolved MTBE plume extends off site from its source with the potential to adversely impact very large areas and volumes of groundwater. MTBE has also been reported to migrate vertically downward through the saturated zone to depths ranging from 10 to 80 ft below the ground surface. In addition, its tendency to biodegrade is slight as evidenced by the large size of these plumes.

1.2 THE USE OF ORGANIC COMPOUNDS IN SOCIETY

Unlike petroleum hydrocarbons, organic compounds in general followed a different evolutionary path. Chlorinated solvents are a common group of organic compounds, and are also the most frequently encountered contaminant in groundwater. Common industrial chemicals that are characterized as chlorinated solvents include trichloroethene (TCE), 1,1,1-trichloroethane (TCA), tetrachloroethene (PCE) or perchloroethylene, chlorofluorocarbon (Freon)-113 (i.e., 1,1,2-trichloroethane or 1,2,2-trifluoroethane), and methylene chloride. In 1997, the EPA reported the presence of TCE and PCE in 852 of 945 groundwater supply systems throughout the United States and in 771 of 1420 Superfund sites.

Worldwide production of chlorinated solvents commenced shortly after World War II, and gradually increased in use, primarily for military purposes, through the 1950s and 1960s. Since then, these compounds have been used extensively in many industries: aerospace, dry-cleaning, semiconductor, photography, and pharmaceutical, among others. TCE was formulated by Fisher in 1864 and used worldwide for about 50 years. During this period, TCE, along with TCA, was used as a common metal-surface degreasing agent. TCA was formulated in 1840, and has also been used extensively as a dry-cleaning agent. Worldwide use of TCA ceased in 1996 because of its adverse atmospheric effect on ozone depletion. PCE was formulated by Faraday in 1821, and has also been used extensively by the dry-cleaning industry since the late 1930s. As of 1992, about 28,000 out of 34,000 dry-cleaning businesses, or 82% of them, in the United States used PCE. About 25,000 of these commercial dry cleaners use about 120,000 metric tons annually. Freon-113 is a common electronic parts defluxing agent. Methylene chloride (or dichloromethane) is an active ingredient in many paint removers, thinners, strippers, etc. Introduced over 60 years ago, methylene chloride was used to replace other, more flammable solvents.

Large regional-scale dissolved contaminant plumes have been reported in many parts of the country. In California, many of the alluvial basin aquifers (i.e., major groundwater basins) were found to be impacted by chlorinated solvents during the period between 1979 to 1981. Beneath one valley a dissolved TCE plume incorporated the upper 50 ft of the saturated zone of a water table aquifer, was over 3 miles wide, and extended hydraulically downgradient over 14 miles.

Not all subsurface occurrences of contaminated groundwater are the direct result of accidental leaks and spills. Society in the mid-20th century showed an inability to foresee how newly introduced chemicals and technologies would significantly affect groundwater quality, regardless of the fact that acceptable waste disposal practices were being followed at the time. The reasons for this oversight by the regulatory and technical community are varied and insightful. A combination of interrelated factors contributed to this situation, including:

- The perception that chlorinated solvents and related compounds prior to the 1970s were potential occupational hazards rather than potential groundwater contaminants;
- The lack of understanding of the fate and transport of petroleum hydrocarbons and organic compounds in the subsurface, which did not appear in print until the mid-1980s with the works of Schwille and others (1984, 1985, and 1988), and not in book form until 1990 (Schwille, 1990);
- The absence of a mechanism and accompanying monitoring network to sample groundwater hydraulically downgradient from potential source areas; and
- The inability to analyze groundwater samples, especially to the parts per billion level, which was not developed and implemented until the 1970s.

1.3 THE ROLE OF THE PROFESSIONAL ENVIRONMENTALIST

Never before has the interest in a healthy environment been such a strong stimulant to the development of a particular branch of scientific practice, as has been the case with groundwater science. The occurrence of subsurface water was traditionally studied to determine the most efficient means of recovering and managing groundwater resources for a source of potable or irrigation water, and for removal of water from mines and construction sites. The traditional approach to aquifer evaluation has recognized the complexity of subsurface materials and conditions as a multidisciplinary science. Contributions to its development have come from many disciplines including geology, civil engineering, agricultural science, chemistry, and a variety of associated fields of interest.

As aquifer restoration began to be recognized as an important aspect of environmental management in the 1980s, the diversity of scientific involvement increased even further. The change from a descriptive science (identifying existing subsurface characteristics, and providing the necessary engineering), to a prescriptive science (modifying subsurface conditions in a controlled manner) for the purpose of aquifer restoration provided a better understanding of all the operating parameters; physical, chemical, and biological processes are all interactive in a dynamic setting.

At specific sites where the contaminants are petroleum hydrocarbon–related products, the spectrum of necessary professional expertise has been greatly expanded. Subsurface recovery and production of multiphase petroleum hydrocarbons has been the primary goal of petroleum engineers. Sorption of organic materials on soil particles has been principally studied by agronomists. Biodegradation of organic chemicals fell into the realm of the microbiologist and wastewater engineers. The quantitative analysis of chemicals and an understanding of their respective reactions and by-products are within the purview of the chemist. This incomplete listing could easily be expanded to include many other scientific, engineering, or health science disciplines.

The motivating force for the professional development and practice of effective and efficient remediation of subsurface materials has been public opinion, which has been expressed in the form of regulations. Almost every level of governing authority has enacted some form of environmental control. Interpretation of these regulations falls, in part, within the bailiwick of the attorney, and, therefore, this group of professionals joins the scientific community as an important member of the remediation team. As a developing technology, innovative new procedures continue to be developed or borrowed from other technical fields. What is important is that the remediation effort is reasonable and ultimately successful.

1.3.1 The Role of the Environmental Geologist

Environmental geology as defined in the American Geological Institute's *Glossary of Geology* is as follows:

> A specialty of geology concerned with earth processes, earth resources, and engineering properties of earth materials and relevant to (1) the protection of human health and natural ecosystems from adverse biochemical and/or geochemical reactions to naturally occurring chemicals or to chemical compounds released into the environment by human activities and (2) the protection of life, safety, and well-being of humans from natural processes, such as floods, hurricanes, earthquakes and landslides, through land-use planning.

Geology incorporates such subdisciplines as hydrogeology, hydrology, structural geology, stratigraphy, geophysics, geochemistry (including aqueous geochemistry), among others. The role of the geologist in dealing with petroleum impacts to the environment is significant. The geologist is commonly requested to specify the type of subsurface information required for a certain project, and evaluate the best means to obtain, analyze, and interpret the data. The geologist through training provides a thorough understanding and knowledge of the subsurface environment and its various characteristics, which are essential to assessing the likely hazards posed by a particular contaminant or waste, and its behavior and fate in the subsurface environment over time. In addition, the geologist also participates in the regulatory and legal process, often being required to translate and interpret information for regulators, attorneys, interested parties, and the general public.

The geologist also participates in subsequent mitigation activities providing input into the design, implementation, and monitoring of a remediation strategy, which is a significant role. Projected costs over the years for assessing the need for and the conduct of remediation have significantly increased. This increase has commonly been attributed to the increase in the number of sites being discovered and associated treatment costs. However, the perceived increase also reflects premature and poorly thought-out design for a remediation strategy before the site has been properly characterized and/or lack of proper understanding of the subsurface environment. In many cases, there is a strong tendency toward emphasizing site remediation at the expense of subsurface characterization. This syndrome of "analysis paralysis" can easily result in diminished quality and effectiveness of the remediation system, escalated costs, and increased liability exposure.

Today's environmental geologists are involved in property transactions, the subsurface characterization of soil and surface and groundwater conditions, assessment of such conditions as it pertains to soil and water quality, determination of the significance of risk associated with a particular impact, and development of technologies and strategies for remediation of soil and groundwater. Geologists also involve themselves in litigation assistance and support, regulatory compliance and guidance, and related public policy issues. Since every segment of society generates waste and disturbs the land to some degree, the environmental geologist today is more visible to the public at large, and is involved in a multitude of activities that relate directly to waste and environmental issues.

1.3.2 The Role of the Environmental Engineer

Environmental engineers like environmental geologists are also trained in the sciences and mathematics, and both disciplines apply scientific principles for the common good of society. But while scientists in general are trained in interpreting scientific phenomena, engineers are trained with a focus on practical application. The main difference between scientists and engineers is, thus, the general approach.

One major role of the environmental engineer is to ensure that technology is designed and operated in a manner compatible with the environment. The environmental engineer is accustomed to solving problems systematically and economically, within the context of a physical system. As with the geological sciences, the traditional world of the engineer has also changed to coincide with society's current needs. Today, the engineer involved in environmental science and professional practice must determine specific applications of scientific knowledge for practical purposes. The engineer must be environmentally knowledgeable in the fate of treated materials and development of upstream processes and products that minimize or eliminate the quantity and toxicity of waste materials. Today's engineer must also maintain several skills including the ability to write and manipulate equations that form the basis of design and prediction, and become more knowledgeable about the overall environment. In the context of this book, the environment is the subsurface.

Typically, application of science involves prediction of function such as determining at what rate a well must be pumped to create a suitable capture zone. What period of time will be required to biodegrade a mass of contaminant within a plume? How much activated carbon will be required to treat the discharged vapor? What will be the cost of electricity to power the remediation system? Engineers are more likely capable of designing a balanced remediation system that has flow rates matched to reaction times or water–air contact rates. Tank sizes, power consumption, and similar rate–time-related calculations also fall within the specialty of the engineer.

In practice, there is rarely a clear division between scientists and engineers, or environmental geologists and environmental engineers, and their professional paths frequently cross. Geologists are familiar with rates of groundwater flow, the principles of structural settlement, general low-temperature geochemistry, and similar disciplines. A civil or environmental engineer with a geotechnical specialty is also capable of calculating groundwater flow, while applying the principles of ion exchange and attenuation to remedial design. Microbiologists use the same calculation procedures as wastewater treatment engineers. In fact, some of the most successful biodegradation projects were designed based on microbial studies performed at health science centers. What is important to keep in mind is that a successful project team comprises individuals who have the balance of education and experience required for the task at hand.

1.4 DEFINING THE ENVIRONMENTAL CHALLENGE

Once an environmental impact has occurred, the significance of the impact can become very difficult to evaluate until years later. Developing an efficient and cost-effective remediation strategy to reduce the impact in a way that satisfies all the parties who share an interest in the outcome can be very challenging. Many of the environmental problems encountered over the past few decades were inherited, in the sense that they reflected antiquated facilities and infrastructures and involved operational and waste disposal practices that are no longer in use or considered acceptable. Understanding of the relationship between groundwater protection and proper waste disposal in an industrialized society goes back at least to the 1940s. Since the publication in 1962 of Rachel Carson's *Silent Spring*, society has steadily progressed toward environmental awareness. Over the past couple of decades, understanding of the subsurface behavior and effect of petroleum hydrocarbons and organic compounds, of the health risks associated with these constituents, and of ways to mitigate in an efficient and cost-effective manner has increased dramatically. The U.S. Congress, through environmental legislation, and industry have allotted significant dollars to this purpose. Furthermore, significant technological advances

FIGURE 1.6 Al-Rawdhatayn oil field showing the impact from crude oil on the desert environment during the Gulf War in March, 1991.

have been made, and new operational procedures implemented, to provide safe and user-friendly products for many years to come.

A variety of recovery and aquifer restoration approaches and strategies are available to the practitioner, and the prudent selection of the right approach or combination of approaches requires a clear understanding of the regulatory and project objectives and a comprehensive assessment of the geologic and hydrogeologic environment, the subsurface occurrence and behavior of hydrocarbons and their various forms, and site-specific factors such as wastewater-handling capabilities and options, time constraints, and numerous other factors. Monitoring the effectiveness and efficiency of such recovery and restoration operations allows for cost-effective system modifications and expansion, and enhanced performance.

The environmental impact as a result of our society's dependence on oil and petroleum products can range from no more than a localized issue, as with the release of gasoline from a UST at one's retail service station, to more regional impacts as a result of large dissolved contaminant plumes, oil spills, or poor air quality as experienced in many of the urbanized and industrialized areas of the country. During the Gulf War, the global population watched as 700 of Kuwait's 1500 oil wells released oil into the environment, with about 600 of them aflame. When this act of environmental terrorism was over, an estimated 11 million barrels of oil either burned or spilled each day, resulting in massive air pollution and the formation of extensive pools of oil on the land surface (Figure 1.6) and extensive oil slicks that migrated into the Persian Gulf.

While society remains dependent upon petroleum both for energy and for economic stability, petroleum hydrocarbons and organic compounds in general continue to be chief environmental concerns in regard to overall air and surface and water quality. The past two decades have clearly demonstrated that a significant number of sites will require, or continue to require, some degree of assessment, monitoring, and remediation. Eventual cleanup of these sites to an acceptable level is an important factor in the protection and preservation of groundwater resources. It is thus only natural to be concerned over the environmental impact from its use. No one is especially interested in an inhospitable environment. Regulations meant to maintain and improve the environment can be financially burdensome to the private-sector to the point that they undermine economic incentives. In these times, it seems that the public-at-large demands high environmental quality, as exemplified by its willingness to pay for environmentally sound options through the purchase of low-energy, recycled, biodegradable products.

Future trends as the new millennium begins will encompass many significant changes for the environmental industry. As remedial strategies become more efficient, they will focus more on containment, removal, and risk assessment and abatement. Site closure will be based more on risk-based corrective action, and complex challenges in the remediation and redevelopment of industrial and significantly impacted sites (brownfields) will be faced. Aquifer restoration will certainly advance beyond simple hydraulic containment with pump-and-treat–type strategies, with more sophisticated techniques, or combinations of techniques, being used. And in regard to modeling, more-sophisticated transport models that incorporate the various chemical and biological reactions that take place in the subsurface will be developed.

As we increase our knowledge and hope to learn from our past errors, it becomes important that we as a professional group servicing society continue our efforts to (1) minimize and reduce the uncontrolled and accidental release of petroleum hydrocarbons and organics into the environment, (2) improve our understanding of the overall behavior of these compounds in the subsurface, and the health risks associated with their presence, and (3) continue to develop sound strategies for the recycling, remediation, and restoration of impacted soil, water, and air.

REFERENCES

Bowie, C. P., 1918, Oil-Storage Tanks and Reservoirs with a Brief Discussion of Losses of Oil in Storage and Methods of Prevention: U.S. Bureau of Mines Bulletin No. 155, Petroleum Technology Report No. 41, 76 pp.

Colton, C. E., 1991, A Historical Perspective on Industrial Wastes and Groundwater Contamination: *Geographical Review*, Vol. 81, No. 2, pp. 215–228.

Colton, C. E. and Skinner, P. N., 1996, *The Road to Love Canal: Managing Industrial Waste before EPA*: University of Texas Press, Austin, 217 pp.

Harmon, B., 1941, Contamination of Ground-Water Resources: *Civil Engineering*, Vol. 11, No. 6, pp. 345–347.

Jackson, R. E., 1999, Anticipating Ground-Water Contamination by New Technologies and Chemicals: The Case of Chlorinated Solvents in California: *Environmental & Engineering Geoscience*, Fall, Vol. V, No. 3, pp. 331–338.

Kerfoot, H. B. and Rong, Y., 1998, Methyl Tertiary Butyl Ether Contamination of Soil and Groundwater: *Environmental Geosciences*, Vol. 5, No. 2, pp. 79–86.

Mackay, D., 1995, Environmental Engineering: A Profession in Transition: In *50th Purdue Industrial Waste Conference Proceedings*, Ann Arbor Press, Chelsea, MI, pp. 1–6.

Morrison, R. D., 2000, *Environmental Forensics — Principles and Applications*: CRC Press, Boca Raton, FL, 351 pp.

Odencrantz, J. E., 1998, Implications of MTBE for Intrinsic Remediation of Underground Fuel Tank Sites: *Remediation*, Vol. 8, No. 3, pp. 7–16.

Pankow, J. F., Feenstra, S., Cherry, J. A., and Ryan, M. C., 1996, Dense Chlorinated Solvents in Groundwater: Background and History of the Problem: In *Dense Chlorinated Solvents and Other DNAPLs in Groundwater* (edited by J. F. Pankow and J. A. Cherry), Waterloo Press, Portland, OR, pp. 1–52.

Pickett, A., 1947, Protection of Underground Water from Sewage and Industrial Wastes: *Sewage Works Journal*, Vol. 19, No. 3, pp. 64–72.

Rong, Y., 1999, Groundwater Data Analysis for Methyl Tertiary Butyl Ether: *Environmental Geosciences*, Vol. 6, No. 2, pp. 76–81.

Sackett, R. L. and Bowman, I., 1905, *Disposal of Strawboard and Oil-Well Wastes*: U.S. Geological Survey Water-Supply and Irrigation Paper, No. 113, 52 pp.

Schneider, W. A., 1948, Industrial Waste Disposal in Los Angeles City: *Water & Sewage Works*, January, pp. 37–39.

Schwille, F., 1985, Migration of Organic Fluids Immiscible in Water in the Unsaturated and Saturated Zones: In *Proceedings of Second Canadian/American Conference on Hydrogeology* (edited by B. Hitchon and M. Trudell), Banff, Alberta.

Schwille, F., Bertsch, W., Linke, R., Reif, W., and Zauter, S., 1988, *Dense Chlorinated Solvents in Porous and Fractured Media: Model Experiments*: Translated from the original German report of 1984 by J. F. Pankow, Lewis Publishers, Chelsea, MI.

Sellers, C., 1994, Factory as Environment: Industrial Hygiene, Professional Collaboration and the Modern Sciences of Pollution: *Environmental History Review*, Spring, Vol. 18, pp. 55–83.

Wolf, K., 1992, Case Study — Pollution Prevention in the Dry Cleaning Industry: A Small Business Challenge for the 1990s: *Pollution Prevention Review*, Summer, pp. 311–330.

2 Regulatory Framework

"Despite our good intentions, it is the regulations that drive the environmental industry."

2.1 INTRODUCTION

Historically, some of the earliest environmental regulations involving crude and petroleum products, such as the Migratory Bird Treaty Act (MBTA) and, later, the Endangered Species Act (ESA) and the Federal Oil and Gas Royalty Management Act (FOGRMA), were established to protect wildlife in private and industrial environments. Since the mid-1950s, there has been the Outer Continental Shelf Lands Act (OCSLA) to protect submerged lands adjacent to the United States; the Clean Air Act (CAA) to protect air quality; and the Clean Water Act (CWA), Safe Drinking Water Act (SDWA), Spill Prevention Control and Countermeasures (SPCC), Oil Pollution Act (OPA), and Federal Water Pollution Control Act (FWPCA) to protect water quality. There is also the Resource, Conservation and Recovery Act (RCRA) to manage, control, and reuse and recycle hazardous wastes; the Toxic Substance Control Act (TSCA) to regulate handling and use of chemical substances; and the Comprehensive Environmental Response, Compensation, and Liability Act (CERCLA) and Superfund Amendments and Reauthorization Act (SARA) to assure that any releases of hazardous waste to the environment are cleaned up. There is the Pipeline Safety Act (PSA) and the Act to Prevent Pollution from Ships (APPS), and other regulations under the Department of Transportation (DOT) to provide safe transportation of hazardous liquids and materials, and the Hazardous Communication Standard (HAZCOM) and Hazardous Waste Operations and Emergency Response Standard (HAZWOPER) to protect workers in industrial environments.

There are also numerous state and local regulations and ordinances that restrict certain activities in sensitive areas. The federal government in some cases has turned over certain regulatory responsibilities and authority to the states, when the opportunity has presented itself, and in many cases state regulations are or can be more stringent than those imposed at the federal level. The states are, however, subject to federal intervention in the case where the state does not effectively enforce the statutes. Some of these more important federal regulations that apply to the prudent control and management of oil and petroleum hydrocarbons, and organic products in general, as applicable to their exploration, development, production, transportation, refining, marketing, and use, are presented in chronological order in Table 2.1.

TABLE 2.1
Chronological Summary of Major Regulations Pertaining to Petroleum Hydrocarbons and Other Organics

Regulation	Year Enacted	Description
Migratory Bird Treaty Act (MBTA)	1918, 1936, 1969, 1972, and 1976	Authorizes fines and imprisonment for operators who allow migratory birds to become injured in pits or open-topped tanks that contain oil, oil products, caustic materials, and certain contaminants
Outer Continental Shelf Lands Act (OCSLA)	1953	Authorizes Secretary of the Interior to grant mineral leases and to regulate oil and gas activities on outer continental shelf lands by maintaining an oil and gas leasing program
National Environmental Policy Act (NEPA)	1969	Provides basic national charter for the protection of the environment by providing a mechanism to interact with government on proposed actions that may affect the environment
Clean Air Act (CAA)	1970	Addresses airborne pollution that may be potentially hazardous to public health or natural resources by setting air quality standards and regulating air emissions
Endangered Species Act (ESA)	1973	Conserves threatened and endangered species, and the ecosystems on which those species depend, by maintaining guidelines for placement of wildlife and plant species on a list, preventing removal of the species and habitat, and providing a mechanism to ensure federal actions will not impair or jeopardize protected species and their habitats
Spill Prevention, Control, and Countermeasures (SPCC)	1974	Encourages pollution prevention and cleanup of waters by requiring plans from onshore facilities that can potentially discharge oil and other pollutants in harmful quantities into or on U.S. navigable waters
Safe Drinking Water Act (SDWA)	1974	Protects underground drinking water sources from toxic contamination by regulating drinking water systems and injection wells by requiring testing of water wells, and subsequent cleanup if necessary
Resource Conservation and Recovery Act (RCRA)	1976	Provides cradle-to-grave management of hazardous waste including regulation of hazardous waste generators, transporters, and treatment, storage, and disposal facilities by tracking hazardous waste from its point of generation to its ultimate disposal; includes regulations for above- and underground storage tanks
Clean Water Act (CWA)	1977	Protects surface and groundwater quality to maintain "beneficial uses" of water by regulating the discharge of toxic and nontoxic pollutants into surface waters of the United States

Toxic Substances Control Act (TSCA)	1977	Regulates the manufacture, processing, distribution, use, and disposal of new and existing chemical substances by requiring a chemical inventory regulating certain chemicals considered to present an unreasonable risk to human health and the environment
Comprehensive Environmental Response, Compensation, and Liability Act (CERCLA)	1980	Provides government authority to regulate location, assessment, and cleanup of contaminated (Superfund) sites, to provide emergency response, to require reporting of releases of hazardous chemicals, and to identify liability and pursue financial reimbursement
Title III of the Superfund Amendments and Reauthorization Act (SARA)	1986	Requires transmittal of information from users on specific chemical hazards for federal and local governments to use to cooperate in providing emergency preparedness
Federal Water Pollution Control Act Amendments (FWPCA)	1987	Prohibits discharge of oil or hazardous substances from any vessel, from any onshore or offshore facility, into or upon the navigable waters of the United States, adjoining shorelines, or the waters of the contiguous zone; also prohibits a discharge that would cause a visible sheen upon the water or adjoining shorelines or cause a sludge or emulsion to be deposited beneath the surface of the water
Pipeline Safety Act (PSA)	1987	Provides for safe transportation of hazardous liquids
Hazard Communication Standard (HAZCOM)	1989	Guarantees employees the right to know about chemical hazards on the job, and how to protect themselves from those hazards, by requiring employers to develop a system to inform and train their employees in the handling of chemical hazards in the workplace
Hazardous Waste Operations and Emergency Response Standard (HAZWOPER)	1990	Requires employers to protect the health and safety of those workers involved in emergency response–related activities by requiring the employer to develop an emergency response plan
Oil Pollution Act (OPA)	1990	Prevents discharges of oil into federal waters from vessels and facilities, and ensures that in such event owners and operators will have the resources to clean them up, by requiring response plans to contain and recover from even a worst-case spill
Department of Transportation (DOT)		Provides for safe transportation of hazardous materials and wastes
California Regional Water Quality Control Board (CRWQCB)	1985	Requires characterization of subsurface hydrogeologic conditions, delineation and chemical characterization of LNAPL product pools, recovery of LNAPL product and overall aquifer restoration, and soil remediation
Act to Prevent Pollution from Ships (APPS)		Regulates the discharge of oil, noxious liquid substances, and garbage generated during the normal operations of vessels, and implements the international treaty on prevention of pollution by ships, known as MARPOL, and prohibits ships from discharging plastic wastes

There now exists a myriad of federal, state, and local legislation and regulations that affect just about every part of our lives, as well as affecting those industries and entities that discover, recover and produce, transport, market, and use petroleum products. From a regulatory perspective, the environmental emphasis was initially directed toward the larger industries and operations that generate large volumes of pollution. These included mainly businesses and industrial companies and areas that were easily identifiable, resulting in standards and limits being set forth by government and industry. These actions were manifested in permit requirements for certain operations, with tracking and reporting requirements, and resulting fines and penalties should companies fail to comply. More recent trends have been toward more regulations that affect not only the actions and operations of industry, but also the actions of individuals. Smog controls on vehicles and requirements on the handling of certain materials like used oil are recent examples of this trend. The ever-changing and evolving regulations over the past few decades continue to undergo significant expansion in scope, economic impact, and enforcement provision and consequences — a trend anticipated only to increase in the future.

Some regulatory programs are designed to detect and correct problems (i.e., detection and compliance monitoring), while others are aimed at groundwater protection and cleanup. Much regulatory attention has been aimed at the ubiquitous occurrence of accidental leaks and spills of hydrocarbon product from pipelines, aboveground storage tanks (ASTs) and underground storage tanks (USTs), transport vehicles, and operational-related activities. Some of these tightly and rigorously regulated areas, as promulgated under the Environmental Protection Agency (EPA), include the UST program, UIC program under the SDWA (Title 40 Code of Federal Regulations, or CFR, Part 144), and the National Pollutant Discharge Elimination System (NPDES) under the CWA (40 CFR Part 122). Presented in this chapter is discussion of agency responsibilities, and those federal, state, and other programs pertinent to some of the more important regulations relating to groundwater impacted by petroleum and organic compounds and its subsequent remediation.

2.2 AGENCY RESPONSIBILITIES

At the federal level, the primary agencies responsible for addressing groundwater and waste-related activities in order of importance are the EPA and the DOT. Other federal agencies that may have regulatory influence under certain circumstances include the U.S. Geological Survey (USGS) and U.S. Department of Agriculture (USDA). Of these, the EPA, USGS, and USDA have the primary responsibility for groundwater activties.

2.2.1 Environmental Protection Agency

The EPA was established from a variety of agencies in 1970. Through its ten regional offices, the EPA administers regulatory programs associated with air and water pollution, pesticides, solid waste, noise control, drinking water, wetlands, and hazardous and toxic wastes. Many of the regulatory programs discussed in this chapter

(CWA, CERCLA, RCRA, etc.) are administered and enforced by the EPA. From a groundwater-quality perspective, the EPA establishes regulations dealing with groundwater contamination from storage tanks, accidental releases and spills, disposal of solid and hazardous waste, and treatment, disposal, and storage facilities (TSDF) operations. Notably, the EPA regulates the cradle-to-grave management of hazardous waste, and associated programs such as underground injection control (UIC), and USTs. The EPA also maintains major research responsibilities, and sponsors an emergency response center, which is funded under the Superfund program and provides immediate information on hazardous materials.

2.2.2 Department of Transportation

The DOT came into being in 1967, and is responsible for the safe transportation, regardless of means, of hazardous liquids and materials. The safety code, made effective in 1970, drew extensively on the voluntary code of the industry. Charged with the responsibility of pipeline safety, standards concerned with the design and operation of pipelines that were ultimately incorporated into the regulations reflected the composite experience of individuals, companies, and professional societies over several years. These efforts were supplemented by research, tests, studies, and investigations. In 1976, the DOT reported that 83% of pipeline accidents were traceable to the line pipe with other accidents traceable to tank farms, pumping stations, and delivery points with successively decreasing percentages. However, the regulations emphasized safety in regard to injuries, not necessarily contamination.

In addition to statistics developed for USTs, in 1984 the EPA Office of Technology Assessment also reported that 16,000 spills occur annually during transport, while the DOT reported that of the 4112 accidents that occurred between 1968 and 1981, 1372 were associated with corrosion and 1101 with pipeline ruptures. Prior to the DOT regulating transport of liquids by pipeline in 1971, 308 interstate pipeline accidents were documented resulting in a loss of about 245,000 barrels of liquid. Increasing DOT involvement may have been the reason for a decline in incidents to 275 in 1980, and 198 in 1981.

2.2.3 Other Federal Agencies

Other federal agencies that may be involved with the release of petroleum products on the land or in the subsurface include the USGS, the USDA, the U.S. Department of Energy (DOE), and the Interstate Oil and Gas Compact Commission (IOGCC). The USGS is a bureau within the Department of the Interior (DOI). Established in 1879 to conduct investigations of a geological nature, the USGS role is one that essentially is responsible for characterizing regional hydrogeologic settings, and administering groundwater-protection research projects. The USDA is primarily responsible for groundwater protection under certain farming and agricultural programs. The DOE was established to complete and balance the national energy plan. The DOE accomplished this objective through coordination and administration of the energy functions of the federal government. In regard to subsurface releases of petroleum products, the DOE has over the past several years been involved in research concerning the subsurface presence and behavior of MTBE. The IOGCC comprises the governors of oil-producing states. Although the IOGCC has no regulatory writing or enforcement authority, it does work cooperatively with the EPA to assist states in improving their regulatory programs.

2.2.4 State Agencies

States have traditionally regulated water use and withdrawal, permitting of water wells, among other activities. State and local agencies and departments typically play the lead role when addressing environmental issues associated with the release of petroleum products to the environment, and may in fact negotiate lead status with federal agencies such as the EPA. These agencies may include regional water boards, environmental quality and health departments, irrigation districts, and local fire and wastewater treatment departments. In regard to petroleum exploration and production wastes, these materials are typically regulated by state oil and gas agencies.

2.3 THE FEDERAL REGULATORY PROCESS

Federal regulations are complicated and to the novice difficult to follow. The process begins with Congress drafting congressional acts and enacting them into law. The actual authority to enforce and administer the law is frequently delegated to a variety of government agencies. The designated agency likely writes its own regulations to enforce or implement the directives specified in the congressional act. For example, where Congress made the RCRA general law, the EPA was charged with writing specific regulations and procedures, and enforcing that law in specific cases. The agency regulations also have the force of law, but operate on a different level from the congressional acts, and tend to be subject to more changes and updates.

Congressional acts are found in the U.S. Code (USC). The USC is a compilation of all the laws passed by Congress that are not exclusively for agencies. A subject index is provided in the USC that gives references by general topics, and the appropriate volume and title (i.e., 16 USC 1531 for the ESA). Copies of new laws not published in the USC because they were passed after the latest edition of the USC, can be retrieved from the Congressional Desk in Washington, D.C. Another source for regulatory information is the *Federal Register.* The *Federal Register* is published daily and contains changes and proposed changes in agency rules and regulations. The *Federal Register* also contains other government actions such as presidential proclamations, and a monthly notice of changes in existing rules and regulations. Each year, all the proposed changes and rules published in the *Federal Register* that actually become final rules and regulations are published in a set of volumes called the Code of Federal Regulations (CFR). The CFR represents all rules from all agencies. The CFR is divided into 50 titles, which represent broad areas subject to federal regulation, with each title divided into chapters that usually bear the name of the issuing agency. Each chapter is divided into parts covering specific regulatory areas, with each part further divided into sections. For example, UST regulations are contained in 40 CFR Parts 280–281.

2.4 PERTINENT FEDERAL REGULATIONS

Those federal regulations of interest and importance for addressing subsurface environmental issues in chronological order of establishment include the National Environmental Policy Act (NEPA), Spill Prevention, Control and Countermeasure (SPCC), the Safe Drinking Water Act (SDWA), the Resource, Conservation, and Recovery Act (RCRA), the Clean Water Act (CWA), the Toxic Substance Control Act (TSCA), the Comprehensive Environmental Response, Compensation, and Liability Act (CERCLA), the Superfund Amendments and Reauthorization Act (SARA), the Federal Insecticide, Fungicide, and Rodenticide Act (FIFRA), and the Petroleum Safety Act (PSA). These regulations are discussed below.

2.4.1 National Environmental Policy Act

NEPA was enacted in 1969, and signed into law in 1970. NEPA is viewed as national policy and is the basic national charter for the protection of the environment; it serves as the umbrella for all other environmental regulations. NEPA states "to use all practicable means and measures, including financial and technical assistance, in a manner calculated to foster and promote the general welfare, to create and maintain conditions under which man and nature can exist in productive harmony, and fulfill the social, economics, and other requirements of present and future generations of Americans" (NEPA Section 101(a); 42 USC Section 4331(a)). While NEPA does not directly regulate assessment and remediation-related actions, some of the processes within NEPA can provide important access for interested individuals or organizations to interact with government on proposed actions that may affect the environment. NEPA also requires that groundwater quality be considered in any environmental impact statement (42 USC 4321 et seq.).

2.4.2 Spill Prevention, Control and Countermeasures

The SPCC plan was enacted in 1974 and required the development, implementation, and maintenance of such plan by owners and operators with large oil storage facilities, with "large" defined as storage facilities containing 1320 gal aboveground or 42,000 gal belowground (40 CFR Part 112). SPCC plans typically include provisions for installation of containment structures, regular inspections, and other preventative measures.

More stringent vessel safety and accident and spill prevention measures are regulated by the Coast Guard as promulgated under the Port and Tanker Safety Act of 1978 (33 USC, Sections 1201–1231), Safety of Life at Sea (SOLAS) Convention of 1973, and the MARPOL Convention of 1978. Financial responsibility to meet potential liability under the CWA must also be met.

2.4.3 Safe Drinking Water Act

The SDWA was enacted in 1974, with amendments promulgated in 1986 and 1996. The SDWA expanded the role of the federal government to establish national standards for the levels of contaminants in drinking water, creating state programs to regulate underground injection wells, and protecting sole source aquifers (SSA). This dramatic expansion of the role of the federal government reflected information suggesting that (1) chlorinated organic chemicals were contaminating major surface and underground drinking water supplies, (2) major aquifers were being threatened by underground injection operations, notably, within the Edwards aquifer which supplies water to much of Texas, and (3) the overall antiquated, underfunded, and understaffed public water supply systems posed an increasing threat to public health. Amendments promulgated in 1986 increased the pace with which EPA could issue standards and implement other provisions of the act by:

- Mandating the issuance of standards for 83 specific contaminants by June, 1989, with standards for 25 additional contaminants to be issued every 3 years thereafter;
- Increasing the EPA enforcement authority;
- Providing increased protection for SSAs and wellhead protection areas; and
- Regulating the presence of lead in drinking water systems.

The SDWA Amendments of 1996 signed by President Clinton on August 6, 1996, required that consumers receive more information about the quality of their drinking water supplies and about what is being done to protect them. The Amendments also provide new opportunities for public involvement and an increased emphasis on protecting sources of local drinking water (groundwater). Water suppliers must thus promptly inform the public through the media if drinking water has become contaminated such "that can have serious adverse effects on human health as a result of short-term exposure." Should a violation occur, information regarding the potential adverse effects on human health, the steps that are being taken to correct the violation, and the need for an alternative water supply (boiled water, bottled water, etc.) until the problem is corrected must be provided through the media. Should a lesser impact occur, the public can be informed of the violation as part of the next water bill sent out to the customer, in an annual report, or by mail within a year. Structured in 11 parts, the most pertinent parts are summarized in Table 2.2.

TABLE 2.2
Hydrogeologic Issues Addressed under the SDWA

Title	Description	Part
Maximum contaminant levels (MCLs)	Provides primary and secondary MCLs for inorganic, organic, turbidity, and radioactive constituents	141–143
Underground injection control program (UIC)	Provides criteria for identifying criteria and standards for the federal and state UIC programs	144–147
Sole source aquifers (SSA)	Provides criteria for identifying critical aquifer protection areas and review of projects affecting the Edwards Underground Reservoir	149

2.4.3.1 Drinking Water Standards

The SDWA requires the EPA to identify contaminants in drinking water that may have an adverse effect on public health, and to specify a maximum contaminant level (MCL) for each contaminant, where feasible. The MCL is defined as the maximum permissible level of a contaminant in water that is delivered to any user of a public water supply system and is under the protection of SSA. These standards require water supply system operators to come as close as possible to meeting the standards by using the best available technology that is economically and technologically feasible. MCLs established under the SDWA have also been used as groundwater cleanup standards under CERCLA and other programs, and have the potential to be used as standards not only for drinking water but also any groundwater regardless of its end use, if used at all.

A total of 83 contaminants are currently required to be regulated under the SDWA of 1986, and 77 substances or classes of substances make up a priority list of contaminants for regulation after the EPA completes standards for the initial list of 83. Although the EPA was required to establish MCLs and maximum contaminant level goals (MCLGs) for 25 of the contaminants by January, 1991, and every 3 years thereafter, the EPA has not kept up with this pace.

2.4.3.2 Underground Injection Control Program

Underground injection is the subsurface emplacement of fluids through a well or hole whose depth is greater than its width. Injected fluids can include hazardous waste, brine from oil and gas production, and certain mining processes. An injection well can include a septic well, cesspool, or dry well. The SDWA allows regulation of UIC to protect usable aquifers from contamination, with additional restrictions and controls under the 1984 Amendments to RCRA.

The SDWA and its Amendments in 1980 mandated that primacy states regulate injection wells to protect underground sources of drinking water (USDW), with additional restrictions and control under the RCRA 1984 Amendments. USDW is defined as aquifers or portions of aquifers with a total dissolved solids value of less than 10,000 mg/l and which is capable of supplying a public water system. About 23 states do not have full UIC authority. These states and Indian lands are regulated by the EPA. Not regulated by the EPA as part of the UIC program are injection of brine from oil and gas production, gas storage requirements, or secondary or tertiary oil and gas recovery, unless public health is directly threatened.

There are five classes of injection wells (Class I through V). Class I disposal wells are used for the disposal of industrial and hazardous waste streams, and may also be subject to certain RCRA, Subtitle C hazardous waste management regulations. Class II wells are defined as those wells used in conjunction with oil and gas production activities. Class III wells are defined as those wells that inject fluids:

- That are brought to the surface in connection with conventional oil or natural gas production and may be commingled with wastewaters from

gas plants that are an integral part of production operations, unless those waters are classified as a hazardous waste at the time of injection;
- For enhanced recovery of oil or natural gas; and
- For storage of hydrocarbons that are liquid at standard temperature and pressure.

The injection of produced water and other oil field fluids into wells started as early as 1928. In 1976, more than 300 industrial waste disposal sites were in operation throughout the country. By 1986, approximately 60 million barrels of oil field fluids were injected through 166,000 injection wells within the conterminous United States. These volumes are anticipated to increase significantly in the future as producing fields continue to be depleted. Thus, construction requirements as listed in 40 CFR 146.22 are an essential prerequisite to the safe disposal/injection of fluids and the prevention of contamination of USDW.

2.4.3.3 Sole Source Aquifers

SSA protection is one of the most significant aspects of SDWA. These aquifers are situated in some of the most populated areas, with the majority of them located in the northeastern part of the United States. Regulations pertaining to SSAs allow communities to petition the EPA to designate their area an SSA which bars federal assistance grants to projects that may threaten the aquifer (i.e., waste disposal sites, water treatment works, etc.). As of January, 1991, 51 SSAs have been designated, most after the 1996 Amendments reflecting the possibility of federal funding for local governments. Delegation of authority was not determined until 1987. To qualify as an SSA, the aquifer must supply 50% of the people in the region with drinking water, have no readily available and reasonably priced alternative water supply, and the aquifer would present a significant risk to public health if contaminated.

SSA boundaries are typically geologic or hydrologic in nature, in lieu of political or geographic boundaries. Designation of an SSA is relatively easier beneath islands than it is within inland areas due to their ease in delineation, localized vertical and lateral extent, high dependency, and high degree of vulnerability. Inland aquifers, on the other hand, are more difficult to designate, reflecting a poor correlation between their overall lateral extent and the population centers, presence of multi-aquifer systems, and whether they are hydraulically connected. SSAs are also situated in less populated resort areas including Cape Cod, Nantucket, and Block Island due to the acute vulnerability of coastal groundwater and the relative ease of aquifer identification.

Also included as part of the overall SSA program is wellhead protection. A wellhead protection area is "the surface and subsurface area surrounding a water well or well fields, supplying a public water system, through which contaminants are reasonably likely to move toward and reach such well or wellfield." The importance of the SSA and wellhead protection program cannot be overstated due to the large populations that can be potentially affected, and the ability of these programs to set the stage for more elaborate state and federal groundwater protection legislation.

2.4.4 Resource Conservation and Recovery Act

Regulatory programs under RCRA address potential groundwater quality issues relating to TSDF operating standards, underground injection (see Section 2.4.3.2), and USTs. All three of these programs are designed to protect groundwater from contamination by leaked or spilled materials. The Land Ban rules under RCRA further protect groundwater resources by prohibiting the land disposal of most hazardous wastes. The 1984 Amendments to the RCRA required the EPA to develop federal regulations dealing with leakage of hydrocarbon product or substances from underground storage tanks designated as hazardous, as defined under CERCLA of 1980, Section 101(14).

2.4.4.1 Treatment, Disposal, and Storage Facilities

Standards for owners and operators of hazardous waste TDSFs are regulated under RCRA. Prior to 1976, disposal of hazardous waste into the atmosphere or bodies of water was restricted. Economics in conjunction with the prevailing regulations favored land disposal and deep-well injection of hazardous waste. In 1976, RCRA was promulgated to provide cradle-to-grave management of hazardous waste, in turn providing the first comprehensive federal regulatory program for protection of the environment including air, surface water, groundwater, and the land.

Under RCRA, interim-status and permitted TDSFs (i.e., landfills, surface impoundments, land treatment facilities, waste piles, and underground injection wells) must fulfill certain obligations for (1) leachate minimization and control through design and operating requirements and (2) groundwater monitoring and response requirements should leachate reach the groundwater. Although the specific monitoring requirements for each TSDF are specific to its respective permit, all generally include procedures for sampling and chemical testing, evaluation of groundwater data, and appropriate response procedures if a release is detected. In addition, groundwater monitoring has to continue for 30 years after the facility closes. As part of the permitting process, operators have to develop a plan that will detect contamination in the event that a release occurs (detection monitoring). A corrective action plan is also developed to address steps that the facility will take to remediate problems if such release impacts groundwater quality.

As part of detection monitoring, two categories exist: detection monitoring and compliance monitoring. Detection monitoring includes monitoring surface and groundwater quality for certain parameters, establishment of a groundwater monitoring network (compliance points), establishment of background levels, accounting for seasonality, determination of groundwater flow rate and direction, characterization of the uppermost aquifer, adherence to EPA methods and procedures for sampling and analysis, and determination of whether a statistically significant increase over background levels has occurred. Compliance monitoring may be implemented following corrective action and involves periodic monitoring of certain constituents, usually for 3 years, at prescribed compliance points.

2.4.4.2 Underground Storage Tanks

In the public's mind, petroleum contamination is commonly associated with USTs. Millions of USTs were installed in the 1950s and 1960s. The EPA estimates that out of an estimated 2.5 to 3 million USTs throughout the United States, more than 400,000 have leaked or are leaking petroleum hydrocarbons, with more expected to develop leaks in the future. All states have UST programs that reflect the federal regulations under RCRA; however, most states can only provide an estimate of the volume of petroleum-contaminated soil generated, and how petroleum-contaminated soil from UST sites is handled, with only a select few having actual statistics. If one assumes that an average estimated amount of contaminated soil as a result of a leak is on the order of 50 to 80 yd^3, then the volume of contaminated soil solely attributed to USTs is on the order of 20 to 32 million yd^3. This number is very conservative, since in many cases much larger volumes have been reported, ranging up to thousands of cubic yards per site and reflecting significant releases over time. In addition, thousands of USTs remain unrecorded, notably in rural areas, and their individual impacts on the subsurface remain unknown.

The historical usage of USTs for the bulk storage of petroleum products was instigated by the need to store flammable petroleum products underground to reduce fire risk and to increase financial savings in respect to space that could be utilized for other purposes. The use of USTs for petroleum storage made excellent sense; however, in the early 1980s petroleum leakage problems of enormous magnitude were being reported by the general public all across the United States. The problems resulted from leakage of petroleum products from USTs. These releases of stored liquids had contaminated both soils and groundwater. Some releases had contaminated nearby private and municipal drinking water supply wells. The release of petroleum products into soils created problems aside from the obvious soil contamination. Explosive concentrations of gasoline vapors could migrate through porous soils and accumulate in sewer and utility conduits as well as in the basements of buildings. These vapors could potentially pose a substantial fire and explosion hazard. In one case in Colorado in 1980, 41 homes were purchased at over twice their appraised value at a cost of $10 million to the responsible party as a result of a UST leak. In another instance, $5 to 10 million as of May 1984 was spent for cleanup and damages resulting from 30,000 gal of gasoline that leaked from a UST into a New York community's groundwater supply in 1978.

Following the report of the U.S. General Accounting Office in 1984 that UST leaks had been reported in all 50 states, many states in 1985 claimed that USTs were the leading cause of underground contamination. As of 1984, there were an estimated 2.5 million USTs in the United States of which about 97% were used to store petroleum products. By 1986, the EPA identified 12,444 incidents of release that had occurred as of 1984.

A comprehensive study was undertaken by the EPA in 1987 to gather additional facts about releases from USTs. This study resulted in the following findings:

- Most releases do not originate from the tank portion of UST systems, but pipeline releases occur twice as often as tank releases.

- Spills and overfills during refueling are the most common cause of releases.
- Older steel tanks fail primarily because of corrosion, but new cathodically protected steel, fiberglass-clad steel, and fiberglass tanks have nearly eliminated external corrosion failure.
- Corrosion, poor installation, accidents, and natural events are the four major causes of piping failure.
- When pressurized piping fails, significantly larger releases can occur.

This study was very important in the formulation of federal UST regulations. UST systems are located at over 700,000 facilities nationwide. Over 75% of the existing systems are made of unprotected steel, a type of tank shown by studies to be the most likely to leak. Most of these facilities are owned and operated by small "Mom & Pop" enterprises that do not have easy access to the financial resources for potential cleanup costs and that are not accustomed to dealing with complex regulatory requirements. Current innovations and technologies associated with USTs are changing rapidly, and keeping abreast of them is sometimes difficult and immensely time-consuming. In response to these unique aspects associated with USTs, the EPA identified several key objectives that it would have to incorporate into the regulations to be formulated for USTs. These objectives were as follows:

1. The regulatory program must be based upon sound national standards that protect human health and the environment.
2. The regulatory program must be designed to be implemented at a state and local level.
3. The regulations must be kept simple and easily understood and implemented.
4. The regulations must not impede technological developments.
5. The regulations must retain some degree of flexibility.

With these key objectives in mind, the EPA formulated the Federal UST Regulations, as described in 40 CFR, Parts 280 and 281, effective December 22, 1988, with financial responsibility requirements effective January 24, 1988. It should also be noted that fire safety matters are not emphasized in the EPA UST regulations. The National Fire Protection Association Code 30 (NFPA Code 30) relating to the storage and handling of flammable and combustible liquids is used by the majority of local regulatory communities in the United States in regulating fire safety tanks that are used to store petroleum products. On November 8, 1984, President Reagan signed into law the Hazardous and Solid Waste Amendment Act of 1984 (HSWA). Subtitle I of HSWA amended the RCRA of 1976 to address specifically the regulation of USTs. Under Section 9003 of RCRA, the EPA had to establish regulatory requirements for the following:

- Exemptions
- UST Systems Design, Construction, Installation, and Notification
- General Operating Requirements

- Release Detection
- Release Prevention
- Release Reporting, Investigation, and Confirmation
- Release Resource and Corrective Action
- UST Closures
- State Program Approval
- Enforcement and Penalties
- Financial Responsibility

2.4.5 Clean Water Act

The basic framework for federal water pollution and ocean-dumping control regulation was promulgated in 1972 by the enactment of the Federal Water Pollution Control Act (FWPCA) and the Marine Protection, Research, and Sanctuaries Act (MPRSA), respectively. The FWPCA was renamed the CWA in 1977, and focused upon rigorous control of toxic water pollution. The CWA protects groundwater by establishing surface water quality standards. This is accomplished in part by the establishment of procedures for dredge-and-fill operations affecting groundwater quality (40 CFR 232–233, Section 404). CWA also establishes water quality standards for the concentration of constituents in sludge for monofiling (40 CFR 257–258). Spills of petroleum products continue to be regulated primarily under Section 311 of the CWA. Spills at deep-water ports or from outer continental shelf oil and gas operations are also governed in part under the Deep Water Ports Act of 1974 and Outer Continental Shelf Act Amendments of 1978. Although oil and hazardous substances are traditionally covered under the same CWA provision, they have been treated separately by the EPA.

Section 311 of the CWA also governs the discharge of hazardous substances. Approximately 300 substances were designated as hazardous (40 CFR Part 116). In addition, the EPA designated quantities of these substances that may be considered harmful (i.e., reportable quantity) (40 CFR Part 117). Five categories (X, A, B, C, and D) were designated:

- X substance is 1 pound;
- A substance is 10 pounds;
- B substance is 100 pounds;
- C substance is 1000 pounds; and
- D substance is 5000 pounds.

Discharges made in compliance with a Natural Pollutant Discharge Elimination System (NPDES) permit are excluded, reflecting the purpose of the amendments to limit "classic" hazardous substance spills, not chronic discharge of designated substances if the discharge complies with an NPDES permit. In addition, the facility has the option to regulate intermittent anticipated spills (i.e., into plant drainage ditches) under its existing NPDES permit. Discharges from industrial facilities to a publicly owned treatment works (POTW) are not currently regulated. However, the regulations can apply to all discharges of reportable quantities of hazardous substances to POTWs by a mobile source such as trucks under certain circumstances (40 CFR Part 117). Reporting requirements for spills have been significantly supplemented by CERCLA and Title III of the SARA of 1986.

2.4.5.1 Storm Water Permitting Program

One important aspect of the CWA is that it prohibits the discharge of pollutants into or adjacent to waters in the United States unless an NPDES permit has been applied for by the owner or operator of the facility. The initial NPDES storm water permitting program was developed by the EPA to regulate discharges of pollutants to surface waters of the United States, and focused on industrial and municipal discharges. After several years of monitoring, overall water quality was not perceived to be improving rapidly enough. The reason seemed to be the significant impact on water quality from non-point-source runoff such as sediment loading from agricultural areas, and contaminated water from industrial and construction sites. The CWA was amended in 1990 by the Oil Pollutant Act (OPA) in direct response to several oil spills including the Exxon *Valdez* incident off Prince Williams Sound in Alaska.

Storm water is defined as storm water runoff, snowmelt, and surface water runoff and drainage, all originating from a point source. A point source (40 CFR Section 122.2) is "any discernable, confined, and discrete conveyance, including but not limited to, any pipe, ditch, channel, tunnel, conduit, well, discrete fissure, container, rolling stock, concentrated animal feeding operation, landfill leachate collection system, vessel or other floating craft from which pollutants are or may be discharged." A storm water discharge permit is required for most discharges of storm water from industrial facilities, construction sites of more than 5 acres, and for discharges into certain municipal storm sewer systems. As part of the NPDES program, certain regulatory standards and requirements must be met. NPDES permit applications usually include a description of the control technology to be used to minimize the discharge of pollutants, possibly limitations on materials and water quality criteria that can be discharged into surface waters, monitoring, sampling, and analysis requirements, and record-keeping and reporting requirements.

2.4.6 Toxic Substance Control Act

TSCA was enacted in 1976 to provide the EPA the authority to require testing of both old and new chemical substances and mixtures of substances entering the environment and to regulate them, where necessary, when such substances might present an unreasonable risk of injury to human health or the environment. TSCA has two main regulatory components: (1) acquisition of sufficient information by the EPA to identify and evaluate potential hazards from chemical substances and (2) regulation of the production, use, distribution, and disposal of such substances when necessary. Under TSCA, a mechanism exists for the EPA to monitor and generate data on the occurrence, migration, and transformation of toxic substances in groundwater (40 CFR 704).

2.4.7 Comprehensive Environmental Response, Compensation, and Liability Act

CERCLA, or Superfund, was enacted in 1980, and amended in 1986, for the basic purpose of providing funding and enforcement authority to clean up any site where there is a past unremedied release of a hazardous substance or hazardous substance spill. Such sites are typically characterized as areas where hazardous waste or materials have been disposed of improperly, with little if any responsible action being taken to mitigate the situation. Standards for financial responsibility were promulgated by the SARA of 1986 which further amended Section 9003 of RCRA and mandated that the EPA establish financial responsibility requirements for UST owners and operators to guarantee cost recovery for corrective action and third-party liability caused by accidental releases of USTs containing petroleum products.

The RCRA and SDWA groundwater quality standards are considered "applicable, relevant and appropriate requirements" for purposes of CERCLA cleanups. Thus, the EPA may require CERCLA sites to be cleaned up to the degree that groundwater quality around the site meets the RCRA and SDWA groundwater quality standards.

Petroleum, natural gas, and synthetic fuels are excluded from the definition of a hazardous substance, and the definitions of pollutants and contaminants under CERCLA; this is known as the Petroleum Exclusion. Although the EPA has the authority to regulate the release or threatened release of a hazardous substance, pollutant, or contaminant, the release of petroleum, natural gas, and synthetic fuels from active or abandoned pits or other land disposal units is currently exempt from CERCLA. Such sites cannot use Superfund dollars for cleanup, nor can the EPA enforce an oil and gas operator, landowner, or other individual to clean up a release under CERCLA. Substances exempt include such materials as brine, crude oil, and refined products (i.e., gasoline and diesel fuel) and fractions.

A petroleum product is excluded even if it contains other listed hazardous substances provided that those substances were not added to the oil after the refining process, and are found at concentrations normally detected in crude oil or refined petroleum products. Just what is a petroleum product is important since such determination will decide if it meets the criteria for exclusion. The two issues are (1) is the material released a petroleum product, and (2) does it contain hazardous substances otherwise listed by the EPA, which limits the applicability of the exclusion to the release in question. The EPA in 1985 interpreted the petroleum exclusion to pertain to

> materials such as crude oil, petroleum feedstock, and refined petroleum products, even if a specifically listed or designated hazardous substance is present in such products. However, EPA does not consider materials such as waste oil to which listed CERCLA substances have been added to be within the petroleum exclusion. Similarly, pesticides are not within the petroleum exclusion, even though the active ingredients of the pesticide may be contained in a petroleum distillate.

In March 1996, the EPA further determined that used oil that contains hazardous substances at levels which exceed those normally found in petroleum are, however, subject to CERCLA. If these spilled materials meet the criteria for hazardous waste

TABLE 2.3
Department of Transportation Code of Federal Regulations

CFR Regulation	Part Description
40 CFR, Part 190	Pipeline safety program procedures
40 CFR, Part 191	Transportation of natural and other gas by pipeline; annual reports and incident reports
40 CFR, Part 192	Transportation of natural and other gas by pipeline; minimum federal safety standards
40 CFR, Part 193	Liquefied natural gas facilities; federal safety standards
40 CFR, Part 195	Transportation of hazardous liquids by pipeline

under RCRA, they are then subject to all applicable hazardous waste regulations and are managed accordingly. Furthermore, if they are not spilled during exploration- and production-related activities but rather spilled during transportation, they are no longer excluded.

2.4.8 Federal Insecticide, Fungicide, and Rodenticide Act

The Federal Insecticide, Fungicide, and Rodenticide Act (FIFRA) indirectly protects groundwater resources by the regulation of the use of pesticides. FIFRA establishes pesticide registration procedures that involve groundwater assessment by the EPA. FIFRA is mostly applicable to manufacturers, distributors, and suppliers of pesticides. Pesticide registration can be suspended or denied once pesticides have been shown to impact groundwater quality adversely.

2.4.9 Pipeline Safety Act

The PSA of 1987 is currently promulgated by the DOT Office of Pipeline Safety under the Natural Gas Pipeline Safety Act of 1968 and the Hazardous Materials Transportation Act as summarized in Table 2.3. Petroleum releases are addressed under 40 CFR Part 195 which is subdivided into six subparts (Subparts A through F). Although these regulations apply to pipeline facilities and the transportation of hazardous liquids associated with those facilities in or affecting interstate or foreign commerce, they do not apply to transportation of hazardous liquids via the following:

- Gaseous state;
- Pipeline by gravity;
- Pipeline stress level of 20% or less of the specified minimum yield strength of the live pipe;
- Transportation in onshore gatherlines in rural areas;
- Transportation in offshore pipelines located upstream from the outlet flange of each facility on the Outer Continental Shelf where hydrocarbons are produced or where produced hydrocarbons are first separated, dehydrated, or processed (whichever facility is farther downstream);

- Transportation via onshore production (including flow lines), refining, or manufacturing facilities, or storage or in-plant piping systems associated with such facilities; and
- Transportation by vessel, aircraft, tank truck, tank car, or other vehicle or terminal facilities used exclusively to transfer hazardous liquids between such modes of transportation.

The PSA of 1987 provides regulations establishing federal safety standards, including requirements for release detection, prevention, and correction, for the transportation of hazardous liquids and pipeline facilities, as necessary to protect human health and the environment. These regulations apply to each person who engages in the transportation of hazardous liquids or who owns or operates pipeline facilities. Of notable importance is the lack of regulations pertaining to pipelines associated with refining or manufacturing facilities, or associated storage or in-plant piping systems that are land bound and situated above aquifers of beneficial use.

2.5 STATE PROGRAMS AND REGULATIONS

The EPA has encouraged states to follow a voluntary approach called the Comprehensive State Ground Water Protection Program (CSGWPP), which is designated to increase a state's overall capacity for groundwater protection by identifying gaps and barriers, helping to set state and agency priorities, and better directing resources based on the relative use, value, and vulnerability of groundwater resources. States have traditionally regulated water use and withdrawal, permitting of water wells, designation and classification of beneficial use waters, among other activities, in addition to maintaining compliance with federal programs. Many states and local regions have developed and implemented more stringent regulations governing these programs and subsequent remedial actions where hydrocarbons have been released to soil and groundwater.

2.5.1 Voluntary Cleanup Programs

Voluntary cleanup programs (VCPs) began in 1989 in Michigan and have become quite popular since about 1995. VCPs typically apply to those sites that are not on the CERCLA NPL or State Minifund lists, or are not handled under the RCRA permitting process. With VCPs, a responsible entity with funding is set forth solely for remediation to put the site to some beneficial use.

2.5.2 Brownfield Initiative

Brownfields are contaminated and derelict land typically in urban settings in which federal funds, limited to $200,000, have been made available. These funds are often associated with significant tax benefits, and serve as seed money for state-granted assistance to selected private-sector developers. The objective of the brownfield initiative, like that of the VCPs, is to assist in the rebuilding of distressed urban areas, to create jobs, and to strengthen cities infrastructure. Under this alternative, the EPA passes the funding responsibility to the cities, who in turn rely on the state legislature, with little assistance from the federal government.

2.5.3 Underground and Aboveground Storage Tanks Program

Underground and aboveground storage tanks are regulated by a variety of federal regulations, depending on what substance is being stored: petroleum products, hazardous waste, or hazardous chemicals. States may obtain approval from the EPA to regulate storage tanks. Many have done so and, in some cases, have developed more stringent regulations. About 23 states currently have approved programs that they regulate. These states are Arkansas, Connecticut, Delaware, Georgia, Iowa, Kansas, Louisiana, Maine, Maryland, Massachusetts, Mississippi, Montana, Nevada, New Hampshire, New Mexico, North Dakota, Oklahoma, Rhode Island, South Dakota, Texas, Utah, Vermont, and Washington. In regard to ASTs, no single comprehensive regulation covers ASTs. Federal regulations covering ASTs can be found under the CWA, OPA, CAA, and RCRA. In addition, certain states may maintain more stringent requirements.

In California, for example, the UST program was established in 1983, prior to establishment of the federal UST regulations in 1988. Since the California UST program has not been approved by the EPA, UST owners are subject to both the federal and state requirements. The regulatory responsibilities and potential liability of operators and owners of USTs are set forth primarily in the California UST Law (Health and Safety Code Section 25280), California UST Regulations (California Code of Regulations, Subchapter 16, Title 23), and numerous local codes and ordinances. In the case of the UST program, the basic requirements of tank owners under the California Underground Storage Tank Regulations (CUSTR) provide provisions for existing tanks including:

- Determining whether past leaks have occurred from the tank system;
- Determining whether the tank system is currently leaking; and
- Implementing a leak detection and monitoring program to detect future releases.

However, as straightforward as this may seem, there are 23 local health and fire departments that have adopted their own version of this bill, making compliance increasingly difficult.

Regulations exist that prohibit the ownership or operation of a UST used for the storage of hazardous substances unless a permit for its operation is issued in addition to imposing various design, installation, monitoring, and release reporting requirements for USTs. However, there are many ASTs currently in service that were constructed 20 or more years ago. This has allowed sufficient time for metal corrosion and stress-fatigue cracking to weaken and perhaps breach the tank floors. The resultant tank floor leaks often go unnoticed by facility personnel during the performance of their daily routines. These leaks can be costly, both in terms of product loss and potential liability for environmental pollution. Until recently, techniques for tank floor integrity testing were generally limited to visual inspection or still-gauging. Visual inspections required the tank to be emptied, cleaned, and carefully examined, putting the tank out of service for weeks or more. Even then, some small leaks could be overlooked.

Still-gauging methods are adequate for only the largest leaks. The accuracy of most tank-installed liquid level gauges is usually $^1/_8$ in. at best. A product loss reflected by a 0.062-in. level drop for a 100-ft-diameter tank translates to a 306 gal/day leak. If this $^1/_{16}$-in. drop in product level is not discernible from the masking effects of fluid expansion, losses in excess of 2650 barrels annually will go undetected. At, for example, \$20 per barrel, this loss amounts to over \$53,000 for a single tank. Most importantly, the associated liability risks of groundwater contamination could involve much greater potential costs.

In 1989, California enacted the Aboveground Petroleum Storage Act. Unique to California, this law regulates the storing of crude oil and petroleum products in liquid form and requires each owner or operator of an aboveground storage tank at a facility to establish and maintain a monitoring program within 180 days of preparing a spill prevention control and countermeasure plan. Second, if the tank facility has the potential to impact surface waters or sensitive ecosystems as determined by the reviewing agency, based on the tank location, tank size, characteristics of the petroleum being stored, or the spill containment system, the facility owner will need to perform tasks similar to that for UST owners. These requirements would include either of the following:

- Install and maintain a system to detect releases into surface waters or sensitive ecosystems; and/or
- If any discharge from a tank facility flows, or would reasonably be expected to flow, to surface waters or a sensitive ecosystem, allow for a drainage valve to be opened and remain open only during the presence of an individual who visually observes the discharge.

Third, if because of tank facility location, tank size, or characteristics of the crude oil or its fractions being stored (16° API gravity or lighter), a facility that has the potential to impact the beneficial uses of the groundwater, and that is not required to have a groundwater monitoring program at the tank facility pursuant to any other federal, state, or local law, shall do any of the following:

- Install a tank facility groundwater monitoring system that detects releases of crude oil or its fractions into the groundwater;
- Install and maintain a tank foundation design that will provide for early detection of releases of crude oil or its fractions before reaching the groundwater;
- Implement a tank water bottom monitoring system and maintain a schedule that includes a log or other record that will identify or indicate releases of crude oil or its fractions before reaching the groundwater; and/or
- Use other methods that will detect releases of crude oil or its fractions into or before reaching the groundwater.

All positive findings from the detection system, excluding tanks whose exterior surface (including connecting piping and the floor directly beneath the tank) can be monitored by direct viewing, must be reported to the appropriate agency within 72 h after learning of the finding.

Federal legislation for ASTs is also in the process of being mandated under RCRA. This bill would include requirements for development and implementation of a release prevention plan, a tank system that is capable of containing 110% of the content of the tank and preventing off-site release, and inspection of tank systems by a qualified professional engineer. Provisions will also include evidence of financial responsibility and cleanup of product releases. There is little doubt that legislation for ASTs will be more stringent in the years to come.

2.5.4 State Programs Affecting the Petroleum Refining Industry

Up to about December 1970, certain sectors of the petroleum industry, notably the exploration, development, and production activities and operations, have remained essentially unregulated from a subsurface environmental perspective. For the past two decades, growing concerns regarding actual and anticipated groundwater problems in these areas have persisted. These concerns reflect in part over 50 years of operations and activities in the petroleum exploration, production, and refining industry. Although many favorable changes have come about in the way these operations and activities are conducted today, in recent years increasing environmental regulations are being felt throughout the industry. The industry has been fortunate to date in that it has temporarily escaped the designation and regulation of its wastes as hazardous. But with concerns about future cleanup costs, an increase in the level of regulatory attention to the conduct of oil and gas exploration, production, and refinery operations is anticipated.

An excellent example of the juxapostion of oil field development and production and urban growth is in southern California. Southern California has a rich history of oil and gas exploration and production going back to 1876, the first year of commercial production. Of 56 counties in California, 30 are known to produce oil and gas. The price of real estate is also at a premium in areas such as Los Angeles County, where much of the available land that remains to be developed is oil field property. These oil field areas are rapidly undergoing environmental pressures as they reach the end of their productive lives. In addition, the conversion of manufacturing-related land to services-related land use and significant increases in land value have resulted in an increasing number of property transfers involving oil field production and storage areas.

Several refineries, tank farms, and other petroleum-handling facilities are included on the EPA National Priorities List (U.S. EPA, 1986) or are regulated under RCRA. These facilities presented several potential subsurface environmental concerns, including:

- Soils containing elevated hydrocarbon concentrations such that the soil may be considered a hazardous waste;

- LNAPL pools that have leaked from reservoirs, tank farms, and pipelines and that serve as a source of soil contamination and a continued source for groundwater contamination;
- Dissolved hydrocarbons in groundwater that may adversely affect water-bearing zones considered of beneficial use or as drinking water supplies; and
- Accumulation of vapors that could pose a fire or explosion hazard.

The highly visible presence and potential hazard to public health, safety, and welfare prompted a minimum of 17 oil refineries and tank farms to be designated as health hazards by the California Department of Health Services (DOHS). This designation reflected the potential and actual subsurface occurrence of leaked hydrocarbon product derived from such facilities that during the course of their operations had migrated into the subsurface, resulting in the presence of LNAPL pools overlying the water table and regional groundwater contamination. Although several of these refineries were listed as hazardous waste sites with remediation being required under RCRA, the majority of such facilities fell under the Los Angeles Region of the California Regional Water Quality Control Board Order No. 85-17 adopted in February 1985 for assessment and subsequent remediation. This order was the first to address large-scale regional subsurface environmental impacts by the petroleum refinery industry. This order required, in part, assessment of the subsurface presence of hydrocarbons and other associated groundwater pollutants that may affect subsurface soils and groundwater beneath such facilities. Specifically, the following items were included:

- Characterization of subsurface geologic and hydrogeologic conditions;
- Delineation of LNAPL pools including chemical characterization and areal extent and volume;
- Implementation of LNAPL recovery;
- Overall aquifer restoration and rehabilitation of dissolved hydrocarbons and associated contaminants (this would also include assessment of the potential of dissolved chlorinated hydrocarbons and DNAPLs); and
- Eventual soil remediation of residual hydrocarbons.

A more in-depth discussion of the subsurface hydrogeologic setting, areal extent of LNAPL and dissolved hydrocarbons in groundwater, remedial strategy, and current status is presented in Chapter 12 (LNAPL Recovery Case Histories).

2.6 RISK-BASED CORRECTIVE ACTION OR "REBECCA"

Not all remedial strategies result from the federal or state regulatory environment. The focus of corrective action programs through the 1980s and early 1990s was to reduce the amount of contaminants to a regulatory acceptable level, with the eventual objective of reaching background levels or below some water quality criteria such as MCL, independent of site characteristics. The problem with this approach was that significant financial resources were being spent at sites of low risk, or under conditions where reaching such goals was not technically feasible. Another approach was to base the level of remedial effort expended depending on current and reasonable potential risks to the environment and human health. In the latter case, the goal was to reduce the current risk or potential adverse impact to below some acceptable level.

Risk-Based Corrective Action (RBCA) or "Rebecca" evolved from the American Society for Testing and Materials (ASTM) in 1994 with the development of Emergency Standard ES-38 ("Emergency Standard for RBCA Applied at Petroleum Release Sites"). This emergency standard followed the October 31, 1994 EPA release that more than 270,000 leaking UST sites had been reported nationwide and that more than 1000 newly discovered sites were being documented each week. Coordinated with an expressed need from industry and regulatory agencies for a standardized approach to RBCA for petroleum release sites, ASTM ES-38 "Guide for Risk-Based Corrective Action at Petroleum Release Sites" was approved as an emergency standard by the ASTM Committee on Standards. RBCA was a means to assist in the remediation process for low-toxicity, highly biodegradable petroleum hydrocarbons. With the EPA anticipating 400,000 new leaking UST cases over the next several years, the ASTM Rebecca process was quickly embraced by responsible parties.

When considering the potential risk from an impacted site, subsurface geologic and hydrogeologic conditions play a significant role, along with how beneficial the water is considered, the chemical(s) of concern, the potential receptors and exposure levels, among other factors. Some states such as Illinois require consideration of the ASTM RBCA as a suitable means of selection of cleanup action levels for virtually all sites of uncontrolled chemical contamination. Although considered sound for its initial intended application, RBCA supporters have extended its use past simple single-contaminant, petroleum hydrocarbon origin, and not necessarily by fully characterizing the site and waste materials.

REFERENCES

American Petroleum Institute, State Relations Department, 1984, *Special Report: State Underground Petroleum Products Storage Mandates*, American Petroleum Institute, Washington, D.C., September 21.

Arbuckle, J. G., et al., 1989, *Environmental Law Handbook*: Government Institutes, Inc., Tenth Edition, Rockville, MD, 664 pp.

Arscott, R. L., 1989, New Directions in Environmental Protection in Oil and Gas Operations: In *Environmental Concerns in the Petroleum Industry* (edited by S. M. Testa), Pacific Section American Association of Petroleum Geologists Symposium Volume, pp. 217–227.

Artz, N. S. and Metzler, S. C., 1985, Losses of Stored Waste Oil from Below-Ground Tanks and the Potential for Groundwater Contamination: In *Proceedings of the National Conference on Hazardous Wastes and Environmental Emergencies*, May, pp. 60–65.

Bennett, P. D., Brumbach, B., Farmer, T. W., Funkhouser, P. L., and Hatheway, A. W., 1999, Remedy Selection for Cleanup of Uncontrolled Waste Sites: *Practical Periodical of Hazardous, Toxic, and Radioactive Waste Management*, Vol. 3, No. 1, January, pp. 23–34.

Buonocore, P. E., Ketas, G. F., and Garrahan, P. E., 1986, New Requirements for Underground Storage Tanks: In *Proceedings of the National Conference on Hazardous Wastes and Hazardous Materials*, March 4–6, pp. 246–250.

California Code of Regulations, Title 23, Chapter 3, Subchapter 16, Article 4.

Dezfulian, H., 1988, Site of an Oil-Producing Property: In *Proceedings of the Second International Conference on Case Histories in Geotechnical Engineering*, Vol. 1, pp. 43–49.

Doll, B. E., 1947, Formulating Legislation to Protect Ground Water from Pollution: *Journal AWWA*, October, pp. 1003–1009.

Eger, C. K. and Vargo, J. S., 1989, Prevention: Ground Water Contamination at the Martha Oil Field, Lawrence and Johnson Counties, Kentucky: In *Environmental Concerns in the Petroleum Industry* (edited by S. M. Testa), Pacific Section of the American Association of Petroleum Geologists Symposium Volume, pp. 83–105.

Fenster, D. E., 1990, *Hazardous Waste Laws, Regulations, and Taxes for the U.S. Petroleum Refining Industry*: PenWell Books, Tulsa, OK, 215 pp.

Garcia, D. H. and Henry, E. C., 1989, Environmental Considerations for Real Estate Development of Oil Well Drilling Properties in California: In *Environmental Concerns in the Petroleum Industry* (edited by S. M. Testa), Pacific Section American Association of Petroleum Geologists Symposium Volume, pp. 117–127.

Government Institutes, Inc., 1989, *Environmental Statutes*: Government Institutes, Inc., Rockville, MD, 1169 pp.

Gregston, T. G., 1993, *An Introduction to Federal Environmental Regulations for the Petroleum Industry*: Petroleum Extension Service, Division of Continuing Education, University of Texas, Austin, 194 pp.

Hansen, P., 1985, L.U.S.T.: Leaking Underground Storage Tanks: In *Proceedings of the National Conference on Hazardous Waste and Environmental Emergencies*, May 14–16, pp. 66–67.

Henderson, T., 1989, Assessment of Risk to Ground Water Quality from Petroleum Product Spills: In *Proceedings of the National Water Well Association and American Petroleum Institute Conference on Petroleum Hydrocarbons and Organic Chemicals in Ground Water: Prevention, Detection and Restoration*, NWWA, Houston, TX, pp. 333–345.

Jones, S. C. and O'Toole, P., 1989, Increasing Environmental Regulation of Oil and Gas Operations: In *Environmental Concerns in the Petroleum Industry* (edited by S. M. Testa), Pacific Section American Association of Petroleum Geologists Symposium Volume, pp. 209–215.

Lovegreen, J. R., 1989, Environmental Concerns in Oil-Field Areas during Property Transfers: In *Environmental Concerns in the Petroleum Industry* (edited by S. M. Testa), Pacific Section American Association of Petroleum Geologists Symposium Volume, pp. 129–158.

McFaddin, M. A., 1996, *Oil and Gas Field Waste Regulations Handbook*: PenWell Publishing Company, Tulsa, OK, 390 pp.

Meyer, C. F., 1973, *Polluting Ground Water: Some Causes, Effects, Controls and Monitoring*: U.S. EPA Environmental Monitoring Series, EPA-600/4-73-001b.

Michie, T., 1988, Oil and Gas Industry Water Injection Well Corrosion Study: In *Proceedings of the UIPC Summer Meetings*, pp. 47–67.

Moore, E. B., 1997, *The Environmental Impact Statement Process and Environmental Law*: Battelle Press, Columbus, OH, 148 pp.

Mulligan, W. S., 1988, Waste Minimization in the Petroleum Industry: In *Proceedings of the Fourth Annual Hazardous Materials Management Conference/West*, November 8–10, 1988, pp. 540–543.

Office of Technology Assessment, 1984, *Protecting the Nation's Ground Water from Contamination,* Vol. I and II: OTA, Washington, D.C., No. OTA-0-233, October, 503 pp.

Orszulik, S. T. (editor), 1997, *Environmental Technology in the Oil Industry*: Blackie Academic and Professional, New York, 400 pp.

Patrick, R., Ford, E. and Quarles, J., 1988, *Groundwater Contamination in the United States*: University of Pennsylvania Press, Second Edition, Philadelphia, PA, 513 pp.

Pearson, G. and Oudijk, K. G., 1993, Investigation and Remediation of Petroleum Product Releases from Residential Storage Tanks: *Ground Water Monitoring & Remediation*, Summer, pp. 124–128.

Reinke, D. C. and Swartz, L. L. (editors), 1999, *The NEPA Reference Guide*: Battelle Press, Columbus, OH, 267 pp.

Robb, A. E., Moore, T. J., and Simon, J. A., 1998, Recent Developments in Cleanup Technologies: *Remediation*, Vol. 8, No. 3, pp. 133–140.

Rowley, K., 1986, The Rules of the Games in Ground-Water Monitoring: In *Proceedings of the Second Annual Hazardous Materials Management Conference West*, December 3–5, pp. 365–374.

Savini, J. and Kammerer, J. C., 1961, *Urban Growth and the Water Regime*: U.S. Geological Survey Water-Supply Paper, No. 1591-A, 43 pp.

Sittig, M., 1978, *Petroleum Transportation and Production — Oil Spill and Pollution Control*: Noyes Data Corporation, Park Ridge, NJ, 360 pp.

Syed, T., 1989, An Overview of the Underground Injection Control Regulations for Class II (Oil and Gas Associated) Injection Wells — Past, Present and Future: In *Environmental Concerns in the Petroleum Industry* (edited by S. M. Testa), Pacific Section American Association of Petroleum Geologists Symposium Volume, pp. 199–207.

Testa, S. M., 1989, Regional Hydrogeologic Setting and Its Role in Developing Aquifer Remediation Strategies: In *Proceedings of the Geological Society of America, 1989 Annual Meeting Abstracts with Programs*, Vol. 21, No. 6, p. A96.

Testa, S. M., 1990, Light Non-Aqueous Phase Liquid Hydrocarbon Occurrence and Remediation Strategy, Los Angeles Coastal Plain, California: In *Proceedings of the International Association of Hydrogeologists, Canadian National Chapter, on Subsurface Contamination by Immiscible Fluids*, April, in press.

Testa, S. M., 1994, *Geologic Aspects of Hazardous Waste Management*: CRC Press/Lewis Publishers, Boca Raton, FL, 537 pp.

Testa, S. M. and Townsend, D. S., 1990, Environmental Site Assessments in Conjunction with Redevelopment of Oil-Field Properties within the California Regulatory Framework: In *Proceedings of the National Water Well Association of Groundwater Scientists and Engineers Cluster of Conferences.*

Testa, S. M., Henry, E. C., and Hayes, D., 1988, Impact of the Newport-Inglewood Structural Zone on Hydrogeologic Mitigation Efforts — Los Angeles Basin, California: In *Proceedings of the National Water Well Association of Ground Water Scientists and Engineers FOCUS Conference on Southwestern Ground Water Issues*, pp. 181–203.

U.S. Accounting Office, 1984, *Federal and State Efforts to Protect Groundwater*: Washington, D.C., February, 80 pp.

U.S. Environmental Protection Agency, 1968, *Code of Federal Regulations*, Title 40, Parts 190, 191, 192, 193, and 194.

U.S. Environmental Protection Agency, 1980, *Code of Federal Regulations*, Title 40, Section 101(14).

U.S. Environmental Protection Agency, 1984, *Code of Federal Regulations*, Title 40, Parts 280 and 281.

U.S. Environmental Protection Agency, 1985a, *National Water Quality Inventory 1984 National Report to Congress*: Office of Water Regulations and Standards, Washington, D.C., EPA-440/4-85-029, 173 pp.

U.S. Environmental Protection Agency, 1985b, *Code of Federal Regulations*, Title 40, Parts 124, 144, 145, 146, and 147.

U.S. Environmental Protection Agency, 1985c, *Code of Federal Regulations*, Title 40, Parts 124, 144, 145, 146, and 147.

U.S. Environmental Protection Agency, 1986a, Amendment to National Oil and Hazardous Substances Contingency Plan National Priorities List, Final Rule and Proposed Rule: *Federal Register,* Vol. 51, No. 111, June 10, pp. 21053–21112.

U.S. Environmental Protection Agency, 1986b, *Summary of State Reports on Releases from Underground Storage Tanks*: Office of Solid Waste, Washington, D.C., EPA-600/M-86-020, July, 95 pp.

Wolbert, G. S., 1979, *U.S. Oil Pipe Lines — An Examination of How Oil Pipe Lines Operate and the Current Public Policy Issues Concerning Their Ownership*: American Petroleum Institute, Washington, D.C., 556 pp.

3 Hydrogeologic Principles

"The science of hydrology would be relatively simple if water were unable to penetrate the Earth's surface."

3.1 INTRODUCTION

Successful subsurface characterization, detection monitoring, and ultimate remediation is predicated on a conceptual understanding of the subsurface environment. Factors that affect the fate and transport of contaminants, determination of possible adverse risks to public health, safety, and welfare, and the degradation of groundwater resources are largely controlled by regional and local subsurface conditions. Successful resolution of issues pertaining to the occurrence and mitigation of subsurface contamination requires adequate geologic and hydrogeologic characterization to gain insight into the subsurface distribution and preferential migration pathways of contaminants, potential receptors, and development of an appropriate remedial strategy.

From a geologic perspective, an understanding of both the regional and local geologic setting is essential. Evaluation of regional geologic features might describe a fault-bounded structural basin containing sediments representing several depositional environments both laterally and vertically. Each depositional environment might be composed of rocks containing some of the same building blocks, gravels, sands, silts, and clays, in different configurations and juxtapositions. The distribution of these sediments on a regional scale may be dependent on a number of interrelated controls: sedimentary processes, sediment supply, climate tectonics, sea level changes, biological activity, water chemistry, and volcanism. The relative importance of these regional factors can vary between different depositional environments. On the local and sublocal scale, the porosity and permeability of a particular sandstone or unconsolidated sediment beneath a specific site in that basin might depend on such factors as individual grain size, grain size sorting, primary and secondary cementation, and lateral and vertical facies changes. It is these factors that largely account for widely disparate contaminant concentrations, up to several orders of magnitude, and large fluctuations in chemical data with time.

The largest percentage of environmentally contaminated sites in the world is situated on alluvial and coastal plains consisting of complex interstratified sediments. Lithology and stratigraphy are the most important factors affecting contaminant movement in soils and unconsolidated sediments. Stratigraphic features including geometry and age relations between lenses, beds, and formations, and lithologic characteristics of sedimentary rocks, such as physical composition, grain size, grain packing, and cementation, are among the most important factors affecting groundwater and contaminant flow in sedimentary rocks. Igneous rocks, which include volcanic rocks such as basalts (or lava flows) and tuffs, and plutonic rocks such as granites, are produced by cooling of magma during emplacement, or deposition. Metamorphic rocks are produced by deformation after deposition or crystallization. Groundwater and contaminant flow in igneous and metamorphic rocks is most affected by structural features such as cleavages, fractures, folds, and faults.

Overall understanding of the regional geologic and hydrogeologic framework, characterization of regional geologic structures, and proper delineation of the relationship between various aquifers are essential to assessing soil and groundwater vulnerability and to implementing reasonable short- and long-term aquifer restoration and rehabilitation programs. Fundamental geologic and hydrogeologic principles are the focus of this chapter, which will set the foundation for later discussion. Geologic aspects including porosity, permeability, and diagenesis are discussed, in addition to the concepts of sedimentary sequences and facies architecture. From a hydrogeologic perspective, discussion of relationships that govern migration of fluids, contaminants, and gases in the subsurface, including the flux equation, Darcy's law, and the behavioral manner of gases and vapors is presented. The types of aquifers, occurrence and migration of water within the saturated system under steady-state and non-steady-state flow, unsaturated systems, and capillary barriers are also discussed.

3.2 POROSITY, PERMEABILITY, AND DIAGENESIS

Preferred fluid migration pathways are influenced by porosity and permeability, sedimentary sequences, facies architecture, and fractures. Porosity is a measure of pore space per unit volume of rock or sediment and can be divided into two types: absolute porosity and effective porosity. Absolute porosity (n) is the total void space per unit volume and is defined as the percentage of the bulk volume that is not solid material. The equation for basic porosity is listed below:

$$n = \frac{\text{bulk volume} - \text{solid volume}}{\text{bulk volume}} \times 100 \tag{3.1}$$

Porosity can exist as individual open spaces between sand grains in a sediment or as fracture spaces in a dense rock. A fracture in a rock or solid material is an opening or a crack within the material. Matrix refers to the dominant constituent of the soil, sediment, or rock, and is usually a finer-sized material surrounding or filling the interstices between larger-sized material or features. Gravel may be composed of large cobbles in a matrix of sand. Likewise, a volcanic rock may have large crystals floating in a matrix of glass. The matrix will usually have properties different from the other features in the material. Often, either the matrix or the other features will dominate the behavior of the material, leading to the terms matrix-controlled transport, or fracture-controlled flow.

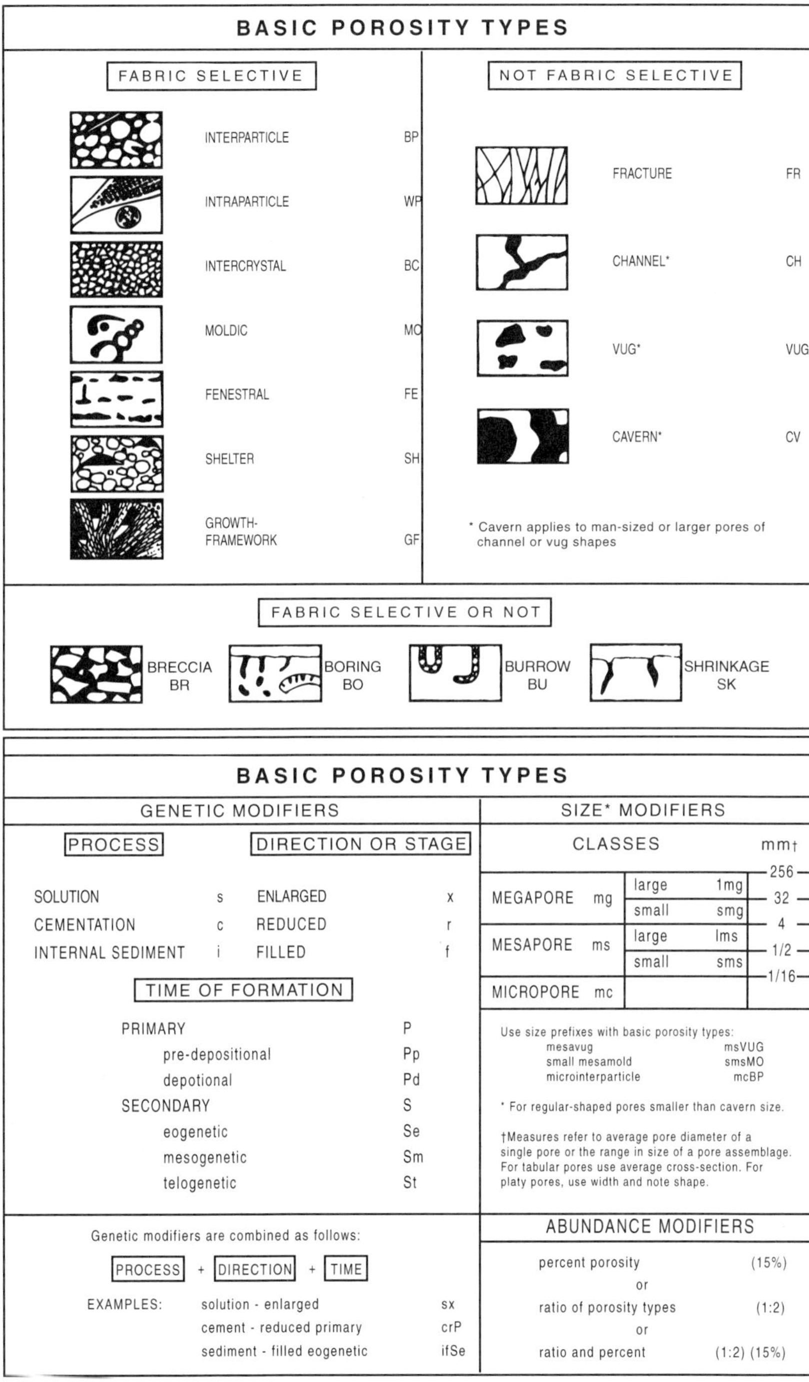

FIGURE 3.1 Classification of porosity types. (After Choquette and Pray, 1970.)

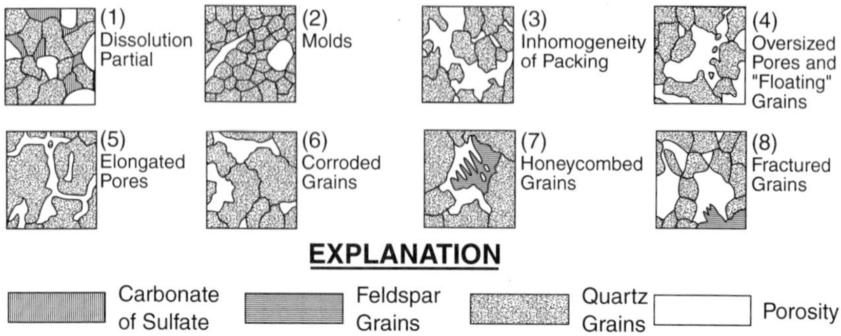

FIGURE 3.2 Petrographic criteria for secondary porosity. (After Schmidt et al., 1977.)

Effective porosity (N_e) is defined as the percentage of the interconnected bulk volume (i.e., void space through which flow can occur) that is not solid material. The equation for effective porosity is listed below:

$$N_e = \frac{\text{interconnected pore volume}}{\text{bulk volume}} \times 100 \tag{3.2}$$

Effective porosity (N_e) is of more importance and, along with permeability (the ability of a material to transmit fluids), determines the overall ability of the material to store and transmit fluids or vapors readily. Where porosity is a basic feature of sediments, permeability is dependent upon the effective porosity, the shape and size of the pores, pore interconnectiveness (throats), and properties of the fluid or vapor. Fluid properties include capillary force, viscosity, and pressure gradient.

Porosity can be primary or secondary. Primary porosity develops as the sediment is deposited and includes inter- and intraparticle porosity (Figure 3.1). Secondary porosity develops after deposition or rock formation and is referred to as diagenesis (Figure 3.2).

Permeability is a measure of the connectedness of the pores. Thus, a basalt containing many unconnected air bubbles may have high porosity but no permeability, whereas a sandstone with many connected pores will have both high porosity and high permeability. Likewise, a fractured, dense basaltic rock may have low porosity but high permeability because of the fracture flow. The nature of the porosity and permeability in any material can change dramatically through time. Porosity and permeability can increase, for example, with the dissolution of cements or matrix, faulting, or fracturing. Likewise, porosity and permeability can decrease with primary or secondary cementation and compaction.

The nature of porosity and permeability in any material can change with time through diagenesis. Diagenetic processes are important because secondary processes can significantly affect the overall porosity and permeability of a sediment (Tables 3.1 and 3.2). Once a sediment is deposited, diagenetic processes begin immediately and can significantly affect the overall porosity and permeability of the unconsolidated materials. These processes include compaction, recrystallization, dissolution,

TABLE 3.1
Processes Affecting Permeability of Sediment Following Deposition

Depositional Processes	Diagenetic Processes
Texture	Compaction
Grain size	Recrystallization
Sorting	Dissolution
Grain slope	Replacement
Grain packing	Fracturing
Grain roundness	Authigenesis
Mineral composition	Cementation

replacement, fracturing, authigenesis, and cementation. Compaction occurs by the accumulating mass of overlying sediments, called overburden. Unstable minerals may recrystallize, changing the crystal fabric but not the mineralogy, or they may undergo dissolution and/or replacement by other minerals. Dissolution and replacement processes are common with limestones, sandstones, and evaporites. Authigenesis refers to the precipitation of new mineral within the pore spaces of a sediment. Lithification occurs when cementation is sufficient in quantity such that the sediment is changed into a rock. Examples of lithification include sands and clays changing into sandstones and shales, respectively (Figure 3.3).

The most important parameters influencing porosity in sandstone are age (time of burial), mineralogy (i.e., detrital quartz content), sorting, the maximum depth of burial, and, to a lesser degree, temperature. Compaction and cementation will reduce porosity, although porosity reduction by cement is usually only a small fraction of the total reduction. The role of temperature probably increases above a geothermal gradient of 4°C/100 m. Uplift and erosional unloading may also be important in the development of fracture porosity and permeability. Each sedimentary and structural basin has its own unique burial history and the sediments and rocks will reflect unique temperature and pressure curves vs. depth.

TABLE 3.2
Diagenesis Processes and Effects of Secondary Porosity

Diagenesis Process	Effects
Leaching	Increase ϕ and K
Dolomitization	Increase K; can also decrease ϕ and K
Fracturing; brecciation	Increase K; can also increase channeling
Recrystallization	May increase pore size and K; can also decrease ϕ and K
Cementation by calcite, dolomite, anhydrite, pyrobitumen and silica	Decrease ϕ and K

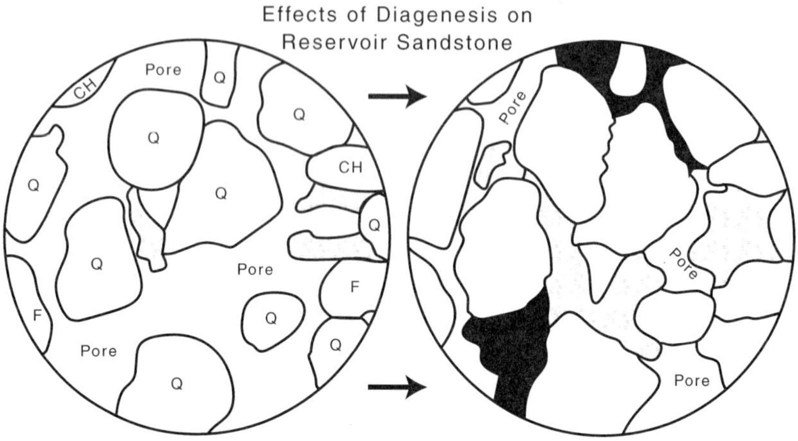

FIGURE 3.3 Reduction in porosity in sandstone as a result of cementation and growth of authigenic minerals in the pores which affects the amount, size, and arrangement of the pores. (Modified after Ebanks, 1987.)

3.3 SEDIMENTARY SEQUENCES AND FACIES ARCHITECTURE

A three-dimensional perspective of the geologic setting beneath a site begins with an understanding of the environment in which the subsurface materials were formed. This process begins with the characterization of subsurface materials, correlation of similar materials, and development of conceptual facies models. Analysis of sedimentary depositional environments is important since groundwater resource occurrence and usage are found primarily in unconsolidated deposits formed in these environments. The more important depositional environments include deltaic, eolian, fluvial, glaciofluvial, lacustrine, and shallow marine. These are important because aquifers, which are water-bearing zones often considered of beneficial use and warranting protection, occur in these environments. Thus, most subsurface environmental investigations conducted are also performed in these types of environments.

The literature is full of examples where erroneous hydraulic or contaminant distribution information has been relied upon. In actuality, however, (1) the wells were screened across several high-permeability zones or across different zones creating the potential for cross-contamination of a clean zone by migrating contaminants from impacted zones, (2) inadequate understanding of soil-gas surveys and vapor-phase transport in heterogeneous environments allowing for an ineffective vapor extraction remedial strategy prevailed, (3) wells screened upward-fining sequences, or (4) the depositional environment was erroneously interpreted.

Geologic heterogeneities control contaminant movement in the subsurface. Heterogeneities within sedimentary sequences can range from large-scale features associated with different depositional environments that further yield significant large- and small-scale heterogeneities via development of preferential grain orientation

FIGURE 3.4 Photograph showing small-scale sedimentary structures that can significantly affect migration pathways of contaminants.

(Figure 3.4). This results in preferred areas of higher permeability and, thus, preferred migration pathways of certain constituents considered hazardous.

To characterize these heterogeneities adequately, it becomes essential that subsurface hydrogeologic assessment include determination of the following:

- The depositional environment and facies of all major stratigraphic units present;
- The propensity for heterogeneity within the entire vertical and lateral sequence and within different facies of all major stratigraphic units present; and
- The potential for preferential permeability (i.e., within sand bodies).

The specific objectives to understanding depositional environments as part of subsurface environmental studies are to (1) identify depositional processes and resultant stratification types that cause heterogeneous permeability patterns, (2) measure the resultant permeabilities of these stratification types, and (3) recognize general permeability patterns that allow simple flow models to be generated. Flow characteristics in turn are a function of the types, distributions, and orientations of the internal stratification. Since depositional processes control the zones of higher permeability within unconsolidated deposits, a predictive three-dimensional depositional model to assess potential connections or intercommunication between major zones of high permeability should also be developed. A schematic depicting the various components of an integrated aquifer description has been developed (Figure 3.5).

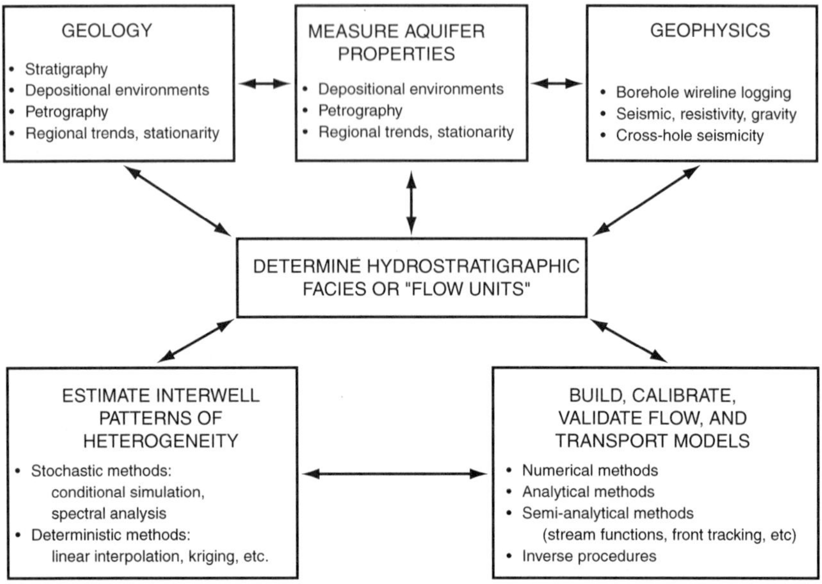

FIGURE 3.5 Schematic depicting the various components of an integrated aquifer description.

3.3.1 Hydrogeologic Facies

Understanding of facies architecture is extremely important to successful characterization and remediation of contaminated soil and groundwater. Defining a hydrogeologic facies can be complex. Within a particular sedimentary sequence, a hydrogeologic facies can range over several orders of magnitude. Other parameters such as storativity and porosity vary over a range of only one order of magnitude. A hydrogeologic facies is defined as a homogeneous, but anisotropic, unit that is hydrogeologically meaningful for purposes of conducting field studies or developing conceptual models. Facies can be gradational in relation to other facies, with a horizontal length that is finite but usually greater than its corresponding vertical length. A hydrogeologic facies can also be viewed as a sum of all the primary characteristics of a sedimentary unit. A facies can thus be referred to with reference to one or several factors, such as lithofacies, biofacies, geochemical facies, etc.

The importance of facies cannot be understated. For example, three-dimensional sedimentary bodies of similar textural character are termed lithofacies. It is inferred that areas of more rapid plume migration and greater longitudinal dispersion correlate broadly with distribution and trends of coarse-grained lithofacies and are controlled by the coexistence of lithologic and hydraulic continuity. Therefore, lithofacies distribution can be used for preliminary predictions of contaminant migration pathways and selection of a subsurface assessment and remediation strategy. However, caution should be exercised in proximal and distal assemblages where certain layered sequences may be absent due to erosion and the recognition of cyclicity is solely dependent on identifying facies based simply on texture. Regardless, the facies reflects deposition in a given environment and possesses certain characteristics of that environment. Sedimentary structures also play a very important role in deriving permeability-distribution models and developing fluid-flow models.

3.3.2 Hydrostratigraphic Models

Hydrostratigraphic analysis is conducted in part by the use of conceptual models. These models are used to characterize spatial trends in hydraulic conductivity and permit prediction of the geometry of hydrogeologic facies from limited field data. Conceptual models can be either site specific or generic. Site-specific models are descriptions of site-specific facies that contribute to understanding the genesis of a particular suite of sediments or sedimentary rocks. The generic model, however, provides the ideal case of a particular depositional environment or system. Generic models can be used in assessing and predicting the spatial trends of hydraulic conductivity and, thus, dissolved contaminants in groundwater. Conventional generic models include either a vertical profile that illustrates a typical vertical succession of facies or a block diagram of the interpreted three-dimensional facies relationships in a given depositional system.

Nearly all depositional environments are heterogeneous, which for all practical purposes restricts the sole use of homogeneous-based models in developing useful hydraulic conductivity distribution data for assessing preferred contaminant migration pathways and developing containment and remediation strategies. Much discussion in the literature exists regarding the influences of large-scale features such as faults, fractures, significantly contrasting lithologies, diagenesis, and sedimentalogical complexities (Figure 3.6). Little attention, however, has been given to internal heterogeneity within genetically defined sand bodies caused by sedimentary structures and

POINT BAR MODEL

ROCK	STRUCTURE	HORIZONTAL PERMEABILITY
• Siltstone, very fine grained, muddy sandstone	Horizontal Apple bedded	Very low
• Silty, fine grained sandstone, poorly sorted	Ripple bedded, parallel bedded	Low to moderate
• Fine to medium grained sandstone, well sorted	Cross bedded	Moderate to high
• Medium to coarse grained sandstone and conglomerate, poor to moderate sorting	Massive of cross bedded	Low to moderate

FIGURE 3.6 Point-bar geologic model showing the influence of a sequence of rock textures and structures in a reservoir consisting of a single point-bar deposit on horizontal permeability, excluding effects of diagenesis. (Modified after Ebanks, 1987.)

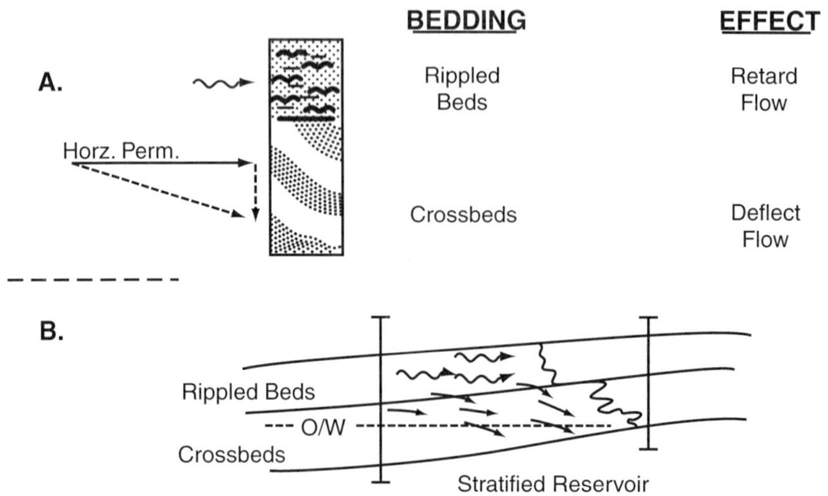

FIGURE 3.7 Diagram showing effects of sedimentary structures and textures on the flow of fluids in a point-bar sandstone reservoir. The cross-bedded unit is coarser grained and is inferred to have better reservoir properties (i.e., permeability) than the overlying rippled unit (A). Uneven advance of injected fluids illustrating permeability variations results from differences in reservoir quality. (Modified after Ebanks, 1987.)

associated depositional environment and intercalations. In fact, for sand bodies greater variability exists within bedding and lamination pockets than between them.

An idealized model of the vertical sequence of sediment types by a meandering stream shows the highest horizontal permeability (to groundwater or contaminants) to be the cross-bedded structure in a fine- to medium-grained, well-sorted unconsolidated sand or sandstone (Figure 3.7). The various layers illustrated affect the flow of fluids according to their relative characteristics. For example, in a point-bar sequence, the combination of a ripple-bedded, coarser-grained sandstone will result in retardation of flow higher in the bed, and deflection of flow in the direction of dip of the lower trough cross beds.

3.3.3 Sequence Stratigraphy

In lieu of correlating specific lithologies, one can also consider correlating lithology sequences that share a common ancestry. This method may have merit when addressing regional-scale hydrogeologic issues and is referred to as sequence stratigraphy. Sequence stratigraphy is a method of relating the depositional environment of sediments and sedimentary rocks to the stratigraphic framework in which they were deposited. A sequence for this purpose is a fundamental stratal unit for stratigraphic analysis, and can be defined as a relatively conformable, genetically related succession of strata bounded by unconformities; it may include all genetically related lithologies (lithofacies) between the defining conformities and regionally extensive aquifers and aquitards lithologies.

Hydrostratigraphy is the study of the stratigraphic framework of a distinct hydrogeologic system consisting of aquifers and aquitards. The hydrostratigraphic model defines the aquifer architecture, where *aquifer* is defined as water-yielding lithologies or systems of water-yielding lithologies that do not necessarily include all genetically related lithologies. This state-of-the-art method has been used by the petroleum industry for over 25 years as a standard procedure for the study of strata in the subsurface, and for predicting and delineating reservoir geometry and continuity. Only recently has it been used as a tool in the environmental industry to define contaminant migration pathways and hydrostratigraphy in certain areas where significant contaminant impact exists or could occur.

3.4 STRUCTURAL STYLE

In consolidated (bedrock) geologic environments, structural elements such as faults, fractures, joints, and shear zones can significantly impact groundwater flow and, thus, contaminant migration. Faults are usually of less importance in most site-specific situations, whereas fractures, joints, and shear zones can play a significant role. Regional geologic processes that produce certain structural elements, notably secondary porosity (or fracture porosity), include faulting (seismicity), folding, uplift, and erosional unloading and compressional loading of strata. The characterization of fractured media is very important because certain hydrocarbon and organic constituents such as solvents and chlorinated hydrocarbons, which are denser than water, are likely to migrate vertically downward through these preferred pathways, and may even increase the permeability within these zones.

Fracture classification is summarized in Table 3.3. Tectonic and possibly regional fractures result from surface forces (i.e., external to the body as in tectonic fractures); contractional fractures are of varied origin resulting from desiccation, syneresis, thermal gradients, and mineral phase changes. Desiccation fractures develop in clay- and silt-rich sediments upon a loss of water during subaerial drying. Such fractures are typically steeply dipping, wedged-shaped openings that form cuspate polygons of several nested sizes. Syneresis fractures result from a chemical process involving dewatering and volume reduction of clay, gel, or suspended colloidal material via

TABLE 3.3
Classification of Fractures

Fracture Type	Classification	Remarks
Experimental	Shear	
	Extension	
	Tensile	
Natural	Tectonic	Due to surface forces
	Regional	Due to surface forces (?)
	Contractional	Due to body forces
	Surface-related	Due to body forces

tension or extension. Associated fracture permeability tends to be isotopically distributed since developed fractures are caused by the cooling of hot rock (i.e., basalts). Mineral phase-change fracture systems are composed of extension and tension fractures related to a volume reduction due to a mineral phase change. Characterized by poor geometry, phase changes such as calcite to dolomite or montmorillonite to illite can result in about 13% reduction in molar volume.

Surface-related fractures develop during unloading, release of stored stress and strain, creation of free surfaces or unsupported boundaries, and general weathering. Unloading fractures or relief joints commonly occur during quarrying or excavation activities. Upon a one-directional release of load, the rock relaxes and spalls or fractures. Free or unsupported surfaces can develop both extension and tensional fractures. These types of fractures are similar in morphology and orientation to unloading fractures. Weathering-derived fractures are related to mechanical and chemical weathering processes such as freeze–thaw cycles, mineral alluation, diagenesis, small-scale collapse and subsidence, and mass-wasting processes.

Faults are regional structures that can serve as barriers, partial barriers, or conduits to groundwater flow. The influence and effect of faults on subsurface fluid flow depend on the physical properties of the rocks and/or sediments, and orientation of the strata within their respective fault blocks. The influence of regional structures such as faults and, to a lesser degree, folds can have a profound effect on groundwater occurrence, regime, quality, and usage, and delineation and designation of water-bearing zones.

Most fractured media consist of rock or sediment blocks bounded by discrete discontinuities (Figure 3.8). The aperture can be open, deformed, closed, or a combination. The primary factors to consider in the migration of subsurface fluids within fractured media are fracture density, orientation, effective aperture, and the nature of the rock matrix. Fracture networks are complex three-dimensional systems. The analysis of fluid flow through a fracture media is difficult since the only means of evaluating hydraulic parameters is by the performance of hydraulic tests. These tests require that the geometric pattern and/or degree of fracturing be known. Fracture density, or the number of fractures per unit volume of rock, and fracture orientation are the most important factors in assessing the degree of interconnection of fracture sets. Fracture spacing is influenced by intrinsic and environmental properties. Intrinsic properties include load-bearing framework, grain size, porosity, permeability, thickness, and previously existing mechanical discontinuities. Environmental properties include net overburden, temperature, time (strain rate), differential stress, and pore fluid composition. Fracturing can also develop under conditions of excessive fluid pressures.

Clay-rich soils and rocks, for example, are commonly used as an effective hydraulic seal. The integrity of this seal, however, can be jeopardized should excessive fluid pressures be induced, resulting in hydraulic fracturing. Hydraulic fracturing in clays is a common feature in nature at hydrostatic pressures ranging from 10 kPa up to several mega pascals. Although hydraulic fracturing can significantly decease the overall permeability of the clay, the fractures are likely to heal in later phases due to the swelling pressure of certain clay minerals.

FIGURE 3.8 Photograph of near vertical, irregular, clay-filled fault in sedimentary siltstone and sandstone strata.

3.5 THE FLUX EQUATION

The average macroscopic flow velocity of a fluid is the average of all the microscopic flow velocities, is usually observed by tracking some chemical tracer or fluid flow front through the material, and is a bulk property of the media. There are several bulk properties of the media that are important and that are more readily determined than their microscopic components. These bulk properties describe the system itself and can be used to compare different systems under similar conditions and the same system under different conditions. Certain bulk properties describe the migration of fluids, dissolved molecules, and gases through the porous media and are called the transport parameters. The transport equations are the mathematical relationships used to relate the transport parameters, among other things, to the overall migration of a species of interest.

The flux equation is one of the most powerful relationships for describing migration through a conducting medium, e.g., water through a soil, electricity through a wire, molecules through a membrane, heat through a wall, gas through a pore space, etc. The flux equation works for those situations where the flow rate is proportional to the driving force. The flux equation has many manifestations depending upon what is moving, and is consequently given different names, like Fick's law for diffusing molecules or Darcy's law for flowing water. But the three essential components of the equation are always the same:

$$\text{Flux Density} = \text{Conductivity} \times \text{Driving Force}$$

The flux design is the flow rate per unit of cross-sectional area, or the amount of whatever is flowing per unit time divided by the area through which it is flowing. For a garden hose it can be described as the number of gallons per minute per square inch of hose cross-sectional area. For a thermos, it might be the number of joules or calories transferred per second per square centimeter of surface area. The conductivity is the parameter that describes the bulk property of the system with respect to whatever is flowing and describes how well or how poorly that material conducts that water, or that heat, or that electricity. For flowing water, it is called the hydraulic conductivity, for molecular diffusion it is the diffusion coefficient, for heat it is the thermal conductivity. The driving force is usually a potential gradient. It might be a pressure gradient, a gravitational acceleration, a temperature gradient, or a voltage.

The two flux equations of importance to subsurface transport are Darcy's law for the advective flow of water and other liquids and Fick's law for the diffusive flow of molecules and gases. These laws are independently discussed below.

$$\text{Darcy's law:} \quad q = -K\tilde{N}h \tag{3.3}$$

$$\text{Fick's law:} \quad J = -D\tilde{N}C \tag{3.4}$$

where q is the flux density of water, K is the hydraulic conductivity, $\tilde{N}h$ is the potential gradient (hydraulic gradient when the fluid is water), J is the flux density of molecules, D is the diffusion coefficient, and $\tilde{N}C$ is the concentration gradient of the molecular species. The negative sign means that flow will occur in the direction of decreasing potential (i.e., flow will occur downhill or downgradient).

3.5.1 Darcy's Law

The hydraulic conductivity, K, is a parameter that includes the behavior of both the porous media and the fluid. It is often desirable to know the behavior of just the porous media or its intrinsic permeability, k, which is theoretically independent of the fluid. The relationship between the two parameters is given by

$$K = k\frac{rg}{h} \tag{3.5}$$

where r is the fluid density, g is the gravitational acceleration, and h is the viscosity of the fluid. The proportionality between K and k is the term rg/h, called the fluidity. Equation 3.3 can be used to determine the Darcy behavior of one fluid in a porous media, such as oil, from the measured behavior of another fluid, such as water, in the same porous media.

It is often easy to measure the flux density, e.g., using a flowmeter, and then determine the hydraulic conductivity or diffusion coefficient by dividing the flux by the driving force. One of the most difficult problems is determining how to represent the driving force. The symbol ∇ is called an operator, which signifies that some mathematical operation is to be performed upon whatever function follows. ∇ means to take the gradient with respect to distance. For Darcy's law under saturated conditions in the subsurface, this can often be easy because the potential gradient can be described by the hydraulic head gradient. The hydraulic head, h, is the height, z, of a free water surface above an arbitrary reference point plus any additional pressures on the system, Ψ, such that $h = \Psi + z$. The hydraulic head gradient, designated $\nabla h/\nabla l$, is the difference between two hydraulic heads with respect to the distance between them. Therefore, in Equation 3.3, ∇h can be given by $\nabla h/\nabla l$. For mathematical convenience in working with functions and infinitesimal changes, the differences are symbolized as a differential, dh/dl.

3.5.2 Fick's Law

Fick's law relates mainly to diffusion of dissolved chemicals with respect to differences in concentration gradients in the material or fluid, or migration of gases or vapors in response to differences in gas pressure. However, before discussing Fick's law, it is necessary to define the term *ion*. An ionic species, or an ion, is defined as a species that carries a charge (i.e., Na^+, Cl^-, Fe^{2+}, Sr^{2+}, NO^{3-}, CO_3^{2-}, etc.), as opposed to dissolved SiO_2 and H_2CO_3, which are not ionic species, but rather are neutral species, sometimes designated as SiO_2^0. A monovalent ion has a single charge or a charge number of one (i.e., Na^+ or Cl^-). A divalent ion has two charges or a charge number of two (i.e., Ca^{2+} or SO_4^{2-}). A trivalent ion has three charges or a charge number of three (i.e., Ce^{3+}, PO_4^{3-}, etc.). Often, shorthand notations are used to depict complicated ions and use an abbreviation along with an important aspect or functional group (i.e., HOAc means acetic acid and OAc^- is the acetate ion formed by loss of H^+). Ionization is the process by which a neutral species becomes an ion, or a preexisting ion becomes further ionized. The energy required to ionize a molecule or species is called the ionization potential, and must be exceeded to remove or add electrons to produce an ion. There are as many ionization mechanisms as there are energy sources, such as chemical reactions, energy deposited by light or high-energy particles in the case of ionizing radiation, thermal ionization using heat, etc.

Keeping in mind Fick's law and Equation 3.4, consider diffusion of chloride ion (Cl^-) through a 0.0001-cm membrane separating fresh water with a Cl concentration of 100 mg/l (0.1 g/1000 cm^3 or 0.0001 g/cm^3), from salt water with a Cl concentration of 10,000 mg/l (10 g/1000 cm^3 or 0.01 g/cm^3). The driving force is the concentration gradient, $\nabla C/\nabla l$, or [(0.01 – 0.0001), 0.0001] g/cm^3/cm or 99 g/cm^4. If the diffusion coefficient of the membrane is 0.00001 cm^2/s, then the flux is 0.00001(99) = 0.00099 g/cm^2s, meaning that 0.00099 g of Cl will pass through each square centimeter of membrane surface per second.

3.5.3 Gases and Vapors

For vapors and gases, the driving force is a pressure gradient in the gas, often resulting from changes in temperature in different regions of the subsurface. Mixtures of gases can behave more or less ideally, and each gas can be described by its partial pressure (i.e., the pressure that the gas would exert if it alone occupied the total space of the mixture). Therefore, all the individual partial pressures of each gas in a gas mixture add up to the actual pressure of the whole mixture. Partial pressures are very sensitive to temperature. A common reference pressure for gases is the atmospheric pressure at sea level, called an atmosphere (atm), which is equal to approximately 14.5 psi. This is the basal pressure generated by a column of air the height of the atmosphere at sea level, a column of mercury 760 mm high, or a column of water 1020 cm high. Ordinary air is 78% nitrogen, 21% oxygen, and 0.03% carbon dioxide, with the rest argon and some trace gases. The partial pressures for each gas are, therefore, $P_{N_2} = 0.78$ atm, $P_{o_2} = 0.21$ atm, and $P_{co_2} = 0.0003$ atm. Alternatively, pressure can be given in the mass of gas per volume of air (g/cm^3).

Consider a shallow waste disposal vault made of cement buried in a sandy loam zone. The vault contains highly alkaline, liquid low-level defense waste with a pore water composition equivalent to a 3 *M* $NaNO_3$ solution. The vault is surrounded by 30 cm of a coarse gravel. The pore water of the soil is normal dilute groundwater with a composition of about 0.005 *M* $NaHCO_3$. The partial pressure of water vapor decreases as the solute concentration in the water increases. The partial pressure of water vapor in the soil water at 21°C is 1.56×10^{-5} g/cm^3. Therefore, a pressure gradient, $DP = 0.19 \times 10^{-5}$ g/cm^3, exists that will drive water from the soil through the gravel toward the vault, where it can condense out on the wall of the vault and potentially leach out contaminants. If the vapor diffusion coefficient of the gravel is 0.081 cm^2/s, then the flux of water vapor according to Fick's law is $(0.081)(0.19 \times 10^{-5}) = 5.1 \times 10^{-9}$ g/cm^{2}s or 0.16 g of water diffusing through each cross-sectional square centimeter of gravel per year.

Of course, more complicated situations and conditions will require more sophisticated mathematical treatment, especially for the driving force, but the basic flux relationships are similar for any liquid and gas migration through the subsurface. If the hydraulic conductivities and diffusion coefficients are known for the materials and each migrating fluid of interest, then predictive computer models can often handle the difficult calculations associated with multiple fluids, multiple pressures, and multiple types of materials.

3.6 SATURATED SYSTEMS

Water entering the subsurface through precipitation enters a complex three-dimensional system which is controlled by a wide variety of physical and chemical processes. In general terms, water entering the subsurface will continue to migrate in the direction of lowest pressure or be retained by the soil until energy equilibrium conditions are met. Forces controlling water movement include gravity, adhesion to soil particles, viscosity, and surface tension of the water itself. Each of these forces will vary in importance depending on several factors, such as the dissolved content of particulate matter in the water, the presence of minerals, soil pore size, and temperature.

Groundwater can be found in the traditional sense at the water table below which the soil pore spaces are essentially saturated and the water is free to move, and in the unsaturated zone (or vadose zone) above the water table. It is possible for water to migrate through both of these zones, transporting dissolved components (or contaminants). The interaction of the various forces involved will determine the direction and rate of migration.

Water table is a traditional term that describes the level at which water surface will stabilize when it freely enters a boring (or well) at atmospheric pressure. Below this level the water occupies all the pore spaces between mineral grains, cracks, and other openings (i.e., fractures, joints, cavities, etc.), and the formation is considered saturated.

All voids in the subsurface medium are classified as porosity. When pore spaces are interconnected so that water can flow between them, the medium is said to be permeable. The actual openings that permit water flow are referred to as effective porosity. Effective porosity is calculated as the ratio of the void spaces through which water flow can occur to the bulk volume of the medium (expressed as a percentage) as follows:

$$n = \frac{V_t - V_s}{V_t} = \frac{V_v}{V_t} \tag{3.6}$$

where V_t = total volume of soils, V_s = volume of solid, and V_v = volume of voids.

Porosity that includes the voids between mineral (or soil) grains is referred to as primary porosity. When the porosity is the result of cracks, fractures, or solution channels, it is known as secondary porosity. The porosity of soft clay is often over 50%, but clay typically has low permeability because the pores are either not interconnected or are too small to permit easy passage of water. On the other extreme, nonfracture igneous rock often has a porosity of less than 0.1% but, again, low permeability.

Within the two extremes, porosity is dependent upon the organization of the mineral or soil grains. If perfect spheres (i.e., marbles) of uniform size are packed into a cube, the open volume (porosity) can vary from a maximum of 48% to a minimum of 26% depending on how the spheres are organized (Figure 3.9). If several different sizes of spheres are placed in the cube, the porosity will be dependent on the sphere size distribution and the packing arrangement. In nature, porosity values are also dependent upon cementing of the grains by minerals such as carbonates or silicates. Typical total porosity values for various soils and rocks are listed in Table 3.4; density values are provided in Table 3.5.

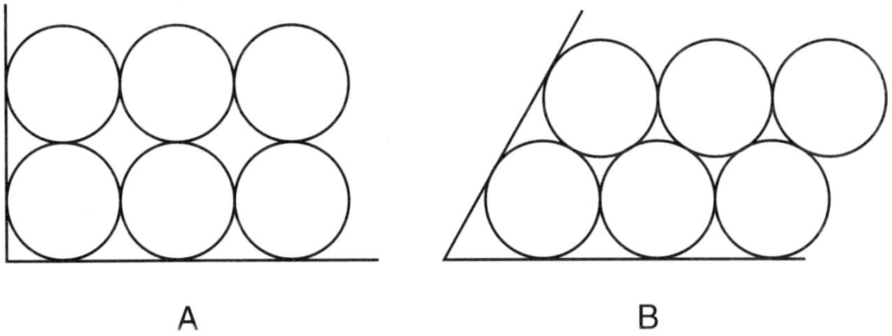

FIGURE 3.9 Cubic packing of spheres with a porosity of 47.65% (A) vs. rhombohedral packing of spheres with a porosity of 25.95% (B).

TABLE 3.4
Selected Values (% by volume) of Porosity, Specific Yield, and Specific Retention

Geologic Material	Porosity	Specific Yield	Specific Retention
Soil	55	40	15
Clay	50	2	48
Sand	25	22	3
Gravel	20	19	1
Limestone	20	18	2
Sandstone (semiconsolidated)	11	6	5
Granite	0.1	0.09	0.01
Basalt (recent)	11	8	3

The quantity of water that can be retrieved from a medium is related to size and shape of the connected pore spaces within that medium. The quantity of water that can be freely drained from a unit volume of porous medium is referred to as the specific yield. The volume of water retained in the medium by capillary and surface active forces is called the specific retention. The sum of specific retention and specific yield is equal to the effective porosity (see Table 3.4). Neither term has a time value attached. Drainage can occur over long periods (i.e., weeks or months).

The rate at which a porous medium will allow water to flow through it is referred to as permeability. Henry Darcy was the engineer who performed the first time-rate studies of water flowing through a sand filter. Darcy determined that, for a given material, the rate of flow is directly proportional to the driving forces (head) applied (hence, Darcy's law).

TABLE 3.5
Densities (g/cm^3) of Sediments and Sedimentary Rocks

Rock Type	Range	Average (wet)	Range	Average (dry)
Alluvium	1.96–2.0	1.98	1.5–1.6	1.54
Clays	1.63–2.21	2.21	1.3–2.4	1.70
Glacial drift	—	1.80	—	—
Gravels	1.7–2.4	2.0	1.4–2.4	1.95
Loess	1.4–1.93	1.64	0.75–1.6	1.20
Sand	1.7–2.3	2.0	1.4–1.8	1.60
Sand and clay	1.7–2.5	2.1	—	—
Silt	1.8–2.2	1.93	1.2–1.8	1.43
Soil	1.2–2.4	1.92	1.0–2.0	1.46
Sandstone	1.61–2.76	2.35	1.6–2.68	2.24
Shale	1.77–3.2	2.40	1.56–3.2	2.10
Limestone	1.93–2.90	2.55	1.74–2.76	2.11
Dolomite	2.28–2.90	2.70	2.04–2.54	2.30

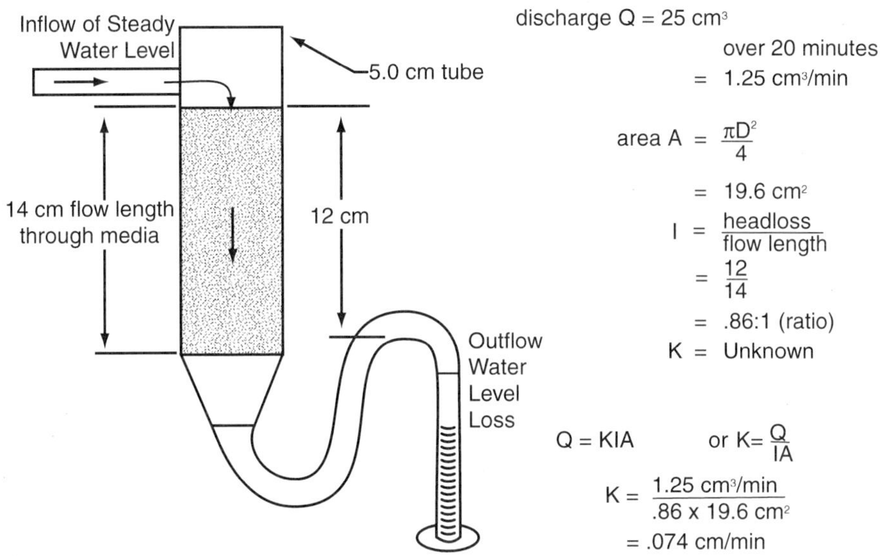

FIGURE 3.10 Example of determining seepage velocity using a permeameter.

Hydraulic conductivity is defined as volume units per square unit of medium face per unit of time under a unit hydraulic gradient (often expressed as units^3/units^2/time). However, many variations of this definition are used for convenience. For example, in the United States hydraulic conductivity is referred to in terms of gallons per day per square foot or, by the U.S. Geological Survey, as square feet per day.

Darcy's work was confined to the quantity of water discharged from a sand filter. Four examples of the application of Darcy's law as applied through a sand filter (actually a permeameter) is shown in Figure 3.10. Notice that the orientation of the cylinder has no effect on permeability. An example calculation of hydraulic conductivity (K) is presented in Figure 3.10 using the equation below:

$$K = \frac{Q}{IA} \tag{3.7}$$

where Q = discharge (units^3/units^2/time), I = hydraulic gradient (units/units), and A = cross-sectional area (units^2). Later studies found that other factors were also involved, such as viscosity and density of the liquid. However, for these purposes, one can assume that the density and viscosity of underground fresh water are relatively uniform the world over. To reduce term confusion, the ability of a porous medium to convey water flow will be referred to as hydraulic conductivity for the remainder of this chapter. Typical ranges of hydraulic conductivity for various soil and rock types are presented in Figure 3.11.

In practical fieldwork, the idea of expressing hydraulic conductivity in units of square area is cumbersome. A more common convention is to recognize that the

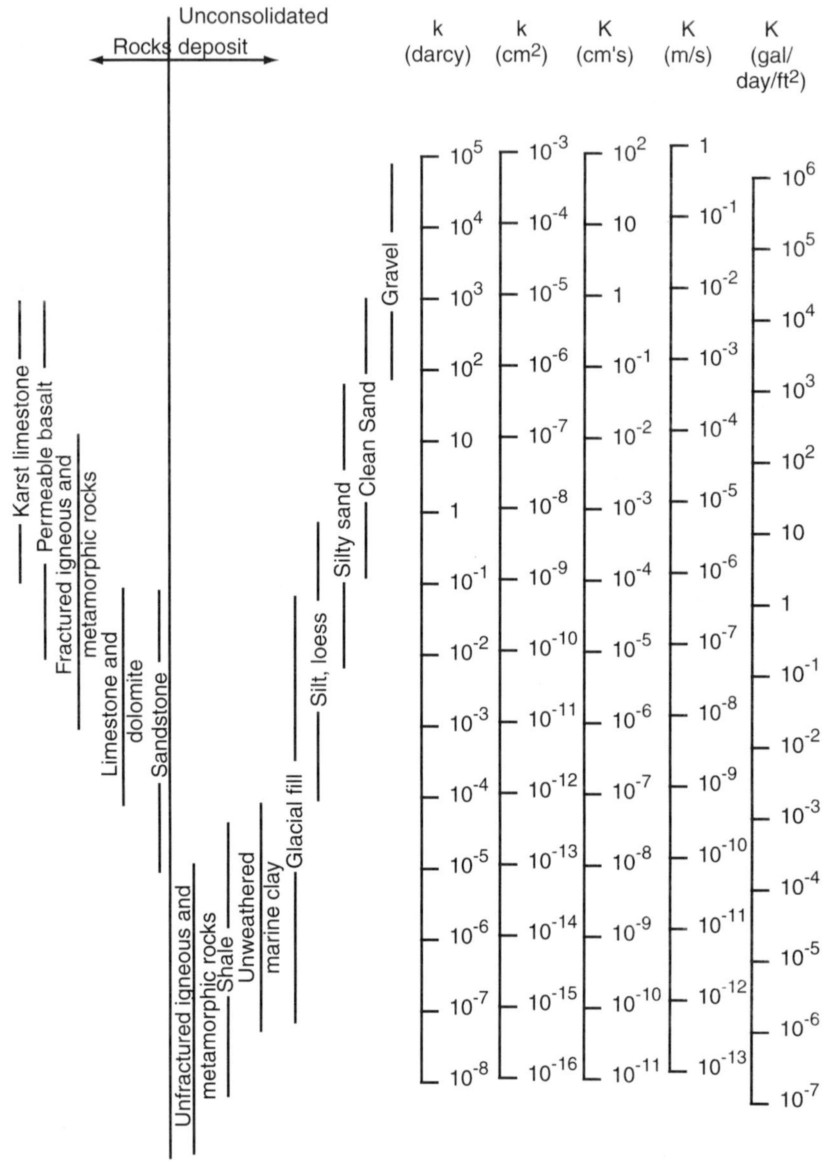

FIGURE 3.11 Range of hydraulic conductivity and permeability values. (After Freeze and Cherry, 1979.)

entire thickness of the water-bearing formation will allow water flow. If the aquifer is divided into vertical unit width segments (i.e., 1 ft wide), each segment then has a water-conducting capacity of the hydraulic conductivity times the thickness, which is called transmissivity (formerly transmissibility), which may be expressed in units such as gallons per day per foot or, in USGS terms, square feet per day (actually, $ft^3/ft^2/day \times ft = ft^2/day$).

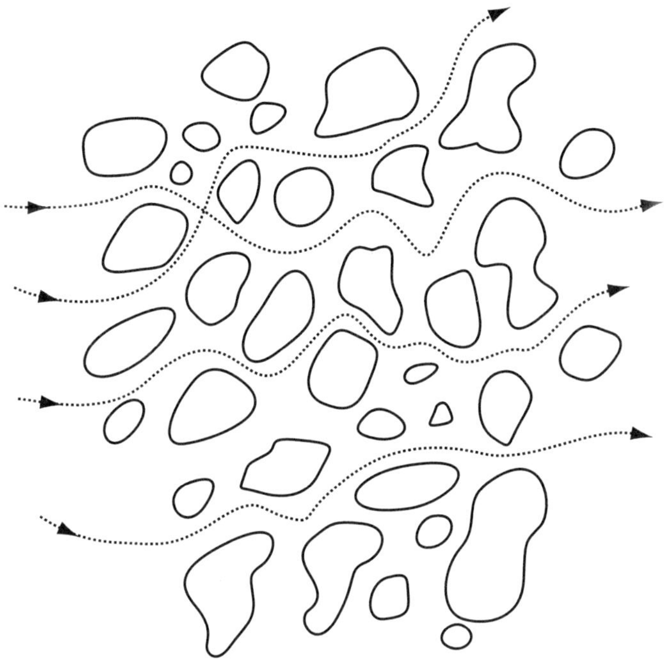

FIGURE 3.12 Tortuous flow paths.

Although hydraulic conductivity is often expressed in length units per unit time (cm/s or ft/day), these units do not express the actual linear flow velocity through the porous medium. Because hydraulic conductivity is measured as a discharge rate (i.e., cm^3/cm^2, or ft^3/ft^2), it would only represent the linear velocity of flow through a unit area of the medium if it consisted of 100% porosity. In reality, a cross section of a porous medium usually exhibits less than 30% open area; therefore, the actual linear velocity of water molecules must be greater in order to supply the discharge from the entire surface area. The "tortuous path" that the water must travel to reach the discharge surface is shown in Figure 3.12. Determination of the actual linear velocity can be calculated by dividing the hydraulic conductivity value by the effective porosity.

Determination of the actual velocity of the water through the permeameter used in Figure 3.13 is presented below:

$$V = \frac{KI}{n} \tag{3.8}$$

If $K = .074$ cm/min, $I = 0.86$, v = velocity (actual), and effective porosity (n) = 0.30, then

$$V = \frac{(0.074 \text{ cm / min})(0.84)}{0.30} = 0.21 \text{ cm / min}$$

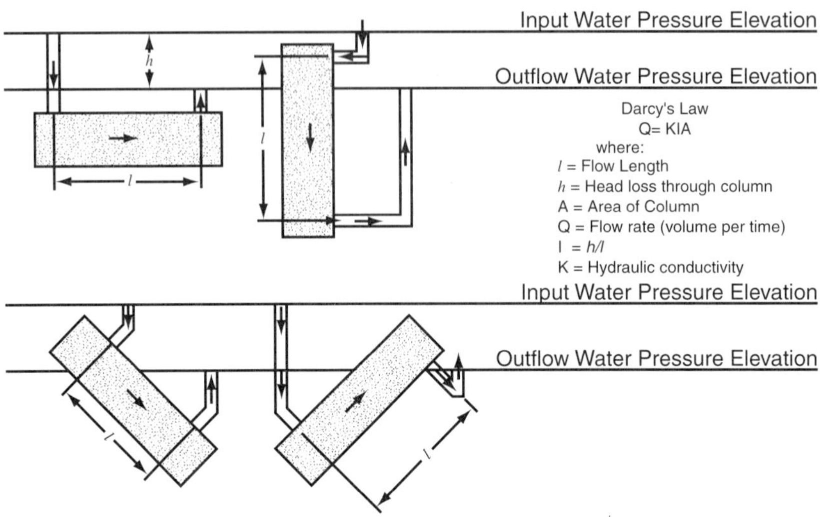

FIGURE 3.13 Examples of the application of Darcy's law for the movement of water through a sand filter or permeameter.

Thus, the actual forward movement of water is 0.21 cm/min.

The horizontal migration rate of water at contaminated sites is a primary concern for evaluation. Example 2 describes the procedure for determining the actual migration rate of water at a site. The hydraulic conductivity and effective porosity are the same as presented in Figure 3.13. The gradient (difference in height/length of flow) is determined from the water table contour map as shown in Figure 3.14.

Two hydraulic head gradients determined from wells or piezometers are shown in Figure 3.15. Piezometers are basically pipes or wells put in the ground which are

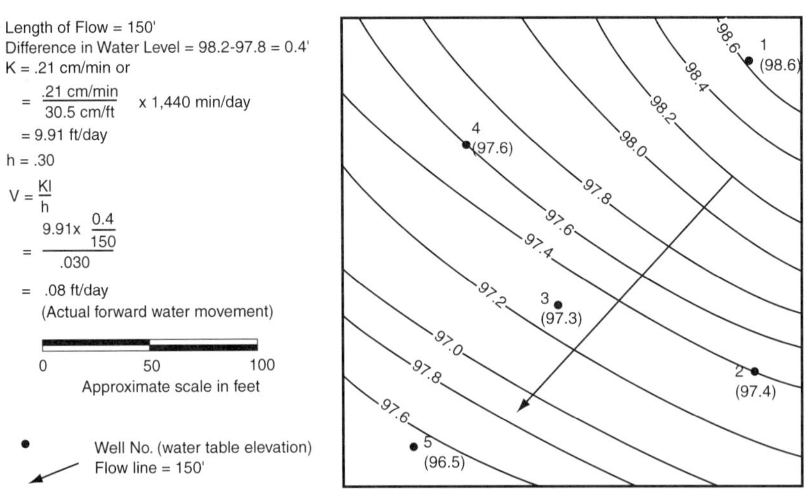

FIGURE 3.14 Determination of hydraulic gradient from a water table map.

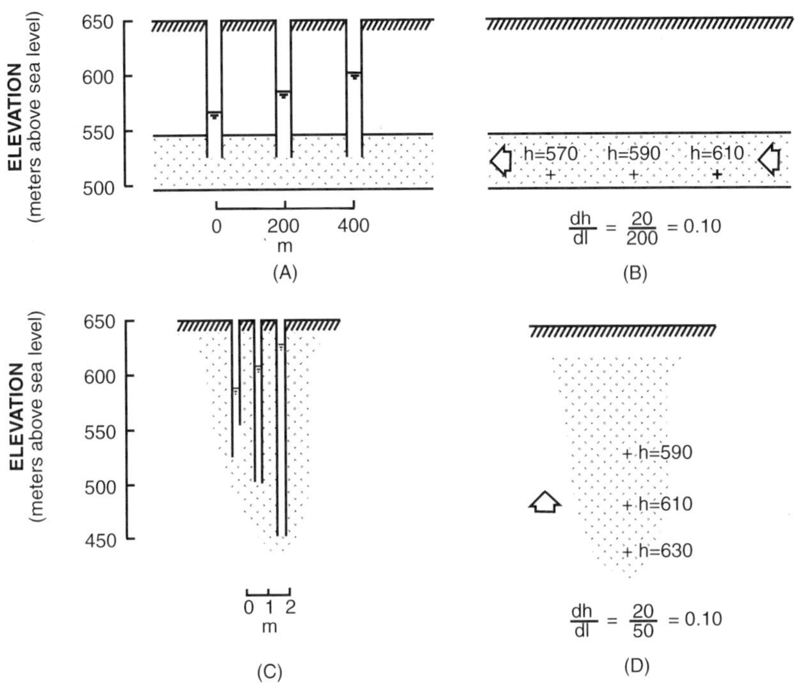

FIGURE 3.15 Determination of hydraulic gradient from wells. (After Freeze and Cherry, 1979.)

sealed along their lengths and open to water flow at the bottom and open to atmospheric pressure at the top. In Figure 3.15A and B, the gradient determined from three widely spaced piezometers is seen to be 0.1 and flow will occur to the left, downgradient. In Figure 3.15C and D, the gradient is 0.4, determined from three closely spaced, or nested, piezometers that are tapping three different depths, probably three different geologic formations. These piezometers show that the flow is actually directed upward, which, even so, is still downgradient. Unusual subsurface conditions can cause unusual flow patterns, but flow will always occur downgradient.

3.6.1 Types of Aquifers

Water existing below the water table has the ability to migrate in the subsurface reflecting the hydraulic conductivity of the formation it is found in and the hydraulic gradient (driving force) of its position. A formation that can supply a "reasonable" quantity of water to a borehole or well within a fairly short period of time for sampling purposes is called an aquifer. An aquitard is a geological unit that is permeable enough to transmit water in significant quantities over large areas or over long periods of time, but of sufficiently low permeability to restrict its use for commercial or public use. An aquiclude is an impermeable geologic unit that does not transmit water in significant quantities, and differs in permeability relative to adjacent units by several orders of magnitude.

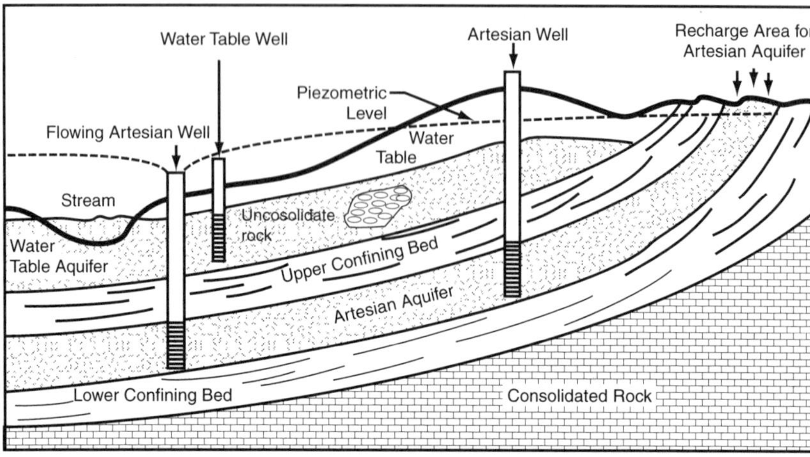

FIGURE 3.16 Schematic showing various aquifer types.

Aquifers exist in a wide variety of forms and can be classified based on lithic characteristics or hydrogeologic behavior (Figure 3.16). Aquifers classified on the basis of lithic character are referred to as either unconsolidated or rock (or consolidated) aquifers.

Unconsolidated aquifers are generally composed of sand and gravel deposit with subordinant or minor amounts of silt and clay. Although compacted to some degree, lithification due to cementation is minor or absent, with much of the original intergranular porosity retained. Groundwater flow is controlled and channeled through zones of relatively higher permeability. Consolidated or rock aquifers comprise grains that are cemented or crystallized into a firm and cohesive mass. Groundwater movement in consolidated aquifers is primarily via secondary porosity. In igneous and some metamorphic rocks, secondary porosity occurs as highly vesicular zones, joints, and fractures. In carbonate rocks, secondary porosity is prevalent by chemical dissolution. Clastic sedimentary rocks may maintain some primary porosity, but it tends to be much lower than that of unconsolidated aquifers.

Unconfined or water table aquifers maintain a saturated surface that is exposed directly to the atmosphere. These are often similar to a bathtub full of sand or gravel to which water has been added. A well drilled through the water table would fill with water to the common water elevation in the tub. Thus, the potentiometric head in the aquifer is at the elevation of the water table. Unconfined aquifers are also characterized by a fluctuating water table, which responds seasonally. With unconfined aquifers, the water table is at atmospheric pressure, and only the lower portion of the aquifer is saturated. Recharge to a water table aquifer comes from rainfall that seeps downward to the water table. The water table level in this type of aquifer rises in direct proportion to the effective porosity. If the equivalent of 2 in. of rainfall seeps into the water table (actually reaches the water table) in an aquifer with an effective porosity of 0.3, the water table would rise 6.7 in. Alternatively, if the same water is pumped and removed from a well, the water table aquifer is then derived from the "storage" in the formation in the immediate vicinity of the well. Natural

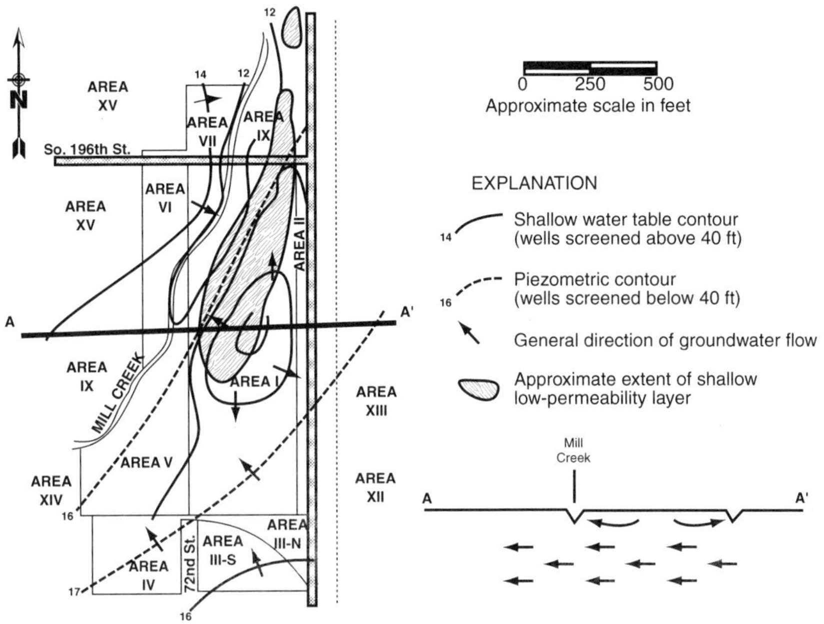

FIGURE 3.17 Example of water table contour map and hydrogeologic cross section showing two groundwater flow regimes.

migration through a water table aquifer is toward a lower elevation discharge. Most streams derive their base flow from water table aquifers (Figure 3.17).

A confined aquifer is a saturated, hydrogeologic unit that is bounded top and bottom by relatively low permeability or impermeable units through which groundwater flow is nonexistent or negligible. All voids within the aquifer are filled with water at a pressure greater than atmospheric; thus, the potentiometric head in the confined aquifer is at a level higher than the top of the aquifer. The classic example is the sand layer that is sandwiched between two clay layers. When a well penetrates through the upper confining layer, the water rises in the borehole above the top of the sand layer to the elevation which is equal to the pressure of water in the sand formation. This type of aquifer is also called an artesian aquifer. Recharge to a pure artesian aquifer occurs at some location where the confining layer is nonexistent and the aquifer formation is exposed to infiltration at a higher elevation. In many cases, the recharge area is a water table aquifer at a higher elevation. The driving force that causes water flow through a confined aquifer is the pressure caused by the standing water at the higher elevation. Any disturbance to water pressure in an artesian aquifer (such as starting a pump) is felt very quickly as a shock wave over a wide area of aquifer. This rapid pressure response is due to the pressurized nature of the system. If water is pumped from an artesian aquifer, only a very small quantity of the water retrieved from the well is derived from "storage" near the well. Water flows in toward the well from a wide area to replace that removed by the well.

Semiconfined or leaky aquifers are similar to confined aquifers with the exception that the overlying and/or underlying units are not impermeable and some leakage occurs. Groundwater is free to migrate upward or downward. If the water level in a well tapping a semiconfined aquifer equals that of the water table, then the aquifer is considered to be in hydrologic equilibrium.

More complex arrangements of aquifers, aquiclude, and aquitards, notably in deep sedimentary basins, are systems of interbedded geologic units of variable permeability. These systems are referred to as a multilayered aquifer system. Such systems are considered more of a succession of semiconfined aquifers separated by aquitards.

Above the water table, groundwater can also occur in perched aquifer conditions. In these instances, groundwater occurs in relatively permeable soil that is suspended over a relatively low permeability layer of limited lateral extent and thickness at some elevation above the water table. Perched groundwater occurrences are common within the vadose zone; high-permeability zones overlie low-permeability zones of limited lateral extent in unconsolidated deposits. However, perched conditions can also occur within low-permeability units overlying zones of higher permeability in both unconsolidated and consolidated deposits. In the latter case, for example, a siltstone or claystone overlies jointed and fractured bedrock such that groundwater presence reflects the inability of the water to drain at a rate that exceeds replenishment from above.

In subsurface characterization studies and detection-monitoring design, defining the horizontal (K_h) vs. vertical (K_v) hydraulic conductivities is of prime importance in detecting and monitoring dissolved constituents in groundwater. Aquifers can thus be characterized as homogeneous, heterogeneous, isotropic, and/or anisotropic.

Aquifers are rarely homogeneous and isotropic; that is, hydraulic conductivity is the same throughout the geologic unit and is the same in all directions, respectively (i.e., $K_h = K_v$). Since most geologic units vary both horizontally and vertically in their physical and structural characteristics, they are seldom considered homogeneous but rather heterogeneous. For example, geologic units rarely maintain uniform thickness, and individual layers may pinch out, grain size will vary, and stratification and layering is common — such that the horizontal hydraulic conductivity is greater than the vertical by several orders of magnitude (i.e., $K_h > K_v$). Such heterogeneity is commonplace in nature. When the hydraulic conductivity is significantly greater in one direction relative to another, as within fractured rock where K is greater parallel to the fracture rather than normal to it, this aquifer is considered anisotropic (i.e., in lieu of isotropic). When a well is drilled into a water-bearing formation, its purpose is to remove water from that formation. Wells are constructed to be holes below the water table that allow easy entrance of water, but prevent inflow of formation solids. When a well is pumped, the standing water level (static level) in the well is lowered, and the well becomes a low-pressure area (Figure 3.18). Water in the surrounding area naturally flows in the direction of lowest pressure, thereby recharging the well. This same basic phenomenon occurs whether the producing formation is under artesian or water table conditions. If the pumping rate is increased, the lowered pumping level will increase the inflow.

Darcy's law ($Q = KIA$) is the operating factor in well flow. The only significant difference in the application of this law to wells rather than to a laboratory permeameter is that the water flows to the well from a radial pattern rather than via linear flow as shown in the permeameter. The flow paths found in both artesian and

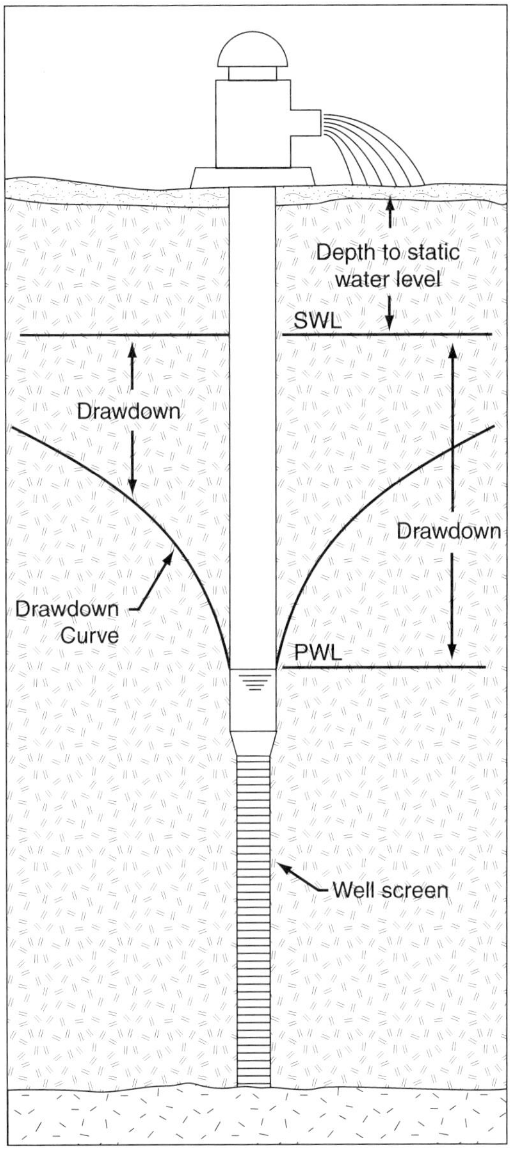

FIGURE 3.18 Terms relating to well performance. (After Driscoll, 1986.)

water table wells are shown in Figure 3.19. The circular lines ("potential") are really contour lines drawn to represent the elevation of the surface of the water table. The surface of the water table is the elevation that water would rise in a well drilled at that location and thus represents the "driving force" that causes water flow to the well.

At a given pumping rate, a more permeable formation will have less drawdown as less effort is required to move the water. Low-permeability formations often offer so much resistance to flow that it is not possible for the well to produce water at a

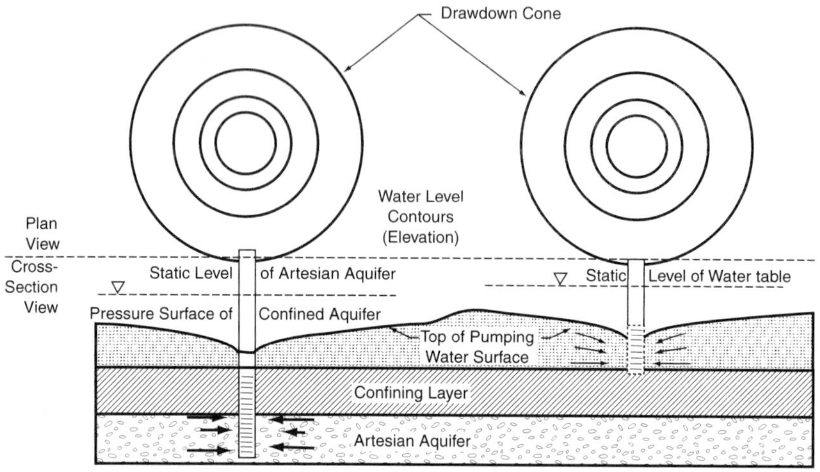

FIGURE 3.19 Schematic showing flow paths encountered in artesian and water table wells.

rate equal to the pump capacity, and the well goes "dry." After a period of recovery, the static level of water in the well will rise to the original level.

3.6.2 Steady-State Flow

If a well is pumped continuously at a constant, reasonable rate for a sufficiently long period of time, the water level in the well will stabilize with the inflow rate equal to the pumping rate. Alternatively, the radius of influence (as measured by observation wells at some distance from the pumped well shown in Figure 3.20) will expand until adequate recharge is captured to balance the pumping rate of the well. When the water level in a well reaches a stable level while the pumping rate remains constant, the well is said to be at "steady state." The mathematical relationships that describe a classic steady-state relationship among the aquifer characteristics, the pumping rate, and the well construction are presented in Figure 3.21.

Aquifer test analyses can be conducted by one or several methods depending upon whether the solution is applied to a specific aquifer type and with certain assumptions being made on the hydrogeologic nature of the aquifer and nature of the test. The initial analytical methods employed assume that the aquifer is

- Confined
- Homogeneous
- Isotropic
- Areally infinite
- Uniform in thickness

and that

- Monitoring and observation wells fully penetrate the aquifer and are fully screened.

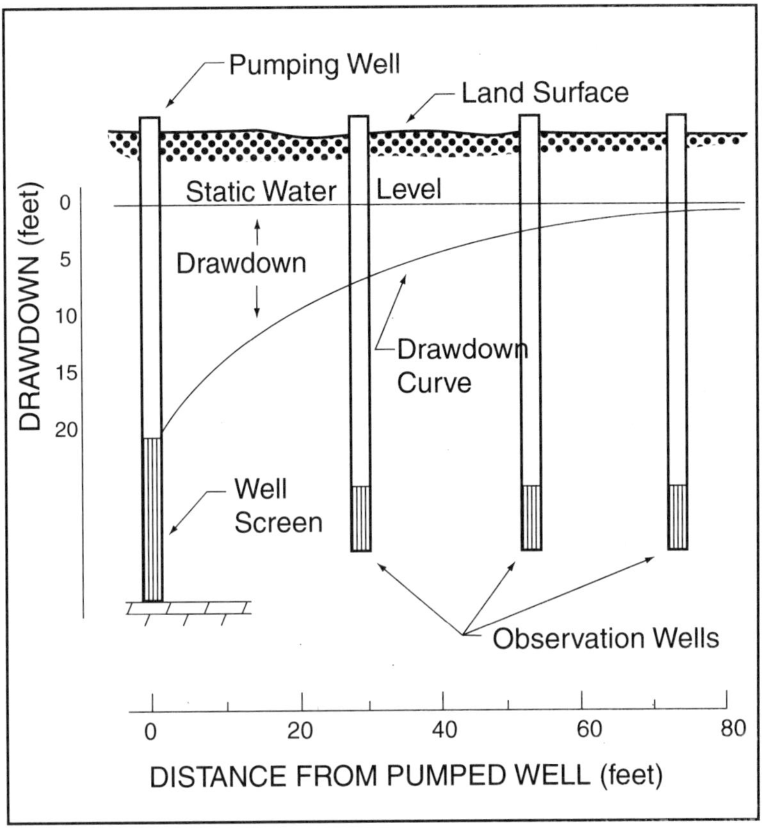

FIGURE 3.20 Trace of half a cone of depression showing variations in drawdown with distances from a pumping well.

- Pumping rate is constant.
- Water storage is negligible.
- Potentiometric surface prior to pumping is horizontal.
- Water removed from storage is immediately discharged once the head declines.

Use of these assumptions is necessary to limit the number of variable parameters that must be considered in the equations. Calculation of the response of an aquifer that is not homogeneous, isotropic, or infinite in extent becomes very complex. Many complex situations are better suited to sophisticated computer simulations.

Certain factors must, however, be considered in choosing the appropriate analytical solution: unconsolidated vs. consolidated conditions, fully vs. partially penetrating wells, variable discharge rules, delayed yield, and aquifer boundaries. Most methods are best suited for unconsolidated aquifers with well-defined overlying and underlying boundaries, whereas with consolidated aquifers, the effective aquifer thickness is uncertain. A pumping well that fully penetrates a confined aquifer (i.e.,

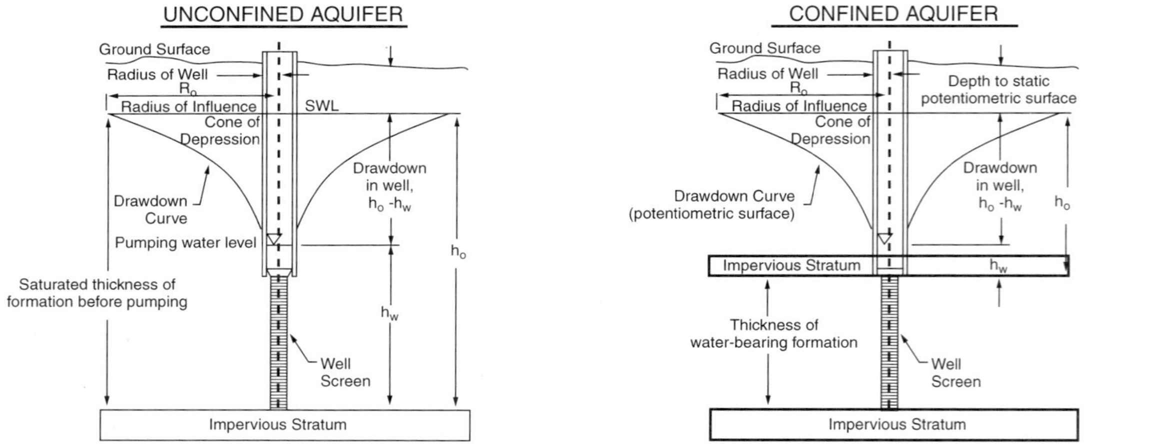

Unconfined aquifer

General Equation
Standard Units

$$Q = \pi k \frac{h_o^2 - h_w^2}{\ln (R_o/r_w)}$$

NOTE: Use consistent units
ln= natural log

American Units

$$Q = \frac{k(h_o^2 - h_w^2)}{1055 (\log_{10} R_o/r_w)}$$

WHERE: Q= Well yield in gpm
k = Hydraulic conductivity in gallons per day/ ft²
h_o= Static head measured from bottom of aquifer (ft)
h_w = Depth of water in the well while pumping
log = log to base 10
r_w = Radius of well (ft)

Confined aquifer

General Equation
Standard Units

$$Q = \pi k \frac{h_o^2 - h_w^2}{\ln (R_o/r_w)}$$

NOTE: Use consistent units
ln= natural log

American Units

$$Q = \frac{k(h_o^2 - h_w^2)}{1055 (\log_{10} R_o/r_w)}$$

WHERE: Q= Well yield in gpm
k = Hydraulic conductivity in gallons per day/ ft²
h_o= Static head measured from bottom of aquifer (ft)
h_w = Depth of water in the well while pumping
log = log to base 10
r_w = Radius of well (ft)

FIGURE 3.21 Mathematical relationships for steady-state confined and unconfined conditions.

screened through 80% or more of the saturated thickness of the aquifer) would be analyzed differently from that which is partially penetrating. In the latter case, vertical flow in the aquifer may affect water levels in nearby observation wells, whereas with fully penetrating wells, groundwater flow toward the pumping well is assumed to be horizontal. To avoid problems associated with vertical flow near the well screen, observation wells should be at a distance of generally 1.5 times the saturated thickness of the aquifer from a partially penetrating well. It is difficult to achieve a constant pumping rate during aquifer testing; however, if the pumping rate does not vary more than 10% during conduct of the test, then the rate can be considered constant.

Delayed yield is the process wherein water in an unconfined aquifer is yielded to the pumping well after the water level has declined and before steady state has been reached. Delayed yield of water is reflected in a flattening of the drawdown curve simulating a steady-state condition; however, once over, water levels may continue to drop (Figure 3.22). Boundary conditions (i.e., boundary of an aquifer) can be characterized as no-flow or constant-head boundaries. A no-flow boundary (i.e., impermeable fault zone, pinching-out of aquifer, etc.) contributes no groundwater flow and is reflected by a sudden water-level decline in one or more observation wells, whereas a constant-head boundary (i.e., perennial stream or lake) may contribute significant recharge to the aquifer (thus lessening drawdown) or sudden stabilization of water levels (Figure 3.23). A listing of several of the most common analytical solutions for aquifer analysis in unconsolidated, fractured rock, and karst environments is presented in Tables 3.6 and 3.7, respectively.

In consolidated or fractured rock aquifers, analytical techniques focus on a dominant characteristic such as a well-developed fracture or void. Boundaries and other aquifer properties are assigned constant values. Some approaches as presented in Table 3.7 assume that the aquifer (i.e., fracture or void) is important for groundwater movement, but is relatively insignificant as a reservoir of groundwater storage. Other approaches consider groundwater storage in both the fractures and in the aquifer matrix. Some of the methodologies listed address specific features associated with fractured rock aquifers such as anisotropy, storage-release effects, and transmissivity contrasts of the bulk aquifer matrix and fractures. Such an approach is applicable where test conditions do not significantly compromise boundary conditions as specified via conventional analysis of transmissivity and storativity. Other methods noted address phenomena typically observed in fractured rock aquifers. For example, double-porosity models emphasize the role of fractures and the aquifer matrix as sources of groundwater storage. The time-drawdown response of groundwater released from these sources is similar to the delayed-yield response of unconfined aquifers. The single-fractured models emphasize the interaction between the aquifer matrix and a fracture penetrated by a production well. Time-drawdown responses for wells located on a fracture or fracture system often exhibit a diagnostic half-slope (0.5) on a log–log plot.

Fractured rock aquifers may be analyzed in a manner similar to unconfined aquifers, providing the responses exhibit the same general characteristics. During the initial testing, the fracture contributes water to the well, followed by primarily

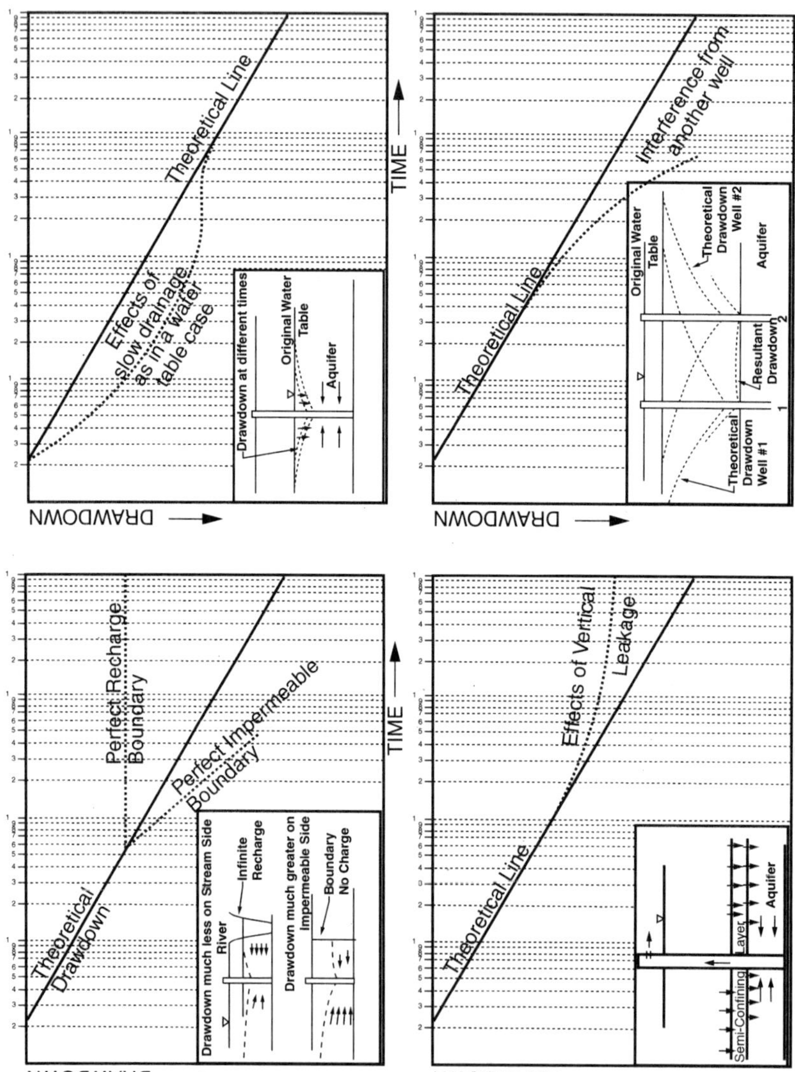

FIGURE 3.22 Typical curves showing effects of delayed yield, vertical leakage, no-flow boundary, and constant-head boundary.

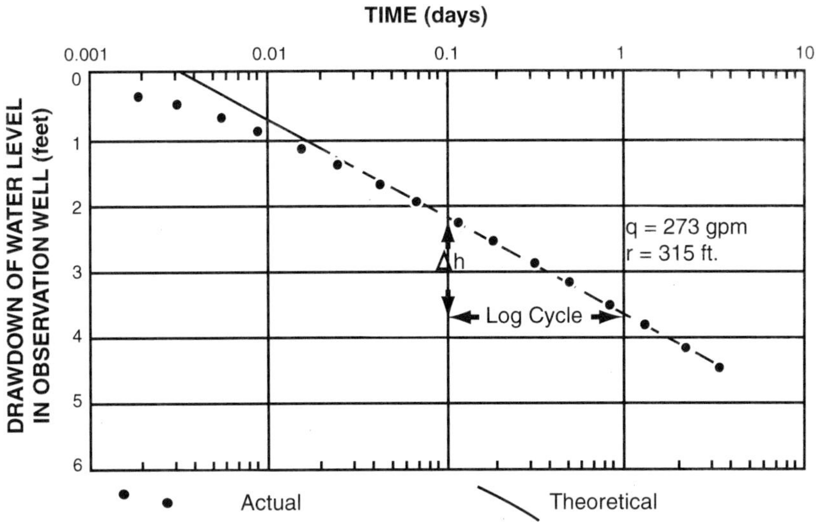

FIGURE 3.23 Time-drawdown graphs for a pumping test.

pores and smaller fractures being dewatered, exhibiting the effects of delayed yield. Groundwater migrates to the well from fractures situated farther away during the latter part of the test.

Carbonate aquifers (i.e., limestone and dolomite) are typically characterized by cavernous zones as a result of chemical dissolution. Some analytical methods that emphasize long, well-developed fractures may thus be applicable to solution-channeled aquifers. Notably, the block-and-fissure model applicable to fractured rock aquifers may also be useful in carbonate aquifers where significant solution channel development has occurred. Carbonate rocks can be highly anisotropic and nonhomogeneous on a localized scale, but behave more homogeneously on a regional scale. As such, analytical methods that recognize water table and/or leaky artesian conditions generally may also be useful in evaluating aquifer characteristics in fractured and solution-channeled carbonate rock aquifers.

Important to any aquifer restoration program is the radius of influence or capture zone to be anticipated during pumping. The radius of influence (r_o) for a vertical pumping well can be calculated using the following equation.

$$r_o = 300(h_o - h_w)\sqrt{K} \tag{3.9}$$

where h_o = thickness of aquifer at radius of influence, h_w = height of water in well (or water equivalent), and K = hydraulic conductivity.

The zone of influence of a horizontal well can be calculated either by the rectangular or elliptical approach. The rectangular approach assumes the radius of influence is constant along the length of the well and is equal to the extent of the

TABLE 3.6
Types of Aquifer Test Analyses for Unconsolidated Environments

Aquifer	Type of Solution	Name of Solution	Method of Solution	Assumption
Confined	Steady state	Thiem	Calculation	a
	Unsteady state	Theis	Curve fitting	
		Chow	Nomogram	
		Jacob	Straight line	
		Theis recovery	Straight line	
Semiconfined	Steady state	De Glee	Curve fitting	a
		Hantush Jacob	Straight line	
		Ernst modification of Thiem method	Calculation	
	Unsteady state	Walton	Curve fitting	a
		Hantush I	Inflection point	
		Hantush II	Inflection point	
		Hantush III	Curve fitting	
Unconfined, with delayed yield	Unsteady state	Boulton	Curve fitting	a
Semi-unconfined, with delayed yield	Unsteady state	Boulton	Curve fitting	a
Unconfined	Steady state	Thiem-Dupuit	Calculation	Solutions for the confined, unsteady-state case can be applied to the unconfined, unsteady-state case only if the drawdown is modified by an appropriate factor
	Unsteady state	Thiem		
		Theis		
		Chow		
		Jacob		

Aquifer type	Flow	Method	Procedure	Remarks
Confined or unconfined	Steady state	Dietz	Calculation	Aquifer crossed by one or more fully penetrating recharge or barrier boundaries
	Unsteady state	Stallman	Curve fitting	
		Hantush image	Straight line	
Confined or unconfined	Unsteady state	Hantush	Calculation	Aquifer homogeneous, anisotropic, and of uniform thickness
		Hantush-Thomas	Calculation	
Semiconfined	Unsteady state	Hantush	Calculation	
Confined	Unsteady state	Hantush	Curve fitting	Aquifer homogeneous and isotropic but thickness varies exponentially
Unconfined	Steady state	Culmination point	Calculation	Prior to pumping the potentiometric surface slopes
	Unsteady state	Hantush	Curve fitting	
Confined or unconfined	Unsteady state	Cooper-Jacob	Straight line	Discharge rate variable
		Aron-Scott	Straight line	
		Sternberg	Straight line	
		Sternberg recovery	Straight line	
Confined	Steady state	Huisman correction I and II	Calculation	Partially penetrating pumping well
		Jacob correction	Calculation	
Semiconfined	Steady state	Huisman correction I and II	Calculation	
Unconfined	Steady state	Hantush correction	Calculation	
Confined	Unsteady state	Hantush modification of Theis	Curve fitting	
		Hantush modification of Jacob	Straight line	
Semiconfined	Steady state	Huisman-Kemperman	Nomograph and curve fitting	Two-layered aquifer with semi-pervious dividing layer
		Bruggeman	Straight line	

Note: a = Aquifer is homogeneous, isotropic, areally infinite, and of uniform thickness; pumping and observation wells fully penetrate and screen the aquifer; prior to pumping the piezometric surface is horizontal; discharge rate is constant and storage in the well can be neglected; water removed from storage is discharged instantaneously with decline of head.

Source: Modified after Kruseman and DeRidder, 1979.

TABLE 3.7
Summary of Analytical Solutions for Anisotropy in Fractured Rock and Karst Environments

Aquifer Type(s)	Phenomenon Modeled	Method of Solution	Wells Required for Calculations	Remarks	Ref.
Confined, homogeneous	Two-dimensional anisotropy	Curve fitting, or straight line with calculation	4	Horizontal aquifer[a]	Papadopoulos (1964)
Leaky and nonleaky, homogeneous	Two-dimensional anisotropy	Curve fitting, or straight line with calculation	4	Horizontal aquifer[a]	Hantush (1966)
			3	Horizontal aquifer[a]	Neuman et al. (1984)
Homogeneous and heterogeneous, horizontal and vertical anisotropy, partial penetration	Three-dimensional anisotropy	Curve fitting with calculation	4	Horizontal aquifer[a]	Way and McKee (1982)
Confined, homogeneous, isotropic	Double porosity block and fissure storage	Curve fitting	1	Fractured rock or karst aquifers	Boulton and Streitsova (1977)
Unconfined, homogeneous, isotropic	Double porosity block and fissure storage	Curve fitting	1	Fractured rock or karst aquifers	Boulton and Streitsova (1978)
Confined, matrix is homogeneous and isotropic fracture and aquifer system strongly anisotropic	Pumping well penetrates vertical fracture or horizontal fracture	Curve fitting	1	Analysis for pumping well data only	Gringarten and Witherspoon (1972); Gringarten (1962)
Confined, matrix is homogeneous and isotropic fracture and aquifer system strongly anisotropic	Pumping well penetrates vertical fracture	Straight line	1	Analysis for hydraulic diffusivity (T/S), estimate of S from other methods needed to solve for T	Jenkins and Prentice (1982)

[a] For application in fractured rock aquifers, it is assumed the aquifer behavior approximates that of a porous medium. Standard methodologies and their applicable assumptions are used to obtain values of transmissivity and storage, from which anisotropy is calculated.

zone of influence from the end points of the well measured in the plane containing the well. Roughly rectangular, the area of drainage (A_r) is

$$A_r = 2LR_{ev} + \pi R_{ev} \tag{3.10}$$

where L = length of well screen and R_{ev} = effective drainage radius of the well calculated from the pumping tests conducted in a vertical well.

The elliptical approach assumes that the zone of influence is elliptical (A_e) with the well end points constituting the face of the ellipse, and the minor semiaxis equal to the radius of influence. The area of drainage (A_e) is

$$A_e + \pi R_{ev} c/2 \tag{3.11}$$

where c = minor axis of the ellipse (screen length or L) and the radius of influence of the well is

$$C = 2\sqrt{[(L/2)^2 + R_{ev}{}^2]} \tag{3.12}$$

3.6.3 Nonsteady Flow

In many situations, only very limited geologic information is available for a site that is under investigation. The published literature may include only limited or approximate hydrologic information. At these sites, the site investigation will necessarily include a drilling program that is designed to sample subsurface materials. An important part of this field program is an evaluation of the hydraulic properties of the subsurface formation.

After the borings are drilled and the wells installed, pumping tests or slug tests can be conducted to evaluate the aquifer parameters. These tests disrupt steady-state flow equilibrium and, thus, are nonsteady state. Pumping tests are conducted by pumping a single well at a uniform rate and measuring the time-rate of water-level decline in the pumped well and/or adjacent wells. A diagram of time drawdown for a pumping test is shown in Figure 3.23. Comparison of field data with theoretical data results in the definition of aquifer characteristics. Pumping tests are best suited for permeable formations that are capable of producing more than 1 gal of water/min. Slug tests are made by "instantaneously" adding (or withdrawing) a measured quantity of water from a well and monitoring the time-rate of water-level recovery in the same well. Because of the mechanics of measuring the water-level response in a timely manner, slug tests are best suited for less-permeable formations. Interpretation of the results of either a slug or pumping test is theoretically a rather straightforward process. In practice, however, the task often becomes more of an art than science.

3.7 UNSATURATED SYSTEMS

There are some important differences between the behavior and flow of water in the unsaturated zone (vadose zone) above the water table and the saturated zone below the water table. The surface tension of water or other fluids becomes important when there is a gas phase in contact with the fluid phase and the solid phase. If the total volume of a porous medium (V_T) is divided into the volume of the solid portion (V_s), the volume of the water (V_w), and the volume of the air or gas (V_a), then the volumetric water content, θ, is equal to V_w/V_T. Like the porosity (n), the volumetric water content may be reported as a fraction or as a percent. For saturated flow, $\theta = n$ and all pores are filled with liquid; for unsaturated flow, $\theta < n$ and $n = V_w + V_a$.

As previously defined, the pressure head, Ψ, can be either a positive hydraulic pressure as it would be below a standing column of water, or a negative pressure that would be generated by the suction of a dry soil. Therefore, $\Psi > 0$ below the water, $\Psi = 0$ at the water table, and $\Psi < 0$ above the water table in the vadose zone. This negative pressure in the vadose zone results from the surface tension of water at the air/water interface, often called the capillary pressure or the matrix suction. When $\Psi < 0$, Ψ is called the tension head or suction head. Therefore, because $h = \Psi = z$, the gradient can become complicated in the vadose zone with various negative pressures or suctions operating in addition to gravity. Some important aspects of the vadose zone/saturated zone characteristics are illustrated in Figure 3.24. Note in Figure 3.24D that the pressure head, Ψ, varies from some negative value (*–ve*) in the vadose zone, through zero at the water table, to some positive value (*+ve*) below the water table. The hydraulic head, h, is the pressure head plus the height above the arbitrarily chosen reference point, called the datum. Because flow occurs down gradient, Figure 3.24E shows that flow is downward for this normal situation. Remember that the negative capillary pressure or matrix suction operates in three dimensions, whereas gravity operates vertically in only one direction.

Darcy's law under saturated conditions uses the saturated hydraulic conductivity (K_s) for K in the above equation. However, K is a strong, nonlinear function of the volumetric water content, $K(\theta)$, meaning that the hydraulic conductivity decreases drastically as the water content decreases. The $K(\theta)$ relationship is often called the characteristic behavior of the sample with respect to hydraulic conductivity. That several types of sediments have different characteristic behaviors, along with a saturated caliche (a calcite–cement soil that in this case has become a rock) that shows different behavior along each of its three orthogonal directions, is shown in Figure 3.25. For example, the sample from a depth of 154.2 ft in the Ringold Formation has a saturated K of 10^{-4} cm/s at a 31% volumetric water content, but the same material has a K of only 10^{-8} cm/s at a 12% volumetric water content, a four order of magnitude difference! The exact shape and magnitude of the $K(\theta)$ relationship depends upon the pore-size distribution, which results mainly from the distribution of grain sizes in aggregate materials (soils and sediments) and upon fracture aperture sizes in fractured media (rock).

Hydraulic steady state is achieved in a porous media when the flux density and driving forces are unchanging with time. For unsaturated systems, the volumetric water content will also be constant and unchanging in a homogeneous material. However, under the same hydraulic steady state, each different material will attain its own steady-state water content depending upon the texture and pore-size distribution in the material. Because there is no suction gradient ($\nabla\Psi = 0$) at unsaturated hydraulic steady state, the only driving force is gravity, a situation referred to as unigradient conditions.

Water movement in the unsaturated zone is dependent on the factors discussed previously — driving force, hydraulic conductivity, viscosity, and density — but is

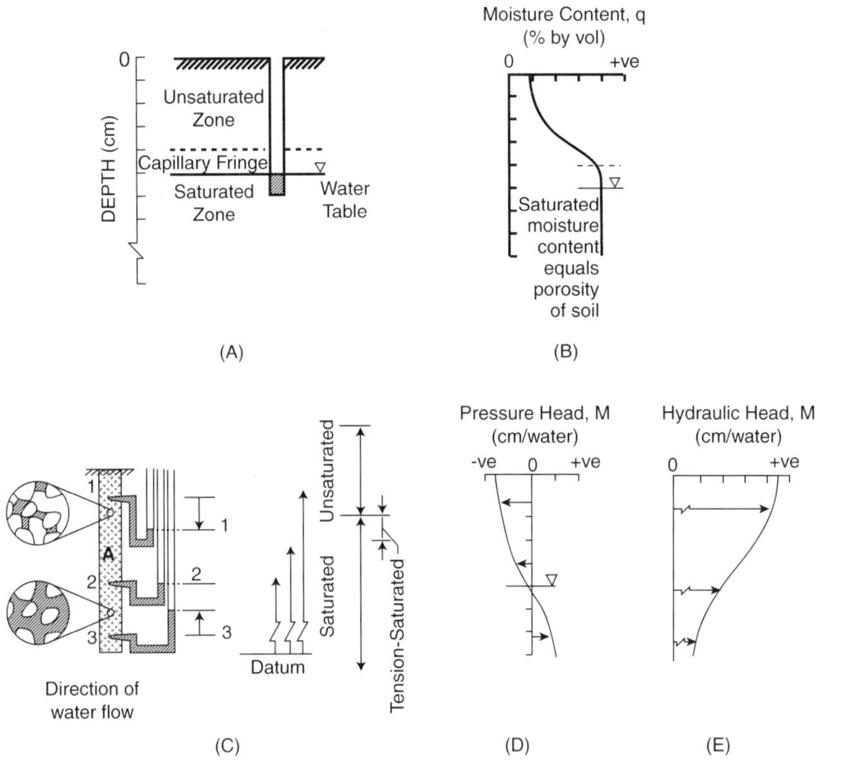

FIGURE 3.24 Groundwater conditions near the ground surface. Saturated and saturated zones (a), profile of moisture content vs. depth (b), pressure head and hydraulic head relationships; insets = water retention under pressure heads less than (top) and greater than (bottom) atmospheric (c), profile of pressure head vs. depth (d), and profile of hydraulic head vs. depth (e). (After Freeze and Cherry, 1979.)

also strongly influenced by additional forces including surface tension and adhesion. The basic atomic organization of the water molecule and electrical charge of the oxygen and hydrogen are responsible for these additional factors.

A water molecule is composed of one oxygen and two hydrogen atoms. These three atoms are organized so that the two hydrogens are at an angle of 105° from each other (Figure 3.26). The weak attraction between positive (oxygen) and negative (hydrogen) atoms of adjacent molecules are the forces that hold water together in a liquid state. Without this attraction, water would exist only as a gas. The impact of this angular arrangement is most evident when water freezes. As water is cooled to the temperature where the molecules begin to assemble into crystalline form, the only structure that they can readily assume is one that has a void in the center. This void causes the bulk ice to have a density of less than water, and thus it floats. In its liquid state, water has a definite volume but no definite form. At the boundary of the water mass, the three-dimensional attraction between molecules is no longer possible. Water molecules at the uncontained surface are forced to contort themselves into a two-dimensional form. The result is a molecular organization that is in tension —

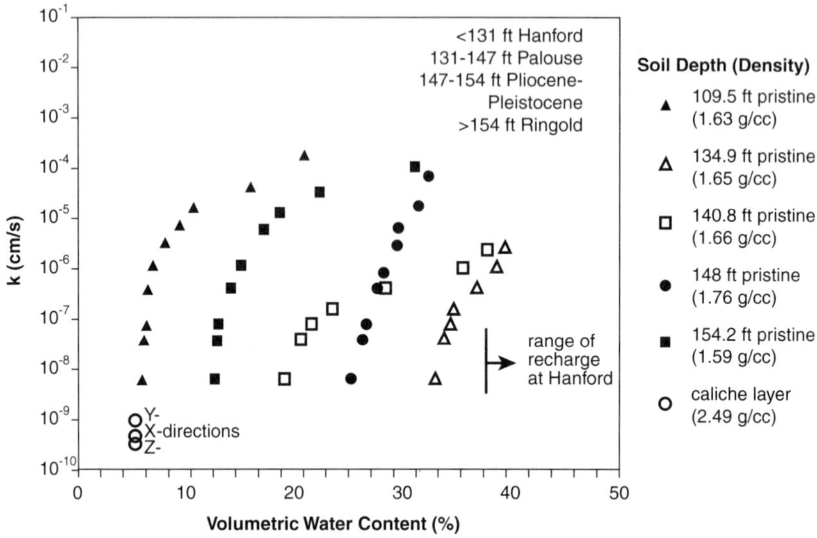

FIGURE 3.25 Graphs of hydraulic conductivity vs. volumetric water content showing characteristic curves for different sediments from the VOC-arid site integrated demonstration at the Hanford site.

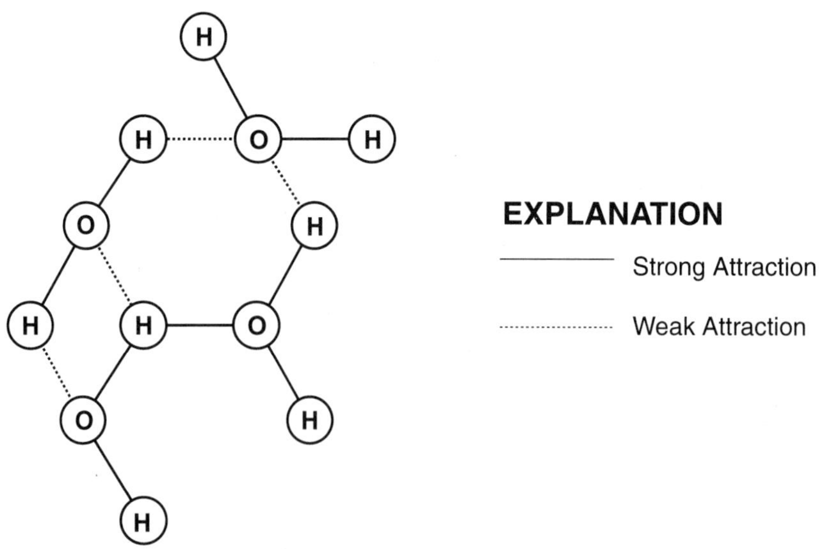

FIGURE 3.26 Attraction of water molecules.

TABLE 3.8
Surface Tension of Water

Temperature (C°)	Surface Tension lb/ft	Surface Tension g/cm
0	0.0052	0.076
10	0.0051	0.074
20	0.0050	0.073
30	0.0049	0.071
40	0.0048	0.070

"surface tension." Pushed to its extreme, the molecules try to associate into the form with the least surface area per volume, which is a sphere. A raindrop falling in a vacuum would hold a spherical form. A smaller radius results in a more stressed molecular organization. The surface tension of pure water is summarized in Table 3.8.

The addition of a small quantity of electrolyte (such as minerals dissolved from adjacent soil particles) increases surface tension. A small quantity of soluble organic compound (alcohol, soap, or acid) decreases the surface tension. The addition of glycerine to water reduces surface tension and thus makes it possible to "stretch" water film into bubbles, as with a child's bubble-blowing game.

A second important concept is that of adhesion. Attractive forces between hydrogen and oxygen in the water molecules are also effective between similar molecules in other substances. For example, the beading of raindrops onto a windshield is the result of the attraction of atoms of oxygen and hydrogen to the silicon and oxygen atoms in the glass. Solid surfaces that exhibit adhesion to water molecules are said to be hydrophilic (water attracting). Other molecules, such as paraffin, which do not attract water are hydrophobic (water repelling). The relationship between water and hydrophobic and hydrophilic surfaces is shown in Figure 3.27. The size of the alpha angle indicates the preference. If the angle is <90°, the water is attempting to maintain its spherical form, although somewhat squashed by the force of gravity. Quartz and most mineral particles in soil have a contact angle of >90°. Crystalline quartz and water containing some impurities are about 25°, which indicates a rather tight bond. In the subsurface, cohesive forces between soil particles and water are often so large that great energy (such as evaporation) is required to separate water from the soil.

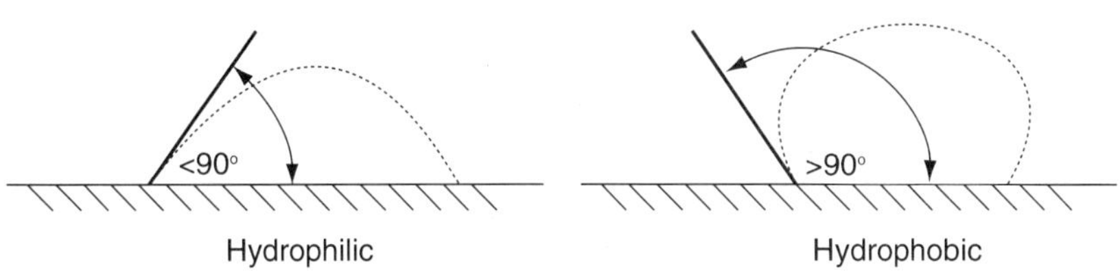

FIGURE 3.27 Relationship between water and hydrophobic and hydrophilic surfaces.

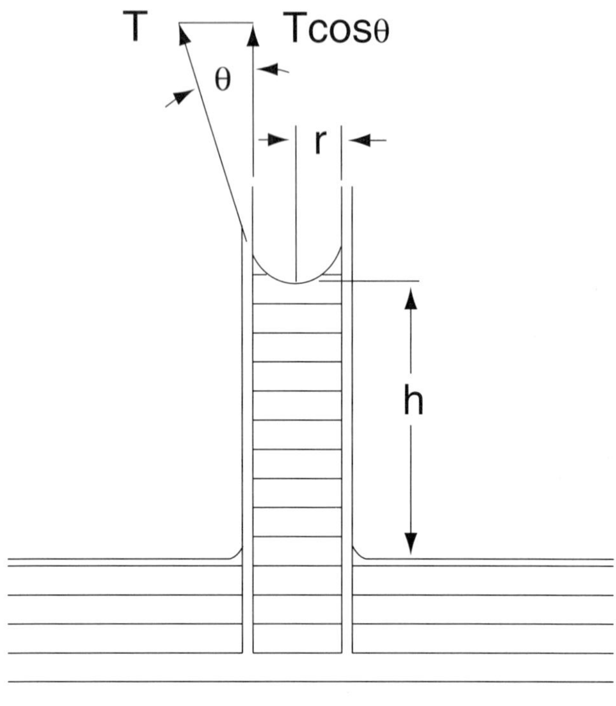

FIGURE 3.28 Rise of liquid in a capillary tube.

A combination of adhesion and surface tension gives rise (pardon the pun) to capillary action. By its adhesion to the solid surface of the soil particles, the water wants to cover as much solid surface as possible. However, by the effect of surface tension, the water molecules adhering to the solid surface are connected with a surface film in which the stresses cannot exceed the surface tension. As water is attracted to the soil particles by adhesion, it will rise upward until attractive forces balance the pull of gravity (Figure 3.28). Smaller-diameter tubes force the air–water surface into a smaller radius, with a lower solid-surface-to-volume ratio, which results in a greater capillary force. Typical heights of capillary rise for several soil types are presented in Table 3.9. The practical relationship between "normal" subsurface water and capillary rise is presented in the following equation.

$$h = \frac{2T\cos\theta}{rpg} \tag{3.13}$$

where h = rise in centimeters; T = surface tension in cynes; θ = well–liquid interface angle; p = density of fluid; g = acceleration of gravity; and r = radius of tube in centimeters.

In the subsurface, capillary forces can be active in all three dimensions depending on the location of the water source. Regardless of the direction of movement, the water is moving in the direction of least pressure. This phenomenon is often called

TABLE 3.9
Capillary Rise in Various Soils

Soil Type	Average Grain Size (mm)	Capillary Rise (m)
Sand	2–0.5	0.03–0.1
	0.5–0.2	0.1–0.3
	0.2–0.1	0.3–1.0
Silt	0.1–0.05	1.0–3.0
	0.05–0.02	—
Silt/clay	0.02–0.006	3–10
	0.006–0.002	10–30
Clay	<0.002	30–300

"soil suction." Dry soil particles always have a tendency to retain water molecules on their surface.

Even after a large supply of water has migrated downward through a soil zone, under conditions where gravity is the dominant force, some water will be retained on and between the soil particles as "residual saturation." The relationship between several soil horizons in the unsaturated zone is shown in Figure 3.29. Water in each of these zones is held according to the local conditions of soil suction. It is important to note that water in liquid form cannot be held by soil if the soil suction is >0.7 atm (10 psi). The zones of unsaturated soil are listed below:

- *Residual saturation* (hygroscopic water or pendular saturation): Water is held under tension of 31 to 10,000 atm force.
- *Funicular saturation*: Similar to residual saturation, except that each grain is surrounded by a water film and has a large air content.
- *Capillary fringe*: The zone immediately above the water table which is essentially saturated; this is the height that can support saturated conditions at negative pressure (due to soil suction).

Water flow through unsaturated soil is controlled by the same forces as capillary action and water retention (i.e., adhesion, gravity, and surface tension). Flow can occur only when the water phase is continuous from pore to pore. If gravity is the controlling force, downward flow will occur according to Darcy's law in direct proportion to the percentage of water-filled, connected pores. For example, if only half the pores in a cross section are water filled, the flow through that section will be half of that predicted by $Q = KIA$.

When water enters the subsurface from the surface, a great percentage initially flows through macropores (cracks, rootholes, and worm bores). After these larger openings are filled (or plugged by washed-in debris, or soil swelling), the primary flow paths become the soil pores.

If the source of water is relatively unlimited (long-term flooding), the downward migrating water will displace air from the pores and migrate according to Darcy's

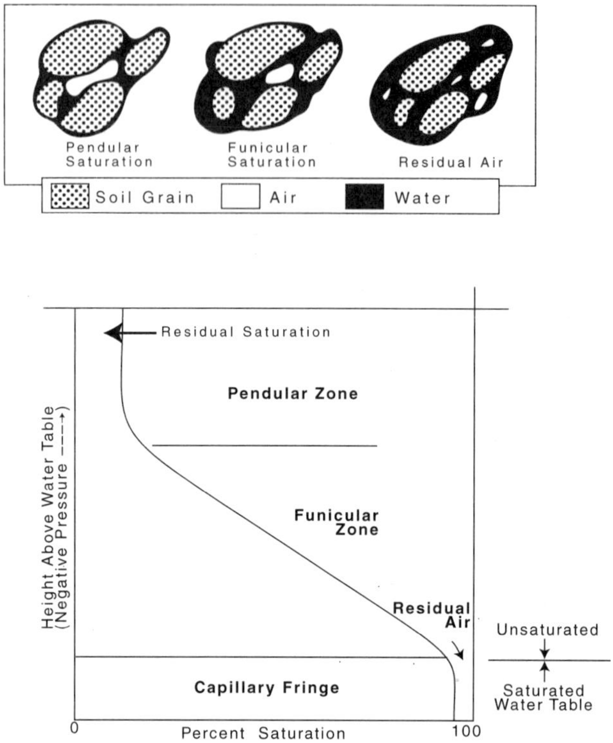

FIGURE 3.29 Soil zones defined by percent saturation. (Modified after Abdul, 1988.)

law. As the percentage of saturated pores increases, the downward flow will increase. The limit of this process is saturated downward flow.

During normal rainfall events, the quantity of water available for infiltration is finite. As water moves from pore to pore, some is retained along the pore walls by adhesion and surface tension in the throat between the pores. After these forces are satisfied, and most of the air displaced, the flow can continue to the next pore. If the unsaturated zone is thick, it is possible that all the water will be retained as residual saturation and none will reach the water table. The retention capability is often observed at the surface when a garden sprinkler sprays only a small quantity of water on sandy soil. The water will "bead up" on the surface until sufficient water has been applied to "wet" the surface soil fully; then infiltration will start. Following a major rainstorm, when the quantity of percolating water is greater than can be absorbed by the soil, a downward-moving "wetting front" occurs. This phenomenon is the result of the water moving downward, pore by pore. When the wave has passed, some water is retained as residual saturation. A graphic representation of the moving wetting front is shown in Figure 3.30.

Actual migration of a wetting front is highly dependent on the nature and size of the soil pore spaces. If all of the pores are relatively similar, the migration rate of the wetted front is uniform. If the pore spaces become smaller, the capillary forces

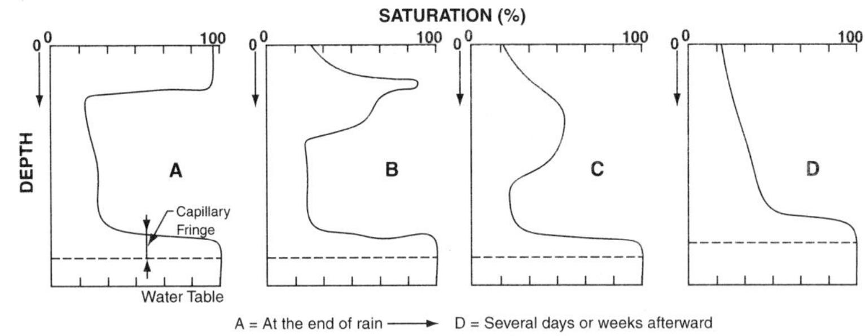

FIGURE 3.30 Downward moving wave of infiltrating water through the vadose zone.

are greater and the rate of movement faster. When significantly larger pores are encountered (such as a sand lens) the capillary forces may not be sufficient to cause flow into the lens. In effect, it is possible for a highly permeable sand to act as a confining layer that restricts downward water flow.

A final thought regarding unsaturated systems relates to the concept of capillary barriers. Under unsaturated conditions, water will not move from a smaller pore into a larger pore unless the smaller pore is saturated. In the previously discussed gravel/vault example, water will not flow from the sand across the boundary with the gravel unless the sand is saturated. If the boundary (called a capillary barrier) is horizontal, saturation is likely to occur as recharge pools in the sand until it finally saturates and flows into the underlying gravel. However, if the boundary is at an angle, even as little as 5°, then the water will not flow into the underlying gravel, but will flow along the sand/gravel boundary remaining with the sand, unless recharge is high enough to saturate the sand. The greater the difference in relative grain size, the better the barrier will perform, although the capillary barrier is not a vapor diffusion barrier.

As a corollary to this, the larger pores will empty first during desaturation of a sample. This is especially important for unsaturated and desaturating fractured rock: the larger fractures will desaturate first while the smaller fractures will still be completely saturated. This situation can be complicated when a contaminated fluid encounters many layers of different fracture/pore-size materials having different water contents. Knowledge of these kinds of unsaturated behavior can be utilized to engineer innovative and effective barriers to contaminant migration in the vadose zone.

REFERENCES

Anderson, M. P., 1997, Geologic Setting: In *Subsurface Restoration* (edited by C. H. Ward, J. A. Cherry, and M. R. Scalf), Ann Arbor Press, Chelsea, MI, pp. 17–25.

Back, W., Rosenshein, J. S., and Seaber, P. R. (editors), 1988, *Hydrogeology, The Geology of North America*: Geological Society of America, Boulder, CO, Vol. 0–2, 524 pp.

Bear, J., 1979, *Hydraulics of Groundwater*: McGraw-Hill International, New York, 567 pp.

Bentall, R. (compiler), 1963a, *Methods of Determining Permeability, Transmissibility and Drawdown*: U.S. Geological Survey Water-Supply Paper 1536-I, pp. I243–I341.

Bentall, R. (compiler), 1963b, *Shortcuts and Special Problems in Aquifer Tests*: U.S. Geological Survey Water-Supply Paper 1545-C, pp. C1–C117.

Boulton, N. S. and Streltsova, T. D., 1977, Unsteady Flow to a Pumped Well in a Fissured Water-bearing Formation: *Journal of Hydrology,* Vol. 35, pp. 257–269.

Boulton, N. S. and Streltsova, T. D., 1978, Unsteady Flow to a Pumped Well in an Unconfined Fissured Aquifer: *Journal of Hydrology,* Vol. 37, pp. 349–363.

Bouwer, H., 1978, *Groundwater Hydrology*: McGraw-Hill, New York, 480 pp.

Choquette, P. W. and Pray, L. C., 1970, Geologic Nomenclature and Classification of Porosity in Sedimentary Carbonates: *American Association of Petroleum Geologists, Bulletin,* Vol. 54, pp. 207–250.

Clark, L., 1979, The Analysis and Planning of Step Drawdown Tests: *Quarterly Journal of Engineering Geology*: Vol. 10, No. 2, pp. 125–143.

Darcy, H., 1856, *Les Fontaines Publiques de la Ville de Dijon*: V. Dalmont, Paris, 674 pp.

Driscoll, F. G., 1986, *Groundwater and Wells*: Johnson Division, St. Paul, MN, 1088 pp.

Ebanks, W. I., Jr., 1987, Geology in Enhanced Oil Recovery: In *Reservoir Sedimentology* (edited by R. W. Tillman and K. J. Weber): Society of Economic Paleontologists and Mineralogists, Special Publication No. 40, pp. 1–14.

Ehman, K. D. and Cramer, R. S., 1996, Assessment of Potential Groundwater Contaminant Migration Pathways Using Sequence Stratigraphy: In *Proceedings of the National Ground Water Association 1996 Petroleum Hydrocarbon and Organic Chemicals in Ground Water: Prevention, Detection, and Remediation Conference*, Nov. 13–15, Houston, TX, pp. 289–304.

Fetter, C. W., Jr., 1980, *Applied Hydrogeology*: Charles E. Merrill, Columbus, OH, 488 pp.

Fogg, G. F., 1989, Emergence of Geologic and Stochastic Approaches for Characterization of Heterogeneous Aquifers: In *Proceedings of the U.S. Environmental Protection Agency, Robert S. Kerr Environmental Research Laboratory Conference on New Field Techniques for Quantifying the Physical and Chemical Properties of Heterogeneous Aquifers*, March 20–23, 1989, Dallas, TX, pp. 1–17.

Freeze, R. A., and Cherry, J. A., 1979, *Groundwater*: Prentice-Hall, Englewood Cliffs, NJ, pp. 145–166.

Gregory, K. J. (editor), 1983, *Background to Paleohydrology — A Perspective*: John Wiley & Sons, New York, 486 pp.

Gringarten, A. C., 1982, Flow-test Evaluation of Fractured Reservoirs: In *Recent Trends in Hydrogeology,* Geological Society of America Special Paper 189, pp. 237–263.

Gringarten, A. C. and Witherspoon, P. A., 1972, A Method of Analyzing Pumping Test Data from Fractured Aquifers: In *Proceedings of the Symposium on Percolation through Fissured Rock,* International Society for Rock Mechanics and the International Society of Engineering Geology, Stuttgart, T3, pp. B1–B9.

Grubb, S., 1993, Analytical Model for Estimation of Steady-State Capture Zones of Pumping Wells in Confined and Unconfined Aquifers: *Ground Water,* Vol. 31, No. 1, pp. 27–32.

Hantush, M. S., 1966, Analysis of Data from Pumping Tests in Anisotropic Aquifers: *Journal of Geophysical Research*, pp. 421–426.

Harman, H. D., Jr., 1986, Detailed Stratigraphic and Structural Control: The Key to Complete and Successful Geophysical Surveys of Hazardous Waste Sites: In *Proceedings of the Hazardous Materials Control Research Institute Conference on Hazardous Wastes and Hazardous Materials*, Atlanta, GA, pp. 19–21.

Jenkins, J. D. and Prentice, J. K., 1982, Theory for Aquifer Test Analysis in Fractured Rock under Linear (Non-Radial) Flow Conditions: *Ground Water,* Vol. 20, pp. 12–21.

Keller, B., Hoylmass, E., and Chadbourne, J., 1987, Fault Controlled Hydrology at a Waste Pile: *Ground Water Monitoring Review,* Spring, pp. 60–63.

Kruseman, G. P. and DeRidder, N. A., 1979, *Analysis and Evaluation of Pumping Test Data*: International Institute for Land Reclamation and Improvement, Bull. 11, Wageningen, the Netherlands, 200 pp.

Lohman, S. W., 1972, Ground-Water Hydraulics: U.S. Geological Survey Professional Paper 708, 70 pp.

Love, D. W., Whitworth, T. M., Davis, J. M., and Seager, W. R., 1999, Free-Phase NAPL-Trapping Features in Intermontane Basins: *Association of Engineering Geologists and Geological Society of America, Environmental & Engineering Geoscience*, Spring, Vol. V, No. 1, pp. 87–102.

Neuman, S. P., Walter, G. R., Bentley, H. W., Ward, J. J., and Gonzalez, D. D., 1984, Determination of Horizontal Aquifer Anisotropy with Three Wells: *Ground Water*, Vol. 22, No. 1, pp. 66–72.

Nilsen, T. H., 1982, Alluvial Fan Deposits: In *Sandstone Depositional Environments* (edited by P. A. Scholle and D. Spearing), American Association of Petroleum Geologists, Memoir No. 31, Tulsa, OK, pp. 49–86.

Papadopoulos, I. S., 1965, Nonsteady Flow to a Well in an Infinite Anisotropic Aquifer: In *Proceedings of the Symposium on Hydrology of Fractured Rocks*, IAHS, Dubrovnik, pp. 21–31.

Poeter, E. and Gaylord, D. R., 1990, Influence of Aquifer Heterogeneity on Contaminant Transport at the Hanford Site: *Ground Water*, Vol. 28, No. 6, pp. 900–909.

Potter, P. E. and Pettijohn, F. J., 1977, *Paleocurrents and Basin Analysis*: Springer Verlag, New York, 296 pp.

Reading, H. G., 1978, *Sedimentary Environments and Facies*: Blackwell Scientific Publications, Oxford, England, 557 pp.

Schmelling, S. G. and Ross, R. R., 1989, *Contaminant Transport in Fractured Media: Models for Decision Makers*: U.S. EPA EPA/54014-89/004, 8 pp.

Schmidt, V., McDonald, D. A., and Platt, R. L., 1977, Pore Geometry and Reservoir Aspects of Secondary Porosity in Sandstones: *Bulletin of the Canadian Petroleum Geology*, Vol. 25, pp. 271–290.

Scholle, P. A. and Spearing, D,. 1982, *Sandstone Depositional Environments*: American Association of Petroleum Geologists, Memoir No. 31, Tulsa, OK, 410 pp.

Scholle, P. A., Bebout, D. G., and Moore, C. H., 1983, *Carbonate Depositional Environments*: American Association of Petroleum Geologists, Memoir No. 33, Tulsa, OK, 708 pp.

Selley, R. C., 1978, *Ancient Sedimentary Environments*: Chapman and Hall, London, England, 287 pp.

Sharp, J. M., 1984, Hydrogeologic Characteristics of Shallow Glacial Drift Aquifers in Dissected Till Plains (North-Central Missouri): *Ground Water*, Vol. 22, No. 6, pp. 683–689.

Sheahan, N. T., 1971, Type Curve Solution of Step Drawdown Tests: *Ground Water*, Vol. 9, No. 1, pp. 25–29.

Slough, K. J., Sudicky, E. A., and Forsyth, P. A., 1999, Importance of Rock Matrix Entry Pressure on DNAPLs Migration in Fractured Geologic Material: *Ground Water*, March–April, pp. 237–244.

Soller, D. R. and Berg, R. C., 1992, A Model for the Assessment of Aquifer Contamination Potential Based on Regional Geologic Framework: *Environmental Geology and Water Science*, Vol. 19, No. 3, pp. 205–213.

Strobel, M. L., Strobel, C. J., and Delin, G. N., 1998, Design for a Packer/Vacuum Slug Test System for Estimating Hydraulic Conductivity in Wells with LNAPLs, Casing Leaks, or Water Tables Intersecting the Screens: *Ground Water Monitoring & Remediation*, Fall, pp. 77–80.

Suaveplane, C., 1984, *Planning Test Analysis in Fracture Aquifer Formations: State of the Art and Some Perspectives*: Ground Water Hydraulics, Water Resources Monograph Series No. 9, pp. 171–206.

Testa, S. M., 1994, *Geological Aspects of Hazardous Waste Management*: CRC Press/Lewis Publishers, Boca Raton, FL, 537 pp.

Van Wagoner, J. C., Posamentier, H. W., Mitchum, R. M., Vail, R. M., Sarg, J. F., Loutit, T. S., and Hardenbol, J., 1988, An Overview of Sequence Stratigraphy and Key Definitions: In *Sea Level Changes: An Integrated Approach* (edited by Wilgus et al.), Society of Economic Paleontologists and Mineralogists Special Publication, No. 42, pp. 39–45.

Vroblesky, C. D., Robertson, J. F., and Rhodes, L. C., 1995, Stratigraphic Trapping of Spilled Jet Fuel Beneath the Water Table: *Ground Water Monitoring Review*, Spring, pp. 177–183.

Way, S. C. and McKee, C. R., 1982, In-situ Determination of Three-Dimensional Aquifer Permeabilities: *Ground Water*, Vol. 20, No. 5, pp. 594–603.

Weber, K. J., 1982, Influence of Common Sedimentary Structures on Fluid Flow in Reservoir Models: *Journal of Petroleum Technology*, March, 1982, pp. 665–672.

Whitworth, T. M. and Hsu, C. C., 1999, The Role of Sand/Shale Interfaces in Saturated Zone NAPL Trapping: *Environmental Geosciences*, Vol. 6, No. 1, pp. 25–34.

4 Hydrocarbon Chemistry

"All substances are poisons; there is none which is not a poison. The right dose differentiates a poison from a remedy."

— Paracelcus, 1493–1541

4.1 INTRODUCTION

Petroleum hydrocarbons and other related organic constituents are the most common contaminants found in groundwater. People commonly speak of substances such as gasoline, diesel and other fuels, and solvents, as if they are very familiar with them and clearly understand what they are and how they behave in the subsurface. The truth of the matter is that these substances are very complex mixtures of many organic chemicals, each defined by its own respective physical and chemical properties. It is these specific properties of the original petroleum that allow one to characterize or fingerprint its refined products. The unique circumstances that exist in the formation of petroleum make it quite unlikely for any two distinct petroleum sources (releases) to share the exact same chemistry. In conducting subsurface assessment- and remediation-related activities, it thus becomes important to have a fundamental understanding of the chemistry of these substances, and thus their relative behavior in the environment. It is therefore important at many sites with complex histories and commingled contaminant NAPL pools and dissolved hydrocarbon plumes to differentiate and fingerprint such occurrences spatially and temporally.

This chapter is divided essentially in two parts: hydrocarbon chemistry and forensic chemistry. Discussion of general hydrocarbon chemistry in regard to hydrocarbon type, structure, analytical methods, and transformation and degradation processes is presented. This is followed by discussion of forensic or fingerprinting techniques to distinguish hydrocarbon substances from one another in the subsurface and evaluating source or origin and relative timing of release. Where commingling of NAPL pools or dissolved plumes has occurred, some thoughts regarding relative contribution or, in other words, degree of liability are offered. Discussion of conventional relative age dating isotopic and hydrogeologic techniques as a tool in one's environmental forensic arsenal is also presented.

4.2 DEFINING PETROLEUM

The term *petroleum* is derived from the Latin derivative *petra* for rock and *oleum* for oil. Current usage defines petroleum as any hydrocarbon mixture of natural gas, condensate, and crude oil. Most important to this discussion is crude oil, which is a heterogeneous liquid consisting of hydrocarbons comprising almost entirely the elements hydrogen and carbon in a ratio of about 1.85 hydrogen atoms to 1 carbon atom. Minor constituents, such as sulfur, nitrogen and oxygen, typically make up less than 3% of the total volume. Trace constituents that typically constitute less than 1% in total volume include phosphorus and heavy metals such as vanadium and nickel.

The composition of crude oil may vary with the location and age of an oil field, and may even be depth dependent within an individual well or reservoir. Crudes are commonly classified according to their respective distillation residue, which reflects the relative contents of three basic hydrocarbon structural types: paraffins, naphthenes, and aromatics. About 85% of all crude oils can be classified as either asphalt based, paraffin based, or mixed based. Asphalt-based crudes contain little paraffin wax and an asphaltic residue (predominantly condensed aromatics). Sulfur, oxygen, and nitrogen contents are often relatively higher in asphalt-based crude in comparison with paraffin-based crudes, which contain little to no asphaltic materials. Mixed-based crude contains considerable amounts of both wax and asphalt. Representative crude oils and their respective composition in respect to paraffins, naphthenes, and aromatics are shown in Figure 4.1.

A petrochemical is a chemical compound or element recovered from petroleum or natural gas, or derived in whole or in part from petroleum or natural gas hydrocarbons, and intended for chemical markets.

4.3 HYDROCARBON STRUCTURE

Hydrocarbons in general are simply compounds of hydrogen and carbon that can be characterized based on their respective chemical composition and structure. Each carbon atom can essentially bond with four hydrogen atoms. Methane is the simplest hydrocarbon, as illustrated below:

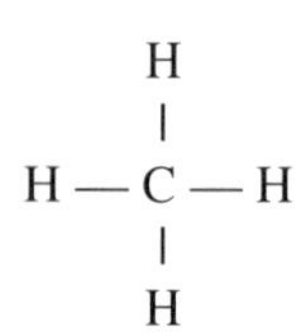

Methane(CH_4)

Each dash represents a chemical bond of which the carbon atom has four and each hydrogen atom one.

More complex forms of methane can be developed by adhering to the simple rule that a single bond exists between adjacent carbon atoms and that the rest of the

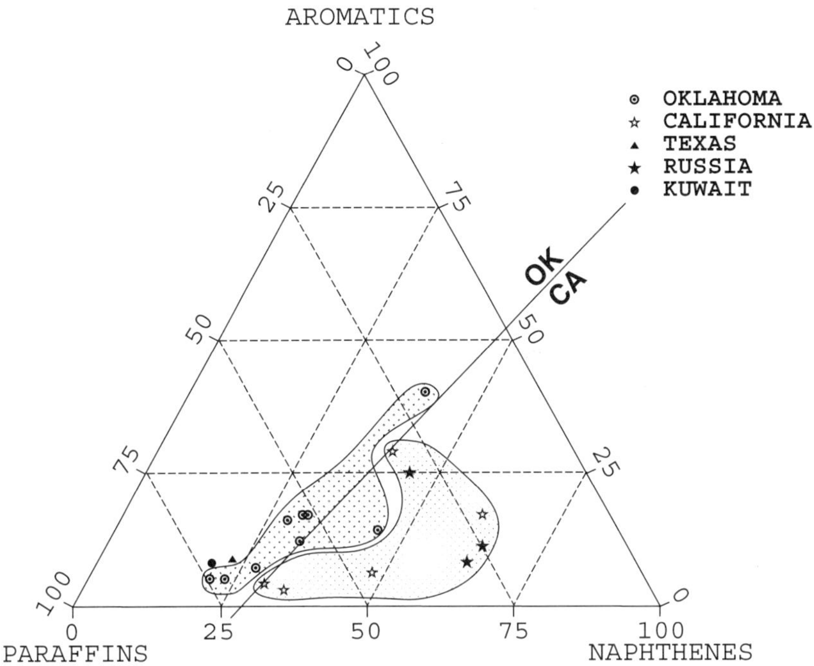

FIGURE 4.1 Ternary diagram showing representative crude oils and their respective composition in respect to paraffins, naphthenes, and aromatics. Note the clustering of the California crudes toward the naphthenes and the Oklahoma crudes toward the paraffins.

bonds are saturated with hydrogen atoms. With the development of more complex forms, thus, there is also an increase in molecular size. Hydrocarbons that contain the same number of carbons and hydrogen atoms, but have a different structure and therefore different properties, are known as isomers. As the number of carbon atoms in the molecule increases, the number of isomers rapidly increases. The simplest hydrocarbon having isomers is butane, as follows:

$$CH_3 - CH_2 - CH_2 - CH_3$$

Normal Butane

$$\begin{array}{l} CH_3 \\ \quad \backslash \\ \quad\quad CH - CH_3 \\ \quad / \\ CH_3 \end{array}$$

Isobutane

Hydrocarbon compounds can be divided into four major structural forms: (1) alkanes, (2) cycloalkanes, (3) alkenes, and (4) arenes. Petroleum geologists and engineers commonly refer to these structural groups as (1) paraffins, (2) naphthenes or cycloparaffins, (3) aromatics, and (4) olefins, respectively; and will be referred to here as such. Olefins are characterized by double bonds between two or more carbon atoms. Olefins are readily reduced or polymerized to alkanes early in diagenesis, and are not found in crude oil and only in trace amounts in a few petroleum products. The following discussions will focus on paraffins, naphthenes or cycloparaffins, and aromatics.

Paraffin-type hydrocarbons are referred to as saturated or aliphatic hydrocarbons. These hydrocarbons dominate gasoline fractions of crude oil and are the principal hydrocarbons in the oldest, most deeply buried reservoirs. Paraffins can form normal (straight) chains and branched-chain structures. Normal chains form a homologous series in which each member differs from the next member by a constant amount, as illustrated in Figure 4.2, where each hydrocarbon differs from the succeeding member by one carbon and two hydrogen atoms. The naming of normal paraffins is a simple progression using Greek prefixes to identify the total number of carbon atoms present. The common names of some normal-chain paraffins along with their respective physical properties are listed in Table 4.1. Branched-chain paraffins reflect different isomers (different compounds with the same molecular formula). Where only about 60 normal-chain paraffins exist, theoretically, over a million branched-chain structures are possible with about 600 individual hydrocarbons identified. Common branched-chain paraffins and their respective physical properties are illustrated in Figure 4.2 and also listed in Table 4.1.

Naphthenes or cycloparaffins are formed by joining the carbon atoms in ring-type structures, the most common molecular structures in petroleum. These hydrocarbons are also referred to as saturated hydrocarbons since all the available carbon atoms are saturated with hydrogen. Typical naphthenes and their respective physical properties are listed in Table 4.2 and shown in Figure 4.3.

Aromatic hydrocarbons usually comprise less than 15% of a total crude oil although often exceed 50% in heavier fractions of petroleum. The aromatic fraction of petroleum is the most important environmental group of hydrocarbon chemicals and contains at least one benzene ring comprising six carbon atoms in which the fourth bond of each carbon atom is shared throughout the ring. Schematically shown with a six-sided ring with an inner circle, the aromatics are unsaturated allowing them to react to add hydrogen and other elements to the ring. Benzene is known as the parent compound of the aromatic series and, along with toluene, ethylbenzene, and the three isomers of xylene (ortho-, meta-, and para-), are major constituents of gasoline. Typical aromatics and their respective physical properties are listed in Table 4.3 and shown in Figure 4.4.

Organic compounds such as the chlorinated solvents also include a wide range of compounds and do not easily fit into the structural classification as described for petroleum hydrocarbons. Chlorinated hydrocarbons are commonly discussed in terms of their relative density (i.e., LNAPL or DNAPL) or degree of halogenation and degree of volatility (i.e., volatile, semi-volatile).

The chemical structure for common chlorinated solvents is shown in Figure 4.5. Chlorinated solvents such as TCE and PCE are composed of double-bonded carbon or ethylene structures with three and four chlorine atoms, respectively. The ethane derivative 1,1,1-TCA has three chlorine atoms. Freon is a chlorofluorocarbon and is also an ethane derivative with four chlorine atoms and three fluoride atoms.

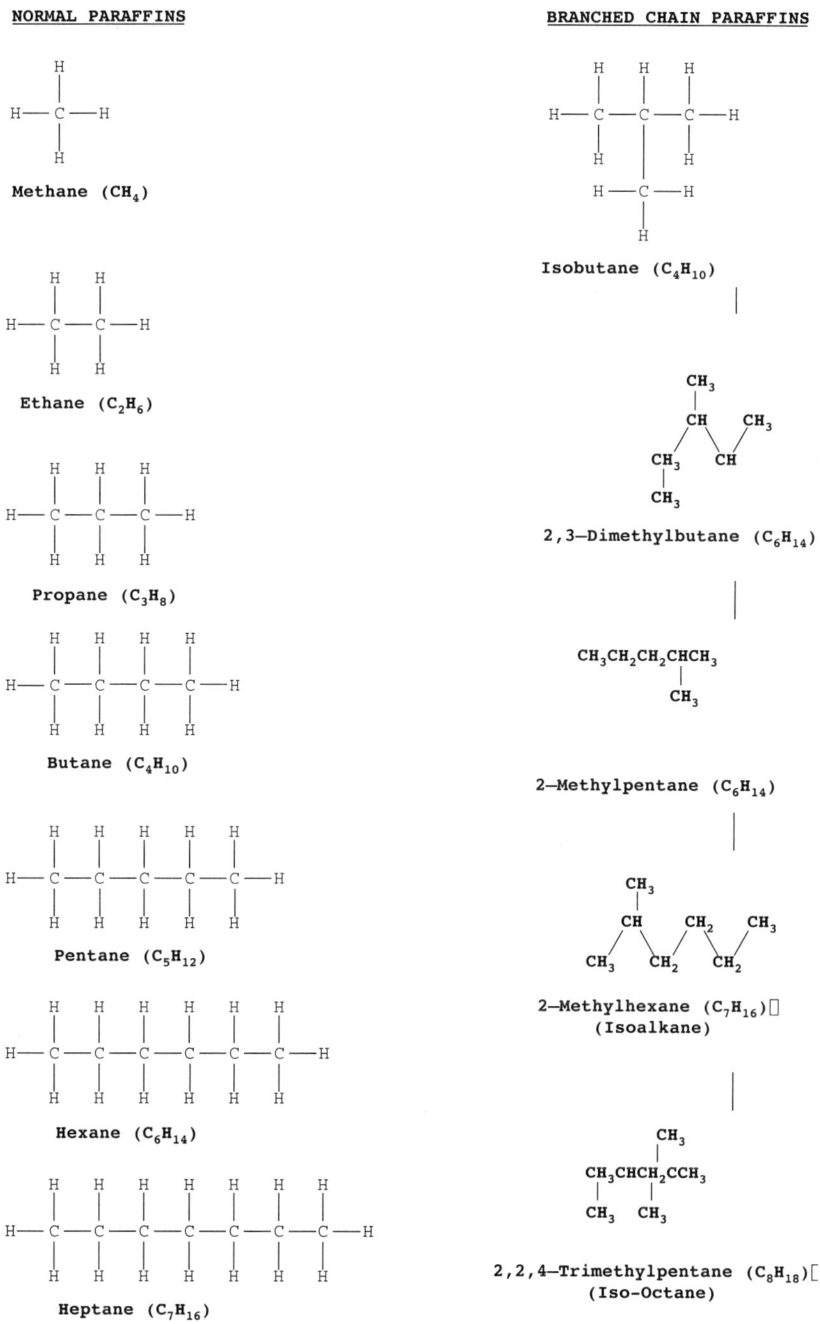

FIGURE 4.2 Structural forms of normal (straight)-chain and branched-chain paraffins.

TABLE 4.1
Chemical and Physical Properties of Common Normal Paraffin and Branched-Chain Paraffins

Compound	Chemical Formula	Molecular Weight (g/mol)	Density (g/cm³)	Solubility (g/m³)	Viscosity (mP)	Boiling Point (°C)	Vapor Pressure (mm)
Normal Paraffins							
Methane	CH_4	16	0.554	—	108.7@20°C	–161	400@–168.8°C
Ethane	C_2H_6	30	0.446	—	98.7@17°C	–89	400@–99.7°C
Propane	C_2H_8	44	0.582	62.4 ± 2.1	79.5@17.9°C	–42	400@–55.6°C
Butane	C_4H_{10}	58	0.599	61.4 ± 2.6	—	–0.5	400@16.3°C
Pentane	C_5H_{12}	72	0.626	38.5 ± 2.0	676,000@25°C	36	426
Hexane	C_7H_{16}	86	0.659	9.5 ± 13	3,260@2°C	69	124
Branched-Chain Paraffins							
Isobutane	C_4H_{10}	58	—	48.9 ± 2.1	—	–12	—
2,2-Dimethylbutane	C_6H_{14}	86	0.649	18.4 ± 1.3	—	50	400
2,3-Dimethylbutane	C_6H_{14}	86	0.668	22.5 ± 0.4	—	58	400
2-Methylpentane	C_6H_{14}	86	0.669	13.8 ± 0.9	—	60	400@41.6°C
2-Methylhexane	C_7H_{16}	100	0.6789	2.54 ± 0.0	—	90	40
3-Methylhexane	C_7H_{16}	100	—	4.95 ± 0.08	—	92	—
2,2,4-Trimethylpentane	C_8H_{18}	114	0.692	2.44 ± 0.12	—	99	40.6

TABLE 4.2
Chemical and Physical Properties of Common Naphthenes

Compound	Chemical Formula	Molecular Weight (g/mol)	Density (g/cm³)	Solubility (g/m³)	Viscosity (mP)	Boiling Point (°C)	Vapor Pressure (mm)
Methylcyclopentane	C_6H_{12}	84	0.749	42 ± 1.6	—	72	—
Cyclohexane	C_6H_{12}	84	0.778	55 ± 2.3	1.02@17°C	81	95
Ethylcyclohexane	C_8H_{16}	112	—	3.29 ± 0.46	—	132	—
1,1,3-Trimethyl-cyclohexane	C_9H_{18}	126	—	1.77 ± 0.05	—	137	—

NAPHTHENES (CYCLOPARAFFINS)

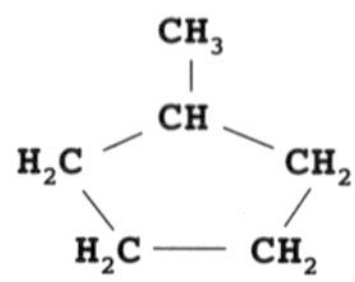

Methylcyclopentane (C_6H_{12})

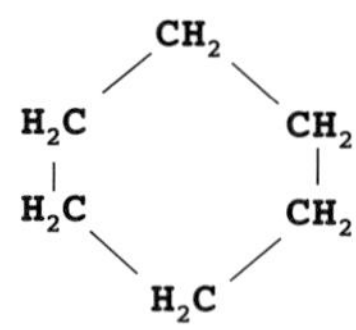

Cyclohexane (C_6H_{12})

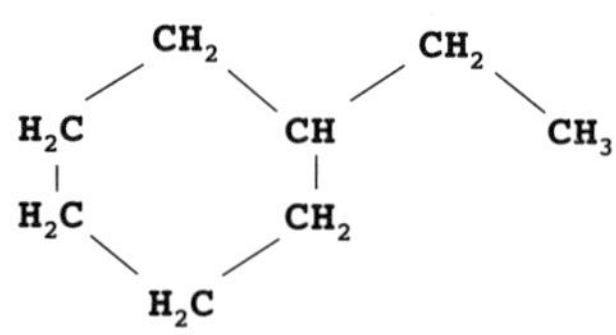

Ethylcyclohexane (C_8H_{16})

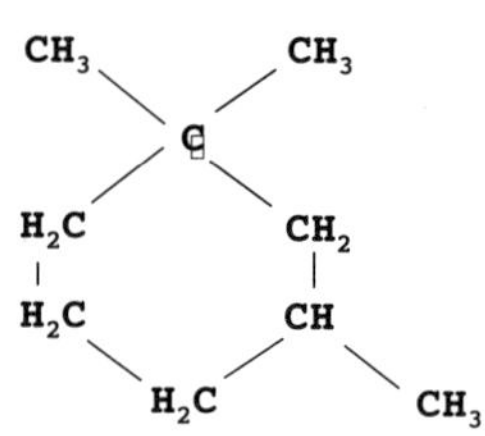

1, 1,3-Trimethylcyclohexane (C_9H_{18})

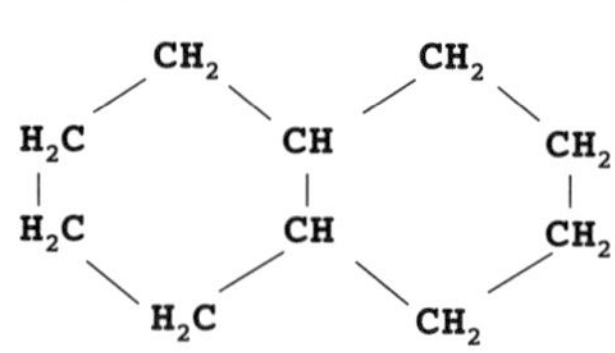

Decalin ($C_{10}H_{18}$)

FIGURE 4.3 Structural forms of typical naphthenes.

Methylene chloride is characterized by what is termed a methylene chlorideane structure and has two chlorine atoms.

4.4 HYDROCARBON PRODUCTS

A refinery converts crude oil and other hydrocarbon feedstocks into useful products and raw materials for other industries (Figure 4.5). Refining involves the separation and blending, purifying and quality improvement of desired petroleum products. The primary products of a refinery are as follows:

- Fuels such as gasoline and diesel fuels, aviation and marine fuels, and fuel oils;
- Chemical feedstocks such as naphtha, gas oils, and gases;
- Lubricating oils, greases, and waxes;
- Butimens and asphalts;
- Petroleum coke; and
- Sulfur.

Transportation fuels are by far the primary product of refineries. The variables that determine gasoline quality include blend, octane level, distillation, vapor pressure ratings, and other considerations, such as sulfur content or gum-forming tendencies. The petrochemical and other industries also utilize refinery products to manufacture plastics, polyester clothing, certain pharmaceuticals, and a myriad of other useful products. The primary fuels, familiar to all, are gasoline, diesel fuel (or Number 2 fuel oil), and fuel oils. Gasoline comprises a mixture of volatile hydrocarbons suitable for use in internal combustion engines. The primary components of gasoline are branched-chain paraffins, cycloparaffins, and aromatics. Diesel fuel is composed primarily of unbranched paraffins with a flash point between 110 and 190°F. Fuel oils are chemical mixtures that can be distilled fractions of petroleum, residuum from refinery operations, crude petroleum, or a mixture of two or more of these mixtures, with a flash point greater than 100°F. These petroleum mixtures basically represent progressive fractions of a distillation column, with the major petroleum hydrocarbons easily represented by increasing the boiling point in a gas chromatograph separating column.

4.4.1 Refining Processes

The refinery uses three primary processes in converting crude oil into these varied fuel products: separation, conversion, and treatment.

4.4.1.1 Separation

The products derived from petroleum are produced in refineries using a fundamental separation process called distillation. The principal method for separating crude oil into useful products is through distillation. Distillation is the heating process whereby hydrocarbons, which make up the crude oil, are converted to vapor form and are

TABLE 4.3
Chemical and Physical Properties for Aromatics

Compound	Chemical Formula	Molecular Weight (g/mol)	Specific Gravity @25°C	Solubility (g/m^3)	Viscosity (mP)	Boiling Point (°C)	Vapor Pressure (mm)	K_{oc}
Benzene	C_6H_6	78.11	0.878	1780 ± 45	0.652	80	76@20°C	1.69–2.00
Toluene	C_7H_8	92.14	0.867	515 ± 17	0.590	111	22@20°C	2.06–2.18
Ethylbenzene	C_8H_{10}	106.17	0.867	161 mg/l@25°C	—	136.2	4.53@25°C	3.05–3.15
Orthoxylene	C_8H_{10}	106.17	0.880	1.75 ± 8	0.810	142	10@25.9°C	2.11
Metaxylene	C_8H_{10}	106	0.867	146 ± 1.6	0.620	138.9	10@28.3°C	3.20
Paraxylene	C_9H_{12}	120	0.860	156 ± 1.6	0.648	138	10@27.3°C	2.31
Isopropylbenzene	C_9H_{12}	120	—	50 ± 5	—	152	—	—
3,4-Benzpyrene	$C_{20}H_{12}$	252	1.351	0.003	—	495	5.49×10^{-9}@25°C	5.81–6.50
MTBE		88.15	0.744	42,000 mg/l	—	—	245	1.035–1.91

AROMATICS

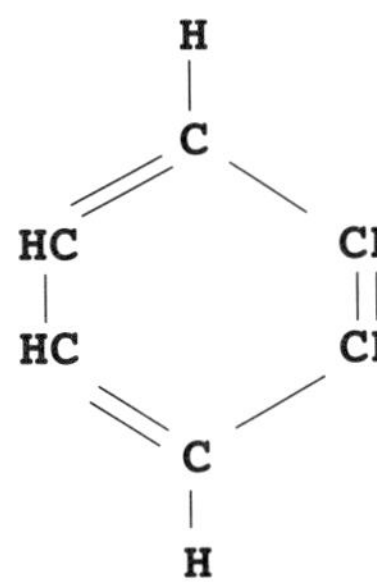

Benzene (C_6H_6)

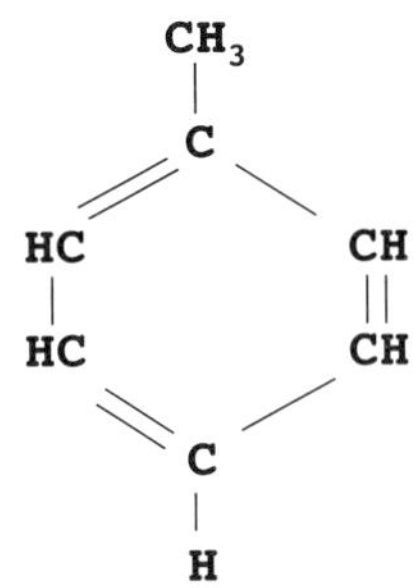

Toluene (C_7H_8)

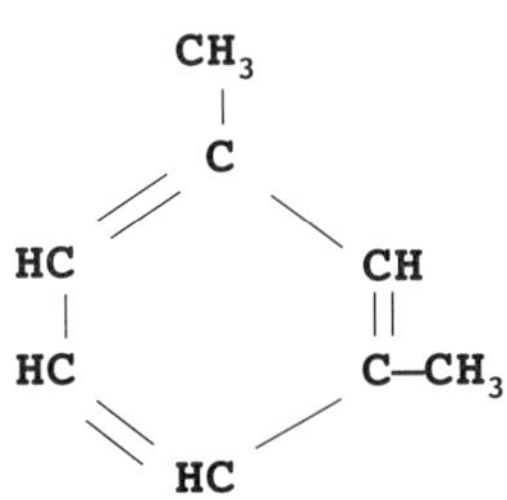

Metaxylene (C_8H_{10})

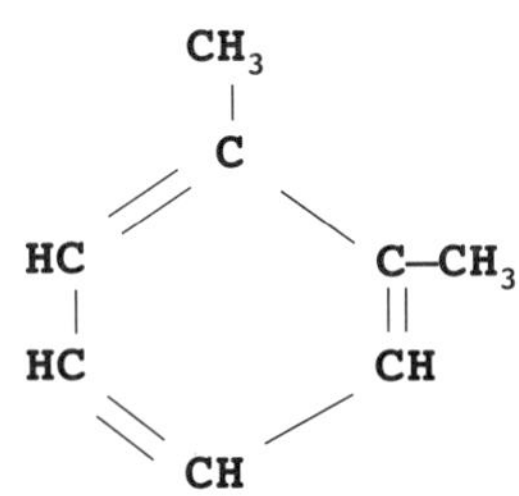

Orthoxylene (C_8H_{10})

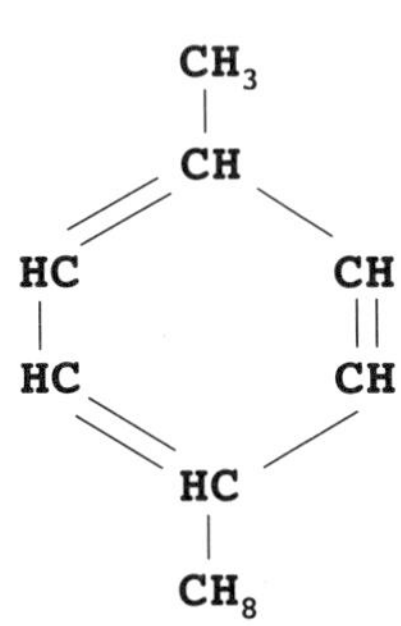

Paraxylene (C_8H_{10})

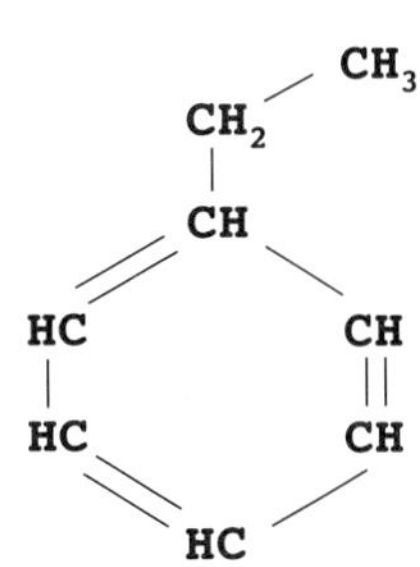

Ethylbenzene (C_8H_{10})

FIGURE 4.4 Structural forms of typical aromatics.

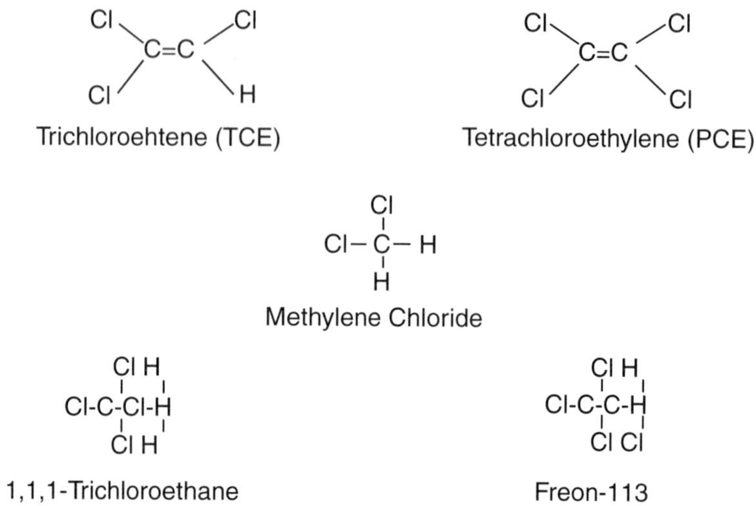

FIGURE 4.5 Schematic showing the distillation process.

released from the oil by heating. The released hydrocarbon vapors are then cooled back to a liquid form. Boiling points of hydrocarbons generally increase with an increase in the number of carbon atoms that comprise the compound. As a crude sample (or any hydrocarbon blend) is heated in increasing increments, the hydrocarbon compounds having a boiling point at or below the current temperature sequentially volatilize. The remaining hydrocarbon compounds in the sample will not volatilize until the temperature is raised to their respective boiling points. By drawing off the hydrocarbons vapors that have similar boiling points, refineries produce a variety of products, such as kerosene, naphtha, gasoline, and butane.

A plot of boiling temperatures (°F) vs. cumulative percent volume removed from the sample is referred to as a distillation curve. The boiling temperatures for various products range from high to low divided into the following product types: residue, heavy gas-oil, light gas-oil, kerosene, naphtha, gasoline, and butanes (Table 4.4).

TABLE 4.4
Major Petroleum Distillation Products

Fraction	Distillation Temperature (°C)	Carbon Range
Butanes	—	C_1 to C_4
Gasoline	40–205	C_5 to C_{14}
Naphtha	60–100	C_6 to C_7
Kerosene	175–325	C_9 to C_{18}
Light gas-oil	275	C_{14} to C_{18}
Heavy gas-oil	275	C_{19} to C_{25}
Residue	—	$>C_{40}$

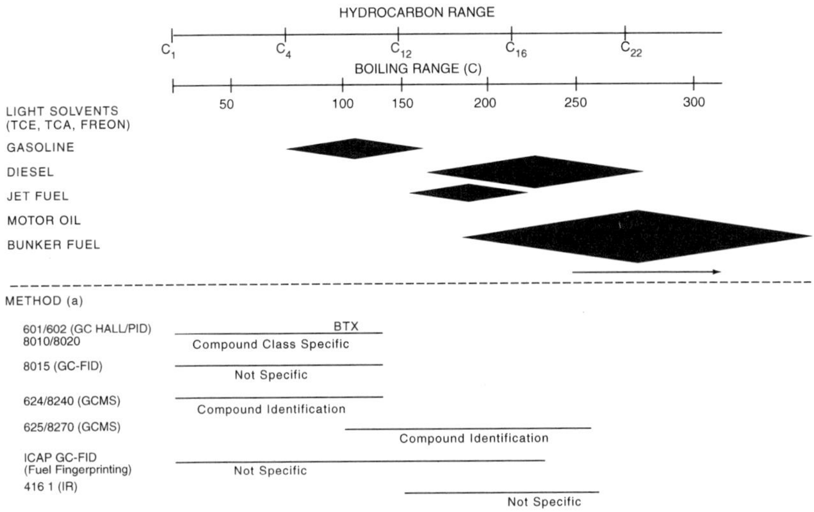

(a) GC = Gas Chromatography
PID = Photoionization
HALL = Electrolytic Conductivity
MS = Mass Spectrometry
ICAP = Inductively Coupled Argon Plasma Spectrophotometry
FID = Flame Ionization
IR = Infared Spectrophotometry

FIGURE 4.6 Hydrocarbon range, boiling point, and analytical method for major petroleum products.

The crude source has a definite effect on the composition of the refined product. A schematic illustrating the distillation process is shown in Figure 4.5. These products correspond to their respective boiling points, carbon ranges, and analytical method as illustrated in Figure 4.6.

Separation can also be accomplished by solvent extraction, adsorption, and crystallization. Solvent extraction is accomplished by selectively dissolving certain hydrocarbon components. Adsorption is similar to solvent extraction but uses a solid to separate out various components selectively based on their tendency to adhere to the surface of the solid adsorbent. Crystallization uses the differing melting points of the components during cooling, which causes some of its compounds to solidify or crystallize, and separate out of the liquid.

4.4.1.2 Conversion

Conversion transforms the chemical composition of the various fractions produced by distillation. The result of conversion is common intermediate components that eventually become parts of other end products. Conversion processes include fluid catalytic cracking, thermal cracking, hydrocracking, coking, and alkylation and reforming. Cracking refers to the process of breaking down large hydrocarbon molecules into smaller ones. Fluid catalytic cracking is the most commonly used conversion process in the United States as it produces about 40% of all domestically refined gasoline. Cracking uses intense heat (1000°F or 358°C), low pressure, and a finely powdered catalyst to convert most heavy hydrocarbons into smaller hydrocarbons within the gasoline range.

One of the oldest conversion processes is thermal cracking where heat and pressure continue to change heavy molecules in fuel oil and asphalt into lighter ones. The use of lower temperatures, greater pressures, and hydrogen to promote chemical reactions is referred to as hydrocracking. This form of conversion is utilized at all refineries, but is most cost-effective in converting medium- to heavy-weight oils into higher-quality products. Coking uses heat and moderate pressure to convert residuum from other conversion processes into lighter products. Coking produces a hard, coal-like substance that is used as an industrial fuel.

The making of materials suitable for blending into gasoline from liquid petroleum gases produced via catalytic cracking is a process referred to as alkylation. This is accomplished by using acidic catalysts to combine hydrocarbon molecules, rather than cracking or breaking apart molecules. Reforming utilizes heat, moderate pressure, and special catalysts to convert nonaromatic hydrocarbons such as naphtha, a relatively low-value fraction, into aromatic hydrocarbons such as high-octane gasoline components (including benzene). Aromatic hydrocarbons are building blocks for many plastics and other products generated by the petrochemical industry.

4.4.1.3 Treatment

A process referred to as hydrotreating removes certain substances that may inhibit refinery processes or result in poor emission performance of fuels. Because of the presence of sulfur and nitrogen, the resulting refining product may still be considered undesirable because these naturally occurring substances create sulfur oxide and nitrogen oxide emissions when the refined product is burned. A process called hydrotreating utilizes hydrogen with a catalyst to react with sulfur compounds and form hydrogen sulfide. Hydrogen sulfur can be easily separated from hydrocarbons and processed to form sulfur.

4.5 DEGRADATION PROCESSES

Degradation processes tend to destroy the paraffins, remove the light ends, and oxidize the remaining fractions of the hydrocarbon. Degradation processes include biodegradation as well as chemical and physical degradation via mechanisms such as oxidation, water washing, and inspissation (the evaporation of the lighter constituents of petroleum leaving the heavier residue behind).

During biodegradation, paraffins in addition to naphthenes and aromatics are all susceptible to microbial decomposition. Over 30 genera and 100 species of various bacteria, fungi, and yeast exist which metabolically utilize one or more kinds of hydrocarbons. Biodegradation results in the selective destruction of specific chemical classes by microbial consumption, starting with the paraffins. The general mechanism for biodegradation is shown in Figure 4.7, where hydrocarbons are essentially oxidized to alcohols, ketones, and acids.

The order in which hydrocarbons are oxidized depends on numerous factors. However, in general small carbon molecules up to C_{20} are consumed before larger

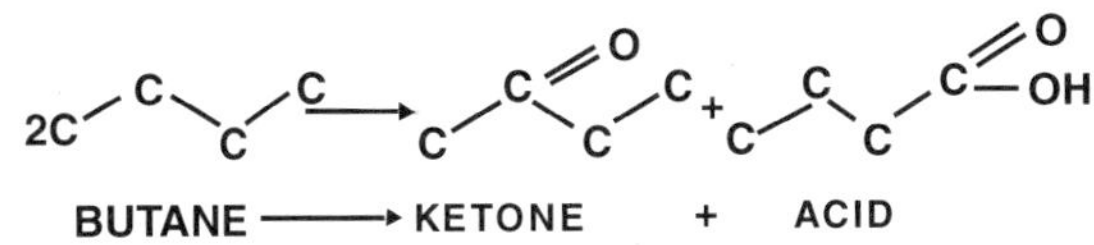

FIGURE 4.7 Conversion of *n*-butane to ketone and acid via microbiological oxidation.

ones. Within the same molecular weight range, the order is usually *n*-paraffins first. These in turn are followed by isoparaffins, naphthenes, aromatics, and polycyclic aromatics. Single-ring naphthenes and aromatics are consumed before isoprenoids, steranes, and triterpanes. The remaining diasteranes and tricyclic terpanes survive heavy biodegradation. Thus, diasteranes and tricyclic terpanes may be useful in chemically fingerprinting biodegraded product (Figure 4.8).

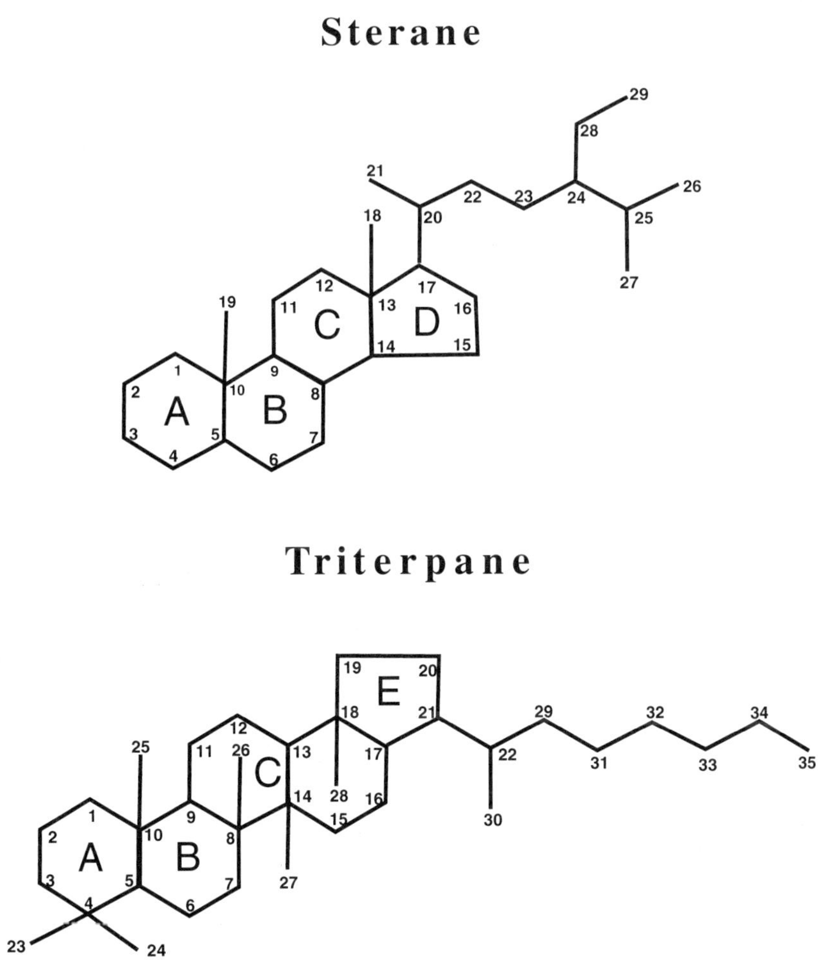

FIGURE 4.8 Sterane and triterpane compound structures.

Other microbiological, chemical, and physical processes also contribute to the degradation of hydrocarbons in the subsurface, and can play a significant role in the characterization and source identification of subsurface hydrocarbons as discussed later in this chapter. The mere migration of hydrocarbons in the subsurface results in the removal of some components, and the retention of others, by soil, sediment, and other subsurface materials. This is most prevalent with the more soluble hydrocarbons such as the light aromatics. Water washing is the preferential dissolving of certain hydrocarbon constituents when in contact with non-steady-state water conditions. Crude hydrocarbon products can undergo oxidation when exposed to dissolved oxygen, certain minerals, and bacteria, resulting in degradation of the lower-molecular-weight paraffinic fractions, notably the *n*-alkanes. Oxidation usually occurs near the surface, resulting in degradation, increased viscosity and specific gravity, and conversion of some liquid components to solids. Deasphalting is the extraction of asphalt components from a petroleum product. For example, propane deasphalting employs propane as a solvent to separate oil and asphalt; the liquid propane dissolves the oil as the asphalt settles out. Biodegradation is the selective destruction of specific chemical classes by microbial consumption. Heavily biodegraded hydrocarbons are typically enriched in nitrogen, oxygen, sulfur, and polycyclic compounds, reflecting either the end product following microbial activity or removal and destruction of less stable constituents.

In addition to these naturally occurring factors are the complexities associated with the historical development (notably industrial) of a particular site or area, property ownership and uses, and transfers over time. The close proximity of crude- and petroleum-handling facilities ranging from the clustering of refineries, bulk storage tank farms, and underground pipelines in industrialized areas, to numerous underground storage tanks associated with many commercial establishments in less-industrialized areas can also complicate matters.

4.6 FORENSIC CHEMISTRY

Environmental forensic chemistry involves the geochemical characterization or "fingerprinting" of leaked crude or refined products. Forensic analysis of spilled hydrocarbons has come into its own over the past decade in distinguishing different substances that may have been released into the environment at different times and commingled, and evaluating the source based on the history of the site in regard to what products it handled and when. This is important in evaluating origin, relative timing, and contribution, thus, liability, in addition to an appropriate remediation strategy. Fingerprinting techniques in some cases are very conventional in nature, while others were specifically developed for use in the petroleum exploration industry in efforts to characterize source rock and crude oil types. These techniques in turn have subsequently been modified and applied to address environmental issues, notably in the identification of fugitive hydrocarbons. With refined petroleum products, similarities and differences between samples can be discerned reflecting differences in the brand of gasoline. Certain product additives of interest and of environmental significance can be qualitatively and quantitatively identified. Such additives include antiknock compounds, blending agents, antioxidants, antirust agents, anti-icing

TABLE 4.5
Dates for Introduction of Certain Chemical Additives

Chemical Additive	Date of Commercial Availability
Chemical Additives	
Carbon tetrachloride	1907
Trichloroethylene (TCE)	1908
1,2-Dichloroethane (DCA)	1922
Tetrachloroethene (PCE)	1925
DDT	1942
Chlordane	1947
Toxaphene	1947
Aldrin	1948
Dibromochloropropane	1955
Bromacil	1955
Gasoline Additives	
Tetraethyl lead (TEL)	1923
Methyl cyclopentadienyl manganese tricarbonyl (MMT)	1957
Tetramethyl lead	1966
Ethylene dibromide (EDB)	pre-1980s
Methyltriethyl lead	pre-1980s
Methyl tert-butyl ether (MTBE)	1980s[a]

[a] Region specific.

agents, or other proprietary purposes (Table 4.5). Common additives include 1,2-dibromoethane (EDB), 1,2-dichloroethane (EDC), isopropyl alcohol, methyl tertiary butyl ether (MTBE), *p*-phenylenediamine and tert-butyl alcohol (TBA).

Current EPA analytical methods do not allow for the complete speciation of the various hydrocarbon compounds. EPA Methods 418.1 and 8015 provide the "total" amount of petroleum hydrocarbons present. However, only concentrations within a limited hydrocarbon range are applicable to those particular methods. Volatile compounds are usually lost, and samples are typically quantitated against a known hydrocarbon mixture and not the specific hydrocarbon compounds of concern or the petroleum product released. By conducting EPA Method 8015 (Modified) using a gas chromatograph fitted with a capillary column instead of the standard, hand-packed column, additional separation of various fuel-ranged hydrocarbons can be achieved.

Several methodologies can be used to identify not only crude or refined product type, but also the brand, grade, and, in some instances, the source crude. The petroleum industry has yielded conventional methods for the characterization of refined products. The simplest is the routine determination of API gravity and development of distillation curves where NAPL is present. More sophisticated methods include gas chromatography and statistical comparisons of the distribution of paraffinic or *n*-alkane compounds between certain C-ranges. With increased degradation and decomposition, the straight-chain hydrocarbons (*n*-alkanes) become less useful. Under these circumstances, molecular compounds such as the sterane and triterpane biomarkers become more important. Under severe biodegradation, the rearranged polycyclic and aromatic compounds may be of use. Chemical testing for trace elements, notably, lead, manganese, nickel, sulfur, and vanadium, may also be useful in distinguishing product types and crude source(s). Determination of certain isotopic ratios of carbon and hydrogen for lighter gasoline-range fractions and ^{15}N and ^{34}S for heavier petroleum fractions can also be utilized for source identification. These techniques are discussed below.

4.6.1 API Gravity

Probably the simplest tool in discriminating between, notably, LNAPL product types is the measurement of API gravity. API gravity is similar to specific gravity but differs by the following conversion:

$$°API = \frac{141.5}{\text{specific gravity}}(-131.5) \tag{4.1}$$

By using this conversion, water has a specific gravity of 1 (unit less) and an API gravity of 10°. The higher the API gravity, the lighter the compound. Typical API gravities for various refined products are listed in Table 4.6. Specific gravity corresponding to API gravity is provided in Appendix B.

The common technique used to measure the API gravity of a liquid sample is with a hydrometer graduated in degrees of API gravity. The hydrometer is inserted in the liquid sample and allowed to float freely. When equilibrium is reached, the API gravity may be read from the exposed portion of the hydrometer.

Another method to evaluate API gravity commonly applied to oil field evaluations is the use of ultraviolet (UV) light having a wavelength of 3600 Å. UV light of this wavelength causes fluorescence of crude oils in differing colors, which is generally characteristic of the gravity of the oil (Table 4.7). Used during drilling, this technique can be used as a quantitative tool for determining API gravity for hydrocarbon-affected soils.

TABLE 4.6
Typical API Gravities

Product Type	Specific Gravity	API Gravity, °
Gasoline	0.74	60
Light crude	0.84	36
Heavy crude	0.95	18
Asphalt	0.99	11

After Leffler, 1979.

TABLE 4.7
Typical Fluorescence Color for Varying API Gravities

API Gravity	Color of Fluorescence
Below 15	Brown
15–25	Orange
25–35	Yellow to cream
35–45	White
Over 45	Blue-white to violet

After Exploration Logging, Inc., 1980.

UV or fluorescence characteristics of hydrocarbons can also be used as a qualitative tool in conjunction with coring or excavation to evaluate the presence of hydrocarbons and their relative abundance in the impacted area. This simple technique can be used as an effective tool to delineate the extent of petroleum hydrocarbons in excavations, locating seeps along the shoreline, and selecting sampling intervals in sediment cores. Most hydrocarbons including light-colored gasolines and jet fuels fluoresce, and although not necessarily detectable under white light, are detectable under UV in contrast to the surrounding uncontaminated media. Combined with solvent extraction, low levels of petroleum in soil can be detected, and relative concentrations assessed. Under UV light, samples whose pore space is filled with air or water will remain dark, while samples whose pore space is filled with hydrocarbon will glow. All other things being equal, the higher percentage of hydrocarbon in total volume will glow the brightest.

4.6.2 Distillation Curves

A distillation curve is essentially a plot of boiling temperature (°F) vs. cumulative percent volume removed from the sample during heating (Figure 4.9). This type of plot can be used to distinguish quickly between one or several product types.

Example 4.1

Typical distillation curves for six LNAPL samples retrieved from beneath a 60-year-old refinery site at a depth of about 50 ft below ground surface, are presented in Figure 4.9. The samples comprise different portions of a single, but coalesced LNAPL pool. Typical distillation curves for primarily light gas-oil (A and C), light gas-oil (B), mixture of light and heavy gas-oil (D), a mixture of primarily kerosene and naphtha (E), and blend of gasoline, naphtha, kerosene, and light gas-oil (F) are shown. What is clearly evident is that the LNAPL pool comprises different petroleum product types, reflecting different releases and volumes, over time.

Example 4.2

Simulated distillation curves produced for two LNAPL samples are shown in Figure 4.10. The two samples depicted were retrieved at a depth of about 10 ft below the

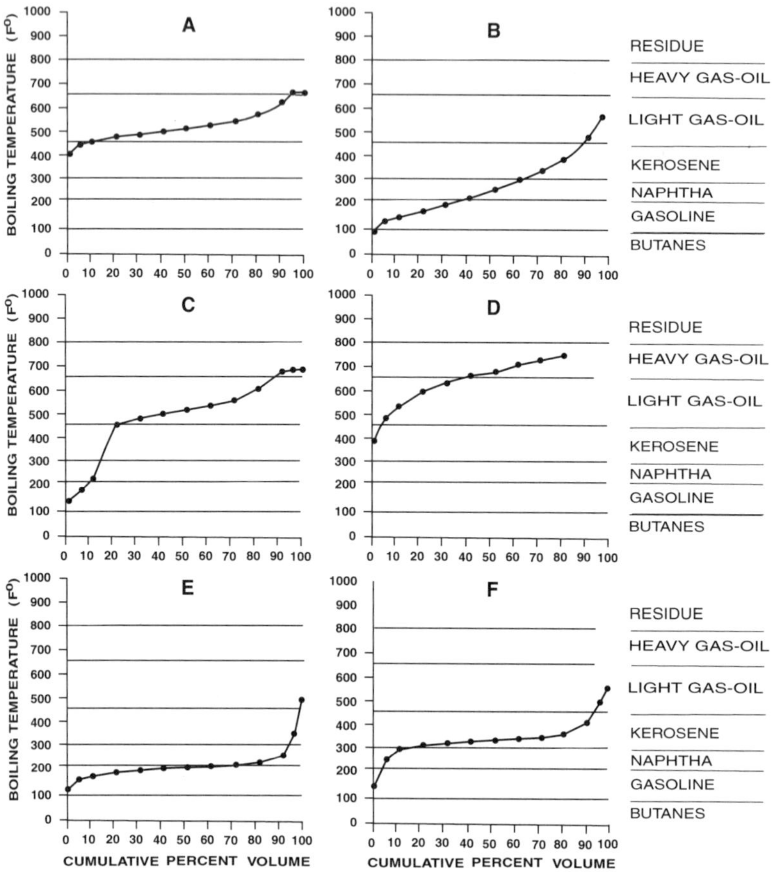

FIGURE 4.9 Typical distillation curves for samples retrieved from a coalesced LNAPL pool from beneath a refinery site.

ground surface from two adjoining sites that formerly had underground storage tanks located on site. LNAPL was released into the subsurface, and although the dissolved plumes have commingled, the LNAPL pools remained separated.

The starting temperature (about 145°F) at which the first hydrocarbon evolves is close for both samples; their distillation curves, however, show different patterns. Monitoring well (MW)-2 shows that at 500°F, at which gasoline-range hydrocarbons (up to C^9) are evolved, about 98% of LNAPL is recoverable. Thus, MW-2 consists almost entirely of gasoline with trace quantities of middle distillate products, likely diesel fuel. For MW-4, only 75% of LNAPL is evolved at the same temperature. The final distillation temperature for MW-4 was very close to 800°F, which is higher than 700°F, whereas the temperature of elution of all diesel-range hydrocarbons is 700°F. Thus, MW-4 in addition to gasoline and diesel fuel also contains a small quantity (5%) of a high distillate product such as heavy residual fuel oil or possibly lubricating oil.

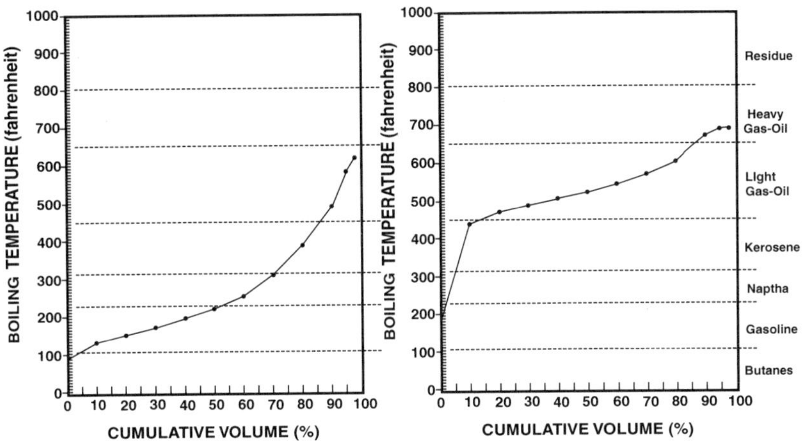

UNIT	Butane	Gasoline	Naptha	Kerosene	Light Gas	Heavy Gas	Residue
Percentage Distillation Concentration	0	50.3	19.5	16.0	14.2	0	0

UNIT	Butane	Gasoline	Naptha	Kerosene	Light Gas	Heavy Gas	Residue
Percentage Distillation Concentration	0	1.7	3.2	6.7	74.4	14.0	0

FIGURE 4.10 Typical distillation curve for (a) predominantly gasoline with trace heavy residual fuel oil and (b) predominantly gasoline with a middle distillate of diesel fuel.

4.6.3 Trace Metals Analysis

Analysis for organic lead (not total) alkyl compounds can sometimes provide useful information in distinguishing between product types, evaluating potential source(s) of gasoline contamination, and relative timing of release (see Table 4.5). Some of these compounds include ethylene dibromide (EDB), tetramethyl lead, trimethylethyl lead, methyltriethyl lead, and tetraethyl lead (TEL). Between 1980 and the present, only TEL was used as an additive in leaded gasoline. Prior to the 1980s, TEL would make up about a third of the total volume to an order of magnitude less. Lead alkyl compounds such as TEL, tetramethyl lead, and EDB were used as antiknock compounds in gasoline. Typically, unleaded gasoline will not contain more than 0.05 g/gal (13 mg/l) while leaded gasoline will not contain greater than 1.1 g/gal (290 mg/l). Organic lead, however, has historically been added to gasoline fuels at concentrations up to 800 mg/l.

The solubility of these compounds lends clues to their origin and timing of release. EDB is more soluble than the lead alkyls. Thus, if both EDB and TEL are present in significant amounts, the release is inferred to be quite recent. If EDB is reduced or absent, it is inferred that some exposure on the order of 5 to 10 years has occurred. Should TEL also be significantly reduced, exposure for a period in excess of 10 years is inferred. When other lead alkyls are present, it is inferred that release took place prior to 1980.

During diagenesis of crude oil source materials, both vanadium and nickel become complexed into a porphyrin ring in addition to other compounds. A porphyrin, such as chlorophyll and heme (of hemoglobin), is a large complex organic ring compound made up of four substituted pyrrole rings, among other rings. Porphyrins are found in plants, as well as carbonaceous shale, crude oil, and coal, and are thus introduced into the crude oil from the source rock, and carry the imprint of the original vanadium–nickel distribution. Therefore, concentrations of vanadium and nickel are several thousand times what would be expected if the metals were not complexed into organic structures. Crude oil concentrations of both vanadium and nickel increase an order of magnitude with a decrease in API gravity of about 10°. Vanadium–nickel ratios are often used for oil field correlation and may also serve as a supportive tool for environmental investigations. Analysis may be conducted using an atomic absorption spectrophotometer and/or inductively coupled plasma-mass spectrometer (ICP). The ICP will tend to be more quantitative, reflecting lower achievable detection limits.

4.6.4 Gas Chromatography Fingerprinting

Crude oil and refined petroleum product can easily be differentiated with the interpretation of gas chromatograms using EPA Method 8015. In this analysis, for example, emitted compounds are scanned using a gas chromatograph within the temperature range from 50 to 300°C at a heating rate of 10°C/min, and then held at 300°C for 5 min. The total run time is 30 min. The chromatograms are then compared with standards, calculated, and reported. Standard chromatograms for gasoline, naphtha, kerosene, JP-5 jet fuel, diesel fuel, and crude oil are shown in Figure 4.11.

Chromatographic fingerprinting of hydrocarbon samples can be conducted utilizing a gas chromatograph with flame ionization detection (FID). The gas chromatograph in this instance is equipped with a capillary column, rather than a standard hand-packed column, to achieve better separation of the compounds as they are eluted. Discussed below is the use of gas chromatography fingerprinting techniques and applications for gasoline, alkane gas, aromatic, and higher petroleum fractions, respectively.

4.6.4.1 Gasoline-Range Hydrocarbons

Results of hydrocarbons in the range of C_4 to C_8 (gasoline range), about 80 individual compounds, can be presented as relative percent (Table 4.8). In the petroleum industry, the ratios of certain compounds or combinations of compounds are used to indicate the degree of maturity, paraffinicity, biodegradation, and water washing, and such applications can be similarly applied to environmental issues. In a refined product such as gasoline, similar ratios are measurable and may be useful in assessing the similarity of differences in samples. These similarities and differences do not have the same significance as with crude samples due to extensive processing. The values are expected to differ as a result of the processing (cracking, reforming, etc.) and blending steps the product has undergone. Therefore, gas chromatograph measurements yield a fingerprint that, in principle, can differ from one brand of gasoline to another. Certain ratios can be compared by plotting such data on a stereonet or ternary diagram.

Example 4.3

The chromatographic results (fingerprint) of five hydrocarbon samples are tabulated in Table 4.8 and graphically illustrated in Figure 4.12. These LNAPL samples were

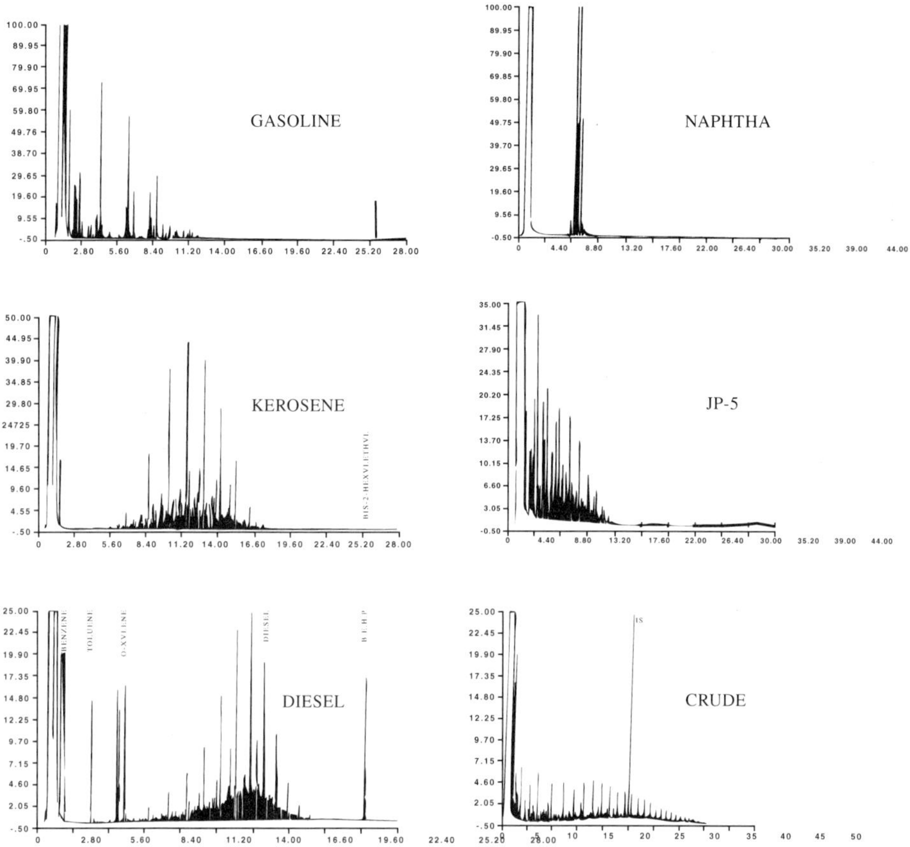

FIGURE 4.11 Standard gas chromatograms for gasoline, naphtha, kerosene, JP-5 jet fuel, diesel fuel, and crude oil.

collected at a depth of about 50 ft below the ground surface beneath a refinery tank farm. Results presented are in the range from *n*-heptane to *n*-octane. The compositional differences between the five samples are tabulated in Table 4.9. As is readily deduced, samples 1 and 4 are similar to one another, as are samples 2 and 3. This similarity is also evident in Table 4.10.

When a ternary diagram is constructed presenting relative percent "normal" alkanes vs. relative percent branched-alkanes (isoalkanes) vs. relative percent cyclic-alkanes, the similarity is even more striking (Figure 4.13). The grouping of sample number 1 with 4 and sample number 2 with 3 is strong, while the dissimilarity of sample number 5 from the other four samples is also clear. Additional aromatic compound distribution can be discerned using gas chromatographs as illustrated in Figure 4.14.

Example 4.4

In Figure 4.15, four LNAPL samples are compared based on data presented in Table 4.11. Six ratios were calculated: T/B, E/B, E/X, X/T, E/T, and X/B. The samples in

TABLE 4.8
Gasoline Range Condensate Hydrocarbon Analysis in Relative Percent

		Sample Number[a]				
	Name	1	2	3	4	5
11	2,3-Dimethylbutane	1.28	0.55	0.17	1.28	3.80
12	2-Methylpentane	4.37	1.54	1.02	5.43	18.23
13	3-Methylpentane	3.38	1.49	1.17	4.54	10.93
14	*n*-Hexane	4.36	1.28	1.16	5.65	6.48
15	2,2-Dimethylpentane					
16	Methylcyclopentane	9.66	5.53	5.04	10.01	10.15
17	1,4-Dimethylpentane	1.22	0.23	0.19	0.74	0.34
18	2,2,3-Trimethylbutane					
19	Benzene		2.87	3.25		3.53
20	3,3-Dimethylpentane		0.04	0.34		
21	Cyclohexane	4.72	3.90	5.01	4.21	2.48
22	2-Methylhexane	5.39	1.83	1.14	4.33	1.73
23	2,3-Dimethylpentane	1.51	1.24	1.16	1.73	0.61
24	1,1-Dimethylcyclopentane		0.93	0.96		0.36
25	3-Methylhexane	3.60	3.09	2.37	3.22	1.48
26	*c*-1,3-Dimethylcyclopentane	4.14	3.25	2.92	4.24	1.40
27	*t*-1,3-Dimethylcyclopentane	3.96	3.18	2.90	3.90	1.46
28	*t*-1,2-Dimethylcyclopentane	8.17	6.34	5.74	8.07	2.50
29	*n*-Heptane	4.37	3.51	2.56	4.73	3.18
30	Methylcyclohexane	10.91	13.04	12.96	11.6	5.07
31	Trimethylcyclopentane(iso?)	1.72	1.94	1.88	1.78	0.79
32	Ethylcyclopentane	2.02	2.10	1.86	1.77	1.02
33	2,5-Dimethylhexane	0.93	0.61	0.46	0.062	
34	2,4-Dimethylhexane	0.96	0.82	0.72	0.89	
35	*t*-1,2-*c*-5-Trimethylcyclopentane	1.90	2.79	2.76	2.09	1.63
36	1,1,2-Trimethylcyclopentane	3.48	5.14	4.65	4.16	1.97
37	2,3,4-Trimethylpentane	1.38	0.00		1.18	
38	Toluene	1.52	0.00		1.13	
39	2-Methylheptane	5.61	10.59	15.25	5.46	6.56
42	Dimethylcyclohexane(iso?)					
43	Dimethylcyclopentane(iso?)	1.10	2.22	15.25	5.46	6.56
44	*n*-Octane	4.03	11.46	12.60	1.77	9.54

[a] Blanks in this table reflect that no peak was detected. Data that are 0.00 were detected at levels less than 0.005%.

Figure 4.15a are LNAPL obtained from two adjacent sites with different histories, and representative unaltered regular unleaded and premium unleaded gasoline samples. High BTEX hydrocarbon concentrations in LNAPL and groundwater are often indicative of gasoline presence. A noticeable depletion in benzene and toluene is evident in the LNAPL samples as compared with unaltered dispensed gasoline,

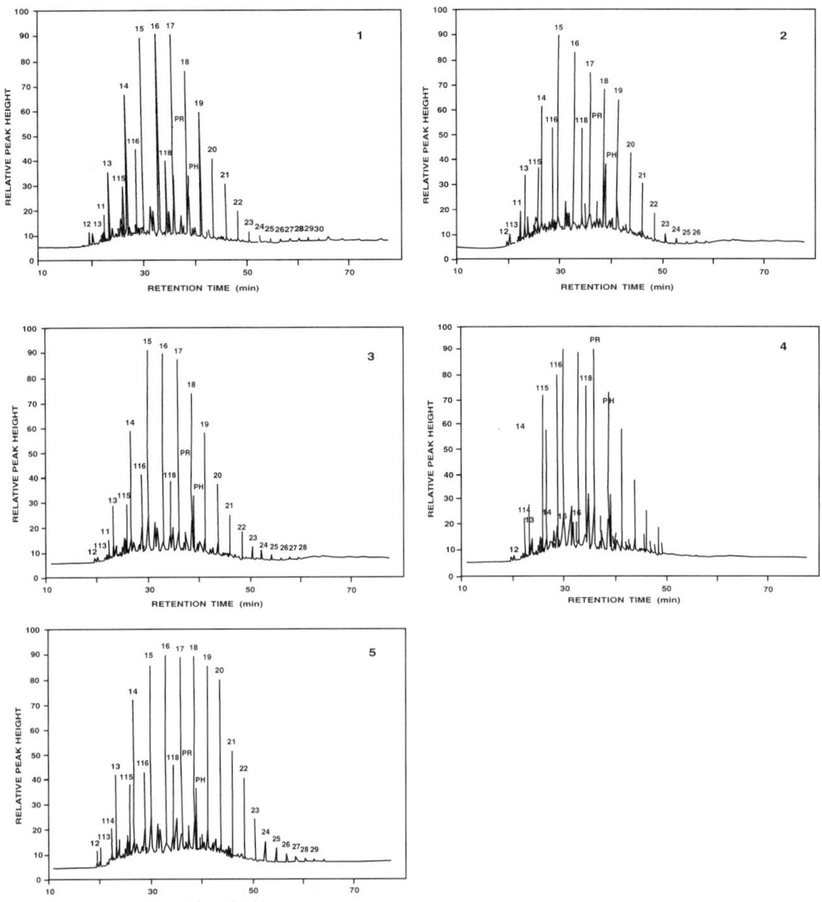

FIGURE 4.12 Gas chromatograph results for crude and refined LNAPL samples retrieved from a monitoring well network within a refinery tank farm.

although ratios of E/B, X/B, E/T, and X/T are not appreciably different from those calculated for gasoline.

The significantly higher benzene and to some extent toluene concentrations relative to those of other BTEX compounds in most of the BTEX-enriched groundwater samples obtained from the same sites (Figure 4.15b) is attributed to their higher water solubility (Figure 4.16). LNAPLs for wells MW-2 and MW-4 are inferred not to be highly altered, and only a relatively small amount of benzene and toluene have been removed by evaporation and/or water washing. The relatively high E/B, X/B, E/T, and X/T ratios for well MW-4 indicate that either (1) LNAPL for MW-4 has been exposed to the environment for a longer period of time relative to MW-2 or (2) MW-4 represents a gasoline release whose composition (formulation) may have been different from that of MW-2, especially because of its low BTEX content.

The presence of dissolved MTBE, a common oxygenated additive blended with gasoline in California to produce reformulated gasoline, in MW-4, but not in MW-2

TABLE 4.9
Gasoline-Range Condensate Hydrocarbon Ratios

Name			Sample Number 1	2	3	4	5
A	19/30	Late maturity index	0.00	0.22	0.25	0.00	0.70
B	38/29	Aromaticity	0.35	0.00	0.00	0.24	0.00
C	(1)	Paraffinity	0.56	0.28	0.21	0.66	1.28
F	29/30	Paraffinity	0.40	0.27	0.20	0.41	0.63
I	(2)	Isoheptane value	0.55	0.39	0.30	0.47	0.60
H	(3)	Heptane value	9.66	9.00	6.99	10.68	16.17
K	(4)	Kerogen type index	10.24	12.96	9.87	14.36	20.34
U	21/16	Naphthene branching	0.49	0.71	1.00	0.42	0.24
R	29/22	Paraffin branching	0.81	1.93	2.24	1.09	1.84
	13/19	Water-washing parameter	0.52	0.36			3.10
	30/38	Water-washing parameter	7.16			10.33	
	13/14	Biodegradation parameter	0.77	1.16	1.00	0.80	1.69
		Relative % normal alkanes	12.95	16.74	16.87	12.29	19.90
		Relative % branched alkanes	34.49	31.42	32.95	33.85	50.23
		Relative % cyclic alkanes	52.55	51.84	50.18	53.86	29.87
		Total	99.99	100.00	100.00	100.00	100.00

(1) (14 + 29)/(21 + 30)
(2) (22 + 25)/(26 + 27 + 28)
(3) [100.0 * 29/(21 + 22 + 24 + 25 + 26 + 27 + 28 + 29 + 30)]

(see Table 4.11), supports alternative 2 above. Because MTBE is more soluble in water relative to hydrocarbons such as BTEX (see Figure 4.16), and thus more easily transferred into groundwater, its absence may signify either its removal from the product during its exposure to water washing, or its absence in the original fuel. No EDC, EDB, and lead alkyl was present. Detection of lead alkyls and the lead scavengers EDC and EDB would imply that leaded gasoline was present, although they could also imply that these compounds were depleted by water washing and/or solution removal.

4.6.4.2 C^{8+} Alkane Gas Chromatography

Aromatic hydrocarbons can be effectively distinguished from nonaromatic hydrocarbons using capillary gas chromatography with photoionization detection (PID) using a UV lamp emitting photons of the appropriate energy level, which is useful for determining the distribution of aromatic compounds throughout the gasoline range. Hydrocarbons of higher molecular weight (less volatile than gasoline), *N*-alkanes, alkylcyclohexanes, and aliphatic biomarkers can be also be assessed by high-resolution gas chromatography–mass spectrometry (GC-MS).

4.6.4.3 Aromatic-Range Hydrocarbons

Differences in polynuclear aromatic hydrocarbons (PAHs) are shown as bar diagrams in Figure 4.17a and b. The high abundance of alkylbenzenes and alkylnaphthalenes

TABLE 4.10
Comparison of Composition Differences in Table 4.9

Name			Differences in Hydrocarbon Ratios[a]
A	19/30	Late maturity index	–0.70
B	38/29	Aromaticity	0.24
C	(1)	Paraffinity	–0.62
F	29/30	Paraffinity	–0.22
I	(2)	Isoheptane value	–0.13
H	(3)	Heptane value	–5.48
K	(4)	Kerogen type index	–5.99
U	21/16	Naphthene branching	0.18
R	29/22	Paraffin branching	–0.74
	13/19	Water-washing parameter	–3.10
	30/38	Water-washing parameter	10.33
	13/14	Biodegradation parameter	–0.89
		Relative % normal alkanes	–7.61
		Relative % branched alkanes	–16.38
		Relative % cyclic alkanes	23.99
Number of ratio differences < 1.0			8
Number of ratio differences < 4.0			3

[a] Differences in hydrocarbon ratios calculated by subtracting hydrocarbon ratio of sample 5 from sample 4.

provide strong evidence for gasoline presence, whereas the presence of fluorenes, biphenyls, phenanthrenes, pyrenes, crysenes, dibenzothiophenes, monoaromatic steranes (MAS), and triaromatic steranes (TAS) in MW-4 (Figure 4.17a) indicates that, in addition to gasoline, small amounts of middle to high distillate products such as diesel fuel and heavy residual fuel oil are also present. This is important because of the presence of MAS and TAS, which are usually absent in light (i.e., gasoline) and middle (i.e., diesel fuel) distillate products, but are present in heavy refined products and/or crude oils in MW-4.

4.6.4.4 Higher-Range Petroleum Fractions

Alkanes distribution, saturated polycyclic isomer biomarker determination, and polynuclear aromatics and their sulfur analogues (thiophenes) distribution can be used under certain circumstances to distinguish among the higher petroleum fractions. Alkanes are measured using capillary gas chromatography. The straight-chain alkanes and a few branched alkanes usually predominate ungraded samples. A dominant paraffin component between C_{15} and C_{24} occurs in mature fossil oil. The more mature the oil, the lower the carbon number that predominates. With extensive biodegradation, there is a complete loss of all paraffins, with the more resistant polycyclic alkanes remaining. The distribution of the sterane and trierpane classes, and polynuclear aromatics and their sulfur analogues, can also be telling.

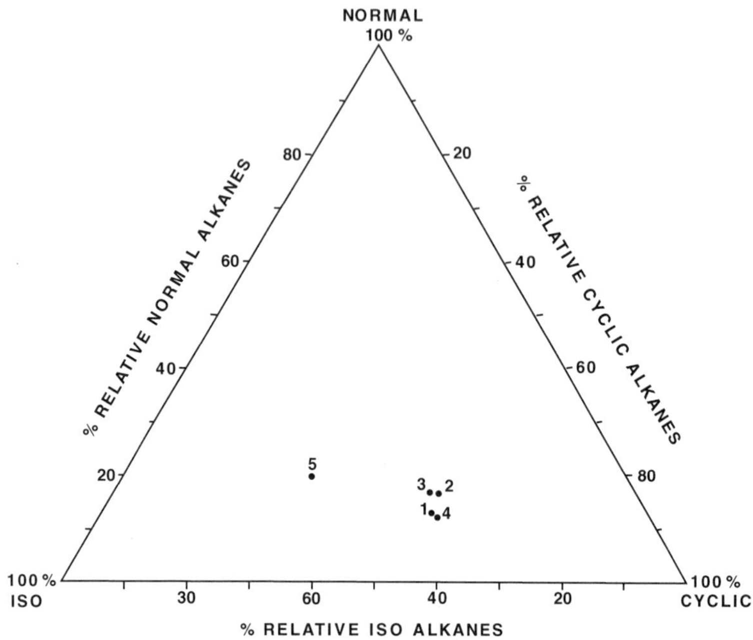

FIGURE 4.13 Ternary diagram showing relative percentages of normal alkanes, branched-alkanes (isoalkanes), and cyclic-alkanes based on gas chromatograph results for leaked LNAPL.

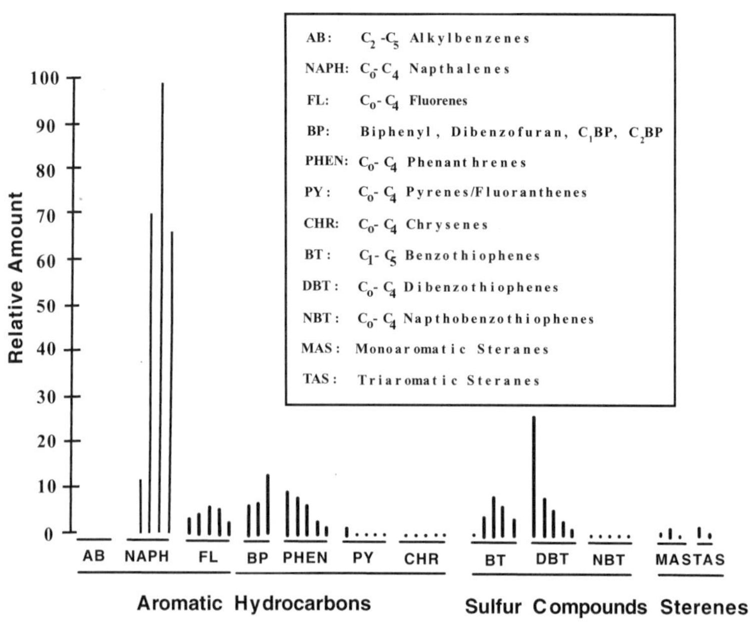

FIGURE 4.14 Additional aromatic compounds, sulfur, and sterene distribution based on gas chromatograph results.

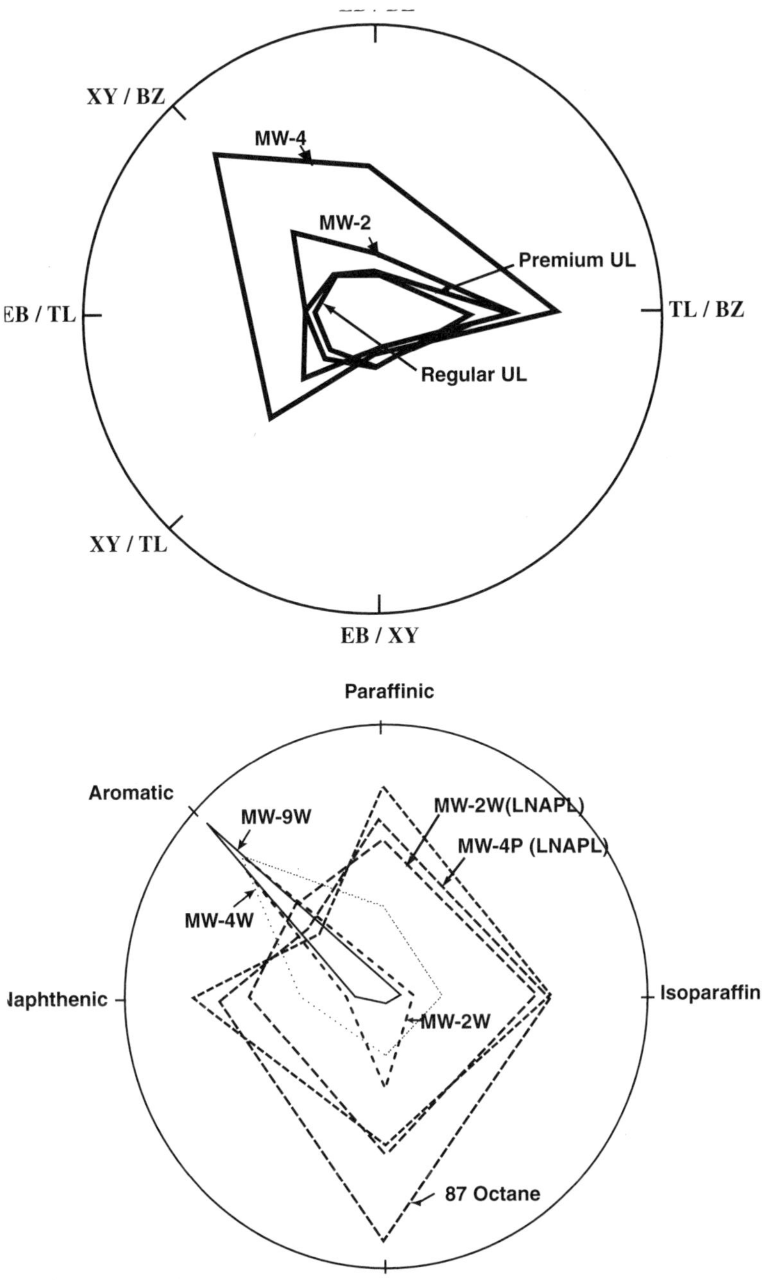

FIGURE 4.15 Star diagram for LNAPL (a) and associated groundwater (b).

TABLE 4.11
Analytical Results for Two LNAPL Samples from Adjacent Sites and Regular and Premium Unleaded Gasoline

Media	Parameter (mg/l)	Site A MW-2	Site B MW-4	Regular Unleaded Gasoline	Premium Unleaded Gasoline
Dissolved phase	TPH-g	150	47		
	TPH-d	ND	37		
	Benzene	14,000	3,200		
	Toluene	15,300	3,600		
	Ethylbenzene	1,330	590		
	Xylenes	10,700	3,280		
	MTBE	3,920	ND		
BTEX ratios	Toluene/benzene	1.09	0.89		
	Ethylbenzene/benzene	0.10	0.18		
	Xylenes/benzene	0.76	1.03		
	Ethylbenzene/toluene	0.09	0.16		
	Xylenes/toluene	0.70	0.91		
	Ethylbenzene/xylenes	0.12	0.18		
LNAPL phase	API gravity	47.9	26.0		
	Benzene	6,840	830	11,900	17,400
	Toluene	25,700	4,060	30,500	61,000
	Ethylbenzene	6,300	1,890	7,800	11,300
	Xylenes	46,300	11,100	3,800	58,000
	MTBE	ND	ND	130,000	128,000
BTEX ratios	Toluene/benzene	3.76	4.89	2.56	3.51
	Ethylbenzene/benzene	0.92	2.28	0.61	0.66
	Xylenes/benzene	6.77	13.37	3.04	3.24
	Ethylbenzene/toluene	0.25	0.47	0.22	0.26
	Xylenes/toluene	1.80	2.73	1.07	1.26
	Ethylbenzene/xylenes	0.14	0.17	0.2	0.2
Isotopes	$\delta^{13}C$	–25.03	–23.33		
	δ	–136	–129		

Note: ND = nondetectable.

4.6.5 Isotope Fingerprinting

Ratios of the stable isotopes of carbon ($^{13}C/^{12}C$) and hydrogen ($^{2}H/^{1}H$) differ both among crude oils by origin and among various fractions within the same oil. Since the isotopic composition does not change to the same extent as the molecular composition, isotope ratios are very useful when used as a tracer for organic materials that have undergone partial degradation through either biological or other processes where the molecular characteristics of the original substance are obscure.

The actual ratio of ^{13}C to ^{12}C is determined using an isotope-ratio-mass spectrometer. The actual ratio is then compared with a standard using the following equation to calculate the ratio difference in parts per thousand:

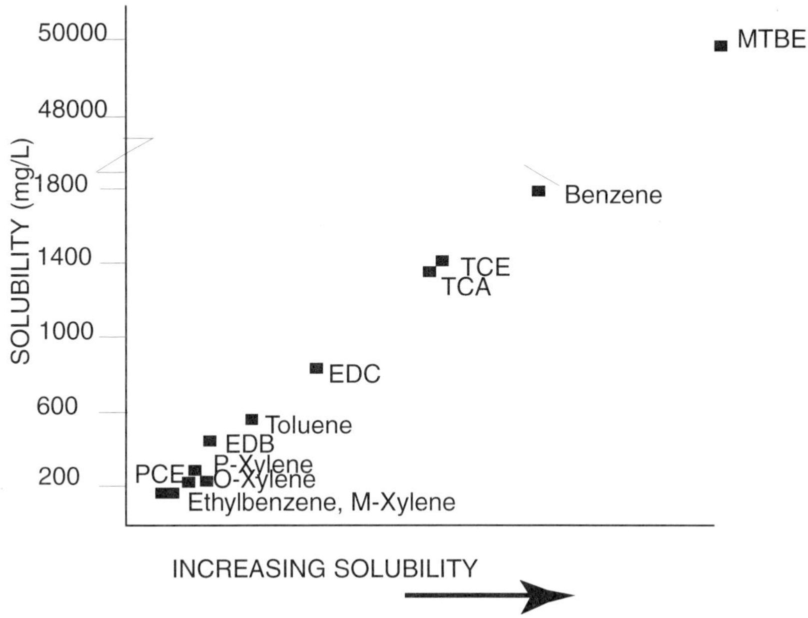

FIGURE 4.16 Relative solubilities of BTEX, MTBE, and the lead scavengers EDC and EDB.

$$d^{13}C° / °° = \frac{(^{13}C / ^{12}C \text{ sample} - ^{13}C / ^{12}C \text{ standard})}{^{13}C / ^{12}C \text{ standard}} \times 1000 \quad (4.2)$$

The most widely used standard is a belemnite from the Peedee Formation in South Carolina (PDB); therefore, some ratios may be expressed as negative values. Most carbon isotope ratio correlations are made on the C^{15+} fraction of crude oil because it is less affected by degradation processes. Valid correlations using carbon isotopes can only be conducted on the same fractions of samples.

Carbon and hydrogen isotopes are measured as a ratio of the heavier to the lighter (most abundant) isotope in a gas introduced into a dual-collecting mass spectrometer. The separated and purified petroleum product is combusted under vacuum with an oxidizing catalyst to produce CO_2 and H_2O. The carbon dioxide is then purified and injected into the mass spectrometer to measure the masses 44($^{12}C^{16}O_2$) and 45($^{13}C^{16}O_2$). The $^{13}C/^{12}C$ is then measured relative to the PDB standard. The water produced during the combustion of the hydrocarbon sample is reacted with zinc metal and converted to H_2 which is then introduced to a mass spectrometer where masses 2(^2H) and 3(DH) are measured. The deuterium/hydrogen ratio thus obtained is compared relative to an international standard.

Example 4.5

A two-component plot of the $^{13}C/^{12}C$ isotope ratio from the saturated and aromatic fractions of five samples previously identified (Table 4.8 and Figure 4.12) is pre-

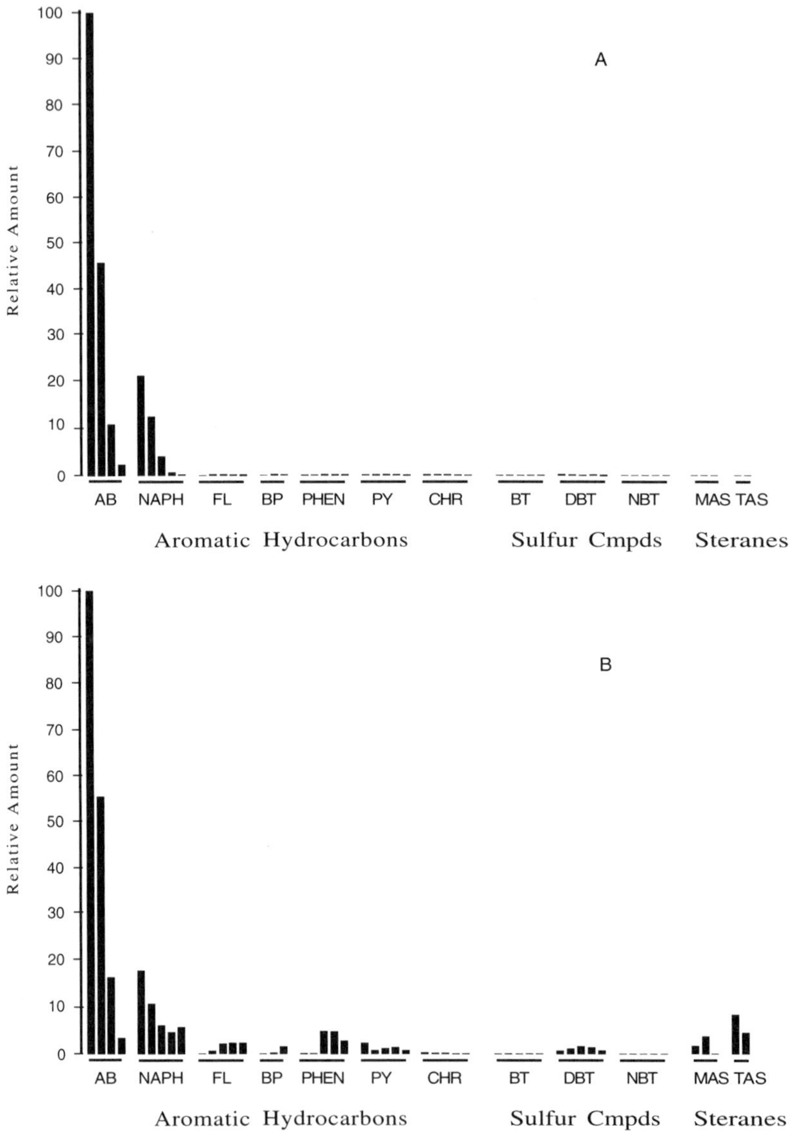

FIGURE 4.17 Additional aromatic compounds, sulfur, and sterene distribution based on gas chromatograph results for two LNAPL samples from adjacent sites.

sented in Figure 4.18. The data indicate that samples 2, 4, and 5 are clearly related to one another, as are samples 1 and 3, indicating similar source crude oils.

Example 4.6

Two LNAPL samples from adjacent sites as previously discussed in Figures 4.15 and 4.17 were analyzed for carbon ($^{13}C/^{12}C$ or $\delta^{13}C$ and hydrogen D/H or δD) ratios.

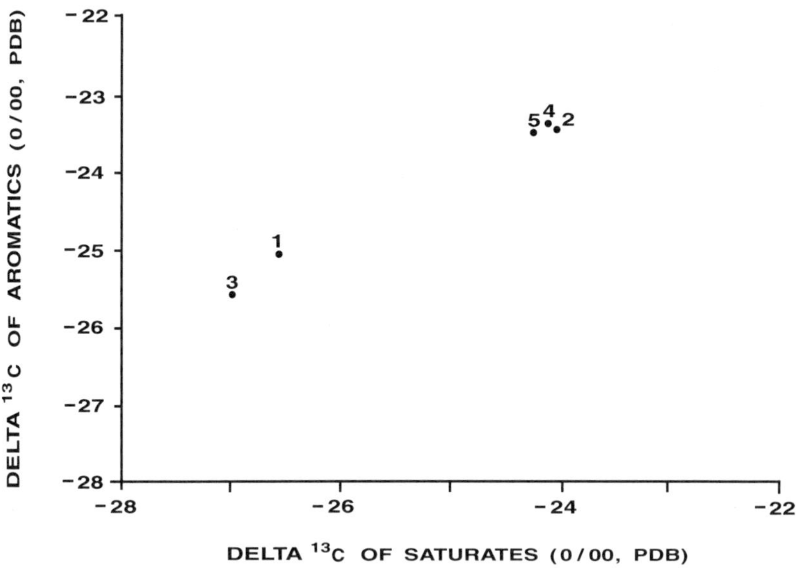

FIGURE 4.18 Two-component plot of $^{13}C/^{12}C$ isotope ratio from the saturated and aromatic fractions in five LNAPL samples.

The results as shown in Figure 4.19 suggest that there is no source relationship between MW-2 and MW-4. The enrichment of heavy ^{13}C carbon isotope in MW-4 may be attributed to the presence of heavy refined product, originating from an isotopically heavy rich (rich in ^{13}C) crude oil feedstock mixture.

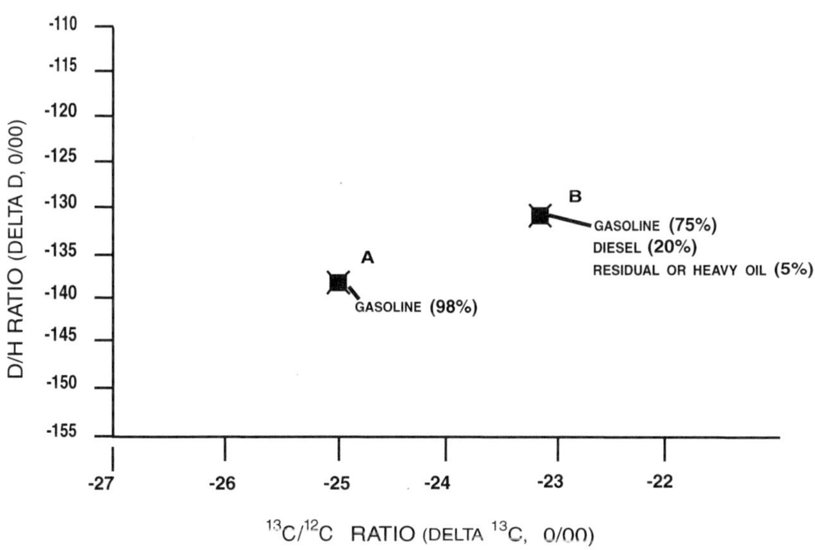

FIGURE 4.19 Two-component plot of the *D*/*H* ratio and $^{13}C/^{12}C$ isotope ratio for two LNAPL samples.

4.7 AGE DATING OF NAPL POOLS AND DISSOLVED HYDROCARBON PLUMES

The determination of the timing of a release of contaminants to the subsurface in years past is likely one of the most difficult challenges facing the environmental geologist, and one of the most often asked questions during litigation, which in turn can have significant liability ramifications. Identification of the responsible party or parties, especially when a particular property has changed ownership and usage with time, the time required for the hydrocarbon to migrate from the ground surface or point of origin to the water table, and the extent of natural attenuation mechanisms and processes make this question all the more difficult to address in a satisfactory manner.

Petroleum hydrocarbons cannot be directly age dated; however, both chemical and hydrogeological approaches can be used to evaluate the relative timing of an NAPL pool or dissolved hydrocarbon plume. Anecdotal information regarding past practices and operations, and chemical testing for certain compounds (i.e., between 1980 and present, only TEL was used as an additive in leaded gasoline in amounts of a third to an order of magnitude less prior to the 1980s; prior to the 1980s, many other leaded alkyls may also have been added by various manufacturers) can be used to establish a relative timing of a release. The most traditional and direct approach is by estimating the velocity of a plume from indirect measurements (refer to Chapter 3), where

$$\text{Velocity} = \frac{\text{hydraulic conductivity} \times \text{hydraulic gradient}}{\text{effective soil porosity}}$$

This traditional approach involves parameters that can be easily measured assuming the understanding that various contaminants or organic compounds do not travel at the same rate in groundwater. Retardation factors for specific constituents can, however, be found in the literature, allowing for a range of possible plume velocities to be determined. Once a velocity is calculated, and the retardation factor for the contaminant of concern determined, the age can be assessed based on the distance between the source and its leading edge for that particular contaminant. Other methods include radioisotope age dating techniques, and those that are more empirical in nature (Table 4.12). The latter include velocity determination by changes in plume movement and configuration over time, changes in the concentration of a contaminant within various portions of the plume over time, and degradation rates and changes in the relative concentration of two contaminants over time. The methods described can be used individually or in concert, which allows for comparison and evaluation of results, giving more weight to those dates that exhibit less variability.

In all these cases, several characteristics of dissolved contaminant plumes and NAPL pools are assumed:

- With a single release event, the leading edge and the center of the plume (i.e., zone of highest concentration) moves increasingly farther from the source over time.

TABLE 4.12
Summary of Age Dating Techniques for Dissolved Hydrocarbon and Organic Plumes

Method Description	Assumptions	Remarks
Obtaining anecdotal information	—	Includes information regarding past operations and practices
Calculating groundwater velocity	Velocity is at a constant rate	Minimum of three data points preferred
Calculating velocity from changes in position of leading edge over time	Single, slug-type release; center of plume traveling from source at constant rate	—
Calculating changes in distance from plume center to its leading edge over time	Single, slug-type release; single point source; plume traveling from source at constant rate	Dispersion processes not important
Calculating changes in areal extent or volume over time	Single point source	Calculates earliest date aquifer is impacted; continued contributions from source not important
Calculating changes in concentrations of contaminant constituents at its center over time	Single, slug-type release; contaminant only partially soluble; no NAPL present	Solubility data derived from literature
Calculating changes in contaminant concentration gradient between center and leading edge of plume over time	Single point source	—
Calculating degradation rate of an organic contaminant	Single, slug-type release; contaminant degrades to more recalcitrant compound	Tendency to underestimate age due to lag time (initial period where no degradation occurs); lag time data derived from literature; likely not applicable to large number of cases
Calculating relative changes in contaminant of two or more contaminants over time	Single, slug-type release; two or more contaminants required	Applicable to large number of cases

- The length of a plume in the direction of groundwater flow increases over time.
- The areal extent and volume of a plume increases over time.
- With a single release event, the concentration of contaminants at the plume center, and the concentration gradient (i.e., change in concentration per unit distance) between the center of the plume and its leading edge, decrease over time.
- Most organic contaminants degrade over time.
- The relative concentration of two or more contaminants may change over time.

4.7.1 Radioisotope Age Dating Techniques

Previously discussed was the role of radioisotopes (i.e., isotope ratios for carbon and hydrogen) in the origin of petroleum hydrocarbons, source identification, and degree of alteration or degradation. From a hydrochemical perspective, radioisotopes have also been used to assess the relative age and origin of surface water and groundwater, estimation of contaminant travel times (i.e., degradation rates of certain alkane components of petroleum hydrocarbons), and potential recharge areas or source areas for groundwater (Table 4.13).

Analysis for such isotopes as carbon and deuterium has been conventionally used to assess the relative age of groundwater, and in evaluating its origin (i.e., meteoric, juvenile, formation, etc.), chemistry, and total salinity. Isotope composition of groundwater and surface water has also been used to correlate between areas of precipitation and groundwater, thus providing an indication of source area(s) of recharge.

Tritium (^{3}H) and cesium (^{137}Cs) are common isotopes used for age dating of groundwater. The rationale behind the use of these two isotopes is the assumption that above background levels are the result of their introduction from the atmospheric testing of thermonuclear devices from about 1953 to 1954. Since 1953 to 1954, tritium and cesium levels increased to about 1963, then declined with secondary increases or peaks around 1973 and 1975 reflecting testing done by China and France.

Radioisotopes are expressed in picocuries (pCi), with naturally occurring radioisotopes ranging from less than a second to billions of years. Tritium is expressed as absolute concentrations referred to as tritium units (TU), where 1 TU is equivalent to one ^{3}H atom per 10^{18} atoms of hydrogen, or 3.2 pCi/kg of water. Tritium is a common short-lived isotope of hydrogen with a half-life of 12.43 years. In other words, a gram of tritium would be expected to decay to one half gram in 12.43 years. Groundwater typically has TUs in the range of <1 to 10 TU. The presence of tritium on the order of 10 TU or less is indicative of the absence of post-1953 "modern or bomb tritium" water. From a practical standpoint, this would indicate that recharge to a water-bearing zone is greater than 45 years. Conversely, greater than the 10 TU

TABLE 4.13
Summary of Isotope Dating Techniques and Half-Life

Media	Age (Years)	Isotope	Half-Life
Modern groundwater	0–45	Tritium (^{3}H)	12.43 years
		Helium (^{6}He)	0.807 s
		Helium (^{8}He)	0.119 s
		Krypton (^{85}Kr)	10.8 years
Submodern groundwater	40–1000	Argon (^{39}Ar)	268 years
		Silicon (^{32}Si)	330 years
Juvenile or old groundwater	1000+	Carbon (^{14}C)	5730 years
Sediment		Cesium (^{137}Cs)	30.2 years
		Cesium (^{135}Cs)	2.3 million years

range indicates a more rapid rate of recharge, and that surface water from recharge, underflow, or release of a spilled fluid has reached the groundwater within the past 45 years. Tritium analysis can also be used to evaluate anticipated recharge rates at potential new waste management sites. ^{137}Ce behaves similarly to tritium, although its half-life is 30.2 years. The presence of cesium was first reported in sediments around 1954, peaking around 1963, with minor increases in 1971 and 1974.

4.7.2 Changes in Configuration of Plume over Time

Determination of the age of a plume can be evaluated by calculating its velocity from changes in the position of its leading edge, the position of its center, the distance from its center and its leading edge, or areal extent or volume, over time. If the leading edge of a plume is migrating from the source at a constant rate, then the age of the plume on any particular date is equal to the distance between the source and location of its leading edge on that date, divided by the rate as follows:

$$\Delta V = 0, \quad \text{then } T_1 = (D_1 - D_0)/V \tag{4.3}$$

where ΔV is the change in velocity, T_1 is the age of the plume on a particular date, D_1 is the location of the leading edge on that same date, D_0 is the location of the source, and V is the velocity of the plume. A graph of data showing distance from source over time can be used to determine whether the plume velocity is linear, thus migrating at a constant rate. Solving for velocity (V) allows solution of Equation 4.3 for the age of the plume:

$$V = (D_2 - D_1)/(T_2 - T_1) \tag{4.4}$$

where V is the plume velocity and $D_2 - D_1$ is a function representing the distance between the leading edge of the plume at two known times represented by the function $T_2 - T_1$. Changing D_1 in Equation 4.3, which represents the location of the leading edge of the plume on date T_1, to D_1 representing the location of the center of the plume on date T_1, one can determine the age of a plume resulting from a single slug-type release (i.e., in lieu of a release from a continuous source).

Changes in the distance from the center of the plume to its leading edge over time can also be useful in determining the age of the plume. Assuming that the plume resulted from a single, slug-type release, that originated at a single point source, and traveled from its source at a constant rate, then the date in which the contaminants entered the water table is equivalent to the age when the distance from the center of the plume to its leading edge is essentially zero:

$$\text{If } R_d \circledR 0, \quad \text{then} \quad T_R = T_{(D \circledR 0)} \tag{4.5}$$

where R_d is the duration of the release, T_R is the date of the release, and $T_{(D \circledR 0)}$ is the date on which the distance from the center to the leading edge of the plume approached zero. This method does not require that the source location be known, and can be used to verify the results from Equation 4.3 or 4.4. Once the distance between the center of the plume and its leading edge is known on two or more dates, one needs only to extrapolate to determine the date upon which this distance would have approached zero.

Another graphical means of estimating the age of a plume is from changes in its areal extent or volume over time. Assuming that a groundwater plume originated from a single point source, the point in which the contaminant entered the water table or aquifer is equivalent to the time in which the areal extent and volume of the plume are essentially zero, as expressed by

$$\text{If } R_N = 1, \quad \text{then} \quad T_R = T_{(S®0)} \tag{4.6}$$

where R_N is the number of sources, T_R is the date of the release, and $T_{(S®0)}$ is the date on which the areal extent or volume of the plume approaches zero. In this case, the plume need not have been derived from a single slug-type release event, although the more closely the initial release approximates a point source, the more accurate the results.

4.7.3 Changes in Concentrations over Time

In place of distance measurements, concentration can be used to determine the age of a plume. Assuming that the plume is derived from a single slug-type release of a contaminant which is only partially soluble in groundwater, and not present as an NAPL, then the date upon which the contaminant enters the water table or aquifer is equivalent to that point in time when the concentration of the contaminant approaches its solubility limit:

$$\text{If } R_d \text{ ® } 0, \quad \text{then} \quad T_R = T_{(C®100\%)} \tag{4.7}$$

where R_d is the duration of the release, T_R is the date of the release, and $T_{(C®100\%)}$ is the date on which the concentration of a particular contaminant approached its solubility limit.

This method is best solved graphically by plotting the maximum concentration measured at the center of the plume on different dates, then extrapolating to determine the date on which the concentrations of the contaminant in question approaches its solubility limit. The solubility limit, corrected for the temperature of the aquifer, for a particular contaminant can be obtained from the literature.

It is assumed that the contaminant enters the water table or aquifer at a concentration near its solubility limit, although there is no practical means to verify this. This method is more favorable when the release occurred as a single, short-term episode. A long-term release from a continuing source would result in a date that more closely represents the last date upon which the contaminant entered the aquifer at or near its solubility limit. Should the contaminant enter the aquifer below its solubility limit, then a date earlier than the actual event would result. Conversely, should the contaminant enter the aquifer as NAPL for a period of time, a date in which all the NAPL dissolved in groundwater would result. If NAPL was present when measurements were obtained, then the zone of highest concentration would approach the solubility of the contaminant and coincide with the source, thus making this method unsatisfactory. This method is also unsatisfactory for multiple events or sources.

Changes in concentration gradient between the center of the plume and its leading edge over time can be expressed as:

$$\text{If } R_N = 1, \quad \text{then} \quad T_R = T_{(Cf=Co)} \tag{4.8}$$

where R_N is the number of sources, T_R is the date of the release, and $T_{(Cf=Co)}$ is the date upon which the concentration at the leading edge of the plume, *Cf*, is the same as the concentration at the center of the plume, *Co*. With this method, a graph of the concentration gradient between the leading edge of the plume and its center (or zone of highest concentration), on a representative date, is first generated. Assuming the log concentration is linear, the slope of the log concentration vs. distance between the leading edge and the center of the plume on each date is calculated. A graph of the changes in this slope over time can then be used to extrapolate the date at which the slope of the graph approaches a straight line (i.e., the date at which the concentration gradient approaches zero).

4.7.4 Degradation Rates

The use of degradation rates to determine the age of a contaminant plume assumes a contaminant from a single, slug-type release, which degrades to a more recalcitrant chemical compound (i.e., trichloroethene degrading to 1,2-dichloroethene), will enter the water table or aquifer at a point in time when none of the daughter product is present in the aquifer:

If *A* degrades to *B*, and *B* is not further degraded, then

$$T_R = T_{(A®100\%)} \tag{4.9}$$

where *A* is the parent compound initially released into the aquifer, *B* is the daughter compound of *A*, T_R is the date of the release, and $T_{(A®100\%)}$ is the date upon which the concentration of *A* approaches 100%, which is equivalent to the date immediately prior to the appearance of compound *B*. This is similar to Equation 4.7, although based on rate of degradation rather than on changes in concentration, which could be due to such processes as dilution and volatilization. Degradation rates are further discussed in Chapter 13.

4.7.5 Changes in Concentration of Two Contaminants over Time

This method assumes that if two or more contaminants were released from a single, slug-type release, the date upon which these contaminants entered the water table or aquifer is equivalent to the time when their relative concentrations were the same as the source material. This method does not require initial concentration information, and is not dependent on changes due to a particular subsurface process (i.e., biodegradation, volatilization, adsorption, etc.). Changes in relative concentration of two contaminants over time are plotted, and trends are identified. The age of the plume is derived by extrapolation to identify the date upon which the slope of the graph is nearly linear.

REFERENCES

Alvarez, P. J., Heathcote, R. C., and Powers, S. E., 1998, Caution Against Interpreting Gasoline Release Dates Based on BTEX Ratios: *Ground Water Monitoring Review*, Fall, 1998, pp. 69–76.

Baker, E. W., 1964, Vanadium and nickel in crude petroleum of South America and Middle East Origin: *Journal of Chemistry and Engineering News*, Vol. 42, No. 15, pp. 307–308.

Bland, W. F. and Davidson, R. L., 1967, *Petroleum Processing Handbook*: McGraw-Hill Book Company, New York.

Bowie, C. P., 1918, Oil Storage Tanks and Reservoirs with a Brief Discussion of Losses of Oil in Storage and Methods of Prevention: *U.S. Department of the Interior Bureau of Mines Bulletin*, No. 155, Petroleum Technology No. 41, 76 pp.

Brookman, G. T., Flanagan, M., and Kebe, J. O., 1985a, Literature Survey: Unassisted Natural Mechanisms to Reduce Concentrations of Soluble Gasoline Components: *American Petroleum Institute of Health and Environmental Sciences Department*, No. 4415, August, 1985, 73 pp.

Brookman, G. T., Flanagan, M., and Kebe, J. O., 1985b, Literature Survey: Hydrocarbon Solubilities and Attenuation Mechanisms: *American Petroleum Institute of Health and Environmental Sciences Department*, No. 4414, August, 101 pp.

California Department of Health Services, 1989, Total Petroleum Hydrocarbons (TPH) Analysis — Gasoline and Diesel: In *California Water Resources Control Board Leaking Underground Fuel Tank (LUFT) Manual*, Appendix C.

Christensen, L. and Larsen, T., 1993, Method for Determining the Age of Diesel Oil Spills in the Soil: *Ground Water Monitoring & Remediation*, Vol. 23, No. 4, pp. 142–149.

Clark, J. H., Clarke, A. N., and Smith, J. S., 1999, Environmental Forensics: *Environmental Protection*, September, pp. 49–54.

Colligan, T. H. and LaManna, J. M., 1993, Using Ultraviolet Light to Investigate Petroleum-Contaminated Soil: *Remediation*, Spring, pp. 193–201.

Donley, J. W. and Jamison, T. P., 1995, Simple Techniques for Estimating the Age of Groundwater Contamination: In *50th Purdue Waste Conference Proceedings*, Ann Arbor Press, Chelsea, MI, pp. 75–86.

Douglas, G., Bence, A., Prince, R., McMillen, S., and Butler, E., 1996, Environmental Stability of Selected Petroleum Hydrocarbon Source and Weathering Ratios: *Environmental Science and Technology*, Vol. 30, pp. 2332–2339.

Dragun, J., 1988, *The Soil Chemistry of Hazardous Materials*: Hazardous Materials Control Research Institute, Silver Springs, MD, 458 pp.

Exploration Logging, Inc., 1980, *Field Geologists Training Guide, An Introduction to Oilfield Geology, Mud Logging and Formation Evaluation*, pp. 5–46 to 5–47.

Fritz, R. and Fritz, J., 1991, Characterizing Shallow Aquifers Using Tritium and 14C: Periodic Sampling Based on Tritium Half-Life: *Applied Geochemistry*, Vol. 6, pp. 17–33.

Greenberg, R., 1997, Plume Dating Combines Science with Good Detective Work: *Soil and Groundwater Cleanup*, February–March, pp. 30–33.

Guthrie, V. B., 1960, *The Petroleum Products Handbook*: McGraw-Hill Publishing, New York.

Haung, W. Y. and Meinschein, W. G., 1979, Sterols as Ecological Indicators: *Geochemica et Cosmochimica Acta*, Vol. 43, pp. 739–745.

Hawley, G., 1981, *The Condensed Chemical Dictionary*, Tenth Edition: Van Nostrand Reinhold Company, New York.

Hodgson, G. W., 1954, Vanadium, Nickel and Iron Trace Metals in Crude Oils of Western Canada: *American Association of Petroleum Geologists Bulletin*, Vol. 38, pp. 2537–2554.

Howard, P. H., 1989, *Handbook of Environmental Fate and Exposure Data for Organic Chemicals,* Volume I, *Large Production and Priority Pollutants*: Lewis Publishers, Chelsea, MI, 574 pp.

Howard, P. H., Boethling, R. S., Jarvis, W. F., Meylan, W. M., Michalenko, E. M., and Printup, H. T., 1991, *Handbook of Environmental Degradation Rates*: Lewis Publishers, Chelsea, MI, 725 pp.

Hunt, J. M., 1979, *Petroleum Geochemistry and Geology*: W. H. Freeman and Company, San Francisco, CA, p. 617.

Johnson, M. D. and Morrison, R. D., 1996, Petroleum Fingerprinting: Dating a Gasoline Release: *Environmental Protection*, September, pp. 37–39.

Kaplan, I. R., 1989, Forensic Geochemistry in Characterization of Petroleum Contaminants in Soils and Groundwater: In *Environmental Concerns in the Petroleum Industry* (edited by S. M. Testa), Pacific Section of the American Association of Petroleum Geologists, Symposium Volume, pp. 159–181.

Kaplan, I. R., 1992, Characterizing Petroleum Contaminants in Soil and Water and Determining Source of Pollutants: In *Proceedings of the American Petroleum Institute Conference on Petroleum Hydrocarbons and Organic Chemicals in Ground Water: Prevention, Detection and Restoration*, pp. 3–18.

Kaplan, I. R., Galparin, Y., Alimini, H., Lee, R., and Lu, S., 1996, Patterns of Chemical Changes during Environmental Alteration of Hydrocarbon Fuels: *Groundwater Monitoring and Remediation*, Vol. 16, No. 4, pp. 113–124.

Kaplan, I. R., Galparin, Y., Lu, S., and Lee, R., 1997, Forensic Environmental Geochemistry: Differentiation of Fuel-Types, Their Sources and Release Time: *Organic Geochemistry*, Vol. 27, No. 5/6, pp. 289–317.

Kraemer, A. J. and Calkin, L. P., 1925, Properties of Typical Crude Oils from the Producing Fields of the Western Hemisphere: *U.S. Bureau of Mines Technical Paper 346*, pp. 1–43.

Kraemer, A. J. and Lane, E. C., 1937, Properties of Typical Crude Oils from Fields of the Eastern Hemisphere: *U.S. Bureau of Mines Technical Paper 401*, pp. 1–169.

Leffler, W. L., 1979, *Petroleum Refining for the Non-Technical Person*: PennWell Books, Tulsa, OK, 159 pp.

Lundegard, P. D., Haddad, R., and Brearley, M., 1998, Methods Associated with a Large Gasoline Spill: Forensic Determination of Origin and Source: *Environmental Geosciences*, Vol. 5, No. 2, pp. 69–77.

McAuliffe, C., 1966, Solubility in Water of Paraffin, Cycloparaffin, Oleofin, Acetylene, Cycloolefin and Aromatic Hydrocarbons: *Journal of Physical Chemistry*, Vol. 70, pp. 1267–1275.

Mair, B. J., 1967, Annual Report for the Year Ending June 30, 1967: *American Petroleum Institute Research Project 6*, Carnegie Institute of Technology, Pittsburgh, PA.

Montgomery, J. H. and Welkom, L. M., 1990, *Groundwater Chemical Desk Reference*: Lewis Publishers, Chelsea, MI, 640 pp.

Morrison, R. D., 1999, *Environmental Forensics — Principles and Applications*: CRC Press, Boca Raton, FL, 351 pp.

Morrison, R. T. and Boyd, R. N., 1973, *Organic Chemistry*: Third Edition, Van Nostrand Reinhold Company, New York, 1258 pp.

Nelson, W. L., 1941, *Petroleum Refinery Engineering,* Second Edition: McGraw-Hill Book Company, New York, 215 pp.

Nyer, E. K. and Skladany, G. J., 1989, Relating the Physical and Chemical Properties of Petroleum Hydrocarbons to Soil and Aquifer Remediation: *Ground Water Monitoring Review*, Winter, pp. 54–60.

Perry, J. T., 1984, Microbial Metabolism of Cyclic Alkanes: In *Petroleum Microbiology* (edited by R. M. Atlas), MacMillan Publishing Co., New York, pp. 61–98.

Petrov, A. A., 1987, *Petroleum Hydrocarbons*: Springer-Verlag, New York, 255 pp.

Rosscup, R. J. and Bowman, J., 1967, Thermal Stabilities of Vanadium and Petroporphyrins: *Preprints of the Division of Petroleum Chemistry,* American Chemical Society, Vol. 12, 77 pp.

Rossini, F. D., 1960, Hydrocarbons in Petroleum: *Journal of Chemistry*, Vol. 37, No. 11, pp. 554–561.

Schoell, M., 1984, Recent Advances in Petroleum Isotope Geochemistry: *Organic Geochemistry*, Vol. 6, pp. 645–663.

Seifirt, W. K. and Moldowan, J. M., 1979, Applications of Steranes, Terpanes and Monoaromatics to the Maturation, Migration and Source of Crude Oils: *Geochemica et Cosmochimica Acta*, Vol. 42, No. 1, pp. 77–95.

Senn, R. B. and Johnson, M. S., 1985, Interpretation of Gas Chromatography Data as a Tool in Subsurface Hydrocarbon Investigations: In *Proceedings of the NWWA/API Conference on Petroleum Hydrocarbons and Organic Chemicals in Groundwater — Prevention, Detection and Restoration*, National Water Well Association, Dublin, OH, pp. 331–357.

Sofer, Z., 1984, Stable Carbon Isotope Composition of Crude Oils: Application to Source Depositional Environments and Petroleum Alternation: *American Association of Petroleum Geologists Bulletin*, Vol. 68, pp. 31–49.

Testa, S. M., 1990, Hydrocarbon Product Characterization: Applications and Techniques: In *Proceedings of the National Water Well Association of Groundwater Scientists and Engineers Fourth Outdoor Action Conference on Aquifer Restoration, Ground Water Monitoring and Geophysical Methods*, May, 1990.

Testa, S. M. and Halbert, W. E., 1989, Geochemical Fingerprinting of Free Phase Liquid Hydrocarbons: In *Proceedings of the National Water Well Association and American Petroleum Institute Conference on Petroleum Hydrocarbons and Organic Chemicals in Ground Water: Prevention, Detection and Restoration*, NWWA, Houston, TX, pp. 29–44.

Testa, S. M., Baker, D. M., and Avery, P. L., 1989, Field Studies on Occurrence, Recoverability and Mitigation Strategy for Free Phase Liquid Hydrocarbon: In *Environmental Concerns in the Petroleum Industry* (edited by S. M. Testa), Pacific Section of the American Association of Petroleum Geologists Symposium Volume, pp. 57–81.

Uhler, A. D., McCarthy, K. J., and Stout, S. A., 1999, Improving Petroleum Remediation Monitoring with Forensic Chemistry: *Soil & Groundwater Cleanup*, April/May, pp. 26–27.

U.S. Environmental Protection Agency, 1986, *Test Methods for Evaluating Solid Waste; Physical/Chemical Methods, SW-846*, Third Edition: Office of Solid Waste and Emergency Response, U.S. EPA, Washington, D.C.

Volkman, J. K., 1986, A Review of Sterol Markers for Marine and Terrigenous Organic Matter: *Organic Geochemistry*, Vol. 9, pp. 83–99.

Younglass, T., Swansinger, J., Danner, D., and Greco, M., 1985, Mass Spectral Characterization of Petroleum Dyes, Tracers and Additives: *Analytical Chemistry*, Vol. 57, pp. 1894–1902.

5 Fate and Transport

"Physical processes acting in the subsurface determine how a chemical partitions in the subsurface media."

5.1 INTRODUCTION

Many processes affect the fate, transport, and transformation of a particular organic chemical or compound in the subsurface. The overall composition of leachates produced reflects the degree of aerobic and anaerobic decomposition as well as retention within the soil matrix. Understanding how this material can become mobilized and what happens to it when it is mobilized is difficult and complex. Simply applying models that account for certain parameters, such as convection, dispersion, adsorption, and retardation, although commonly used is not necessarily the answer. Physical, chemical, and biological properties and processes must be considered in conjunction with site-specific environmental factors. It is often difficult to separate physical processes from chemical and biological processes because they are often coupled within any given system.

Mathematical relationships are quite useful in understanding nonintuitive processes, notably transport; thus, some basic mathematics are required. A key mathematical concept, the gradient, is one of the most important concepts for all aspects of subsurface science as well as any study of the earth. A gradient is a change in the value of one variable with respect to another variable, like a slope. Most of the discussion will include gradients of some property with respect to distance or time, e.g., an elevation (pressure) gradient across the boundary between two soils, a concentration gradient across a membrane, or a thermal gradient across a surface. Gradients are the real driving forces for change in earth systems and in the environment. Obviously, large or steep gradients can result in rapid and dramatic changes, and the ability to restore a contaminated site or successfully dispose of hazardous material often depends upon the ability to minimize particular gradients within the system.

A major emphasis of environmental science today also concerns an assessment of risk and performance of remediation systems. Predictive models in the form of computer programs attempt to predict the behavior and migration of subsurface fluids and contaminants. Many licensing and government regulations depend upon programming and data input. Relevant data is commonly lacking; thus, educated guesses or extrapolations from less-relevant situations are assumed. Conceptual models serve as the basic paradigm of the program, but many currently available conceptual models are usually too simplified to be meaningful because the actual subsurface environment is very complicated, or heterogeneous. The adage "garbage in = garbage out" summarizes the concern that input of unrealistic data into an oversimplified computer model by an inexperienced operator will result in unreliable and overly conservative results.

Modeling of the NAPL migration, and the fate and transport of its various components and other organics, has been attempted by several means. Vertical equilibrium, sharp-interface flow models have been used as a preliminary means to evaluate quickly the extent and volume of hydrocarbons subsurface releases. Two-phase flow models have conventionally been used to evaluate NAPL phase migration after the NAPL reached the saturated zone (below the water table), whereas three-phase flow models have conventionally been used to evaluate NAPL migration within both the unsaturated and saturated zones and recovery of free-phase LNAPLs as part of aquifer restoration activities. Incorporation of interface mass transfer of NAPL component(s) has been used to assess the migration of the soluble components in groundwater and to evaluate certain aquifer restoration technologies such as water flooding, and, with the added incorporation of air-phase dynamics, to evaluate the effectiveness of remedial strategies such as vapor extraction and air sparging. Since a combination or series of remedial actions may be implemented during the life span of a particular project, the use of any model will depend upon the complexity of the site being investigated, and the remedial activity (i.e., pump-and-treat, soil vapor extraction, etc.) being performed at any given time.

The main objective of this chapter is to develop the framework for extending conceptual models in order to design, characterize, and manage remediation systems more responsibly. Discussion of the general characteristics and behavior of LNAPLs and DNAPLs is presented. This is followed by a synopsis of the physical, biological, and chemical processes that play a significant role in the fate and transport of organic chemicals in the subsurface, and discussion of the occurrence and flow of immiscible liquids within the unsaturated (or vadose) zone and saturated zone. Biodegradation processes, although an important process in the fate and transport of petroleum hydrocarbons and organic compounds in the subsurface, is not discussed in this chapter, but rather presented in Chapter 13 (Site Closure).

5.2 NAPL CHARACTERISTICS AND SUBSURFACE BEHAVIOR

NAPLs, referred to in the federal regulations as "free product," can occur in the subsurface as LNAPLs or DNAPLs (Figure 5.1). LNAPLs are lighter than water and have a density less than 1, whereas DNAPLs have densities greater than 1. Typical LNAPLs include most crude oil, used oil, and fuels (such as gasoline, diesel, and jet fuel), Stoddard solvents, and mineral oils. Densities for these substances range from about 0.6 to 1.0 g/ml.

DNAPLs are broadly classified on the basis of certain chemical properties such as density, viscosity, and solubility. Some of the more common DNAPLs are chlorinated solvents (i.e., trichloroethylene, TCE; tetrachloroethylene, PCE; and trichloroethane, TCA), creosote, and coal tar, that is, polycyclic aromatic hydrocarbons

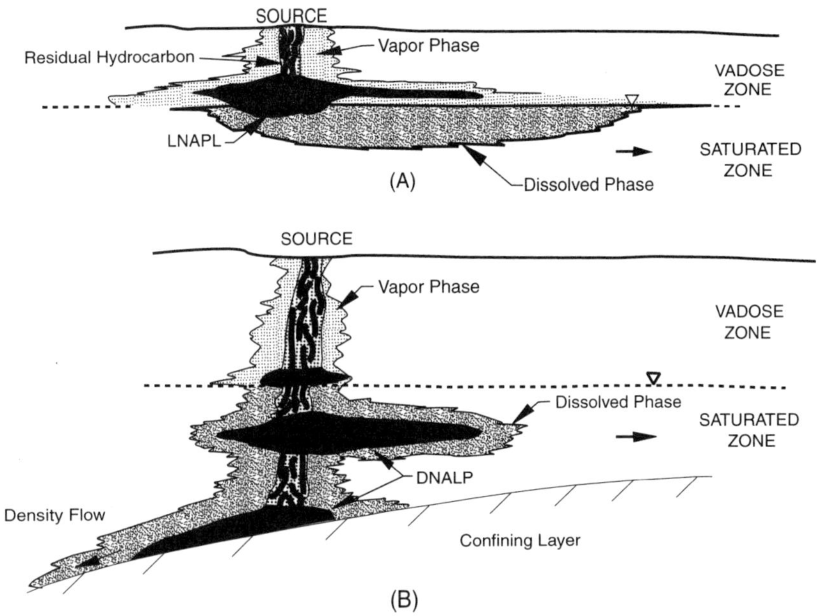

FIGURE 5.1 Schematic showing general distribution of LNAPL and DNAPL in the subsurface.

(PAHs) or polynuclear aromatic hydrocarbons (PNAs) such as anthracene, chrysene, fluorene, naphthalene, phenanthrene, and pyrene. Although many chlorinated solvents are characterized by relatively high densities and low viscosities, creosote and coal tar compounds have relatively low densities and high viscosities. In comparison with water, chlorinated solvents are characterized by relatively high densities, low viscosities, and significantly high specific gravities. Creosote and coal tar compounds have viscosities of only 10 to 20 times greater than that of water, and specific gravities only slightly greater than water. Some of the more common LNAPLs and prevalent DNAPLs reported at CERCLA sites, and their respective physical properties, are presented in Tables 5.1 and 5.2, respectively.

It is these contrasts that result in the subsurface behavior of both the immiscible phase and dissolved phase of DNAPL being different from that of LNAPLs. When released into the subsurface, DNAPLs behave much the same as LNAPLs within the vadose zone; however, once groundwater is encountered, LNAPLs will tend to form a pool or "pancake layer," whereas DNAPLs will tend to continue to migrate vertically downward through the water column until a significant permeability contrast is encountered, provided enough of a volume of DNAPL was released with time. Immiscible chlorinated solvents are relatively mobile and strongly influenced by gravity, while the dissolved phases are also relatively mobile, reflecting sorption properties and significant solubilities of some constituents. The dissolved constituents can thus migrate large distances. Some dissolved compounds derived from DNAPLs, such as creosote and coal tar, are less influenced by gravity, thus less mobile. Lower solubilities and concomitant stronger sorption can also result in lower mobility.

TABLE 5.1
Density and Dynamic Viscosity of Selected Fluids with Specific Gravity Less than 1[a]

Fluid	Density (g/ml)	Dynamic (Absolute) Viscosity (cP)
Water	0.998	1.14
Automotive gasoline	0.729	0.62
Automotive diesel fuel	0.827	2.70
Kerosene	0.839	2.30
No. 6 jet fuel	0.844	—
No. 2 fuel oil	0.866	—
No. 4 fuel oil	0.904	47.20
No. 5 fuel oil	0.923	215.00
No. 6 fuel oil or Bunker C	0.974	—
Norman Wells crude	0.832	5.05
Avalon crude	0.839	11.40
Alberta crude	0.840	6.43
Transmountain Blend crude	0.855	10.50
Bow River Blend crude	0.893	33.70
Prudhoe Bay crude	0.905	68.40
Atkinson crude	0.911	57.30
LaRosa crude	0.914	180.00

[a] All measurements at 15°C.

Source: Modified after API (1996).

DNAPL mobility is influenced by their respective density, viscosity, and interfacial tension with water. Mobility in the soil matrix is influenced by small-scale features, such as soil type, intrinsic permeability, mineralogy, pore size, pore geometry, and micropores, and large-scale features, such as heterogeneities and anisotropic conditions, and structural and stratigraphic features (Figure 5.2). Once released in the subsurface, DNAPLs migrate vertically downward through the vadose zone with some lateral spreading where significant permeability contrasts are encountered. Within the vadose zone, DNAPL residual hydrocarbon becomes trapped in pore space via surface tension, as dissolved constituents in residual soil water, or as vapor. With significant releases, capillary or entry pressures are overcome, and the DNAPL will eventually reach the water table or saturated zone. At this point, DNAPLs continue to migrate downward by the influence of gravity occurring as dissolved constituents and within pore spaces. The pressure head required for penetration increases as grain size decreases. DNAPLs are often gradient controlled following the dip of the interface between the aquifer and the top of the lower confining layer, regardless of groundwater flow directions, filling in topographic lows, or migrating along preferred pathways, such as fractures, bedding planes, and zones of relatively higher permeability. DNAPLs will not necessarily pond at depth above some confining layer. In cases where significant vertical separation exists between the water table and the lower confining layer, the DNAPL may manifest itself in essentially

TABLE 5.2
Most Prevalent Chemical Compounds at U.S. Superfund Sites with a Specific Gravity Greater than 1

DNAPL Compound	Density (g/cc)	Dynamic Viscosity (cP)[a]	Kinematic Viscosity (cS)[b]	Water Solubility (mg/l)		Henry's Law Constant (atm-m^3/mol)		Vapor Pressure (mm Hg)	
			Halogenated Semivolatiles						
1,4-Dichlorobenzene	1.2475	1.2580	1.008	8.0	E + 01	1.58	E – 03	6	E – 01
1,2-Dichlorobenzene	1.3060	1.3020	0.997	1.0	E + 02	1.88	E – 03	9.6	E – 01
Arochlor 1242	1.3850			4.5	E – 01	3.4	E – 04	4.06	E – 04
Aroclor 1260	1.4400			2.7	E – 03	3.4	E – 04	4.05	E – 05
Aroclor 1254	1.5380			1.2	E – 02	2.8	E – 04	7.71	E – 05
Chlordane	1.6	1.1040	0.69	5.6	E – 02	2.2	E – 04	1	E – 05
Dieldrin	1.7500			1.86	E – 01	9.7	E – 06	1.78	E – 07
2,3,4,6-Tetrachlorophenol	1.8390			1.0	E + 03				
Pentachlorophenol	1.9780			1.4	E + 01	2.8	E – 06	1.1	E – 04
			Halogenated Volatiles						
Chlorobenzene	1.1060	0.7560	0.683	4.9	E + 02	3.46	E – 03	8.8	E + 00
1,2-Dichloropropane	1.1580	0.8400	0.72	2.7	E + 03	3.6	E – 03	3.95	E + 01
1,1-Dichloroethane	1.1750	0.3770	0.321	5.5	E + 03	5.45	E – 04	1.82	E + 02
1,1-Dichloroethylene	1.2140	0.3300	0.27	4.0	E + 02	1.49	E – 03	5	E + 02
1,2-Dichloroethane	1.2530	0.8400	0.67	8.69	E + 03	1.1	E – 03	6.37	E + 01
trans-1,2-Dichloroethylene	1.2570	0.4040	0.321	6.3	E + 03	5.32	E – 03	2.65	E + 02
cis-1,2-Dichloroethylene	1.2480	0.4670	0.364	3.5	E + 03	7.5	E – 03	2	E + 02
1,1,1-Trichloroethane	1.3250	0.8580	0.647	9.5	E + 02	4.08	E – 03	1	E + 02
Methylene chloride	1.3250	0.4300	0.324	1.32	E + 04	2.57	E – 03	3.5	E + 02
1,1,2-Trichloroethane	1.4436	0.1190	0.824	4.5	E + 03	1.17	E – 03	1.88	E + 01
Trichloroethylene	1.4620	0.5700	0.390	1.0	E + 03	8.92	E – 03	5.87	E + 01
Chloroform	1.4850	0.5630	0.379	8.22	E + 03	3.75	E – 03	1.6	E + 02
Carbon tetrachloride	1.5947	0.9650	0.605	8.0	E + 02	2.0	E – 02	9.13	E + 01
1,1,2,2-Tetrachloroethane	1.6	1.7700	1.10	2.9	E + 03	5.0	E – 04	4.9	E + 00
Tetrachloroethylene	0.8900	0.8900	0.54	1.5	E + 02	2.27	E – 02	1.4	E + 01
Ethylene dibromide	1.6760	1.6760	0.79	3.4	E + 03	3.18	E – 04	1.1	E + 01

TABLE 5.2 *(continued)*
Most Prevalent Chemical Compounds at U.S. Superfund Sites with a Specific Gravity Greater than 1

DNAPL Compound	Density (g/cc)	Dynamic Viscosity (cP)[a]	Kinematic Viscosity (cS)[b]	Water Solubility (mg/l)		Henry's Law Constant (atm-m³/mol)		Vapor Pressure (mm Hg)	
Nonhalogenated Semivolatiles									
2-Methyl naphthalene	1.0058			2.54	E + 01	5.06	E – 02	6.80	E – 02
o-Cresol	1.0273			3.1	E + 04	4.7	E – 05	2.45	E – 01
p-Cresol	1.0347			2.4	E + 04	3.5	E – 04	1.08	E – 01
2,4-Dimethylphenol	1.0360			6.2	E + 03	2.5	E – 06	9.8	E – 02
m-Cresol	1.0380	21.0	20	2.35	E + 04	3.8	E – 05	1.53	E – 01
Phenol	1.0576		3.87	8.4	E + 04	7.8	E – 07	5.293	E – 01
Naphthalene	1.1620			3.1	E + 01	1.27	E – 03	2.336	E – 01
Benzo(a) anthracene	1.1740			1.4	E – 02	4.5	E – 06	1.16	E – 09
Fluorene	1.2030			1.9	E + 00	7.65	E – 05	6.67	E – 09
Acenaphthene	1.2250			3.88	E + 00	1.2	E – 03	2.31	E – 02
Anthracene	1.2500			7.5	E – 02	3.38	E – 05	1.08	E – 05
Dibenz(*a,hh*) anthracene	1.2520			2.5	E – 03	7.33	E – 08	1	E – 10
Fluoranthene	1.2520			2.65	E – 01	6.5	E – 06	4.8	E – 05
Pyrene	1.2710			1.48	E – 01	1.2	E – 05	6.67	E – 06
Chrysene	1.2740			6.0	E – 03	1.05	E – 06	6.3	E – 09
2,4-Dinitrophenol	1.6800			6.0	E + 03	6.45	E – 10	1.49	E – 05
Miscellaneous									
Coal tar	1.028[c]	18.98[c]							
Cresote	1.05	1.08[c]							

[a] Centipoise (cP); water has a dynamic viscosity of 1 cP at 20°C.
[b] Centistokes (cS).
[c] 45°C (70).

Source: After Huling and Weaver (1991).

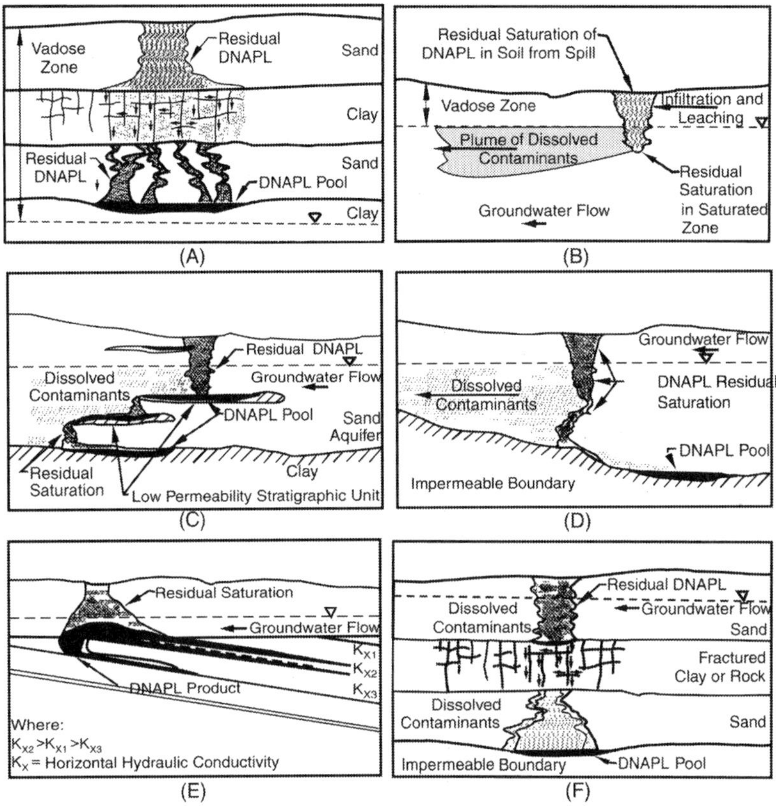

FIGURE 5.2 Schematic showing various scenarios for DNAPL occurrence.

thin lenses or layers within the saturated zone as evidenced at the Savannah River site, Georgia.

NAPLs can be trapped or pooled in the subsurface by large-scale hydrogeologic features such as confining layers and bedrock, in addition to small-scale depositional features. For LNAPL, traps may consist of buried channels, gravel bars, eolian wedges, and lateral and vertical facies changes, among others. For DNAPLs, traps include scoured channel bases, volcanic collapse features, irregular topographic depressions on the upper surface of a confining layer, etc. Interface geometry and grain size are important factors in determining whether interface trapping of NAPLs is possible. To trap NAPLs, the height of the trapping feature (trap-closure height) must be greater than the capillary intrusion of water into the NAPL phase, and the trap boundary must be sufficiently fine-grained to prevent NAPLs from entering its pores. For both LNAPLs and DNAPLs, the closer the fluid density is to that of water, the greater the closure height necessary to retain the NAPL in an interface trap (i.e., sand/shale interface). Likewise, NAPLs with densities that significantly differ from that of water require the least closure for trapping. Trap-closure heights can range from one to several centimeters for coarse-grained material, and 1 to more than 5 m for fine-grained sands.

Where DNAPL enters fractured geologic media, the DNAPL will preferentially enter the larger pore spaces, which are the individual fractures. A small amount of DNAPL can migrate a significant distance should the surrounding rock be characterized by small pores having high displacement or entry pressure. Conversely, the DNAPL may penetrate into the rock matrix should a buildup of DNAPL pressure within the fracture network occur, thus reducing the extent of DNAPL migration within the fracture network.

5.3 SUBSURFACE PROCESSES

Release of organic chemicals can occur under a wide variety of scenarios and environmental settings. The extent of any threat to human health and the environment depends on release-specific conditions. Some of the factors that determine the risk include:

- Type of organic chemicals (contaminants of concern);
- Quantity and age of the release;
- Migration pathways (i.e., soil type, depth to water table, utility trenches);
- Proximity to receptors (groundwater, surface water, basements, direct animal and human exposure, etc.); and
- Attenuation, dilution, and degradation processes.

Once released, the organic chemical interacts physically with soil and water by one or more distinct processes:

- Volatilization — Gaseous state, primarily in the unsaturated zone;
- Sorption — Attached to soil particles and trapped within soil pores (can be above or below the water table);
- Aqueous phase migration — Dissolved in groundwater and soil moisture, advection, dispersion, and diffusion; and
- Liquid phase migration and retention — NAPL held suspended by the water table or capillary fringe or perched above low permeability zones (water wet soil) in the unsaturated zone.

The relationship of these phases with soil zones is shown in Figure 5.3. The phase distribution of a 30,000-gal gasoline spill into medium sand soil within a water table aquifer is presented in Table 5.3. Note the relationship between the percent in each phase (free liquid, residual, and dissolved) compared with the contaminated volume. While 64% of the gasoline remains in free-phase form, it only contaminates 7100 yd^3. The residual phase represents that portion (35%) which remains relatively fixed (immobile) in the soil, but contaminates 250,000 yd^3. Only 1% of the gasoline is in the dissolved phase; however, it is mobile in that it migrates with the groundwater by diffusion and dispersion, and thus contaminates 960,000 yd^3 of aquifer. “Contamination” in this example is defined as exceeding some regulatory

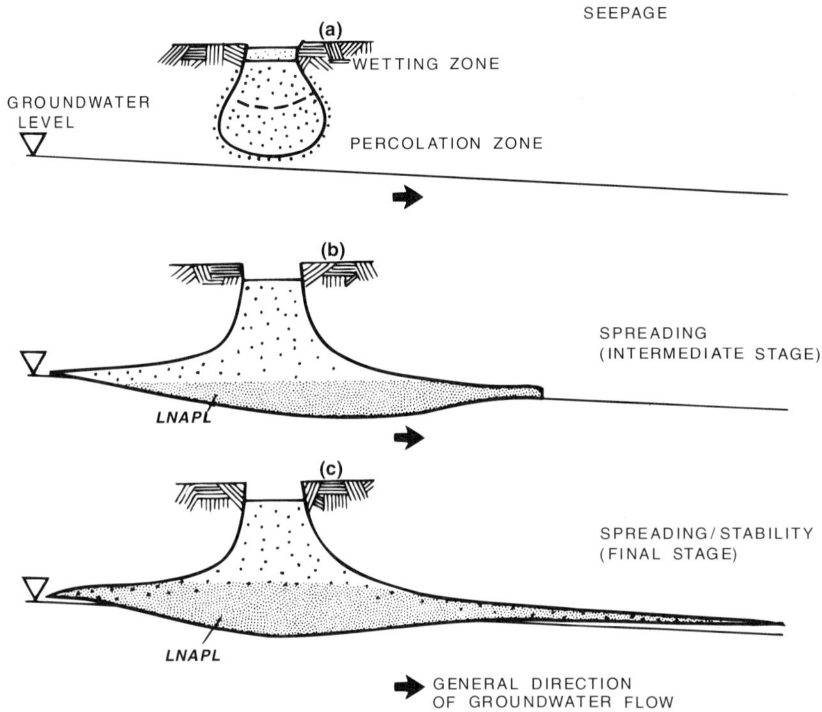

FIGURE 5.3 Stages of subsurface hydrocarbon migration. (After Schwille, 1967.)

limit. Dissolved-phase limits are often in the parts per billion range, and thus a small quantity of hydrocarbon can affect a very large volume of water.

5.3.1 Volatilization

In the liquid and solid state, molecules are held together by a variety of intermolecular forces. When transferring from solid to liquid, liquid to vapor, or solid to vapor, these forces must be overcome by absorbing energy, usually thermal energy from the environment. Volatilization is the process by which this occurs. The more volatile

TABLE 5.3
Phase Distribution at a 30,000 Gasoline Spill Site

Groundwater Media	Phase	Gasoline Release Volume (gal)	% of Total	Contaminated Volume (yd^3)	% of Total
Medium sand	Free-phase LNAPL	18,500	64	7,100	1
	Residual hydrocarbon	10,000	35	250,000	20
	Dissolved hydrocarbons in groundwater	333	1	960,000	79

Source: EPA (1996).

the chemical, the easier it is for molecules to leave the surface. Volatility depends upon the temperature and the chemical composition of the compound. At any particular temperature, a volatile liquid or solid has a characteristic vapor pressure which is the equilibrium partial pressure in the atmosphere or air space surrounding the compound. Therefore, whether in the pore space of the vadose zone at 25°C (normal groundwater) or in the atmosphere above an open disposal trench at 25°C, if the vapor pressure of the highly volatile compound carbon tetrachloride (CCl_4) is 0.15 atm (114 mm Hg), then the partial pressures of CCl_4 vapor in the pore space and in the atmosphere above the trench will both be 0.15 atm. Vapor pressures for certain organic compounds typically encountered in the subsurface are presented in Table 5.4.

If a vacuum is applied to the gas in the vadose zone, to remove the CCl_4 vapor, then more CCl_4 will volatilize to restore the partial pressure to 0.15 atm. This is the operating principle of the vadose zone remediation of volatile contaminants referred to as soil vapor extraction (SVE) in which vapor extraction causes continuous volatilization of the contaminant until little or none remains. Once the compound volatilizes, its vapor behaves like any other gas with respect to the physical transport processes. Obviously, the more volatile the compound, the better this method will work. SVE works well for remediating CCl_4 in the vadose zone because of its high vapor pressure, but will not work as well for the pesticide methyl parathion which has a vapor pressure of only 10^{-11} atm at 25°C. However, after many years, the loss of methyl parathion through volatilization will be an important process, especially if other degradation processes are slow.

The potential for a chemical to volatilize from the water to the air phase can be estimated by Henry's law. This law states that, in a very dilute solution, the vapor pressure of a chemical should be proportionate to its concentration.

$$V_p = K_h C \quad \text{or} \quad K_h = V_p/C \tag{5.1}$$

where V_p = vapor pressure, C = concentration of chemical in water, and K_h = Henry's law constant.

Note: K_h is expressed in different dimensions throughout the literature. Units used are atm-m^3/mol, atm-cm^3/g, or dimensionless. Caution should be used to assure numerical values are compatible.

The presence of organic vapors in the shallow subsurface is an indication that there is (or was) a hydrocarbon that has volatilized. The source of vapors can be an LNAPL pool, hydrocarbon adsorbed onto soil particles (residual hydrocarbons), or dissolved hydrocarbon in groundwater. After the hydrocarbon has volatilized into the soil atmosphere phase, it moves by dispersion gradient from areas of higher to lower concentration. Ultimately, it may reach the surface. If the air above the soil surface is moving, the rate of transfer can be much greater (by the process of convection). When a mechanical air pressure differential is established, as during SVE or air sparging, the vapors will flow along with the moving air. Under natural conditions (without mechanical assistance), the movement of air in the vadose zone is very slow, resulting in a long contact time between the phase, and in establishment of a concentration equilibrium near the phase boundary.

TABLE 5.4
Physical and Chemical Properties of Certain Common Organic Compounds

Compound	Molecular Weight	Melting Point (°C)	Boiling Point (°C)	Vapor Pressure (p, kPa)	Solubility (S, g/m³)	Henry's Law Constant[a] kPa m³/mol		
						Calc.	Exp.	Rec.
Monoaromatics at 25°C								
Benzene	78.11	5.53	80.1	12.7	1780	0.562	0.562	0.550 ± 0.025
Toluene	92.13	–95	110.6	3.80	515	0.68	0.673	0.670 ± 0.035
Ethylbenzene	106.2	–95	136.2	1.27	152	0.887	0.854	0.80 ± 0.07
p-Xylene	106.2	13.2	138	1.17	185	0.671	NA[b]	0.710 ± 0.08
m-Xylene	106.2	–47.9	139	1.10	162	0.721	NA	0.700 ± 0.10
o-Xylene	106.2	–25.2	144.4	0.882	175	0.535	NA	0.50 ± 0.06
Polynuclear Aromatics at 25°C								
Napthalene	128.19	80.2	218	1.01×10^{-2}	34.4	0.0407	0.0489	0.0430 ± 0.004
Fluorene	166.2	116	295	8.86×10^{-5}	1.90	0.00775	0.0101	0.0085 ± 0.002
Anthracene	178.23	216.2	340	1.44×10^{-6}[c]	0.075	0.0034	0.073	0.0060 ± 0.003
Halogenated Alkanes and Alkenes at 25°C								
1,1-Dichloroethane	98.97	–96.98	57.5	30.10	5100	0.585	NA	0.58 ± 0.02
1,1,1-Trichloroethane	133.4	–30.4	74.1	16.53	720	3.06	NA	2.8 ± 0.04
1,1,2-Trichloroethane	133.4	–36.5	113.8	4.04	4420	0.122	NA	0.12 ± 0.02
1,1,1,2-Tetrachloroethane	167.85	–70.2	130.5	1.853	1100	0.283	NA	0.28 ± 0.02
Vinyl chloride	62.5	–153.8	–13.4	344 (20°C)	2700 (20°C)	2.35	NA	NA
1,1-Dichloroethene	96.94	–122.1	37	79.73	400	13.32	NA	NA
1,2-Dichloroethene (*cis*)	96.94	–80.5	60.3	27.46	3500	0.761	NA	NA

TABLE 5.4 *(continued)*
Physical and Chemical Properties of Certain Common Organic Compounds

Compound	Molecular Weight	Melting Point (°C)	Boiling Point (°C)	Vapor Pressure (p, kPa)	Solubility (S, g/m³)	Henry's Law Constant[a] kPa m³/mol Calc.	Exp.	Rec.
1,2-Dichloroethene (*trans*)	46.94	–50	47.5	43.47	6300	0.669	NA	NA
Tetrachloroethene	165.83	–19	121	2.48	140	2.94	1.239	2.3 ± 0.4
Halogenated Aromatics at 25°C								
Chlorobenzene	112.56	–45.6	132	1.581	471.7	0.377	0.382	0.35 ± 0.05
d-Dichlorobenzene	147.01	–17.0	180.5	0.196	145.2	0.198	0.193	0.19 ± 0.01
1,2,3-Trichlorobenzene	181.45	53	218	0.0530[d]	16.6	0.306	0.127	NA
1,2,3,4-Tetrachlorobenzene	215.9	47.5	254	0.00876[d]	4.31	0.261	NA	NA

[a] Modified after MacKay and Shiv (1981).
[b] NA = not available.
[c] Calculated from the extrapolated vapor pressure with a fugacity ratio correction.
[d] Extrapolated from liquid state.

After vapor has dispersed through atmosphere in the soil pores, its fate may include escape to the atmosphere, adsorption by soil particles, destruction by biological activity, or re-solution into percolating water. In highly permeable unsaturated settings, volatilization of light hydrocarbons (i.e., gasoline) can be a major factor in soil remediation. Similarly, where large quantities of gasoline are located on a shallow water table under a building (or adjacent to underground structures), the potential for a fire or explosion hazard is a real concern.

Several soil–vapor monitoring techniques are currently being used to define areas of volatile organic chemical contamination. These procedures usually involve the collection of representative samples of the soil gas for analysis of indicator compounds. Maps marked with concentration contours of these indicator compounds can be used to identify potential sources to delineate the contaminated area. Indicator compounds (usually the more volatile compounds) are selected for each specific situation. For gasoline contamination, the compounds are usually benzene, toluene, ethylbenzene, and total xylene (BTEX). In the case of a fuel oil spill, the most commonly used indicator is naphthalene. Some laboratories have adapted the laboratory procedures used for quality analysis of wellhead condensate (i.e., normal paraffins) to include light-end (<8 carbons) molecular analysis.

This mapping technique can be an effective tool when used in conjunction with a detailed subsurface investigation. However, the unknown presence of confining layers, permeable pathways, or previous spills may lead the investigator to erroneous conclusions. The volatility of some organic chemical products can, however, prove to be a favorable characteristic in developing a remediation strategy. Under favorable geologic conditions strategically placed vapor extraction wells, and high volatility, can be highly effective for the recovery of certain hydrocarbons.

5.3.2 Sorption

Adsorption refers to the attachment of a chemical to the *surface* of a soil particle; absorption infers incorporation of a chemical into the *structure* of the soil particle. As a practical matter, it is difficult to determine the contribution of each process; thus, they are usually combined into the single term *sorption*. In soil, this process is exhibited when a molecule of gas, free liquid, or contaminants dissolved in water is attached to the surface of an individual soil particle (often in the form of organic carbon). This surface attachment can be by any of three general processes:

1. Physical: Very weak, caused by van der Waals forces.
2. Chemical: Much stronger, similar in strength to some chemical bonds; often requires significant effort to separate.
3. Exchange: Characterized by electrical attraction between the *sorbate* and the surface; exemplified by ion-exchange processes.

Molecular structure determines the gross activity of an organic compound, as it is responsible for the molecular volume, water solubility, vapor pressure, density, and electrical charge of the compound. The three-dimensional structure of an organic molecule may have considerable effect on the potential for sorption of that molecule. The larger the planer contact area, the greater the sorption.

The size of a molecule is directly related to its probability of sorption. Larger molecules have a greater tendency to be sorbed. As the surface area of contact increases, the holding forces increase by van der Waals bonding. Some very soluble molecules are held tightly, due to their many contact points with solid surfaces. An example of this is soluble dye added to cotton fabric. A large portion of the soluble dye is sorbed by the cloth fibers. Subsequent washing removes only a small portion of the color with each washing. Recent studies indicate that a number of organic molecules (i.e., pesticides, polynuclear hydrocarbons, and aromatic hydrocarbons) bind easily to large dissolved organic molecules such as humic acid, fulvic acid, and organic matter. These large molecules are not mobile, as they themselves tend to be adsorbed to soil particles.

Some organic molecules prefer attachment to hydrophobic ("water-hating") surfaces instead of aqueous solvents or hydrophilic ("water-liking") surfaces. Molecular fragments of C, H, Br, Cl, and I tend to be hydrophobic, whereas fragments of N, S, O, and P are primarily hydrophilic. Other organic molecules have a built-in positive or negative charge. Examples of these molecules include metal–organic compounds. These molecules are easily adsorbed on to surfaces with the opposite charge such as cation (or anion) exchange sites.

Sorptive reactions of LNAPLs and dissolved organic compounds moving through soil are almost always reversible equilibrium reactions. A concentration equilibrium is established between the concentration of chemical dissolved in water and that which is attached to the soil particles. When concentrations change, the soil may adsorb additional organic molecules or release them to reestablish the equilibrium.

Since sorption is primarily a surface phenomenon, its activity is a direct function of the surface area of the solid as well as the electrical forces active on that surface. Most organic chemicals are nonionic and therefore associate more readily with organic rather than with mineral particles in soils. Dispersed organic carbon found in soils has a very high surface-to-volume ratio. A small percentage of organic carbon can have a larger adsorptive capacity than the total of the mineral components.

Sorption is especially important in sediments containing a high percentage of organic matter (peat bogs, former lake beds, etc.). Mass migration of free-phase hydrocarbon through these soils is greatly retarded. Most soil and aquifers materials, however, contain less than 0.1% organic matter and the rate of sorption is minimal.

Sorption isotherms for relatively low concentrations of hydrophobic compounds (i.e., less than one half solubility) are relatively linear and may be estimated by the following equation:

$$q = K_d C \tag{5.2}$$

where q = mass adsorbed per mass of solid phase, K_d = slope of the adsorption isotherm, and C = equilibrium concentration of the chemical dissolved in water. As this equation indicates, sorption of a chemical from the dissolved phase onto the solid surface is dependent upon the molecule being adsorbed. Less-soluble molecules are more easily adsorbed. Another factor that affects the slope of the isotherm (for any given temperature and pressure) is the molecule and its ability to fit within the pore space between the soil particles.

K_d is difficult to define when dealing with actual field conditions. Laboratory studies have demonstrated that an approximation of K_d can be made based on the more easily determined factors such as organic carbon content of the soil and the octanol–water partitioning coefficient of the compound. For sediment particles <50 μm:

$$K_d = 0.6 f_{oc} K_{ow} \tag{5.3}$$

where K_d = slope of adsorption isotherm, f_{oc} = fractional organic carbon in the solid phase, and K_{ow} = octanol–water partition coefficient. K_{ow} for many compounds has been determined by measuring the ratio of the concentration of a particular compound in octanol (a relatively nonpolar alcohol) to that in water. This ratio is also often used to approximate the partitioning behavior of compounds between soil-containing organic matter and water.

The net result of sorption on organic contaminated soils is to retard the movement of contaminants. When a pollutant is adsorbed onto soil, it can be released only when the equilibrium between it and the passing fluid (water or air) is disrupted. *Retardation* is the term used to describe the apparent discrepancy between the actual migration rate of aquifer water and that of a dissolved organic chemical (somewhat slower). The difference in travel rates is the result of sorption of the chemical onto the aquifer matrix and release into water by the concentration gradient and time of contact. A general equation used for gross estimation of the retardation factor R_f is

$$R_f = 1 + K_d(b/P) \tag{5.4}$$

where R_f = retardation factor, K_d = distribution coefficient, b = bulk density of soil, and P = total porosity. As an example, at a particular site, the relatively soluble gasoline component benzene tends to travel in groundwater with a retardation factor (R_f) of 1.1, indicating that it migrates 1/1.1 or 90% as fast as groundwater. Xylene tends to have an R_f of 2, meaning it migrates at a rate of 50% as fast as the groundwater. If the seepage velocity of uncontaminated water passing through the system is too fast, the water will contain less than the equilibrium concentration of the contaminant. Monitoring wells in dynamic flow settings may demonstrate low concentrations of dissolved contaminants during periods of rapid flow. When slower flow is again present, concentrations will increase to the equilibrium level. Typical retention factors for various LNAPLs and DNAPLs are presented in Table 5.5.

5.3.3 Advection, Dispersion, and Diffusion

The physical transport of dissolved organic compounds through the subsurface occurs by three processes: advection, hydrodynamic dispersion, and molecular diffusion. Together, these three cause the spread of dissolved chemicals into the familiar plume distribution. Advection is the most important dissolved chemical migration process active in the subsurface, and reflects the migration of dissolved chemicals

TABLE 5.5
Typical Residual Saturation Data for Various LNAPL and DNAPL Types

Residual Fluid Type	Hydrogeologic Conditions[a]	Media	Residual Saturation (S_r) or Retention Factor (R_f) (l/m³)	
			S_r	R_f
Water	Unsaturated	Sand	0.01	
		Silt	0.07	
		Sandy clay	0.26	
		Silty clay	0.19	
		Clay	0.18	
LNAPLs				
Gasoline	Unsaturated	Coarse gravel		2.5
		Coarse sand and gravel		4.0
		Medium to coarse sand		7.5
		Fine to medium sand		12.5
		Silt to fine sand		20
		Coarse sand	0.15–0.19	
		Medium sand	0.12–0.27	
		Fine to coarse sand	0.19–0.60	
Kerosene		Stone, coarse sand	0.46–0.59	
		Gravel, coarse sand	5	
		Coarse to medium sand	8	
		Fine to medium sand	15	
		Fine sand and silt	25	
Diesel and light fuel oil		Soil	40	
Light oil and gasoline		Soil	0.18	
Lube and heavy fuel oils		Soil	10.18	
			0.15	
Middle distillates	Unsaturated	Coarse gravel		5.0
		Coarse sand and gravel		8.0
		Medium to coarse sand		15
		Fine to medium sand		25
		Silt to fine sand		40
Fuel oils	Unsaturated	Coarse gravel		10
		Coarse sand and gravel		16
		Medium to coarse sand		30
		Fine to medium sand		50
		Silt and fine sand		80
Crude oils	Saturated	Sandstone	0.35–0.43	
		Sandstone	0.16–0.47	
		Sandstone	0.26–0.43	
DNAPLs				
Benzene	Saturated	Sand (92% sand, 5% silt, 3% clay)	0.24	
Tetrachloroethane	Saturated	Fracture (0.2 aperture)		0.05 1/m²
Tetrachloroethene	Unsaturated	Coarse Ottawa sand	0.15–0.25	

TABLE 5.5 *(continued)*
Typical Residual Saturation Data for Various LNAPL and DNAPL Types

Residual Fluid Type	Hydrogeologic Conditions[a]	Media	Residual Saturation (S_r) or Retention Factor (R_f) (l/m³) S_r	R_f
1,1,1-Trichloroethane	Saturated	Coarse Ottawa sand	0.15–0.40	
Trichloroethene	Unsaturated	Medium sand	0.20	
		Fine sand	0.19	
		Fine sand	0.15–0.20	

[a] Unsaturated refers to NAPL–water–air or water–air systems; saturated refers to NAPL–water systems.

Source: Modified after Mercer and Cohen (1992).

along with groundwater flow, the mass of actual mass movement of a fluid through the porous media.

Hydrodynamic dispersion refers to the tendency of a solute or chemical dissolved in the fluid, to spread out over time (i.e., to become dispersed in the subsurface). The mechanical component of dispersion results from the differential flow of the fluid through pore spaces that are not the same size or shape, and from different flow velocities and the fluid near the walls of the pore where the drag is greatest vs. the fluid in the center of the pore (Figure 5.4).

Molecular diffusion also tends to disperse fluids and dissolved solutes through the random travel of individual molecules. Diffusion results from the thermal kinetic energy of each molecule and the driving force of entropy. Contaminants move from areas of higher to areas of lower concentration. Diffusion is especially important when volatile compounds are present in situations where flow velocities are extremely low. The classic example is the placement of a single drop of food coloring into a stagnant container of water. Some time later, without agitation, the coloring will have diffused throughout the entire container. In computer modeling, the effects of diffusion and dispersion are often combined into a single term *dispersivity*, which is expressed in terms of linear units (cm, m, or ft).

An idealized example of dispersion is shown in Figure 5.4. Molecule diffusion can cause an increased concentration in every direction, including the upgradient,

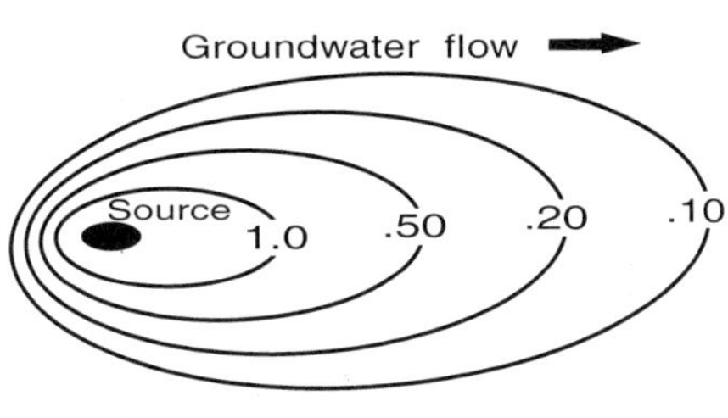

FIGURE 5.4 Idealized dispersion of a single component indicating relative concentration.

especially in slowly moving groundwater. In some cases, where the ion (such as chloride) has an R_f of 1 (no retardation) the ion can migrate (by diffusion) and appear to travel faster than the groundwater. Slower water movement results in more nearly circular diffusion patterns.

The above discussion is most nearly correct when only one contaminating chemical is involved. When contaminants such as gasoline are introduced into the subsurface, the setting becomes more complex. Gasoline is a mixture of mostly small molecules, such as benzene, xylene, toluene, hexane, and other molecules that are mostly less than C_{14}. Diffusion of a mixture of dissolved chemicals involves several factors.

Each individual molecular structure has individual characteristics when dissolving and diffusing through water. Some molecules disperse more easily than others based on the solubility, water–octanol partition coefficient, aquifer grain composition, and temperature. The result is a chromatographic distribution of components throughout a dissolved contaminant plume. For gasoline components, benzene disperses more easily than toluene, which spreads more slowly than xylene. In an idealized plume in a uniform aquifer, benzene would be the most widely dispersed of the three chemicals. It would be observed first in downgradient monitoring wells.

Investigators must exercise professional judgment when interpreting analytical results from monitoring wells. In "real-world" situations, many other factors are involved that may interfere with the ideal. For example, the soil may contain fine particles (clays or free organic carbon) that have an affinity for a particular component or certain microfluora that biodegrade some molecules more easily than others.

Diffusion and dispersion occur in both the liquid and vapor phases. In the vadose zone, liquid-phase dispersion is a major factor following a rainfall (or irrigation) event. Dispersion of vapor phase can also occur. Driving forces for vapor and dispersion are gravity and variations of atmospheric pressure (or SVE/air sparge operations).

5.4 OCCURRENCE AND FLOW OF IMMISCIBLE LIQUIDS

5.4.1 The Unsaturated Zone (above the Water Table)

5.4.1.1 Water Flow through the Unsaturated Zone

Water present in and moving through the unsaturated zone is subjected to several rules of physics in addition to those influencing the water below the water table. The presence of retained moisture above the water table is due to adsorptive forces between the water molecules and soil particles, in addition to surface tension of the water surface.

The term *capillary action* describes the upward movement of a fluid as a result of surface tension through pore spaces. The fluid can rise until the lifting forces are balanced by gravitational pull (see Figure 3.28). The rise of fluid in a small tube above the water table surface, as previously discussed in Chapter 3, can be described using Equation 3.13. Lifting of fluids above the water table is a true negative pressure compared with atmospheric pressure (also described as soil suction). In soil situations, pore spaces are not uniform, elongated cylinders, and, therefore, the actual rise of fluid does not create a smooth, uniform front. Capillary forces can also promote downward or horizontal fluid migration.

These principles in soil above the water table are depicted in Figure 3.29. Immediately above the water table, the majority of the pore spaces are filled with water. This condition extends upward from the capillary fringe, which is virtually saturated. Next in upward progression is the funicular zone, in which the volume of water retained is reduced as the number of directly connected small pores decreases and the percentage of air saturation increases. Above the funicular zone is the pendular zone, in which water is retained as residual saturation in the necks of individual pores. This situation is very stable because adhesive forces per volume retaining the remaining water are much greater than the gravity draining force. Under some circumstances, the adhesive forces are as great as several atmospheres. Water in the liquid state cannot exist under a suction of greater than 0.7 atm.

Although water is the most common liquid fluid found in the unsaturated zone, other liquids that may occur are controlled by the same physical forces. Laboratory testing of diesel fuel and water in sands demonstrates that free-floating oil above water also creates a capillary fringe. Because the surface tension of oil is characteristically less than that of water, the oil-capillary fringe for the diesel fuel used in laboratory experiments was approximately one half as high as that of the water-capillary fringe.

Water flow through the unsaturated zone is controlled by several factors. The primary forces that determine the rate of liquid flow through the unsaturated zone are gravity (acting as a downward force) and a combination of capillary and adhesive forces, called the moisture potential (usually upward). Moisture potential is often described as the tendency of soil to retain moisture. Depending upon the moisture content, either the gravity force or the retention forces may predominate. When saturation is such that the water phase becomes continuous from pore to pore space, flow may be possible. At saturations without continuous water connection between pores, the moisture potential is greater and flow does not occur. As the percentage of pore spaces filled increases, the volume of flow increases.

Darcy's equation can be used to describe flow in this region; however, the value of permeability varies as a function of saturation. Also, the value of moisture potential is a function of saturation. The total potential for flow (hydraulic gradient in Darcy's equation) can be defined as the difference between the moisture potential (minus) and the elevation potential (plus). When the potential for flow is positive, flow can occur.

Field measurements of moisture potential made by tensiometer, in conjunction with moisture content, demonstrate that the moisture potential is substantially different if the soil has been in a wetting or drying phase. Soils that have been in a drying phase prior to testing have a greater degree of saturation at the same moisture potential than a soil that has been in a wetting phase. The reason for this hysteresis is threefold. First, as water reenters a dry narrow channel, a local increase in suction is required. If the pore space is too wide, the interface cannot advance until a neighboring pore is filled, allowing the wall–liquid interface angle to be small enough to cause capillary rise. Second, the contact angle at an advancing interface differs

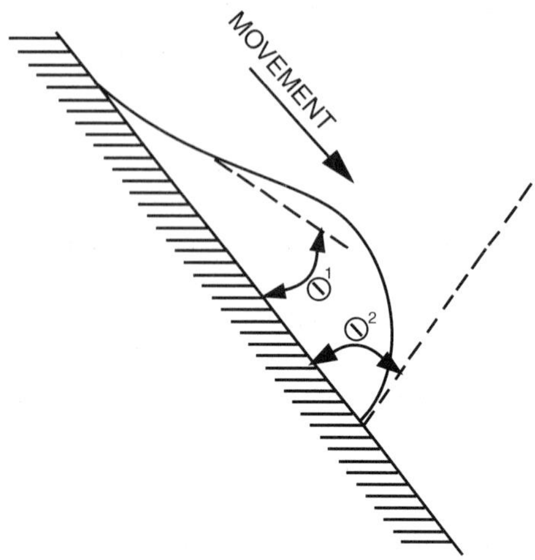

FIGURE 5.5 Liquid–solid interface contact showing advancing and retreating contact angle.

from that of a receding one, as shown in Figure 5.5. Entrapped air is the third factor causing hysteresis.

The permeability of wetting or drying unsaturated soils follows a similar relationship with moisture at the degree of saturation. A graphic demonstration of the similarity of these two functions is shown in Figure 5.6.

5.4.1.2 Multiphase Fluid Flow in the Unsaturated Zone

In addition to water, NAPLs, such as petroleum, oils, tars, and biological fluids, are often present in the subsurface. When more than one fluid is present, there is a need to describe how well they mix, referred to as their *miscibility.* Water and vegetable oil are immiscible fluids. Many of the NAPLs are immiscible with water and will occur as separate fluid bodies, droplets, zones, etc. in the subsurface environment.

Migration of free-phase NAPLs in the subsurface is governed by numerous properties including density, viscosity, surface tension, interfacial tension, immiscibility, capillary pressure, wettability, saturation, residual saturation, relative permeability, solubility, and volatilization. Thc two most important factors that control their flow behavior are density and viscosity.

Density is the mass of a material per unit volume, which is the ratio of the density of a substance to that of some standard, notably water. Water has a density of about 1 g/cm^3, whereas carbon tetrachloride (CCl_4), an important DNAPL contaminant, has a density of about 1.58 g/cm^3 and will tend to sink through the water table.

Viscosity is the ability of a fluid to resist deformation or flow, and is a measure of the tendency of a fluid to flow; for example, molasses has a high viscosity relative to water. Viscosity is highly temperature dependent and has common units of centipoise (cP). Water has a viscosity of 1.00 cP at 20°C, whereas carbontetrachloride has a viscosity of 0.97 cP at 20°C. Therefore, the two fluids will physically flow about the same. However, with respect to flow through porous media, surface tension is extremely important.

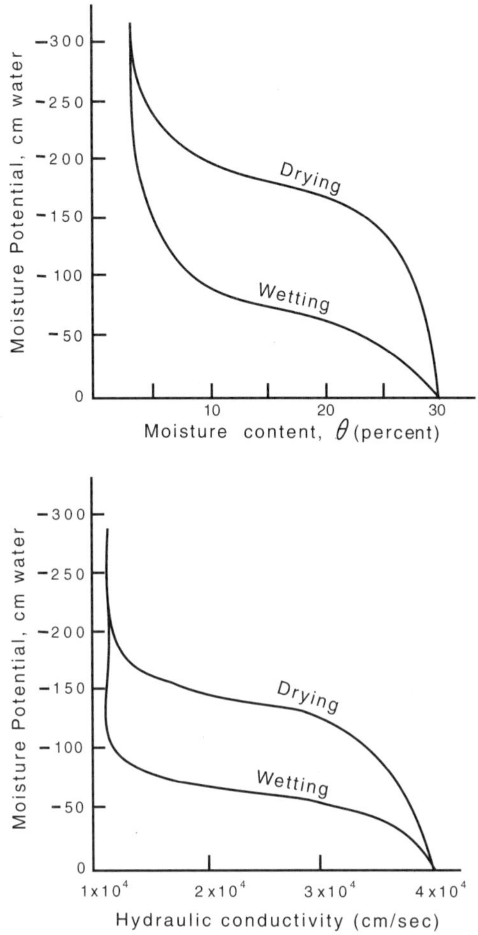

FIGURE 5.6 Graph showing similarities between moisture content vs. moisture potential and hydraulic conductivity vs. moisture potential. (After Fetter, 1980.)

Surface tension is responsible for capillary effects and spreading of the NAPL over the water table. At about 20°C, water has a surface tension of 73.05 dyn/cm, whereas CCl_4 has a surface tension of only 26.95 dyn/cm. Therefore, water will be held in an unsaturated porous media by surface tension to a much greater degree relative to carbon tetrachloride (i.e., the permeability of porous media will be different with respect to each liquid). The ramifications will be important for contaminant transport of mixed wastes.

It should be noted that the properties of any liquid can change as its composition changes. Salt water is much denser than fresh water, and these two fluids will not readily mix without agitation. Likewise, a specific oil composition will determine its density, miscibility, viscosity, and surface tension, as well as other properties.

Wettability refers to the preferential spreading of one fluid over solid surfaces in a two-fluid system and is dependent upon the interfacial tension. The wetting

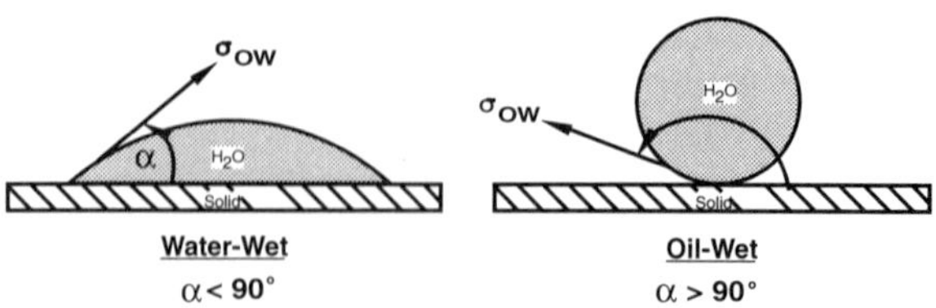

FIGURE 5.7 Configuration of wettability.

fluid (water) will tend to coat the surface of grains and occupy the smaller spaces or pore throats, whereas the nonwetting fluid (oil or NAPL) will tend to occupy the largest openings. The contact angle (α), as measured in the water, determines whether the porous media will be preferentially wetted by water or NAPL varying from 0 to 180°. The system is water-wet if α is less than approximately 70°, and NAPL-wet if α is greater than 110°; neutral conditions persist when α is between 70 and 110° (Figure 5.7). Soils that are initially water-wet tend to remain so; the same is true of oil-wet soil.

5.4.1.3 Saturation Volumes

Saturation (s) is the volume fraction of the total void volume occupied by a specific fluid at a point. Saturation values can vary from zero to 1 with the saturation of all fluids equal to 1. Residual saturation (S_r) is the saturation at which the NAPL becomes discontinuous and immobile due to capillary forces. Residual saturation is dependent upon many factors, including pore size distribution, wettability, fluid viscosity and density ratios, interfacial surface tension, gravity and buoyancy forces, and hydraulic gradients.

The residual saturation capacity of soil is generally about one third of its water-holding capacity. Immobilization of a certain mass of hydrocarbon is dependent upon soil porosity and physical characteristics of the product. The volume of soil required to immobilize a volume of liquid hydrocarbon can be estimated as follows:

$$V_s = 0.2\frac{V_{\text{hc}}}{P(\text{RS})} \tag{5.5}$$

where V_s = cubic yards of soil required to attain residual saturation, V_{hc} = volume of discharged hydrocarbon, in barrels (42 gal = 1 barrel), P = soil porosity, and RS = residual saturation capacity.

The maximum depth of liquid hydrocarbon penetration into the unsaturated zone can be estimated by the following equations:

$$D = \frac{V_s}{A} \tag{5.6}$$

$$D = \frac{KV_{\text{hc}}}{A} \tag{5.7}$$

TABLE 5.6
Typical *K* Values for Various Soil Types

Soil Type	*K* Value		
	Gasoline	Kerosene	Light Gas Oil
Stone to coarse gravel	400	200	100
Gravel to coarse sand	250	125	62
Coarse to medium sand	130	66	33
Medium to fine sand	80	40	20
Fine sand silt	50	25	18

Source: Dietz (1970) and CONCAWE Secretariat (1974).

$$D = 1000 \frac{V_{\mathrm{hc}}}{ARC} \tag{5.8}$$

where D = maximum depth of hydrocarbon penetration in the vadose zone, V_s = cubic yards of soil required to attain residual saturation, V_{hc} = volume of discharged hydrocarbon, A = area of infiltration, K = constant based on soil retention capacity for LNAPL and LNAPL viscosity (Table 5.6), R = soil retention capacity (Table 5.7), and C = approximate correction factor based on LNAPL viscosity (i.e., 0.5 for gasoline to 2.0 for light fuel oil).

5.4.1.4 NAPL Migration

The problems associated with LNAPLs are well documented in the literature, ranging from small releases where just enough LNAPL is present to be a nuisance, to pools ranging up to millions of barrels of LNAPL and encompassing hundreds of acres in lateral extent. Subsurface migration of LNAPL (and DNAPL) are affected by several mechanisms depending upon the vapor pressure of the liquid, the density of the liquid, the solubility of the liquid (how much dissolves in water at equilibrium), and the polar nature of the NAPL.

TABLE 5.7
Typical Oil Retention Capacities for Kerosene in Unsaturated Soils

Soil Type	Oil Retention Capacity (*R*)	
	l/m³	g/yd³
Stone, coarse sand	5	1
Gravel, coarse sand	8	2
Coarse sand, medium sand	15	3
Medium sand, fine sand	25	5
Fine sand, silt	40	8

Source: CONCAWE (1979).

NAPL will migrate from the liquid phase into the vapor phase until the vapor pressure is reached for that liquid. NAPL will move from the liquid phase into the water phase until the solubility is reached. Also, NAPL will move from the gas phase into any water that is not saturated with respect to that NAPL. Because hydraulic conductivities can be so low under highly unsaturated conditions, the gas phase may move much more rapidly than either of the liquid phases, and NAPLs can be transported to wetter zones where the NAPL can then move from the gas phase to a previously uncontaminated water phase. To understand and model these multiphase systems, the characteristic behavior and the diffusion coefficients for each phase must be known for each sediment or type of porous media, leading to an incredible amount of information, much of which is at present lacking.

NAPLs migrate in response to organic liquid pressure gradients, just as water moves in response to the hydraulic pressure gradients. NAPLs are generally less easily adsorbed than water because of their lower surface tensions and lower dipole moments, and will preferentially move in the larger channels, pores, or fractures. This may cause irregular and discontinuous flow patterns relative to the flow of water through the same material when the pore sizes and the degree of water saturation vary. However, some organic NAPLs, such as alkyl phosphates, which are used in many extraction processes, are highly polar and will behave differently.

Relative permeability is the reduction of mobility between more than one fluid flowing through a porous media, and is the ratio of the effective permeability of a fluid at a fixed saturation to the intrinsic permeability. Relative permeability varies from zero to 1 and can be represented as a function of saturation (Figure 5.8). Neither water nor oil is effectively mobile until the S_r is in the range of 20 to 30% or 5 to 10%, respectively, and, even then, the relative permeability of the lesser component is approximately 2%. Oil accumulation below this range is for all practical purposes immobile (and thus not recoverable). Where the curves cross (i.e., at an S_{rw} of 56% and $1 - S_{ro}$ of 44%), the relative permeability is the same for both fluids. With increasing saturation, water flows more easily relative to oil. As $1 - S_{ro}$ approaches 10%, the oil becomes immobile, allowing only water to flow.

5.4.1.5 Three Phase — Two Immiscible Liquids and Air in the Unsaturated Zone

When a petroleum compound is released into the unsaturated zone, it enters a very complex environment. The preexisting condition includes partially saturated (with water) pores that are under strong negative pressure, air-filled pores, and a degree of permeability that is variable depending on whether the soil has been recently in a wetting or draining phase. Additionally, this new fluid has its own properties of density, viscosity, and surface tension, all of which influence its ability to flow. To move through this environment, the LNAPL must satisfy the following:

$$P_{nw} - P_w = \frac{2\sigma\cos\theta}{r} \tag{5.9}$$

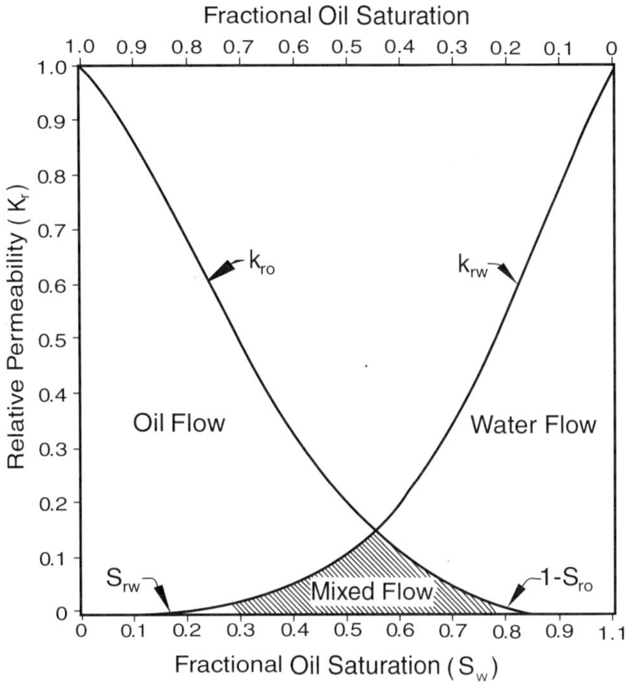

FIGURE 5.8 Representative hydraulic conductivities under highly saturated conditions. (After Testa, 1994.)

where P = fluid pressure, the subscript nw = nonwetting fluid, the subscript w = wetting fluid; r = interface radius; θ = contact angle; and σ = surface tension when one of the fluids is air or other interfacial tension between two liquid phases.

The above equation states that sufficient pressure must be applied to displace the existing fluid (water or air) before the new fluid can enter the pore spaces. In the pendular zone and the upper part of the funicular zone where residual water does not extend completely from grain to grain, the air offers little resistance. As the NAPL migrates through this region, it tends to flow through interconnected pores, which offer the least resistance, especially the larger pores. LNAPL follows the same rules of physics — wetting, being adsorbed, etc. — as does water. However, most petroleum products are nonpolar and have weaker adhesive attraction to the soil grains; therefore, "soil suction" is not as significant. In the pendular and upper parts of the funicular zone, gravity is sufficient to overcome the retention potential and the migration is predominantly downward.

The viscosity of separate LNAPL products varies significantly, ranging from far less to many times that of water. Flow of LNAPL in the unsaturated zone is largely dependent upon viscosity and soil grain size. Finer-grained materials have a higher residual saturation of water, which restricts the number of pores available for LNAPL entry in this region.

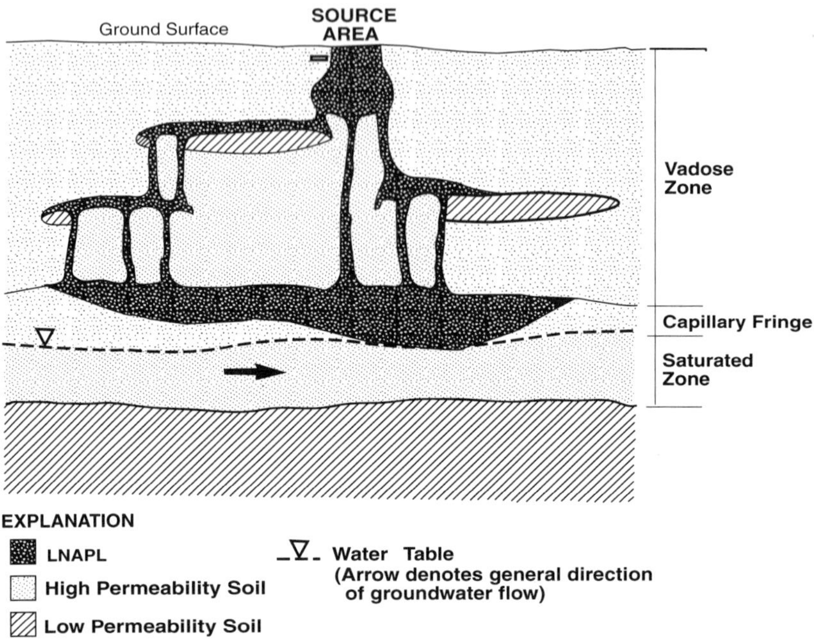

FIGURE 5.9 Relative permeability relationships controlling the flow of two immiscible fluids. (After Leverett, 1939.)

As more LNAPL enters the soils, it increases the pressure on the moving front, which allows increasing displacement of water. If a sufficient quantity of sufficiently mobile oil is available, it can produce a pressure head large enough to displace water through the funicular zone and to form a distinct interface at the top of the capillary fringe.

The migration of LNAPL in the subsurface can be divided into three phases: seepage through the vadose zone, spreading over the water table with development of a pancake layer, and accumulation stability within the water capillary zone (Figure 5.9). During seepage through the vadose zone, downward migration of the hydrocarbon can occur as a bulk product zone of affected soil or by fingering as illustrated in Figure 5.9. Fingering tends to occur under slow velocity conditions such that

$$q < \frac{kpg}{\mu} \tag{5.10}$$

where q = hydrocarbon velocity, k = permeability, p = LNAPL density, g = gravity, and μ = viscosity.

Large capillary forces as anticipated with tight, porous, and nonhomogeneous media will also contribute to fingering. Recognition of conditions favorable for development of fingering is important in adequately assessing subsurface presence and the lateral and vertical extent of hydrocarbon-affected soil. If fingering occurs, the LNAPL (or DNAPL) may not occupy the entire cross-sectional area through which it passes, allowing water to flow through and increasing dissolution. Fingering also results in deeper penetration of the NAPL relative to that of bulk hydrocarbon migration. If conditions are such that

$$q > \frac{kpg}{\mu} \tag{5.11}$$

then fingering will not occur and the hydrocarbon will migrate vertically downward in bulk. If the soil is stratified or contains significant fine-grained materials, some horizontal spreading may occur. To complicate matters, migration directions may vary due to facies architecture, stratigraphic controls imposed by depositional environment (i.e., channeling), bedrock orientation, and fractures.

In unstratified homogeneous soil, the form will be pear-shaped, with the larger part at the bottom. If the soil is stratified or contains significant fine materials, some horizontal spreading will occur. The general shapes that might be expected are shown in Figure 5.10.

When vertically migrating LNAPL hydrocarbon nears the water table, the capillary fringe is initially encountered. This capillary zone rises above the water table to a height dependent upon the grain size distribution of the formation as discussed later in this chapter. Essentially, finer-grained soils such as silt or clay attain a thicker capillary zone than coarser-grained soil, such as sand or gravel. As light hydrocarbon enters the water capillary zone, it begins to fill the pore spaces not occupied by capillary or residual water. Little mixing occurs since the two fluids are immiscible. Upon reaching the top of the water capillary zone, additional light hydrocarbon accumulation begins to spread laterally to form what is referred to as a pancake (see Figure 5.10). The lateral spreading of the pancake layer over the water table can be estimated as follows:

$$S = \left(\frac{1000}{F}\right)\left(V - \left[\frac{Ad}{K}\right]\right) \tag{5.12}$$

where F = thickness of the pancake (mm), V = volume of infiltrating bulk hydrocarbons (m^3), A = area of infiltration (m^2), d = depth to groundwater (m), K = constant dependent upon the soil retention capacity for oil and upon oil viscosity (see Table 5.6). K values generally decrease with decreasing grain size. Typical values for K are summarized in Table 5.6.

Sufficiently large seepage rates typically produce a hydraulic mound that permits limited lateral spreading of hydrocarbon in direction inconsistent with the regional groundwater table gradient, although lateral spreading upgradient could be stratigraphically controlled. The initial stage of spreading is dominated by gravity forces, but as the gravitational potential diminishes, capillary forces tend to control the rate of lateral spread.

Where the source is limited or ceases, capillary spreading eventually slows until further migration is limited and equilibrium is reached. This stable condition is attained when the leading edge of the laterally spreading light LNAPL fails to be

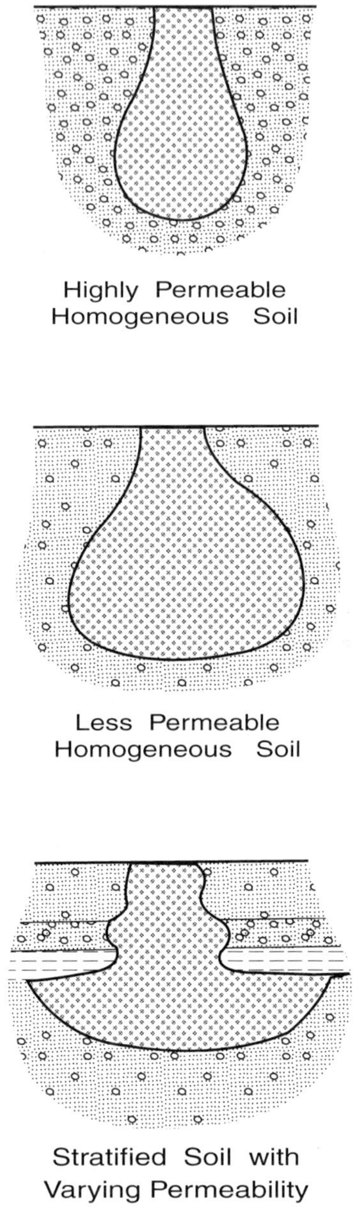

FIGURE 5.10 Schematic showing distribution of LNAPL in the subsurface.

replenished by more hydrocarbon. In this stable condition, the formation reaches its immobile or residual hydrocarbon saturation.

Residual hydrocarbon saturation will exist within both the unsaturated and capillary zone through which the LNAPL phase migrated. As might be expected, residual hydrocarbon saturation tends to be higher as the grain size decreases and the hydrocarbon viscosity increases. For example, hydrocarbons become immobile at a much higher saturation in clay that in sand. As a practical matter, this means that following hydrocarbon recovery efforts, more hydrocarbon is retained in clay formations than in sandy formations.

Once formed, the LNAPL is referred to as a pool that is really a continuous accumulation of LNAPL. Subpools are individual accumulations of relatively uniform free-product types based on geochemistry (API gravity, specific constituents, isotope ratios, etc.) that have coalesced, reflecting multiple-source accumulations with time. The LNAPL pool, once formed, also maintains a capillary fringe. The anticipated capillary rise of LNAPL can be calculated as follows:

$$h = \frac{2T\cos\theta}{r\rho g} \tag{5.13}$$

where h = rise (in cm), T = surface tension (in dynes), θ = wall–liquid interface angle, ρ = density of fluid, g = acceleration of gravity, and r = radius of tube in centimeters. Because the surface tension of LNAPL is characteristically less than water, the height of the LNAPL capillary rise is less than water.

The basic principles governing downward migration of light hydrocarbons as discussed above are applicable to "dense" hydrocarbons (DNAPLs) as well. The difference is that once groundwater is encountered, dense hydrocarbons continue to migrate downward, reflecting a specific gravity or density greater than that of water. Although a pancake may initially form, downward migration occurs once sufficient mass is attained to overcome the pressure necessary to displace water from the soil pores. Fingering can also occur in the saturated zone, reflecting density and viscosity differences. Vertical density flow ceases when a significant permeability contrast is encountered (i.e., clay) or when the DNAPL is retained as residual hydrocarbon saturation (by sorption). The migration of dense hydrocarbon in the subsurface is schematically shown in Figure 5.2.

5.4.2 The Saturated Zone (below the Water Table)

5.4.2.1 Steady-State Saturated Flow — Single Fluid

The basic equation for water flow through saturated porous media was developed by Henry Darcy in 1856 to calculate the flow of water through sand filters. This equation has been found to be valid for the flow of liquids through porous media when adjusted for the viscosity and density of the liquid. The original form of the equation, designed to calculate discharge is as follows:

$$Q = \frac{KIA}{\gamma} \tag{5.14}$$

where Q = discharge (L^3/T), K = permeability (hydraulic conductivity L/T), I = hydraulic gradient (driving force, or pressure, includes liquid density), γ = viscosity of liquid, and A = cross-sectional area of flow channel (L^2). This form of the equation is valid for liquid flow in the subsurface, when hydraulic conductivity has been determined in relation to water. Common units of hydraulic conductivity are cm/s, ft/day, or m/day.

Although permeability is expressed in units resembling a velocity term, the actual travel rate of the fluid through soil is expressed as

$$V = \frac{KI}{\Phi} \tag{5.15}$$

where V = actual travel velocity of fluid through the media, K = hydraulic conductivity (adjusted for viscosity), I = hydraulic gradient (adjusted for liquid density), and Φ = effective porosity of the media. Addition of the effective porosity Φ to this equation is necessary because the actual cross-sectional area of flow channels through the media is only a fraction of the total area of discharge. The flowing fluid must travel faster through the flow channels to maintain the apparent discharge velocity. Effective porosity describes only those connected pores that are large enough to permit free flow of the fluid. A clay, for example, may have an actual porosity of 50%; however, it may not be capable of transmitting a significant quantity of fluid because the pores are too small or not connected.

For practical purposes, saturated flow of a single fluid such as gasoline, kerosene, or another particular petroleum product can be predicted by the use of these equations. Standard units of linear measurement (feet, meters, etc.) and discharge are accommodated for by the corrections for viscosity and density. Field-testing procedures can be conducted using standard water well testing procedures.

5.4.2.2 Flow of Two Immiscible Fluids

The behavior of mixtures of immiscible fluids flowing through aquifers is often much more important than single-fluid flow during aquifer restoration. Recovery of NAPLs usually involves the movement of water through previously oil-saturated soils or the movement of oil through previously water-saturated soils. When these situations occur, the migrating fluid is a mixture of the two liquids. The relative permeability and flow of two immiscible fluids (water and kerosene) in sands, as previously discussed in Section 5.4.1.4 (Figure 5.9), was found to be directly related to fractional water saturation of the aquifer material. The mixed flow region of this graph is relatively small in comparison with either the water or oil flow regions. In situations that have nearly equal saturation levels of both water and oil, the relative permeability of each is only approximately 15%. This indicates that greater driving pressures are needed to cause the same total fluid flow quantities through the aquifer under this mixed-flow condition.

Each restoration situation would have a specific set of curves to define these relationships. For other materials, the curves may vary in shape, but the concept is similar. Mixed flow substantially reduces the relative permeability. This concept can be very important during the removal of NAPLs by the use of a large drawdown in wells or the raising of water levels for flushing purposes, and recovery of DNAPLs.

5.4.2.3 Dispersion from NAPL to Solution

Water and hydrocarbons occurring together, in shallow aquifer systems, may be considered "immiscible" for flow calculation purposes; however, each is somewhat soluble in the other. Since groundwater cleanup is the purpose behind restorations, it receives greater attention. Definition of water quality based on samples retrieved from monitoring wells relies heavily upon the concentration of individual chemical components found dissolved in those samples. An understanding of the processes that cause concentration gradients is important for the proper interpretation of analytical results.

The transfer of chemical molecules from oil to water is most often a surface area phenomenon caused by kinetic activity of the molecules. At the interface between the liquids (either static or moving), oil molecules (i.e., benzene, hexane, etc.) have a tendency to disperse from a high concentration (100% oil) to a low concentration (100% water) according to the functions of solubility, molecular size, molecular shape, ionic properties, and several other related factors. The rate of dispersion across this interface boundary is controlled largely by temperature and contact surface area. If the two fluids are static (i.e., no flow), an equilibrium concentration will develop between them and further dispersion across the interface will not occur. This situation is fairly common in the unsaturated zone.

The intensity with which dissolved chemicals are released from liquid hydrocarbon over time is referred to as source strength. Two approaches have been reported to provide an estimate of source strength. Source strength can be expressed as mass/time per unit contact area such that

$$S = K_m A \tag{5.16}$$

where S = source strength (mass/time/m^2), K_m = mass exchange coefficient (mg/m^2/s), and A = contact area or interface across which mass exchange occurs (m^2). The contact area governs the actual mass that is exchanged within a given period of time. Quantification of the A is very difficult, reflecting the complexity of hydrocarbon distribution in the pore space. Thus, efforts to quantify K_m have been attempted. Estimated K_m for certain products reported by the USGS (1984) are

Gasoline and tar oil	1.0 mg/m^2/s
Fuel oil, diesel, and kerosene	0.01 mg/m^2/s
Lube oils and heavy fuel oil	0.001 mg/m^2/s

These mass coefficient values are considered to be maximum values. Field values could be as much as one to two orders of magnitude lower.

The second approach to estimating source strength is also provided by the USGS (1984) and utilizes what is referred to as the Sherwood number and the Peclet number:

$$\mathrm{Sh} = 0.55 + 0.025\ \mathrm{Pe}^{3/2} \tag{5.17}$$

where Sh = Sherwood number representing a dimensionless transfer coefficient; and Pe = Peclet number representing a dimensionless flow velocity. When Pe < 1, the chemical exchange should be independent of flow velocity and should be essentially diffusion controlled, as anticipated for most soil systems. When Pe > 10, the chemical exchange should be dependent upon flow velocity. In addition, if the rate of diffusion for one chemical is known, then the rate for another chemical of similar structure can be estimated as follows:

$$\frac{D_1}{D_2} = \left[\frac{d_2}{d_1}\right]^{1/2} = \left[\frac{M_2}{M_1}\right]^{1/2} \tag{5.18}$$

where d_1, d_2 = densities of two chemicals and M_1, M_2 = molecular weight of two chemicals. The relationship in Equation 5.18 is derived from Grahms's law of diffusion. Generally speaking, the larger the molecule, the more slowly it diffuses and dissolves. The range of carbon atoms for various petroleum products is shown in Table 4.4.

The mechanics of dissolution of LNAPL into water are directly related to surface area contact and the time of contact. Longer contact times raise concentrations (possibly to saturation concentration). Conversely, large droplets of LNAPL dispersed throughout the upper aquifer, with rapidly moving water, may result in much less concentration, even though large quantities of product are present. Based on this discussion, dissolved concentrations in monitoring wells may (or may not) be useful in interpreting the presence of LNAPLs.

During the later phases of remediation of an aquifer, it has often been suggested that the expense of continuing LNAPL recovery is not worth the effort, considering the remaining small quantities of recoverable product. The implication of this argument has been that the product will eventually go into solution and therefore will be recovered as part of the dissolved contaminant recovery. Experience has demonstrated that this is not usually a viable option, because the time required and treatment costs to accomplish the task are much greater than with respect to LNAPL product recovery efforts.

REFERENCES

Abdul, S. A., 1988, Migration of Petroleum Products through Sandy Hydrogeologic Systems: *Ground Water Monitoring Review,* Fall Issue, Vol. VIII, No. 4, pp. 73–81.

API (American Petroleum Institute), 1972, *The Migration of Petroleum Products in Soil and Groundwater, Principles and Counter-Measures*: American Petroleum Institute, No. 4149.

API (American Petroleum Institute), 1996, *Underground Spill Cleanup Manual*: American Petroleum Institute, No. 1628, 1st Edition.

Bear, J., 1972, *Dynamics of Fluids in Porous Media*: American Elsevier, New York, 764 pp.

Bear, J., *Hydraulics of Groundwater*: McGraw-Hill Book Company, New York, 567 pp.

Bossert, I., and Bartha, R., 1984, The Fate of Petroleum: In *Soil Ecosystems: Petroleum Microbiology* (edited by R. M. Atlas), Macmillan Co., New York.

Brewster, M. L., et al., 1995, Observed Migration of a Controlled DNAPL Release by Geophysical Methods: *Ground Water*, Vol. 33, No. 6, pp. 977–987.

Cherry, J. A., 1984, Groundwater Contamination: In *Proceedings of Mineralogical Association of Canada Short Course in Environmental Geochemistry*, May, 1984, pp. 269–306.

Cherry, J. A., Billham, R. W., and Barker, J. F., 1983, Contaminants in Groundwater: Chemical Processes: Groundwater Contamination: In *Studies in Geophysics*, National Academy Press, Washington, D.C., pp. 46–64.

Chiou, C. T., Peters, L. J., and Freed, V. H., 1979, A Physical Concept of Soil-Water Equilibria for Nonionic Organic Compounds: *Science,* Vol. 206, pp. 831–832.

CONCAWE Secretariat, 1974, *Inland Oil Spill Clean-up Manual*: Report No. 4/74, The Hague, the Netherlands.

CONCAWE, 1979, *Protection of Groundwater from Oil Pollution*: The Hague, the Netherlands, NTIS PB82-174608.

CONCAWE, Water Pollution Special Task Force No. 11 (T. L. de Pastrovitch, Y. Baradat, R. Barthel, A. Chiarelli, and D. R. Fussell), 1979, *Protection of Groundwater from Oil Pollution*: The Hague, the Netherlands.

Corapcioglu, M. Y. and Baehr, A. L., 1985, Immiscible Contaminant Transport in Soils and Groundwater with an Emphasis on Gasoline Hydrocarbons System of Differential Equations vs. Single Cell Model: *Water Science and Technology*, Vol. 17, No. 9, pp. 23–37.

Corapcioglu, M. Y. and Baehr, A. L., 1987, A Compositional Multiphase Model for Groundwater Contamination by Petroleum Products, 1. Theoretical Considerations: *Water Resources Research*, Vol. 23, No. 1, pp. 191–200.

Dietz, D. N., 1970, Pollution of Permeable Strata by Oil Components: In *Water Pollution by Oil* (edited by P. Hepple), Elsevier Publishing Co. and the Institute of Petroleum, New York.

Dragun, J., 1988, *The Soil Chemistry of Hazardous Materials:* Hazardous Materials Control Research Institute, Silver Springs, MD, 458 pp.

EPA, 1993, *Behavior and Determination of Volatile Organic Compounds in Soil: A Literature Review*: Office of Research and Development, EPA/600/R-93/140.

Fan, C.-Y., and Krishnamurthy, S., 1995, Enzymes for Enhancing Bioremediation of Petroleum-Contaminated Soils: A Brief Review: *Air and Waste Management Association*, Vol. 45, June, pp. 453–460.

Farmer, V. E., Jr., 1983, Behavior of Petroleum Contaminants in an Underground Environment: In *Proceedings of the Canadian Environment Seminar on Ground Water and Petroleum Hydrocarbons*, June 26–28, Toronto, Ontario.

Faust, C. R., Guswa, J. H., and Mercer, J. W., 1989, Simulation of Three Dimensional Flow of Immiscible Fluids within and below the Unsaturated Zone: *Water Resources Research*, Vol. 25, No. 12, pp. 2449–2464.

Goodwin, M. J. and Gillham, R. W., 1982, Two Devices for in Situ Measurements of Geotechnical Retardation Factors: In *Proceedings of the Second International Hydrogeological Conference* (edited by G. Ozoray), International Association of Hydrogeologists, Canadian National Chapter, pp. 91–98.

Hendricks, D. W., Post, J. F., and Khairnar, D. R., 1979, Adsorption of Bacteria on Soils: Experiments, Thermodynamic Rationale and Application: *Water, Air, Soil Pollution,* No. 12, pp. 219–232.

Hoag, G. E., Bruell, C. J., and Marley, M. C., 1984, *Study of the Mechanisms Controlling Gasoline Hydrocarbon Partitioning and Transport in Groundwater Systems*: Institute of Water Resources, University of Connecticut, U.S. Department of the Interior Research Project No. G832–066, 51 pp.

Hounslow, A. W., 1995, *Water Quality Data: Analysis and Interpretation*: CRC Press/Lewis Publishers, Boca Raton, FL, 397 pp.

Hunt, J. R., Sitar, N., and Udell, J., 1988, Nonaqueous Phase Liquid Transport and Cleanup, Part 1. Analysis of Mechanisms: *Water Resources Research.,* Vol. 24, No. 8, pp. 1247–1258.

Huyakorn, P. S., Wu, Y. S., and Panday, S., 1992, A Comprehensive Three-Dimensional Numerical Model for Predicting the Transport and Fate of Petroleum Hydrocarbons in the Subsurface: In *Proceedings of the Petroleum Hydrocarbons and Organic Chemicals in Groundwater Conference*, Water Well Journal Publishing Company, Dublin, OH, pp. 239–253.

Huyakorn, P. S., Panday, S., and Wu, Y. S., 1994, A Three-Dimensional Multiphase Flow Model for Assessing NAPL Contamination in Porous and Fractured Media, I, Formulation: *Journal of Contaminant Hydrology*, Vol. 16, pp. 109–130.

Jackson, D., Payne, T. H., Looney, B. B., and Rossabi, J., 1996, *Estimating the Extent and Thickness of DNAPL within the A/M Area of the Savannah River Site*: (U) WSRC-RP-96-0574, Westinghouse Savannah River Company.

Karickhoff, S. W., Brown, S. D., and Scott, T. A., 1979, Sorption of Hydrophobic Pollutants on Natural Sediments: *Water Resources Research*, Vol. 13, pp. 241–248.

Keuper, B. H., and Frind, E. O., 1988, An Overview of Immiscible Fingering in Porous Media: *Journal of Contaminant Hydrol*ogy, Vol. 2, pp. 95–110.

Keuper, B. H., Abbott, W., and Farquhar, G., 1989, Experimental Observations of Multiphase Flow in Heterogeneous Porous Media: *Journal of Contaminant Hydrology,* Vol. 5., pp. 83–95.

Lance, M. D. and Berba, C. P., 1984, Virus Movement in Soil during Saturated and Unsaturated Flow: *Applied Environmental Microbiology,* No. 47, pp. 335–337.

Lee, M. D., Thomas, J. M., Borden, R. C., et al., 1988, Biorestoration of Aquifers Contaminated with Organic Compounds: *CRC Critical Review Environmental Control*, Vol. 18, Issue 1, pp. 29–89.

Lenhard, R. J. and Parker, J. C., 1987, Measurement and Prediction of Saturation-Pressure Relationships in Three-Phase Porous Media Systems: *Journal of Contaminant Hydrology,* Vol. 1, pp. 407–424, Correction, Vol. 2, pp. 95–110.

Lenhard, R. J. and Parker, J. C., 1988, Experimental Validation of the Theory of Extending Two-Phase Saturation-Pressure Relationships to Three-Fluid Phase System for Monotonic Drainage Paths: *Water Resources Research*, Vol. 24, pp. 373–380.

Leo, A., Hansch, C., and Elkins, D., 1971, Partition Coefficients and Their Uses: *Chemical Review,* No. 71, pp. 525–616.

Leverett, M. C., 1939, Flow of Oil-Water Mixtures through Unconsolidated Sands: *Transactions of American Institute of Mining and Metallurgical Engineers*, Vol. 142, Petroleum Development Technology, pp. 149–171.

Love, D. W., Whitworth, T. M., Davis, J. M., and Seager, W. R., 1999, Free-Phase NAPL-Trapping Features in Intermontane Basins: *Environmental & Engineering Geoscience*, Vol. V, No. 1, pp. 87–102.

Lyman, W. J., Reidy, P. J., and Levy, B., 1992, Mobility and Degradation of Organic Contaminants in Subsurface Environments: C. K. Smoley, Inc., Chelsea, MI, 395 pp.

MacKay, D. M., Cherry, J. A., Fryberg, D. L., et al., 1983, Implementation of a Field Experiment on Groundwater Transport of Organic Solutes: In *Proceedings of the National Conference on Environmental Engineering,* ASCE, University of Colorado, Boulder, July.

Maguire, T. F., 1988 Transport of a Non-Aqueous Phase Liquid within a Combined Perched and Water Table Aquifer System: In *Proceedings of the National Water Well Association of Ground Water Scientists and Engineers FOCUS Conference on Eastern Regional Ground Water Issues,* September, 1988.

Martin, J. P. and Koener, R. M., 1984a, The Influence of Vadose Zone Conditions on Groundwater Pollution, Part I: Basic Principles and Static Conditions: *Journal of Hazardous Materials,* No. 8, pp. 349–366.

Martin, J. P. and Koener, R. M., 1984b, The Influence of Vadose Zone Conditions on Groundwater Pollution, Part II, Fluid Movement: *Journal of Hazardous Materials.,* No. 9, pp. 181–207.

Melrose, J. C., 1965, Wettability as Related to Capillary Action in Porous Media: *Society of Petroleum Engineers Journal*, Vol. 5, pp. 259–271.

Mercer, J. W. and Cohen, R. M., 1990, A Review of Immiscible Fluids in the Subsurface — Properties, Models, Characterization and Remediation: *Journal of Contaminant Hydrology*, Vol. 6, pp. 107–163.

Morrow, N. R., 1990, Wettability and Its Effect on Oil Recovery: *Journal of Petroleum Technology*, Vol. 29, pp. 1476–1484.

Newsom, J. M., 1985, Transport of Organic Compounds Dissolved in Ground Water: *Ground Water Monitoring Review*, Spring, Vol. 5, No. 2, pp. 28–36.

Panday, S. and Corapcioglu, M. Y., 1994, Theory of Phase-Separate Multi-Component Contamination in Frozen Soil: *Journal of Contaminant Hydrology*, Vol. 16, pp. 235–269.

Panday, S., Forsyth, P. A., Falta, R. W., Wu, Y., and Huyakorn, P. S., 1995, Considerations for Robust Compositional Simulations of Subsurface Nonaqueous Phase Liquid Contamination and Remediation: *Water Resources Research*, Vol. 31, No. 5, pp. 1273–1289.

Pfannkuch, H. O., 1984, Determination of the Contamination Source Strength from Mass Exchange Processes at the Petroleum-Groundwater Interface in Shallow Aquifer Systems: In *Proceedings of the National Water Well Association and American Petroleum Institute Conference on Petroleum Hydrocarbons and Organic Chemicals in Groundwater — Prevention, Detection and Restoration,* November, Houston, TX.

Raisbeck, J. M. and Mohtadi, F. M., 1974, The Environmental Impacts of Oil Spills on Land in the Arctic Regions: *Water, Air, Soil Pollution*, No. 3, pp. 195–208.

Rau, B. V., Vass, T., and Stachle, W. J., 1992, DNAPL: Implications to Investigation and Remediation of Groundwater Contamination: In *Proceedings of the Hazardous Materials Control Research Institute, HMC-South* '92, New Orleans, LA, pp. 84–94.

Regens, J. L., Hodges, D. G., et al., 1999, Screening Technologies for Soil and Groundwater Remediation: *Soil and Groundwater*, August/September, pp. 24–27.

Roberts, P. V., McCarty, P. L., Reinhard, M., and Schreiner, J., 1980, Organic Contaminant Behavior during Groundwater Recharge: *Journal of Water Pollution Control,* Vol. 2, pp. 161–172.

Sawyer, C. N., 1978, *Chemistry for Environmental Engineering Basic Concepts from Physical Chemistry:* McGraw-Hill Book Company, New York.

Schwartzenbach, R. P. and Westhall, H., 1981, Transport of Nonpolar Organic Compounds from Surface Water to Groundwater, Laboratory Sorption Studies: *Environmental Science and Technology*, Vol. 15, pp. 1350–1367.

Schwille, F., 1984, Migration of Organic Fluids Immiscible with Water in the Unsaturated Zone: Pollutants in Porous Media: In *Ecological Studies* (edited by B. Yarn, G. Dagan, and J. Goldshmidt), Springer-Verlag, New York, pp. 27–48.

Schwille, F., 1985, Migration of Organic Fluids Immiscible with Water in the Unsaturated and Saturated Zones: In *Proceedings of the National Water Well Association Second Canadian/American Conference on Hydrogeology,* Banff, Alberta, Canada, June, pp. 31–35.

Sitar, N., Hunt, J. R., and Udell, K. S., 1987, Movement of Nonaqueous Liquids in Groundwater: In *Proceedings of Geotechnical Practice for Waste Disposal*, ASCE, Ann Arbor, MI, pp. 205–223.

Slough, K. J., Sudicky, E. A., and Forsyth, P. A., 1999, Importance of Rock Matrix Entry Pressure on DNAPL Migration in Fractured Geologic Material: *Ground Water*, Vol. 37, No. 2, pp. 237–244.

Stone, H. L., 1973, Estimation of Three-Phase Relative Permeability and Residual Oil Data: *Journal of Canadian Petroleum Technology*, Vol. 12, No. 4, pp. 53–61.

Thomas, J. M., Clark, G. L., et al., 1988, *Environmental Fate and Attenuation of Gasoline Components in the Subsurface*: Rice University, Department of Environmental Science and Engineering, Houston, TX, 111 pp.

U.S. Geological Survey, 1984, Groundwater Contamination by Crude Oil at the Bemidje, Minnesota, Research Site, U.S.G.S. Toxic Waste-Groundwater Study: *Water Resources Investigation Report,* No. 84-4188.

U.S. Geological Survey, 1996, *Effectively Recover Free Product at Leaking Underground Storage Tank Sites*: U.S. EPA Office of Underground Storage Tanks, OSWER, September, 1996.

Vance, D. B., 1998, Redox Reactions for in-Situ Groundwater Remediation: *Environmental Technology*, September/October, p. 45.

Van Dam, J., 1967, The Migration of Hydrocarbons in a Water Bearing Stratum: In *Joint Problems of the Oil and Water Industries* (edited by P. Hepple), Institute of Petroleum, London.

Van Duijvenbooden, W. and Kooper, W. F., 1981, Effects on Groundwater Flow and Groundwater Quality of a Waste Disposal Site in Noordwijk, the Netherlands: *Science Total Environment,* No. 21, pp. 85–92.

Vanloocke, R., DeBorger, R., Voets, J. P., and Verstraete, W., 1975, Soil and Groundwater Contamination by Oil Spills; Problems and Remedies: *International Journal of Environmental Studies,* No. 8, pp. 99–111.

Villume, J. G. 1985, Investigations at Sites Contaminated with Dense Non-Aqueous Phase Liquids (NAPLs): *Ground Water Monitoring Review,* Spring, Vol. 5, No. 2, pp. 60–74.

Vroblesky, C. D., Robertson, J. F., and Rhodes, L. C., 1995, Stratigraphic Trapping of Spilled Jet Fuel Beneath the Water Table: *Groundwater Monitoring Review*, Spring, pp. 177–183.

Whitworth, T. M. and Hsu, C. C., 1999, The Role of Sand/Shale Interfaces in Saturated Zone NAPL Trapping: *Environmental Geosciences*, Vol. 6, No. 1, pp. 25–34.

Yang, W. P., 1981, Volatilization, Leaching and Degradation of Petroleum Oils in Sand and Soil Systems: Ph.D. thesis, Department of Civil Engineering, North Carolina State University.

6 NAPL Subsurface Characterization

"When in doubt, measure it; when not in doubt, measure it anyway."

6.1 INTRODUCTION

Groundwater contamination as a result of the subsurface presence of nonaqueous phase liquids (NAPLs) is ubiquitous in today's society. Sources of NAPLs include the release of crude oil and refined petroleum-related products from aboveground and underground storage tanks (USTs), pipeline corridors, dry wells, and accidental spills. Since the 1980s, much focus has been placed upon USTs, although over the past few years emphasis has shifted to other industries and operations such as petroleum refining, bulk liquid storage terminals, major pipeline networks, gas production, steel industry, and coking and wood treating. NAPLs are referred to in the federal regulations as "free product," and as previously discussed can occur in the subsurface in two forms: lighter than water (LNAPLs) and denser than water (DNAPLs) (Figure 5.1). As NAPLS migrate vertically downward through the soil column and groundwater is encountered, an LNAPL will tend to form a pool overlying the capillary fringe and water table. DNAPLs behave much the same as LNAPLs in the vadose zone; however, once groundwater is encountered, DNAPLs will tend to continue to migrate vertically downward through the water column until a significant permeability contrast is encountered, providing a sufficient volume of DNAPL was released. The subsurface presence of NAPLs for the most part can range from essentially unnoticed small releases over very long periods of time to episodic releases due to a significant breach of a storage or transportation unit.

Free-phase NAPL refers to NAPL that exists as an independent phase, not as a dissolved component in the pore water or pore atmosphere. The environmental concerns associated with sites affected with free-phase NAPLs revolve around hydrocarbon-impacted soil (residual hydrocarbon), the NAPL itself (which can serve as a continued source for groundwater contamination), dissolved hydrocarbon constituents in groundwater, and hydrocarbon vapors. The detection of free-phase NAPLs in the subsurface presents many challenges. Two questions frequently arise at sites impacted by NAPLs: how much is there and how long will it take to clean up. Before one can address these two questions, assessments of the type and subsurface distribution of free-phase NAPL, realistic prospects of its recoverability, and the time frame required for recovery of what is recoverable are necessary.

A variety of methods and approaches have been developed over the past decade to address these issues. This chapter takes a look at these methods, and focuses on the occurrence, detection, monitoring, and overall characterization of the subsurface lateral and vertical extent of free-phase NAPLs. Problems and limitations associated with the measurement and determination of apparent vs. actual thickness, empirical and field methods, volume determinations, recoverability, and time frame for recovery are discussed.

6.2 FIELD METHODS FOR SUBSURFACE NAPL DETECTION

6.2.1 Monitoring Well Installation and Design

Subsurface geologic and hydrogeologic conditions, as well as the subsurface presence of hydrocarbons, can be directly determined by the drilling of borings and subsequent installation and construction of monitoring wells. Several techniques are available for the drilling and installation of wells regardless of whether their eventual use will be for monitoring, gauging, delineation, injection, or recovery purposes. A summary of these techniques is provided in Table 6.1. A typical monitoring well construction detail is shown in Figure 6.1.

Although well construction details for monitoring and recovery of NAPL are similar to that of conventional monitoring wells, several factors need to be emphasized. Obviously, the well screen must overlap the mobile hydrocarbon interval and be of sufficient length to account for seasonal fluctuations or changes due to recovery or reinjection influences. Filter pack design can also have a bearing on whether hydrocarbon presence is detected or confirmed. Filter packs must be designed to allow not only mobile hydrocarbon but also capillary hydrocarbon to migrate into the well. Otherwise, since hydrocarbon in the formation can exist at less than atmospheric pressure, a poorly designed filter pack can result in capillary hydrocarbon being unable to migrate into the well. The end result is a much broader areal extent of subsurface hydrocarbon than is being accounted for. Third, well design and construction details among a network of wells, including the filter pack design and developing procedures, should remain consistent. With recovery wells, too coarse a filter pack will minimize the ability of the well to attract capillary hydrocarbon to the well. For example, a typical hydrocarbon product with a density of 0.8 g/cm^3 and an interfacial tension with air of 30 dyn/cm will accumulate to a thickness of approximately 25 to 33 cm in a fine sand before exceeding atmospheric pressure. A part of this capillary hydrocarbon is recoverable with a properly designed finer-grained filter pack.

6.2.2 NAPL Detection Methods

A couple of conventional methods are available to measure the apparent LNAPL product thickness in a monitoring well. The apparent thickness of LNAPL product in a well is typically determined using either a steel tape with water-and-oil-finding

TABLE 6.1
Drilling Techniques for the Construction and Installation of Monitoring, Recovery, and Injection Wells

Drilling Technique	Geologic Material[a]	Depth Limitations (ft)	Well Type[b]	Remarks
Geoprobe	U	50	M	Geologic and hydrogeologic characterization only; excellent sampling and analytical capabilities (soil, water, or vapor); 1-in.-diameter well screen capability; accessibility excellent
Hand-augered	U	15	M,R	Accurate sampling; difficult in coarse sediments or loose sand; physically demanding; fluid levels easily detected; borehole variable due to friction of auger; depth limited; inexpensive
Driven	U	25	M	No sampling capability; quick and easy method to detect and monitor shallow fluid levels
Hollow-stem auger	U	180	M,R,I	Accurate sampling; continuous sampling available; diameter limitations; fluid levels easily detected; no drilling fluids required; smearing of borehole walls in fine-grained soils and sediments causing sealing
Jet	U	200	M	Diameter limitations; fluid level (water and NAPL) difficult; sampling accuracy limited; produced fluids require handling (hazardous if NAPL is encountered)
Bucket auger	U	100	R,I	Sampling of borehole wall samples easy; can install large-diameter well; difficult to control caving
Cable tool	U	1000	M,R,I	Satisfactory sampling; fluid levels easily detected; drilling can be slow
Hydraulic rotary	U or C	2500+	M,R,I	Fast; retrieval of accurate samples requires special attention; knowledge of drilling fluids used to minimize plugging of certain formations is critical; good for recovery and injection well construction; continuous coring available; produced fluids require handling (hazardous if NAPL is encountered)
Reverse circulation	U or C	2000+	M,R,I	Formation relatively undisturbed compared with other methods; large-diameter boreholes can be drilled; no drilling mud usually required because of hydraulics associated with this method; good for recovery and injection well construction; produced fluids require handling (hazardous if NAPL is encountered)
Air rotary	U or C	2000+	M,R,I	Fast; cuttings removal rapid; poor sample quality; diameter limitations; formation not plugged with drilling fluids; dangerous with flammable fluids
Air percussion	U or C	2000+	M,R,I	Fast; cuttings removal rapid; good in consolidated formations

[a] U = unconsolidated; C = consolidated.
[b] M = monitoring well; R = recovery well; I = injection well.

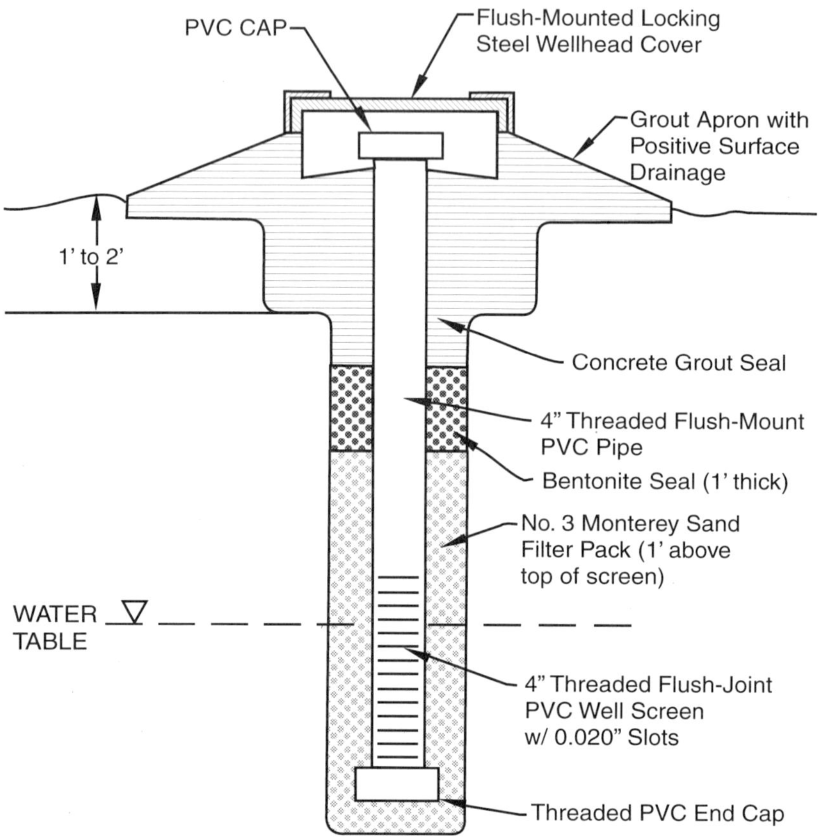

FIGURE 6.1 Typical well construction detail.

paste or a commercially available electronic resistivity probe (Figure 6.2). Either method can provide data with an accuracy to 0.01 ft. However, if the LNAPL product is emulsified or highly viscous, significant error can result. In addition, measurement using electronic resistivity probes can be misleading if the battery source is weak.

The tape-and-paste method involves lowering a weighted steel measuring tape on which hydrocarbon and water-sensitive pastes have been applied into a monitoring well. As the water and liquid hydrocarbons contact the pastes, color changes occur. The area of sharply contrasting colors made on the paste is called the "water cut." The water cut delineates the interface between the floating hydrocarbons and the uppermost surface of groundwater in the well. The top surface of the liquid hydrocarbon is determined by locating the top of the characteristic oily film on the steel tape. The distance between these two markers is recorded from the measuring tape, and the resultant length is equivalent to the apparent thickness of liquid hydrocarbon present in the well.

The interface gauging probe incorporates a measuring tape on a reel, connected to an electronic sensor head. The sensor head contains a float-ball and magnetic relay switch assembly that distinguishes between air and fluids; also contained in

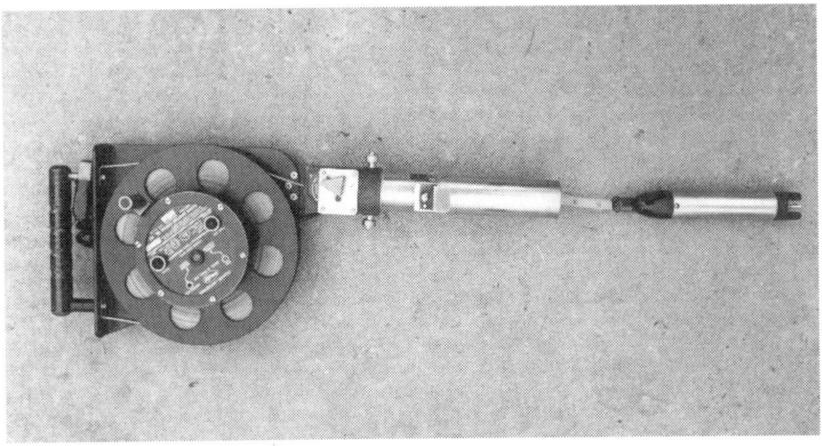

FIGURE 6.2 Photograph of oil–water interface probe.

the sensor head is an electrical conductivity sensor. This sensor consists of two electrodes, between which a small electrical current is passed. This electrical conductivity sensor distinguishes between nonconductive fluids (hydrocarbons) and conductive fluids (groundwater).

The sensor head is lowered into a monitoring well. Upon contact with any fluid, the float ball is raised and a continuous tone emitted from an audible alarm. When the sensor head contacts the interface between LNAPL and groundwater, the change in conductive properties is detected by the electrical conductivity sensor and a beeping tone is emitted. The distances along the tape at which the two changes in the audible alarm occur are recorded as referenced from a presurveyed point on the lip of the monitoring well. The resultant distance is equivalent to the apparent thickness of the LNAPL in the well.

6.3 APPARENT VS. ACTUAL NAPL THICKNESS

6.3.1 LNAPL Apparent vs. Actual Thickness

The subsurface presence of LNAPL can occur under both perched and water table conditions (see Figure 5.1). In addition, occurrence can exist under unconfined and occasionally confined conditions. LNAPL product in the subsurface is typically delineated and measured by the utilization of groundwater monitoring wells. While monitoring wells have provided some insight into the extent and general geometry of the plume, as well as the direction of groundwater flow, difficulties persist in determining actual NAPL thickness and, therefore, the volume and overall NAPL distribution in the subsurface, and ultimately the duration of recovery and aquifer restoration.

One of the more difficult aspects in dealing with the subsurface presence of hydrocarbons, as a practical matter, is that NAPL accumulations in monitoring wells do not directly correspond to the actual thickness in the formation. The thickness of LNAPL product as measured in a monitoring well is an apparent thickness that is not reflective of the actual formation thickness. This relationship can be demon-

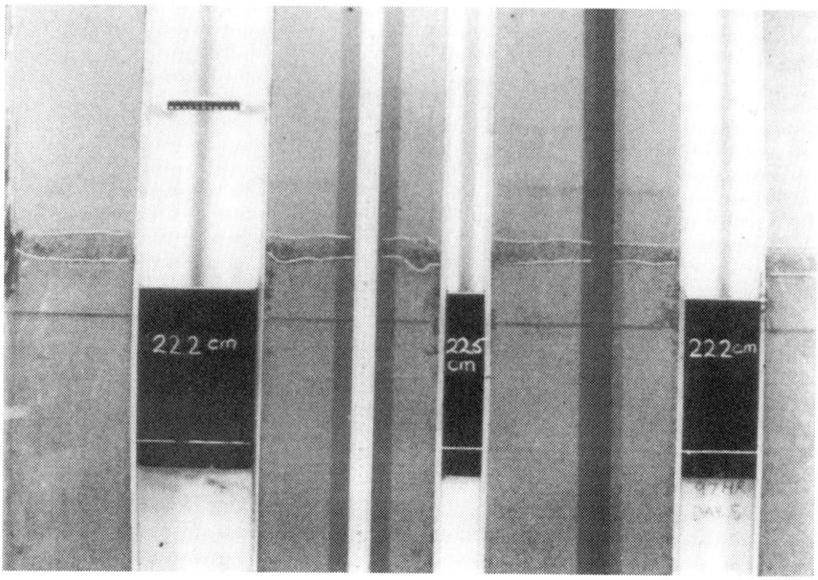

FIGURE 6.3 Sandbox model showing LNAPL overlying capillary fringe, and apparent vs. actual LNAPL thickness. Saturated conditions (water table) are represented by the straight horizontal line.

strated using a sandbox model (Figure 6.3). What is important from a practical perspective is not the actual thickness of LNAPL in the formation, but rather the actual thickness of LNAPL that is considered mobile. The distribution of the various phases that are present in a porous media, notably, water, air, and LNAPL, is illustrated in Figure 6.4. In Figure 6.4, the continuous pore volume is shown to be occupied by one of three fluids:

1. A two-phase zone containing water and air;
2. A three-phase zone containing water, LNAPL and air;
3. A two-phase zone containing water and LNAPL; and
4. A one-phase zone containing solely water.

The formation of a distinct LNAPL layer floating on top of the capillary fringe as illustrated in Figure 6.5, however, violates the fundamental equations that describe the fluid pressure distributions in a porous media and also in the monitoring well under conditions of mechanical equilibrium. This accounts in part for the poor LNAPL yields from spill sites. Thus, the presence of LNAPL overlying the capillary fringe as shown in Figure 6.5 and observed in the laboratory (Figure 6.3) may reflect entrapped air plus water, which over time would be released from this zone providing continuous pore space or connectivity. Excluding the depression of the water level (oil–water interface) in the well, the measured LNAPL thickness in the well also represents an exaggerated volume since the actual hydrocarbon saturation in the formation is at a higher elevation relative to that measured in the well, and the pore spaces are occupied by phases other than LNAPL (i.e., soil particles, air, and water).

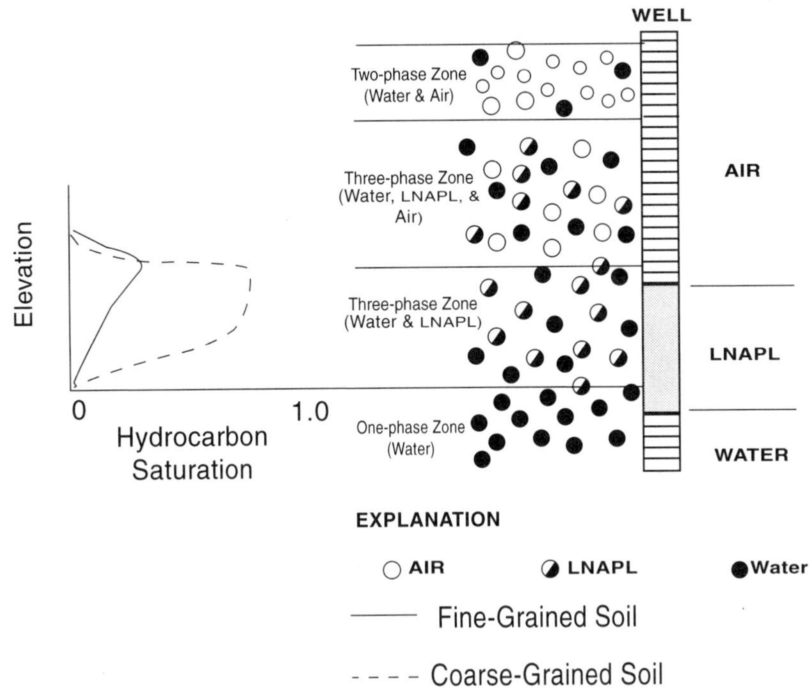

FIGURE 6.4 Conceptual representation of distribution of air, LNAPL, and water in a porous media. (After Farr et al., 1990.)

Since hydrocarbon and water are immiscible fluids, free-phase recoverable LNAPL can simplistically be viewed as being perched on the capillary fringe above the actual water table with the understanding that what is being referred to as actual NAPL thickness is what one could perceive as being equivalent to the approximate thickness of that portion of the zone of hydrocarbon saturation that is considered mobile. The physical relationships that exist are illustrated in Figure 6.5. This discrepancy can be a result of one or a combination of factors or phenomena. Some of the more common factors or phenomena are schematically shown in Figure 6.6 and include:

- Grain size differences reflected in varying heights of the capillary fringe;
- Actual formation thickness of the mobile free hydrocarbon;
- Height of perching layers, if present;
- Seasonal or induced fluctuations in the level of the water table;
- Product types and respective specific gravities; and
- Confining conditions.

The capillary fringe height is grain size dependent as summarized in Table 6.2 and shown in Figure 6.6a. As grain size decreases, the capillary height increases. Coarse-grained formations contain large pore spaces that greatly reduce the height of the capillary rise. Fine-grained formations have much smaller pore spaces which allow a greater capillary height.

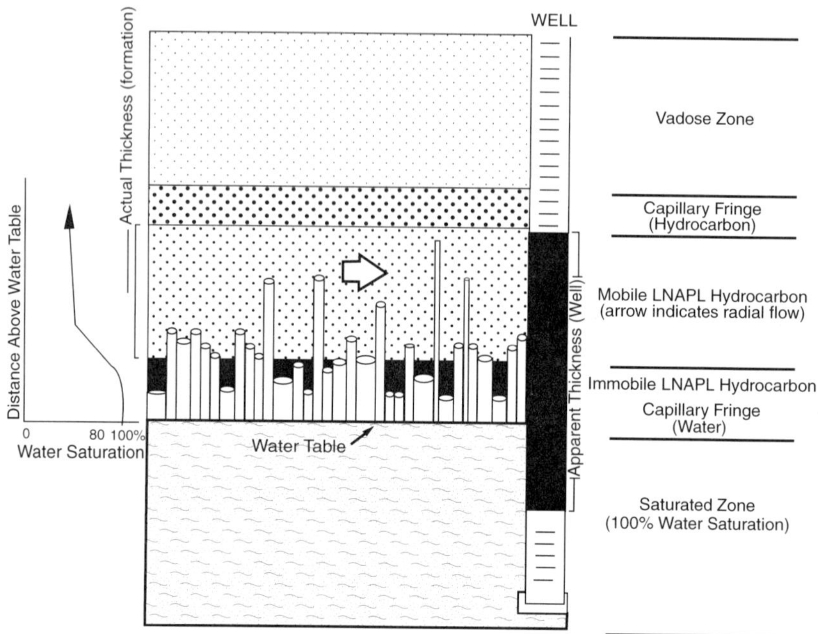

FIGURE 6.5 Schematic showing generalized relationship of apparent vs. actual LNAPL thickness in the well and adjacent formation.

Since LNAPL occurs within and above the capillary fringe, once the borehole or monitoring well penetrates and destroys this capillary fringe, free-phase LNAPL migrates into the well bore. The free water surface that stabilizes in the well will be lower than the top of the surrounding capillary fringe in the formation; thus, mobile hydrocarbons will flow into the well from this elevated position. LNAPL will continue to flow into the well and depress the water surface until a density equilibrium is established. To maintain equilibrium, the weight of the column of hydrocarbon will depress the water level in the well bore. Therefore, a greater apparent thickness is measured than actually exists in the formation. The measured or "apparent" LNAPL thickness is greater for fine-grained formations and less for coarser-grained formations, which may be more representative of the actual thickness.

The measured or "apparent" hydrocarbon thickness is not only dependent on the capillary fringe but also on the actual hydrocarbon thickness in the formation (Figure 6.6b). In areas of relatively thin LNAPL accumulations, the error between the apparent well thickness and actual formation thickness can be more pronounced than in areas of thicker accumulations. The larger error reflects the relative difference between the thin layer of LNAPL in the formation and the height it is perched above the water table. The perched height is constant for thick and thin accumulations; however, a thick accumulation can depress and even destroy the capillary fringe as illustrated in Figure 5.1.

The thickness measured in a monitoring well with LNAPL product situated on a perched layer at some elevation above the water table can produce an even larger

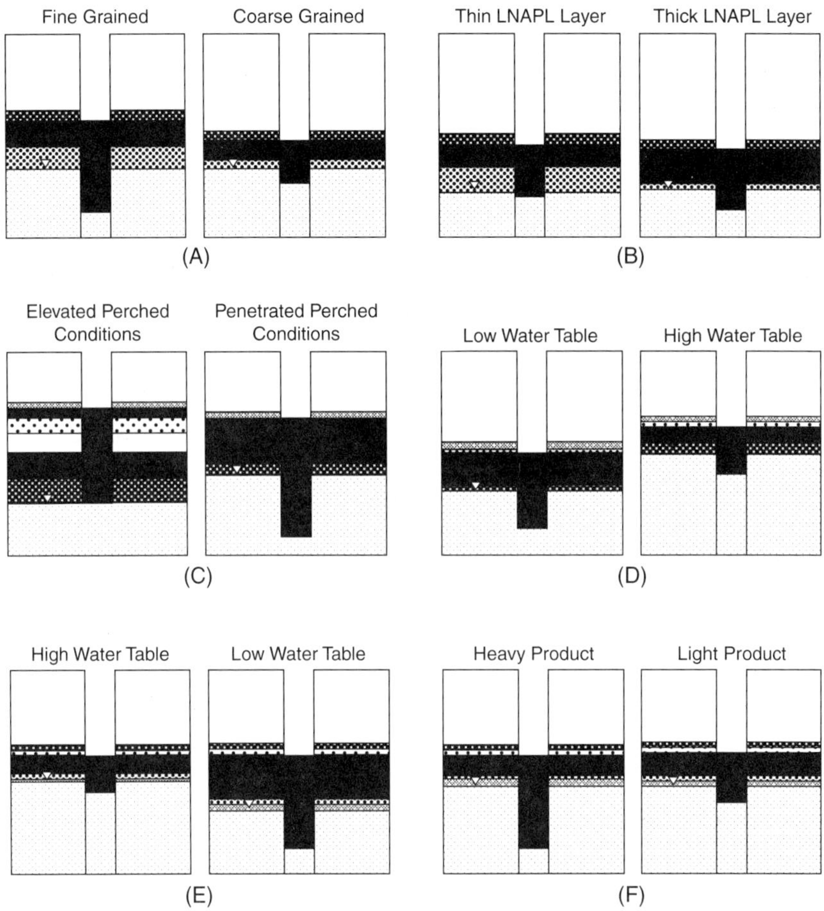

FIGURE 6.6 Cause for discrepancies in apparent LNAPL thickness as measures in a monitoring well in comparison with actual thickness in the formation.

associated thickness error (Figure 6.6c). This commonly occurs when the well penetrates the perched layer and is screened from the perching layer to the water table. LNAPL then flows into the well from the higher or perched elevation. The accumulated apparent thickness is a direct result of the difference of their respective heights. If a situation such as this exists, the difference in respective heights and weights of the column of hydrocarbon should be accounted for in determining the actual thickness.

Additionally, vertical fluctuations in the water table due to recovery operations or seasonal variation in precipitation have a direct effect upon the apparent or measured LNAPL thickness (Figure 6.6d). As the water table elevation declines gradually due to seasonal variations, for example, an exaggerated apparent thickness occurs, reflecting the additional hydrocarbon that accumulated in the monitoring well. The same is true for an area undergoing recovery operations where the

TABLE 6.2
Summary of Equations Relating Apparent LNAPL Thickness (H) to Actual Thickness (h)

Equation	Actual LNAPL Thckness as a Function of: Aquifer Properties	Apparent LNAPL Thickness (H)	Oil Density	Ref.
$h = Z_L - Z\mu$ if $S_o > 0.2$ for $Z_L > Z > Z_\mu$	Yes?	Yes?	Yes?	Van Dam (1967); OILEQUIL
$h_{max} = f$ (aquifer grain size)	Yes	No	No	Dietz (1971)
$h = H + \dfrac{P_c^{OA}}{(\rho_o - \rho_A)g} - \dfrac{P_c^{OW}}{(\rho_w - \rho_o)g}$	Yes	Yes	Yes	Zilliox and Muntzer (1975)
$h \cong H\left(\dfrac{\rho_w - \rho_o}{\rho_o}\right)\dfrac{P_c^{OA}}{P_c^{OW}} \cong \dfrac{H}{4}$	No	Yes	If adjusted	CONCAWE (1979)
$h = H - F$	Yes	Yes	No	Hall et al. (1984)
$h = H - 2\bar{h}_{c,dr} = H - \dfrac{2\lambda h_d}{\lambda - 1}$	Yes	Yes	No	Schiegg (1985)
$h = Z_L\ (S_o = 0.29) - Z_\mu\ (S_o = 0.09)$	Yes	Yes	Yes	Wilson et al. (1988); OILEQUIL
$h = \dfrac{1}{\bar{S}_o}\int_o^{Z_{ow}} S_o dz$	Yes	Yes	Yes	Parker and Lenhard (1987); QUILEQUIL

groundwater elevation is lowered through pumping; thicker apparent thicknesses may be observed.

The reverse of this effect has also been documented at recovery sites (Figure 6.6e). When sufficient recharge to the groundwater system through seasonal precipitation events or cessation of recovery well pumping occurs with the water table at a slightly higher elevation, thinner LNAPL thicknesses may be observed. During this situation, a compression of the capillary zone occurs, lessening the elevation difference between the water table and the free hydrocarbon, which reduces the apparent thickness.

Differences in product types, thus, API gravities, can account for variations in apparent thickness as measured in a monitoring well (Figure 6.6f). Heavier product types of relatively low API gravities will tend to depress the free hydrocarbon/water level interface more than would be anticipated from a product of higher API gravity.

In addition, monitoring wells screened across LNAPL within confined aquifers will exhibit an exaggerated thickness. This exaggerated thickness reflects relatively high confining pressure, which forces the relatively lower density fluid upward within the borehole. Thus, the measured thickness is a function of the hydrostatic head and not the capillary fringe, which has been destroyed by the confining pressures.

It is often assumed that apparent thicknesses are greater near the edge of an LNAPL pool and smaller toward the center. The relationship between apparent

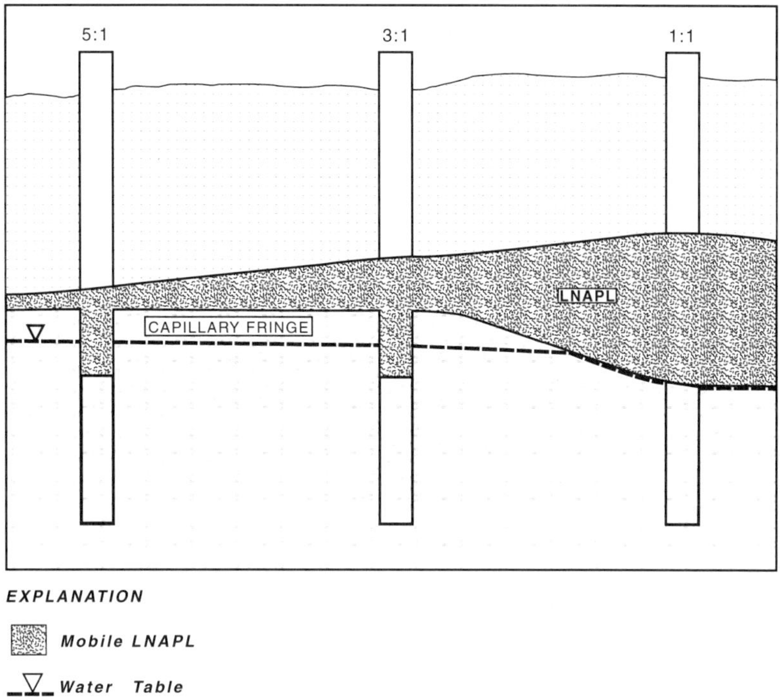

FIGURE 6.7 Cross section of an LNAPL pool showing well-to-formation-LNAPL thickness ratios.

thicknesses as measured in wells and formation thickness across an LNAPL pool is shown in Figure 6.7. Ratios are essentially smaller near the center of the pool, where the thickness of the LNAPL is sufficient to displace water from the capillary fringe. Conversely, where the LNAPL thickness is less, toward the edge of the pool, ratios increase.

Apparent/actual LNAPL thickness ratios can be very high at the perimeter of LNAPL pools, notably, under low permeability conditions. Once a well is installed, it can take several months before the LNAPL migrates from the formation into the well reflecting the presence of low-permeability soils in the zone of LNAPL occurrence. As clearly shown in Figure 6.8, a well screened at the perimeter of a known LNAPL pool initially had no detectable LNAPL until 4 months after installation, whereas upon detection the apparent thickness slowly increased with time up to 15.71 ft.

6.3.2 DNAPL Apparent vs. Actual Thickness

DNAPL once in the subsurface can occur in a variety of geologic scenarios making it difficult to fully assess the occurrences, as well as the lateral and vertical extent of the DNAPL (see Figure 5.2). DNAPL thickness when measured in a monitoring well, as with LNAPL, is not representative of the actual formation thickness. As with LNAPLs, the DNAPL thickness will be exaggerated as illustrated in Figure 6.9a. Limited laboratory studies have shown that the apparent DNAPL thickness is

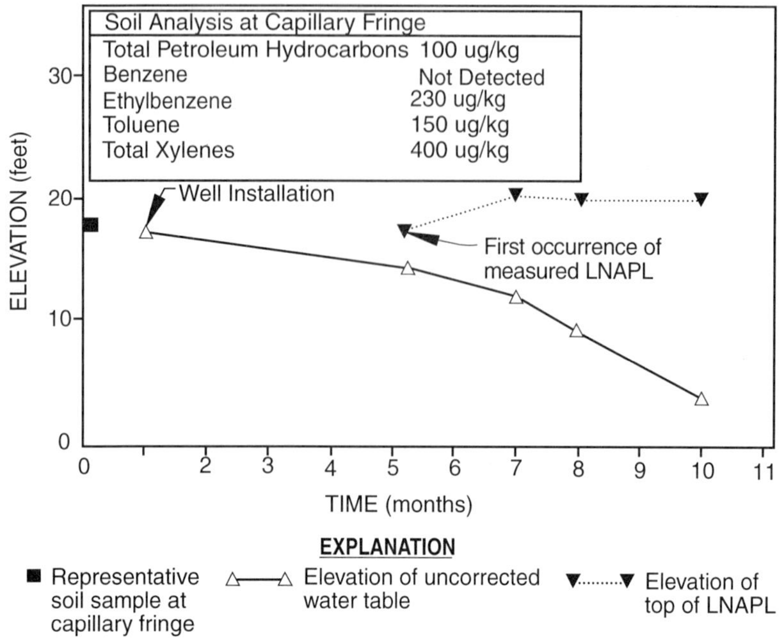

FIGURE 6.8 Graph showing gradual presence of LNAPL in well with time in low-permeability soil along perimeter of LNAPL pool. (After Testa, 1994.)

greater in fine sand than in coarse sand, reflecting variations in grain size. The difference in the apparent vs. actual DNAPL thickness equals the DNAPL-water capillary fringe height. DNAPL thickness in hydrophobic sand, however, is greater than what is measured in a well due to the DNAPL capillary rise. Exaggerated DNAPL thickness in wells can also be due to the well partially penetrating a lower permeability layer (Figure 6.9b) or the presence of thin, low-permeability laminae (Figure 6.9c).

6.4 APPARENT VS. ACTUAL LNAPL THICKNESS DETERMINATION

Various approaches and techniques have been used for the determination of the actual thickness of LNAPL in the subsurface. These approaches essentially fall into two groups: indirect empirical and direct field approaches. Both of these are further discussed in the following subsections.

6.4.1 Indirect Empirical Approach

Several equations and empirical relationships have been formulated in an attempt to relate the apparent LNAPL thickness as measured in a monitoring well to that which actually exists in the adjacent formation. These indirect empirical approaches are summarized in Table 6.3 and discussed below.

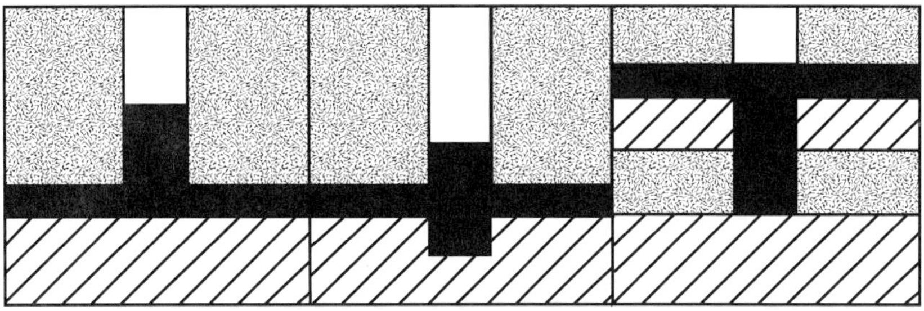

FIGURE 6.9 Causes for discrepancies in apparent DNAPL thickness as measures in a monitoring well in comparison with actual thickness in the formation.

TABLE 6.3
CONCAWE Factor Ranges for Various Petroleum Products

Similar Product Types	Product Density, r_o			CONCAWE Factor
	Relative Density	(°API)	(g/cm)³	$\frac{\rho_o}{\rho_w - \rho_o}$
Gasoline[a]	High	49.6	0.781	3.6
	Low	72.1	0.695	2.3
	Average	61.1	0.735	2.8
JP-4[b]	High	45.7	0.800	4.0
	Low	56.7	0.752	3.0
	Average	53.9	0.763	3.2
#1 Diesel	High	37.8	0.836	5.1
#1 Fuel Oil[c]	Low	47.9	0.790	3.8
Or kerosene	Average	42.4	0.814	4.4
Jet A[b]	High	37.4	0.838	5.2
(JP-5)	Low	47.4	0.791	3.8
(Kerosene)	Average	42.5	0.813	4.4
#2 Diesel[c]	High	22.1	0.920	11.5
#2 Fuel Oil[d]	Low	44.5	0.804	4.1
Or diesel	Average	34.6	0.852	5.8

[a] Derived from NIPER motor gasoline surveys for summers of 1980 and 1981, and winters of 1980–81, 1981–82, and 1985–86.

[b] Derived from NIPER 1981 aviation turbine fuels survey.

[c] Derived from NIPER 1981 diesel fuel survey.

[d] Derived from NIPER heating oil surveys of 1981, 1982, and 1986.

Source: All data derived from various topical reports by the National Institute for Petroleum and Energy Research (NIPER), P. O. Box 2128, Bartlesville, OK 74005; modified after Gruse (1967) and Hampton (1989).

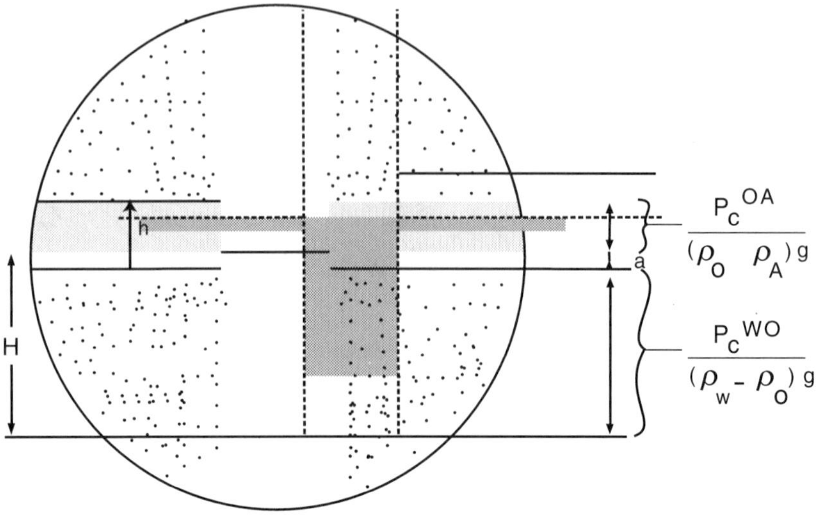

FIGURE 6.10 Oil thickness in a well and adjacent formation. (After CONCAWE, 1979.)

The theory and understanding required to deal quantitatively with spills was initially provided by Van Dam (1967), who also illustrated the physical processes responsible for product accumulation in wells and adjacent porous media. The relationship between actual and apparent thicknesses using a physical laboratory model was developed by Zilliox and Muntzer (1975), who proposed the following equation:

$$\Delta h = H - h = \frac{P_c^{\text{wo}}}{(\rho_{\text{w}} - \rho_{\text{o}})} - \frac{P_c^{\text{oA}}}{(\rho_{\text{o}} - \rho_{\text{A}})g} \tag{6.1}$$

where H = apparent LNAPL thickness, measured in well, h = average LNAPL thickness in soil near well, Δh = difference in thicknesses, P_c^{wo} = pressure difference (capillary pressure) between water and oil at their interface, P_c^{oA} = capillary pressure between oil and air, ρ_{w}, ρ_{o}, ρ_{A} = density of water, oil, and air, and g = gravitational acceleration.

These variables are illustrated in Figure 6.10. Zilliox and Muntzer (1975) also evaluated the effects of falling and rising water tables on Δh in their Plexiglas laboratory model. When the water table fell, the capillary pressures P_c^{wo} and P_c^{OA} were nearly equal, whereas Δh was positive at equilibrium. When the water table rose, they observed that P_c^{oA} varied little and P_c^{wo} decreased considerably. As a result, Δh decreased and could become negative. Hence, the actual thickness could be greater than the apparent thickness.

CONCAWE (1979) presented a modified version of Zilliox and Muntzer's equation as shown in Figure 6.11 and presented below:

$$\frac{H}{h} \approx \frac{H - a}{h - a} = \frac{P_c^{\text{wo}}}{P_c^{\text{oA}}} \frac{(\rho_o - \rho_a)g}{(\rho_w - \rho_o)g} = \frac{P_c^{\text{wo}}}{P_c^{\text{oA}}} \frac{\rho_o}{(\rho_\rho - \rho_\rho)} \approx 4 \frac{P_c^{\text{wo}}}{P_c^{\text{oA}}} \approx 4 \tag{6.2}$$

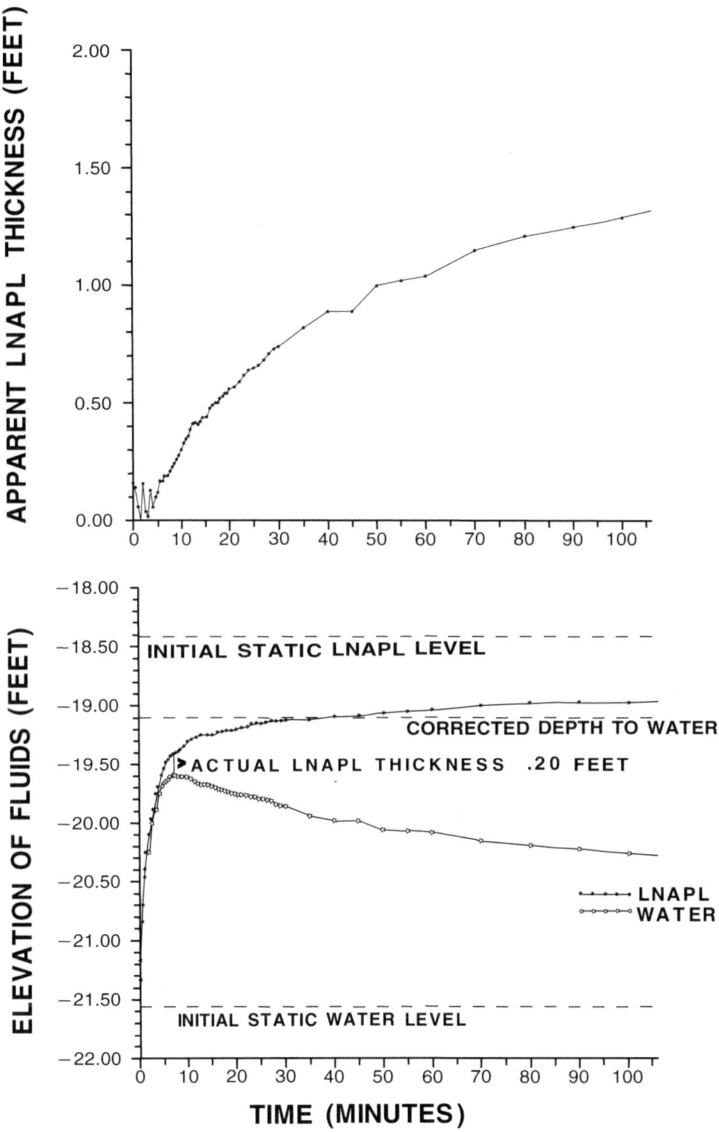

FIGURE 6.11 Representative baildown testing curve results using Gruszenski's (1987) method. Graphs of depth to product and depth to the product–water interface vs. time (a), and product thickness vs. time (b) are produced.

where h = actual LNAPL thickness including LNAPL fringe, and a = LNAPL-saturated portion of h.

In Equation 6.2, the factor 4 results from assuming that $\rho_o = 0.8\rho_w$. This ratio $\rho_o/(\rho_w - \rho_o)$, hereafter called the "CONCAWE factor," actually varies from 2 to 24 as ρ_o varies from $0.67\rho_w$ to $0.96\rho_w$ (Table 6.4). CONCAWE (1979) suggested that "more often than not $P_c^{wo} \approx P_c^{OA}$ from which follows that H may be roughly four

TABLE 6.4
Published Values for *H*/*h*

Spilled Product Type	Reported *H*/*h*	CONCAWE Factor Range	Ref.
Gasoline	2–3	2.3–3.5	Kramer (1982)
Gasoline	2.5–3	2.3–3.5	Yaniga (1982)
#2 Fuel oil	5–10	4.0–11.5	Yaniga and Demko (1983)

times *h*." According to Zilliox and Muntzer, this assumption of equal capillary pressures applies when the water table falls. If all of the CONCAWE assumptions are correct except $\rho_o = 0.8\rho_w$, then the CONCAWE factor should equal the ratio *H*/*h*.

Revisions to this well-known but poorly understood rule of thumb have been suggested. Kramer (1982), after citing CONCAWE (1979) in discussing a gasoline spill, pointed out that "field experiences have shown that the LNAPL thickness in monitoring wells represents two to three times the thickness in the formation." At another gasoline spill into fine glacial sands, Yaniga (1982) observed that the apparent thickness was 2½ to 3 times greater than the real thickness. For a #2 fuel oil spill at one particular site, Yaniga and Demko (1983) claimed that "apparent product thicknesses proved to be five to ten times greater than representative 'actual' values." These figures are actually consistent with Equation 6.2 if the CONCAWE factor of 4 is corrected for product density as in Table 6.4. The comparison is shown in Table 6.5.

Equation 6.2 has been disputed by several investigators. Shepherd (1983) claimed that apparent LNAPL thicknesses in wells may become more exaggerated than accounted for in Equation 6.2 in certain cases. Hall, Blake, and Champlin (1984) described convincing laboratory experiments that contradicted Equation 6.2. Their results can be formalized as follows:

$$h = H - F \tag{6.3}$$

where, h = actual LNAPL thickness, H = apparent LNAPL thickness, greater than minimum value (Table 6.6), and F = formation factor (Table 6.7). Note that F is a

TABLE 6.5
Information Required to Apply Equation 6.3

Classification[a]	Minimum Apparent Thickness, cm	Formation Factor (*F*), cm	Capillary Fringe Height,[b] cm	U.S. Standard Sieve Size Range
Coarse sand	8	5	2–5	#4 to #10
Medium sand	15	7.5	12–35	#10 to #40
Fine sand	23	12.5	35–70	#40 to #200

[a] Based on U.S. Bureau of Reclamation scale using medium grain size.
[b] From Bear (1979).

TABLE 6.6
Estimate of LNAPL from Aquifer Grain Size

Sand Size	Average Grain Diameter (mm)[a]	Corresponding Capillary Zone Thickness (cm)[b]
Extremely coarse to very coarse	2.0–0.5	1.8–9.0
Very coarse to moderately coarse	0.5–0.2	9.0–22.4
Moderately coarse to moderately fine	0.2–0.05	22.4–28.1
Moderately fine to very fine	0.05–0.015	28.1–45.0

[a] Maximum actual oil pancake thickness, h.
[b] Function of aquifer average grain size.

Source: From Dietz (1971).

TABLE 6.7
Direct Field Methods for Measuring Actual LNAPL Thickness

Method	Ref.
Bailer test	Yaniga (1982) Yaniga and Demko (1983)
Continuous core analysis	Yaniga (1984)
Continuous core analysis with ultraviolet light (natural fluorescence)	Hughes et al. (1988)
Test pit	Hughes et al. (1988)
Baildown test	Gruszczenki (1987), Hughes et al. (1988)
Recovery well recharge test	Hughes et al. (1988)
Dielectric well logging tool	Keech (1988)
Optoelectronic sensor	Kimberlin and Trimmell (1988)

function of grain size and that it follows the trend of capillary fringe height for different sand sizes. Equation 6.3 applies only to sands in which H exceeds a minimum value also related to capillary fringe height.

For cases where an aquifer is believed to be contaminated by a product spill but no apparent LNAPL thickness measurements are available, Dietz (1971) presented Table 6.7 relating the maximum expected product thickness to average aquifer grain diameter for various sands. He assumed that the maximum oil pancake thickness equals the capillary zone thickness. For each sand size range, he gave a corresponding range of maximum h values.

Exquisite laboratory experiments in this area have been developed by Schiegg (1985), who formulated another equation for actual thickness:

$$h = H - 2\overline{h}_{c,dr} \tag{6.4}$$

where $\bar{h}_{c,\mathrm{dr}}$ = mean capillary height of saturation curve for water and air during drainage
≈ height of the capillary fringe during drainage

To use Equation 6.4, one must measure or calculate $h_{c,\mathrm{dr}}$ for a given aquifer material. A mathematical formula for the water pressure–saturation curve during drainage is needed to calculate the average water pressure head. The Brooks–Corey model of the drainage curve can be used as follows (Corey, 1986):

$$S_e = \frac{S - S_r}{S_m - S_r} = \left\{\frac{h_d}{h_c}\right\}^1 \tag{6.5}$$

or

$$h_c = h_d S_e^{-1/\lambda}$$

where S_e = effective saturation, S = saturation = phase volume/pore volume, S_r = residual saturation, S_m = maximum field saturation, h_c = capillary pressure head, h_d = air-entry displacement head, and λ = pore-size distribution index.

To find the average capillary head, Equation 6.5 will be inserted into the definition of the average head:

$$\bar{h}_{c,\mathrm{dr}} = \frac{\int_0^1 h_c dS_e}{\int_0^1 dS_e} = \frac{h_d \int_0^1 S_e^{-1/\lambda} dS_e}{1} = \frac{h_d [S_e^{\lambda^{-1+1}}]\big|_0^1}{-\frac{1}{\lambda}+1} = \frac{h_d}{\frac{\lambda - 1}{\lambda}}(1-0) = \frac{\lambda h_d}{\lambda - 1} \tag{6.6}$$

Now that $\bar{h}_{c,\mathrm{dr}}$ is expressed solely in terms of the Brooks–Corey parameters for a given soil, Equation 6.4 becomes

$$h = H - 2\bar{h}_{c,\mathrm{dr}} = H - \frac{2\lambda h_d}{\lambda - 1} \tag{6.7}$$

Parker et al. (1987) have developed constitutive relations describing the various multiphase saturation curves. With these functions, one can then apply the quantitative analyses of Van Dam (1967) to a given soil.

Parker and Lenhard (1989) and Lenhard and Parker (1988) have developed equations that relate the apparent product thickness measured at a well under equilibrium conditions with the product and water saturations in a vertical column of soils adjacent to the well. By integrating the product saturation curve with respect to elevation, an equivalent depth of LNAPL-saturated pores is obtained. This process has been implemented in a computer program called OILEQUIL. The result is reported as a total oil depth in a vertical profile. The water and oil saturation curves with elevation can also be produced and printed in graphical or tabular form.

The results from OILEQUIL can be converted to an active layer thickness in several ways. First, the total thickness of the zone shown to contain product could be used. This would overestimate the thickness of the recoverable product layer, because oil at low saturations is immobile. Second, one could follow Van Dam (1967) in assuming a residual oil saturation of 20% (or another value appropriate to the soil and LNAPL involved) and pick out the thickness of the zone with greater than 20% product saturation from the OILEQUIL tabular output. In equation form,

$$h = Z_L - Z_u \quad \text{and} \quad S_o > 0.2 \quad \text{for} \quad Z_L > Z > Z_u \tag{6.8}$$

where h = actual LNAPL thickness, Z = vertical coordinate, positive downward, S_o = oil saturation, and Z_L = depth to lower boundary at $S_o = 0.2$, which Z_u = depth to upper boundary at which $S_o = 0.2$.

Another similar approach would be to follow Wilson et al. (1988) in using different residual saturations for the vadose and water-saturated zones. Wilson et al. (1988) found s_r values of 9 and 29% in the vadose and saturated zones, respectively. Hence, one could revise Equation 6.8 as follows:

$$h = Z_L(S_o = 0.29) - Z_u(S_o = 0.09) \tag{6.9}$$

This approach has promise, particularly if one can determine the residual product saturation in an independent laboratory experiment. This experiment would consist of acquiring an "undisturbed" sample of the aquifer just above the water table, wetting it with water and allowing it to drain for a few days, and saturating it with water and allowing it to drain for a few days, and saturating it with product and allowing it to drain for a week or two. Then, the LNAPL mass and dry soil bulk density would be determined via Freon extraction and soil drying, respectively. Using these quantities, one could calculate the residual product saturation.

Third, the total oil thickness (or thickness of saturated pores) determined by OILEQUIL could be converted to an active layer thickness using an assumed average oil saturation for the active product layer. One would divide the total oil thickness by the expected average saturation to get the actual thickness. Hence, a 2-cm total oil thickness would be equivalent to a 3-cm-thick active product layer with an average saturation of 66.7%. In equation form,

$$h = \frac{1}{\bar{S}_o} \int_0^{Z_{ow}} S_o dz \tag{6.10}$$

where $\bar{S}_o$ = average oil saturation of active layer, z = depth below land surface, Z_{ow} = oil–water interface depth in monitoring well, and o = land surface datum. Table 6.6 summarizes the equations presented above. Each incorporates various assumptions. These assumptions limit the ability of the equations to account for all of the necessary physical processes. Any equation that does not incorporate all of the important physical processes cannot work, in general. The equations can further be evaluated for the three major factors: measured aquifer properties, apparent thickness, and product density (see Table 6.2). The CONCAWE equation is the only one to omit use of aquifer material properties. Since no soil or rock characterization is required, this method is often used. However, one would not anticipate the same product thicknesses to be found in sand, silt, and clay aquifers for a given apparent thickness. The method of Parker and Lenhard (1989) appears to be the only one amenable to analysis that includes all three major components. Other physical components not shown in Table 6.2 include saturation history, hysteresis, and interfacial tensions.

Superimposed on the difficulty in estimating the actual thickness using LNAPL thickness measurements in monitoring wells is the influence of well diameter on such measurements. Chafee and Weimar (1983) reported that the gasoline thicknesses ranged from 6 to 30 in. in a small-diameter monitoring well and from 1 to 2 in. in a large-diameter well nearby. Similarly, Cummings and Twenter (1986) found 1.27 ft of JP-4 in a 3-in.-diameter well and 0.01 ft of fuel in a 6-in. well about 4 ft away. They deepened and pumped the 6-in. well, but did not significantly reduce fuel thickness in the 2-in. well. Mansure and Fouse (1984) explain these observations using thickness data from a monitoring well network that included wells at the edge of the gravel pack around large-diameter recovery wells. They convincingly demonstrated the difficulty with which fuel migrates from an aquifer into a large-diameter, gravel-packed well. During pumping, product accumulates outside of the gravel pack. Reviews of laboratory results have shown that none of these methods can be trusted without further validation. Although some have provided reasonable estimates, cross-validation and field validation is greatly insufficient.

6.4.2 Direct Field Approach

Several direct field approaches have been reported for the measurement of actual LNAPL product thickness. These approaches are summarized in Table 6.8 and discussed further below.

6.4.2.1 Bailer Test

The bailer test, although not typically used anymore, is a transparent bailer that is lowered into the well. A bailer is a cylindrical device with a check valve on the bottom and a hook for a cord on the top. The bailer must be long enough to ensure

TABLE 6.8
Factors Associated with Recovery of NAPL

Primary Factors	Secondary Factors
Formation permeability	Amount of residual hydrocarbons
Relative permeability[a]	Areal distribution of NAPL
NAPL viscosity	Site-specific constraints
Actual NAPL thickness	

[a] Based on percent water saturation.

that its top will be above the air/LNAPL interface when the check valve is below the LNAPL/water interface. The hydrocarbon thickness measured in a bailer can be slightly greater than that actually present in the well. This is because a volume equivalent to the bailer wall thickness will displace the hydrocarbons. Some of these displaced mobile hydrocarbons will enter the bailer, thus exaggerating the apparent thickness of the LNAPL actually present. The bailer should thus be lowered slowly through the hydrocarbon layer to minimize this discrepancy.

6.4.2.2 Continuous Core Analysis

Continuous dry coring (without water or mud) of the interval immediately above and below the mobile LNAPL layer seems an obvious way of determining the mobile LNAPL thickness. Standard split-spoon samples are commonly used for this purpose. The use of clear acrylic shelby tubes have also been used with favorable results. However, several problems are associated with coring which results in overestimating the actual thickness of the product. These factors include:

- Loss of material each time the sample is driven then extracted;
- Loss of sample in saturated or partially saturated unconsolidated sediments during extraction;
- Depth control;
- Nonrepeatablility; and
- Increased drilling costs.

6.4.2.3 Test Pit Method

The test pit method consists of excavating a pit using a backhoe to a depth where hydrocarbon pools in the bottom of the pit. Soil samples are subsequently obtained above and below the hydrocarbon-affected zone using hand-driven tubes (acrylic small-diameter shelby tubes). Problems associated with this method include:

- Impracticality of excavating test pits due to site-specific factors (i.e., piping, utilities, etc.);
- Coring in unconsolidated sediments; and
- Generation of soils that may be considered hazardous and require special handling.

6.4.2.4 Baildown Test

Baildown testing is a widely used field method to evaluate the "actual" thickness of LNAPL product in a monitoring well. Baildown testing involves the rapid removal of fluids from the well, and subsequent monitoring of fluid levels, both the water level or potentiometric surface (oil–water interface) and NAPL level (oil–air interface), with time. Such testing was originally used as a preliminary field method to evaluate recoverability of NAPLs and thus to determine potential locations for recovery wells. All monitoring wells at a site that exhibited a measurable thickness of LNAPL were typically tested. Whether or not all the LNAPL product could be removed from the well, and the volume of NAPL bailed, were general indicators of areas for "potentially good" recovery.

Although initially used for determining transmissivity, and thus recoverability, baildown tests have been reported in the literature to be useful in the evaluation of the actual thickness of the mobile LNAPL layer in the formation. This involved the estimation of actual LNAPL thickness by the graphical presentation of depth-to-product, depth-to-water, and apparent product thickness vs. time as measured during the fluid recovery period in each well as shown in Figure 6.11. This procedure was considered useful when the mobile LNAPL–water interface is above the potentiometric surface, rate of mobile LNAPL into the well is slow, and sufficient LNAPL thickness is present (greater than 0.1 in.). Baildown testing field procedures are similar to those performed for *in situ* permeability tests and involve the measurement of the initial "apparent" thickness in the monitoring well by an oil–water interface gauging probe or pressure transducer. Two different procedures have been proposed (Gruszczenski, 1987, and Hughes et al., 1988). With Gruszczenski's (1987) method, both water and product are bailed from the well until all of the hydrocarbon is removed or no further reduction in thickness can be achieved. Measured over time are levels of both depth to product and depth to water. Typically, the time increments for measurement follow the same sequence as monitored during an *in situ* rising or falling head permeability test. The test is considered complete when the well levels have stabilized for three consecutive readings or if a significant amount of time has elapsed and the levels have reached 90% of the original measurements.

If the apparent thickness is greater than the actual thickness, and the thickness in the well has been reduced to less than actual during bailing, then at some point during fluid recovery the apparent thickness equals the actual thickness. During recovery of the fluid levels in the well, the top of product in the well rises to its original level. However, the top of water (product–water interface) initially rises and then falls. The fall is due to displacement of water in the well reflecting an over-accumulation of product on the water surface. The point at which the depth-to-water graph changes from a positive to negative slope is referred to as the "inflection point." The distance between the inflection point and the measured stabilized top of product is considered the actual mobile LNAPL thickness in the formation, as shown in Figure 6.11. The difference between the actual product thickness and the static depth to product prior to testing is the height of the capillary fringe. Where no capillary fringe exist, the distance reflects the actual thickness.

Another approach to baildown testing is that proposed by Hughes et al. (1988). With this procedure, only free-phase NAPL (no water) is removed from the well. Hydrocarbon once removed will recharge at a constant rate and then begin to decrease steadily as the well continues to fill. Hydrocarbon removal is accomplished by use of a bailer or skimmer pump. As the rate of recharge changes, this change reflects not the elevation of the mobile LNAPL–water interface, but the base of the mobile LNAPL in the formation. The actual thickness of the mobile LNAPL in the formation is thus the distance between the point of initial rate change and the static depth to the top of the hydrocarbon layer prior to testing. This relationship can be shown on a graph of the top of the mobile hydrocarbon layer as measured in the well vs. time, as presented in Figure 6.12.

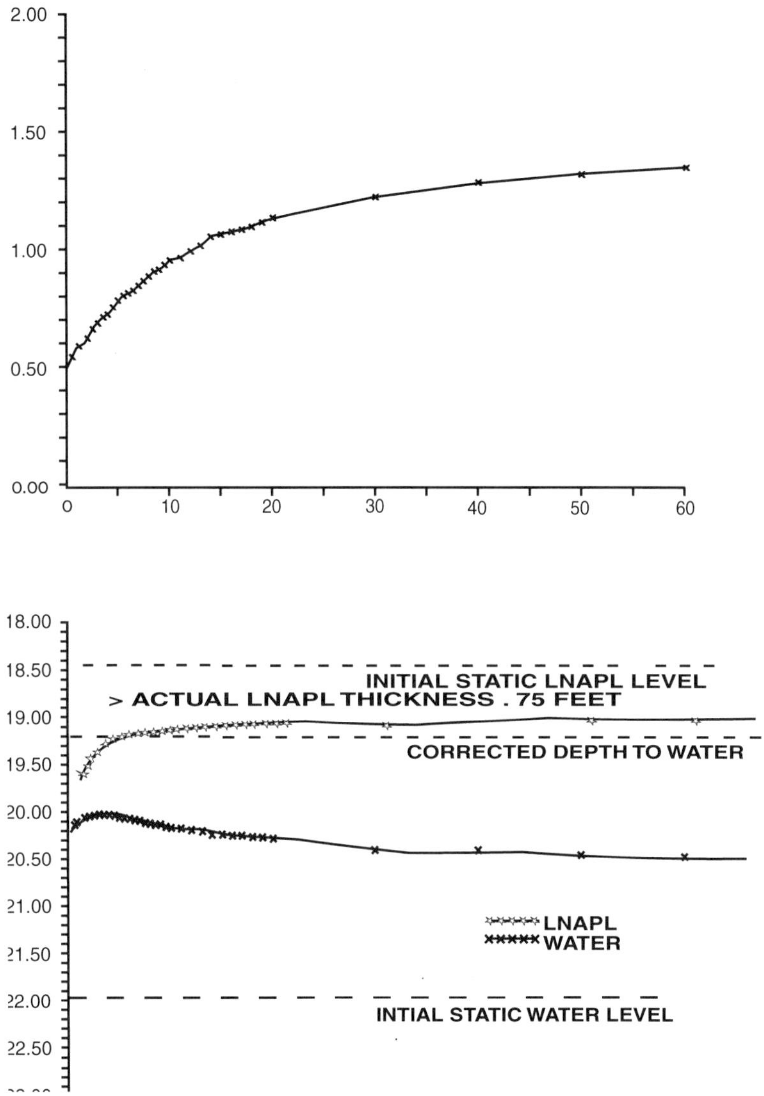

FIGURE 6.12 Representative baildown testing curve results using Hughes et al. (1988) method where depth to top of product layer vs. time is produced.

Baildown test results do not always conform to the theoretical response anticipated. This is evident from Figure 6.12, which shows much fluctuation as a result of borehole effects. In these instances, maximum theoretical values can be determined by subtracting the static depth-to-product from the corrected depth-to-water. Thicknesses provided in this manner are conservative in that true thicknesses must be less than or equal to these values and, thus, this method overestimates the actual thickness by an amount equal to the thickness of the capillary zone.

Although baildown testing is a relatively simple field procedure, the analysis and evaluation of the data remain speculative without a supportive theoretical base. The method contains a number of steps where errors can easily be introduced, and the method does not agree favorably with the hydrocarbon distribution as shown in Figure 6.4. Baildown testing results are, however, relied upon to determine "actual" thickness in a monitoring well and are an initial step and basis for calculating a volume and ultimately a recoverable volume. Baildown tests do not provide usable curves for determining actual thickness when the mobile hydrocarbon–water interface is below the potentiometric surface or where only capillary hydrocarbon exists. Some of the major areas identified where errors can also be introduced include:

- Accuracy of the measuring device used for the initial gauging and recovery of the levels after bailing;
- Operator error in measuring and recording levels with time;
- Inability of operator to collect early recovery data due to rapidly rising well levels;
- Bailing groundwater in addition to LNAPL from a low-yielding formation;
- Lack of a theoretical response or inflection point due to an inordinate length of time for water recovery;
- Variable accumulation rates of LNAPL caused by borehole effects;
- Evaluation of type curves and selection of an inflection point; and
- Lack of cross-validation of methods.

If baildown testing has innate compounding errors, these errors can only be further compounded since the remaining calculations, extrapolations, and evaluations are based upon this initial step. Although this procedure remains essentially unproven, it can be and has been used as a supportive tool in comparing the actual thickness data generated from baildown testing to those derived empirically, thus resulting in a range of values for total mobile LNAPL product volume present, and certainly can be used to evaluate LNAPL transmissivity.

6.4.2.5 Recovery Well Recharge Test

When the elevation of the mobile LNAPL–water interface is not known, recharge testing has been employed to estimate the actual LNAPL thickness in the formation (Hughes et al., 1988). Similar to a conventional pumping test, mobile LNAPL is pumped from the well until steady-state conditions are reached while maintaining the thinnest possible amount of hydrocarbon in the well. No water is pumped from the well. The top of the liquid hydrocarbon surface is recorded as the hydrocarbon recharges. Elevation of the top of the fluid hydrocarbon surface vs. time is plotted. When the recharge rate begins to decline, this corresponds to the elevation of the entry point or inflection point. The actual thickness can then be calculated using the following equation:

$$\mathrm{Tf} = \frac{\mathrm{Tc}}{(1 - \mathrm{SG}_{\mathrm{hc}})} \tag{6.11}$$

where Tf = actual thickness of the mobile LNAPL in the formation
Tc = distance between the top of the mobile LNAPL layer and the buoyancy surface or inflection point
SG_{hc} = specific gravity of the LNAPL product

The recharge test has the advantage of yielding results regardless of the location of the mobile LNAPL–water interface in relation to the potentiometric surface. Although more accurate than a baildown test, it is also more complicated to conduct.

6.4.2.6 Dielectric Well Logging Tool

Innovative ways of determining actual LNAPL thickness in the formation are currently being explored and tested. In conjunction with baildown testing, these techniques include dielectric well logging under unconfined conditions and optoelectronic sensing under confined conditions. Dielectric well logging allows detection of (1) interfaces between dissimilar fluids such as water, hydrocarbon, and air; (2) the relative dielectric constants of the fluids; and (3) well casing (nonperforated PVC casing) reflectance. Resolution is at present 5 to 6 in.; below this thickness a distinct hydrocarbon layer is not consistently discernible although thicknesses as low as 3 in. have been resolved. Thus, this method is not applicable to occurrences of limited relatively small thicknesses, but may prove promising for occurrences of large thicknesses.

6.4.2.7 Optoelectronic Sensor

Optoelectronic sensing is based on the ability to detect rising oil droplets that enter through the well screen openings and rise to the surface in a well. The device used incorporates photosensors into solid-state electronic circuitry. To induce flow of mobile LNAPL into the well, measurement is conducted in conjunction with bailing of the well. The actual LNAPL product thickness (OT_F) is calculated by obtaining a measurement of the apparent thickness in a well bore (OT_A) and then estimating the distance from the top of the product down to the confining layer (Ho). The relationship is as follows:

$$OT_F = (V_w)\, OT_A - Ho \tag{6.12}$$

where OT_F = actual LNAPL thickness, V_w = specific weight of the water gravity, OT_A = apparent LNAPL thickness, and Ho = depth of confining layer.

Potential problems can result if monitoring well construction is such that the gravel pack and screened interval are not discretely terminated at the confining layer and its grouted annulus.

6.5 VOLUME DETERMINATION

Before a volume determination can be estimated, a corrected water table contour map is prepared that depicts the configuration of the water table excluding the depressed nature created by LNAPL presence. In developing water table contour maps when LNAPL is present, since the water level (water table) as measured in the well is depressed by the weight of the hydrocarbon, a corrected depth to water is calculated as follows:

$$\mathrm{DTW}_c = \text{Static DTW}_m - (\mathrm{LNAPL}_{\mathrm{ap}} \times G) \tag{6.13}$$

where DTW_c = corrected depth to water, DTW_m = measured depth to water, $\mathrm{LNAPL}_{\mathrm{ap}}$ = measured apparent LNAPL thickness in well, and G = specific gravity at 60°F. The API gravity data for LNAPL can easily be converted to specific gravity as follows:

$$\mathrm{API} = 145/G - 131.5 \tag{6.14}$$

or

$$G = 145/\mathrm{API} + 131.5 \tag{6.15}$$

The resulting LNAPL thickness is conservative in that it incorporates both the actual thickness of LNAPL in the adjacent formation and the height of the capillary fringe. For most practical purposes, this level of accuracy is sufficient, although the more complex and extensive the site conditions, the more sophisticated approaches may be warranted, as discussed later in this chapter.

Determination of actual LNAPL thickness in the formation is important in providing reasonable estimates of the volume of free product in the subsurface. This, in turn, allows for reasonable estimates of time frame for recovery, as well as provides a mechanism for monitoring the efficiency and effectiveness of a recovery operation. The efficiency and effectiveness of many large-scale recovery operations are monitored by the reduction in volume with time. Thus, the percent reduction with time can easily be viewed as insignificant if exaggerated volumes are used. For example, if one estimates a volume on the order of 100,000 barrels, of which 10,000 barrels have been recovered to date, then a 10% reduction over a certain time interval has been achieved. However, if only 50,000 barrels exist, of which 10,000 barrels have been recovered, then a 20% reduction has actually been achieved.

Prior to initiation of a free hydrocarbon recovery strategy, the total hydrocarbon and recoverable hydrocarbon volume is estimated. Volume estimates are dependent on the determination of actual mobile hydrocarbon thickness, which can be derived empirically and/or directly via field methods such as conducting baildown or recharge tests. A simplistic hydrostatic model is provided by Testa and Paczkowski (1989). Initially, the measurement of apparent LNAPL product thicknesses as measured in monitoring wells is conducted. The data generated are then used to develop an "actual" LNAPL hydrocarbon thickness contour (isopach) map. Once developed, planimetering is performed to derive the areal coverage of incremental apparent hydrocarbon thicknesses. The greater the coverage and number of data points (monitoring wells), the smaller the chosen increment for planimetering. Although apparent thicknesses can vary between monitoring points depending on the thickness of the capillary fringe, calculated thickness values between monitoring points are approximated. Thus, the capillary fringe, hence the apparent thickness, is assumed to be constant between monitoring points. Upon completion of planimetering, the volume of soil encompassed by the LNAPL plume (V_s) is estimated. This value is then multiplied by an "assumed" porosity (∅) value, based on soil types encountered during the subsurface characterization process to calculate grossly the total apparent volume (V_a) present as shown below:

$$V_a = V_s \times \varnothing \tag{6.16}$$

where, V_a = total apparent volume of LNAPL present, V_s = volume of soil encompassed by LNAPL, and ∅ = porosity (assumed). Equation 6.16 overpredicts the hydrocarbon volume, because, in the volume of soil encompassed by LNAPL, finite (and variable) water saturation exists, the minimum value being S_r (the irreducible water saturation). Hence, the maximum value of V_a based on V_s is

$$V_a(\max) = V_s \times \varnothing(1 - S_r) \quad \text{and} \tag{6.17a}$$

$$V_a = V_s \times {}^1\!/_2\varnothing\ S_o \tag{6.17b}$$

In the capillary fringe, a third phase (air) can be present, further reducing the value of V_a.

Since the water table as measured in the well is depressed by the weight of the hydrocarbon, a corrected depth to water is calculated:

$$\text{PT}_{\text{ap}} = \text{DTW} - \text{DTP} \tag{6.18}$$

$$\text{CDTW} = \text{Static DTW} - (\text{PT}_{\text{ap}} \times G) \tag{6.19}$$

where CDTW = corrected depth to water, DTW = depth to water (measured), DTP = depth to LNAPL (measured), PTap = apparent LNAPL thickness, and G = specific gravity of LNAPL at 60°F.

A correction factor is then applied for capillary fringe effects. This factor can be empirically derived reflecting the corrected depth to water as shown below:

$$\text{CF} = (\text{CDTW} - \text{DTP}) - \text{PT}_{\text{ac}} \tag{6.20}$$

where CF = capillary fringe thickness, CDTW = corrected depth to water (calculated), DTP = depth to LNAPL (measured), and PT_{ac} = actual LNAPL thickness (calculated).

Calculation of total apparent volume does not, however, take into consideration the specific yield of the formation. Specific yield is the percentage of the mobile free hydrocarbon product that will drain and be recovered under the influences of gravity. This value is dependent on flow characteristics of the hydrocarbon as well as the characteristics of the formation. Typical values may range from 5 to 20%.

The total apparent volume is multiplied by an "assumed" specific yield for the particular area to obtain the estimated volume of recoverable hydrocarbons:

$$H = S_y \times V_a \tag{6.21}$$

where H = recoverable hydrocarbon, S_y = specific yield, and V_a = total apparent volume of hydrocarbon.

More sophisticated methods for estimating LNAPL volumes have recently been proposed and are based on the general approach that the specific volume of LNAPL can be determined by integrating the hydrocarbon saturation over the vertical distance of its occurrence:

$$V_o = \int_{Z_{ow}}^{Z_{ao}} \varnothing(z) S_o(z) dz \tag{6.22}$$

where V_o is the specific volume (l^3/l^2), $\varnothing$ is the porosity (fraction), S_o is the saturation (fraction), z represents the vertical coordinate, and subscripts ao and ow denote the air–oil and oil–water interfaces, respectively. In turn, the saturation of free-product phase is estimated from the capillary pressure–saturation correlations proposed by Brooks and Corey (1966) or van Genutchen (1980). The above relationships are in the following general form:

a. Brooks–Corey:

$$S_o = f(P_c^{ao}, P_c^{ow}, \lambda, S_r) \tag{6.23}$$

b. van Genutchen:

$$S_o = f(\alpha_{ao}, \alpha_{ow}, ^n, P_c^{ao}, P_c^{ow}, S_r) \tag{6.24}$$

In the above equations, P_c is the capillary pressure, λ is the Brooks–Corey pore-size distribution index, S_r is the residual saturation of the free product phase, α is the van Genutchen fluid/soil parameter, and the superscript n is the van Genutchen soil parameter.

Finally, the capillary pressures between phases are found by:

$$P_c^{ao} = f(\rho_o, D_w^{ao}, P_d^{ao}) \tag{6.25}$$

$$P_c^{ow} = f(\rho_o, \rho_w, D_w^{ow}, P_d^{ow}) \tag{6.26}$$

where D_w is the depth to the interface between two phases in the well, ρ is the phase density, and P_d is the displacement pressure. In Equations 6.23 through 6.26, superscripts a, o, and w denote air, oil (free product), and water phases.

In Equation 6.22, S_o at Z_{ow} is maximum with the limiting value of $1 - S_r$, where S_r is the irreducible aqueous-phase saturation in the porous medium under study.

The value of S_o decreases with increasing elevation. Z_{ao}, the interface between air and the LNAPL phase, may or may not coincide with Z_u, the upper boundary of the aquifer. Typically, the saturation of the LNAPL phase extends over two distinct regions (see Figure 5.10). These are (1) water and LNAPL phase zone, and (2) water, LNAPL phase, and air zone. When a single homogeneous stratum is considered, O can be assumed constant. In a stratified medium, however, saturation discontinuities generally exist due to the variation in soil characteristics, and the determination of LNAPL volume based on Equation 6.22 may become much more involved.

Although the approach is theoretically sound, both the proposed relationships between capillary pressure and saturation (Equations 6.23 and 6.24) are highly nonlinear and limited in practicality by the requirement of multiparameter identification. In addition, due to the inherent soil heterogeneities and difference in LNAPL composition, the identified parameters at one location cannot be automatically applied to another location at the same site, or less so at another site. For example, Farr et al. (1990) has reported the Brooks–Corey and van Genutchen parameters, λ, n, and α_{ao}, for seven different porous media based on least-square regression of laboratory data. The parameters are found to vary about one order of magnitude and do not show any specific correlation for a particular soil type.

The general conclusions of the two studies are listed in the following:

1. The relationship between V_o, the LNAPL volume, and H_o, the apparent thickness of LNAPL in the observation well, is unique, nonlinear, and dependent on the particular porous medium/fluid system.
2. For relatively low values of apparent LNAPL thickness (100 cm or less) observed in the well, the ratio of hydrocarbon volume to apparent product thickness is very low, in the range of 10^{-2} to 10^{-1} or even less.
3. For sufficiently large values of LNAPL thickness, V_o/H_o approaches the limiting maximum value of $\varnothing(1 - S_r)$, which is usually in the range of 0.2 to 0.4.
4. For an unconsolidated porous medium with uniform pore size distribution, the ratio of V_o/H_o approaches its maximum value much more rapidly than that found in a medium where pore size distribution is nonuniform, leading to the significant variation of water saturation in the contaminated zone.

An important assumption in the above approach is that the steady-state (or equilibrium) condition exists among the three phases so that the laws of hydrostatics can be applied in the analysis. In an actual site, the process of attaining a true three-phase equilibrium is rather slow (and is not attained when there is periodic recharge and/or discharge), and a hydrodynamic model may be better suited for parameter identification and subsequent prediction. The limitations of hydrostatic models arise due to the fact that the height of the free-product phase in the observation well is not directly related to the height of the free-product in soil except in the simplest case. When a well is being drilled at a contaminated site, factors affecting the radial flow of the free product phase into the well include the relative permeability of the immiscible phases as a function of phase saturation (and, naturally, the elevation), absolute permeability of the porous medium, capillary pressure as a function of saturation, head difference, and wettability of the porous medium, among others. The LNAPL phase flows into the well only from points in elevation (z-axis) where the LNAPL saturation is sufficiently high, until an equilibrium in pressure in all directions within the observation well is attained.

6.6 RECOVERABILITY

Remediation of petroleum-impacted aquifers where LNAPL is present is typically a phased process. These phases can be simplified as follows:

- Source identification, containment, and/or elimination;
- NAPL pool and dissolved plume delineation and containment;
- Removal of recoverable NAPL;
- Stabilization or removal of dissolved phase to achievable goal;
- Secondary displacement of residual hydrocarbons, and subsequent removal of dissolved and gaseous components.

The time frames associated with the later two phases are dependent on the volume of residual hydrocarbon that remains following initial recovery, hydrogeologic conditions, among other factors, and are further discussed in Chapters 9 and 10, respectively. Important factors such as the presence and extent of residual hydrocarbons, and conceptual understanding of relative permeability and transmissivity as they relate to NAPL recovery, are further discussed below.

6.6.1 Residual Hydrocarbon

The recoverability of hydrocarbon from the subsurface refers to the amount of mobile hydrocarbon available. Hydrocarbon that is retained in the unsaturated zone is not typically recoverable by conventional means. Additional amounts of hydrocarbon that are unrecoverable by conventional methods include the immobile hydrocarbons associated with the water table capillary zone. Residual hydrocarbon is pellicular or insular, and is retained in the aquifer matrix. With respect to recoverability, residual hydrocarbon entrapment can result in volume estimate discrepancies as well as decreases in recovery efficiency. With increasing water saturation, such as when the water table rises via recharge or product removal, hydrocarbons essentially become occluded by a continuous water phase. This results in a reduction of LNAPL and product thickness as measured in the well at constant volume. When water saturation is decreased by lowering the water table (as during recovery operations), trapped hydrocarbons can remobilize, leading to increased recoverability.

In general, as viscosity of the hydrocarbon increases and grain size decreases, the residual saturation increases. Typical residual saturation values for unsaturated, porous soil are tabulated in Table 5.5.

These values are then multiplied by a correction factor to account for hydrocarbon viscosity. Correction factors for different hydrocarbon types are

- 0.5 for low-viscosity products (gasoline);
- 1.0 for kerosene and gas oil; and
- 2.0 for more viscous oils.

The American Petroleum Institute (1980) has presented some similar guidelines for estimating residual saturation. Basing its work on a "typical" soil with a porosity of 30%, the API gives residual saturation values noted as a percentage of the total porosity of the soil as follows:

- 0.18 for light oil and gasoline;
- 0.15 for diesel and light fuel oil; and
- 0.20 for lube and heavy fuel oils.

Similar studies done by Hall et al. (1984) on hydrocarbon of lower API gravities (i.e., gravities between 34 and 40°) show that specific retention for more viscous hydrocarbons can range between 35 to 50% of the pore volume for fine sands with porosities of approximately 30%. The loss due to retention in the aquifer as the hydrocarbon migrates to the recovery well can be significant. Wilson and Conrad (1984) claim that residual losses are much higher in the saturated zone (i.e., capillary zone) than in the unsaturated zone.

Comparisons of the estimated volume to the actual volume recovered prove to be the only reasonable procedure for assessing the recoverable volume considering all the variables involved. These comparisons indicate that the volume of hydrocarbon retained in the aquifer is higher than published residual saturation values. Based on experience for gasoline and low-viscosity hydrocarbons, the recoverable volumes have ranged from 20 to 60% of the pore volume in fine to medium sands.

6.6.2 Relative Permeability

The potential for recovery of LNAPL is governed by several factors including viscosity, density, actual saturated thickness of the hydrocarbon in the formation, residual water saturation, and the permeability of the formation. These factors determine the relative permeability of the formation to the hydrocarbon. The relative permeability is a measure of the relative ability of hydrocarbon and water to migrate through the formation as compared with a single fluid. It is expressed as a fraction or percentage of the permeability in a single fluid system. Relative permeability can be determined experimentally for each formation material and each combination of fluid saturations and fluid properties. During hydrocarbon recovery, their ratios are constantly changing.

Graphs of relative permeability are generally similar in pattern to that shown in Figure 5.10. As shown, some residual water remains in the pore spaces, but water does not begin to flow until its water saturation reaches 20% or greater. Water at the low saturation is interstitial or "pore" water, which preferentially wets the material and fills the finer pores. As water saturation increases from 5 to 20%, hydrocarbon saturation decreases from 95 to 80% where, to this point, the formation permits only hydrocarbon to flow, not water. Where the curves cross (at a saturation of 56% for water and 44% for hydrocarbon) the relative permeability is the same for both fluids. Both fluids flow, but at a level of less than 30% of what the flow of each fluid would be at 100% saturation. With increasing water saturation, water flows more freely and hydrocarbon flow decreases. When the hydrocarbon saturation approaches 10%, the hydrocarbon becomes immobile, allowing only water to flow. For the example given, the hydrocarbon residual saturation is 10% pore saturation limited by the fluid density and viscosity and the formation permeability.

The relations shown in Figure 5.10 have a wide application to problems of fluid flow through permeable material. One of the most important applications for recovery of hydrocarbon is that there must be at least 5 to 10% saturation with the nonwetting fluid, and 20 to 40% saturation with the wetting fluid before flow occurs. Thus, for hydrocarbon (the nonwetting fluid), there must be a minimum of 5 to 10% saturation of the pore space before the fluid can move through the partially saturated or unsaturated formation and accumulate. Conversely, every hydrocarbon accumulation has a quantity of hydrocarbon that is not mobile since it is at or below a saturation of 5 to 10% and is thus not recoverable.

Mixed flow of water and LNAPL to a recovery well at a restoration site is very similar to oil and water flow to a pumping well in a low-pressure water drive reservoir. The equation used by Muskat (1936) to describe flow to a production well is presented below:

$$Q = \frac{2\pi kTP}{\mu \ln(L/r)} \tag{6.27}$$

where Q = rate of flow
k = intrinsic permeability

Note:

$$k = \frac{K\mu}{\surd} \tag{6.28}$$

where $\surd$ = specific weight
μ = viscosity
K = normal permeability
L = radius over which drainage occurs
T = thickness of reservoir
P = pressure drop over distance L
r = well bore radius

Note the similarity to the Dupuit equation for confined aquifers:

$$Q = 2\pi KT\frac{h_o - h_w}{\ln(L/r^2)} \tag{6.29}$$

where K = normal permeability
$h_o - h_w$ = pressure drop (in water head) between L and r

When the Muskat and Dupuit equations are combined and then integrated to approximate steady-state flow to an unconfined aquifer (as found at most remedial sites), the resulting equation takes the following form:

$$Q = \frac{\pi k(h_o^2 - h_w^2)}{\mu \ln(L / r^2)} \tag{6.30}$$

where k = intrinsic permeability
μ = viscosity
h_o = water head equivalent at distance L
h_w = water head equivalent at well

Note: Correction for fluid density is included in the $h_o - h_w$ factor.

If the values for all the factors are known and expressed in common units, this equation will provide a reasonable estimate of flow from recovery wells. The difficulty of actual application of this equation is the determination of the water–oil saturation in the aquifer matrix (thus, the relative permeability). The equation assumes a steady-state flow setting. In a dynamic situation where the oil reserves are being depleted and water–oil mixtures are variable, it is almost impossible (from an economic point of view) to use these equations precisely. However, inclusion of some assumptions based on results of the initial site investigation can often be of assistance in initial spacing of recovery wells and estimating recovery rates.

It should be recognized that observations made in monitoring wells surrounding the recovery wells must be adjusted to reflect the true water head pressure due to varying NAPL thickness in these wells over time. With recovery, the apparent product thickness may increase (or decrease) as the true formation thickness decreases.

6.6.3 LNAPL Transmissivity

Baildown tests have been used for decades during the initial or preliminary phases of LNAPL recovery system design to determine adequate locations for recovery wells and to evaluate recovery rates. Baildown tests involve the rapid removal of fluids from a well with subsequent monitoring of fluid levels, both the LNAPL–water (or oil–water) interface and LNAPL–air (or oil–air) interface, in the well with time. Hydrocarbon saturation is typically less than 1, and commonly below 0.5, due to the presence of other phases in the formation (i.e., air and water). Since the relative permeability decreases as hydrocarbon saturation decreases, the effective conductivity and mobility of the LNAPL is much less than that of water, regardless of the effects induced by increased viscosity and decreased density of the LNAPL.

Determination of transmissivity can be useful in designing LNAPL recovery strategies because the rate of LNAPL recovery is a function of LNAPL transmissivity (T_n).

$$T_n = \int_{z_w}^{z_o} k_{rn} k_i \frac{\rho_n g}{\mu_n} dz \tag{6.31}$$

where T_n = LNAPL transmissivity, k_{rn} = relative permeability of the nonwetting phase, k_i = intrinsic permeability of the soil, ρ_n = density of the LNAPL, g = acceleration due to gravity, μ_n = viscosity of the LNAPL, z_o = elevation of the oil–air interface, and z_w = elevation of the oil–water interface. Modification of conventional slug-test equations can be used to calculate LNAPL transmissivity and the lateral rates of mass flux of LNAPL (Huntley, 2000). A modified version of Bouwer and Rice (1976) analysis of the standard slug withdrawal test can be used when the effective water transmissivity across the screened interval of water is greater than the effective LNAPL transmissivity. When the water level or potentiometric surface does not recover rapidly, a constant drawdown-variable discharge aquifer test approach using a modified Jacobs and Lohman (1952) analysis is suitable.

6.6.4 Other Factors

In addition to factors concerning relative permeability, viscosity, and residual hydrocarbon, areal distribution of the plume and site-specific physical constraints can have a significant impact upon the degree of recoverability. A relatively small NAPL pool in areal extent with concentrated thicknesses is more recoverable, for example, than a thin pool with a large areal distribution. Site-specific physical constraints may have a major impact upon the recoverability of the plume. The problem centers around the difficulty in locating recovery wells in their optimum location without conflicting with the facility layout. Furthermore, most recovery programs generate contaminated groundwater. Depending on the size of the facility and the scale of the recovery project, the recoverability of hydrocarbon and respective time frame may be limited and highly dependent on the amount of water the facility can handle, and the subsequent treatment and disposal options available.

6.7 TIME FRAME FOR NAPL RECOVERY

The length of time required for recovery of LNAPL products is based on the volume of product present in the subsurface and is limited by numerous factors. Most often based on an educated guess, factors regarding volume estimates have innate compounding errors in relation to:

- Accuracy of physical measurement where high-viscosity and emulsified hydrocarbon is encountered;
- Determination of true vs. apparent thickness;
- Validity of baildown tests for estimation of true thickness;
- Extrapolation of geologic and hydrogeologic information between monitoring points;
- Averaging of apparent thickness for planimetering;

- Estimation or assumption for key factors including porosity, specific yield, and retention values;
- Variable product density; and
- Effects of residual trapped hydrocarbons.

After the initial volume estimate has been determined, testing of a pilot recovery system should be initiated to evaluate recovery rates. However, factors that significantly affect recovery rates include the areal distribution and geometry of the free-hydrocarbon product plume, type(s) and design of recovery system selected, and the performance and efficiency of the system with time.

Volume determinations and subsequent time frame for recovery of LNAPL product can be estimated. However, a large number of compounding errors are associated with these calculations. Thus, a reasonable time frame for remediation is clearly an estimate.

The progress of recovery efforts cannot be based confidently on LNAPL product thickness maps. Although these maps provide quantification of overall trends, the numerous factors that impact hydrocarbon thicknesses make accurate quantification difficult. An estimate of effectiveness thus is based on volume recovered to date divided by the total volume that is considered recoverable. Furthermore, as the recovery project progresses and new data are introduced, the volume and time frame for recovery should be continually reevaluated and revised.

Littlefield et al. (1984) provides a method of estimating the volume of recoverable product after recovery has been initiated. Contouring the NAPL thickness in monitoring wells and calculating the apparently oil-saturated soil volume are completed at two separate times. The difference in apparent oil-saturated soil volumes at these times is attributable to the quantity of oil recovered during that period. By using that relationship, an extrapolation can be used to estimate the remaining volume of recoverable oil. This value, in most situations, results in an overstatement of the recoverable volume of LNAPL in place.

In determining total and recoverable volumes of LNAPL, the factor of recharge to the volume is undeterminable, but realistic. Developing a range of total and recoverable volumes is suggested. A valid way to determine this range is a comparison of values generated from the indirect empirical and direct field methods. Also, as additional monitoring well points are incorporated into the project, these new data are coupled with existing data and revised estimates made. Finally, comparisons of the estimated recoverable volumes to the actual volume produced proves to be the only reasonable procedure for estimating the recoverable volume considering all the variables involved.

Recovery of DNAPL is a very slow process that is affected by those factors encountered with LNAPL (i.e., relative permeability, viscosity, residual hydrocarbon pool distribution, site-specific factors, etc.). Dissolution of a DNAPL pool is dependent upon the vertical dispersivity, groundwater velocity, solubility, and pool dimension. Dispersivities for charnolid solvent are estimated for a medium to coarse sand under laboratory conditions on the order of 10^{-3} to 10^{-4} m. Thus, limited dispersion at typical groundwater velocities is anticipated to be slow and may take up to decades or centuries when dealing with concentrations of a few hundred to a few thousand kilograms. To add to this dilemma, velocities are difficult to increase due to limited radial flow conditions from pumping wells, low yields, and limited coproduced water handling (notably where large water volume yields are produced).

Residual hydrocarbons will continue to serve as a source of groundwater contamination; thus, remediation strategies for DNAPLs should emphasize long-term control and management (i.e., source containment, pool control, and recovery) vs. short-term fixes. Regardless of an increased level of effort (i.e., additional wells, increased pumping rates, etc.), the overall time for remediation is not expected to shorten by more than a factor of five.

REFERENCES

Abdul, A. S., 1988, Migration of Petroleum Products through Sandy Hydrogeologic Systems: *Ground Water Monitoring Review*, Vol. 8, No. 4, pp. 73–81.

Abdul, A. S., Kin, S. F., and Gibson, T. L., 1989, Limitations of Monitoring Wells for the Detection and Quantification of Petroleum Products in Soils and Aquifers: *Ground Water Monitoring Review*, Vol. 9, No. 2, pp. 90–99.

Acker, W. L., 1974, *Basic Procedures for Soil Sampling and Core Drilling*: Acker Drill Company, Scranton, PA, 246 pp.

Adams, M. L., Sinclair, N., and Fox, T., 1998, 3-D Seismic Reflection Surveys for Direct Detection of DNAPL: In *Nonaqueous-Phase Liquids, Remediation of Chlorinated and Recalcitrant Compounds* (edited by G. B. Wickramanayake and E. Hinchee), Battelle Press, Columbus, OH, pp. 155–160.

Adams, T. V. and Hampton, D. R., 1992, Effects of Capillarity on DNAPL Thickness in Wells and in Adjacent Sands: In *Proceedings of the IAH Conference on Subsurface Contamination by Immiscible Fluids*, Balkema Publications, Rottendam.

American Petroleum Institute (API), 1980, *The Migration of Petroleum Products in the Soil and Groundwater. Principles and Countermeasures*: API Publication No. 1628.

Anderson, M. R., Johnson, R. L., and Pankow, J. F., 1992, Dissolution of Dense Chlorinated Solvents into Groundwater. Modeling Contaminant Plumes Form Fingers and Pools of Solvent: *Environmental Science and Technology*, Vol. 26, No. 5, pp. 901–907.

Anderson, W. G., 1986a, Wettability Literature Search — Part 2: Rock/Oil/Brine Interactions, and the Effects of Core Handling on Wettability: *Journal of Petroleum Technology*, October, pp. 1125–1149.

Anderson, W. G., 1986b, Wettability Literature Search — Part 2: Wettability Measurement: *Journal of Petroleum Technology*, November, pp. 1246–1262.

Anderson, W. G., 1986b, Wettability Literature Search — Part 3: The Effects of Wettability on the Electrical Properties of Porous Media: *Journal of Petroleum Technology*, December, pp. 1371–1378.

Bear, J., 1979, *Hydraulics of Groundwater*: McGraw-Hill, New York, 569 pp.

Bermejo, J. L., Sauck, W. A., and Atekwana, E. A., 1997, Geophysical Discovery of a New LNAPL Plume at the Former Wurtsmith AFB, Oscoda, Michigan: *Ground Water Monitoring & Remediation*, Fall, 1997, pp. 131–137.

Blake, S. B. and Fryberger, J. S., 1983, Containment and Recovery of Refined Hydrocarbons from Groundwater: In *Proceedings of Groundwater and Petroleum Hydrocarbons — Protection, Detection, Restoration*, PACE, Toronto, Ontario.

Blake, S. B. and Hall, R. A., 1984, Monitoring Petroleum Spills with Wells: Some Problems and Solutions: In *Proceedings of the National Water Well Association of Ground Water Scientists and Engineers*, Fourth National Symposium on Aquifer Restoration and Groundwater Monitoring, pp. 305–310.

Bouwer, H. and Rice, R. C., 1976, A Slug Test for Determining Hydraulic Conductivity of Unconfined Aquifers with Completely or Partially penetrating Wells: *Water Resources Research,* Vol. 12, No. 13, pp. 423–438.

Brooks, R. H. and Corey, A. T., 1966, Properties of Porous Media Affecting Fluid Flow: *Journal of Irrigation Drainage Division*, ASCE, Vol. 92, No. IR2, pp. 61–88.

Chafee, W. T. and Weimar, R. A., 1983, Remedial Programs for Groundwater Supplies Contaminated by Gasoline: In *Proceedings of the National Water Well Association of Ground Water Scientists and Engineers, Third National Symposium on Aquifer Restoration and Groundwater Monitoring*, pp. 39–46.

Charbeneau, R. J., Wanakule, N., Chiang, C. Y., Nevin, J. P., and Klein, C. L., 1989, A Two-Layer Model to Simulate Floating Free Product Recovery: Formulation and Applications: In *Proceedings of the National Water Well Association and American Petroleum Institute Conference on Petroleum Hydrocarbons and Organic Chemicals in Ground Water: Prevention, Detection and Restoration*, November, pp. 333–345.

CONCAWE (T. L. de Pastrovich, Y. Baradat, R. Barthel, A. Chiarelli, and D. R. Fussell), 1979, *Protection of Groundwater from Oil Pollution*: CONCAWE Report No. 3/79, The Hague, the Netherlands, 61 pp.

Corey, A. T., 1986, *Mechanics of Immiscible Fluids in Porous Media*: Water Resources Publications, Littleton, CO, 255 pp.

Corey, A. T., Rathjens, C. H., Henderson, J. H., and Wyllie, M. R. J., 1956, Three-Phase Relative Permeability: *Transactions of the American Institute of Mining and Metallurgy and Petroleum Engineering*, Vol. 207, pp. 349–351.

Cummings, T. R. and Twenter, F. R., 1986, *Assessment of Groundwater Contamination of Wurt Smith Air Force Base, Michigan, 1982–85*: United States Geological Survey Water Resources Investigations Report 86-4188, 110 pp.

Dietz, D. N., 1971, Pollution of Permeable Strata by Oil Components: In *Water Pollution by Oil* (edited by P. Heddle), Institute of Petroleum, London, pp. 127–139.

Driscoll, F. G., 1986, *Groundwater and Wells*: *Second Edition*: Johnson Dividion, St. Paul, 1089 pp.

Farr, A. M., Houghtoalen, R. J., and McWhorten, D. B., 1990, Volume Estimation of Light Nonaqueous Phase Liquids in Porous Media: *Ground Water*, Vol. 28, No. 1, pp. 48–56.

Faust, C. R., Guswa, J. H., and Mercer, J. W., 1989, Simulation of Three Dimensional Flow of Immiscible Fluids within and below the Unsaturated Zone: *Water Resources Research*, Vol. 25, No. 12, pp. 2449–2464.

Feenstra, S., Mackay, D. M., and Cherry, J. A., 1991, A Method for Assessing Residual NAPL Based on Organic Chemical Concentrations in Soil Samples: *Ground Water Monitoring Review*, Vol. 11, No. 23, pp. 128–136.

Folkes, D. J., Bergman, M. S., and Hearst, W. E., 1987, Detection and Delineation of a Fuel Oil Plume in a Layered Bedrock Deposit: In *Proceedings of the National Water Well Association of Ground Water Scientists and Engineers and the American Petroleum Institute Conference on Petroleum Hydrocarbons and Organic Chemicals in Ground Water: Prevention, Detection and Restoration*, pp. 279–304.

Foster, G. D., 1998, Effects of Column Height on DNAPL Behavior: In *Nonaqueous-Phase Liquids, Remediation of Chlorinated and Recalcitrant Compounds* (edited by G. B. Wickramanayake and E. Hinchee), Battelle Press, Columbus, OH, pp. 25–30.

Gruszczenski, T. S., 1987, Determination of a Realistic Estimate of the Actual Formation Product Thickness Using Monitor Wells: A Field Bailout Test: In *Proceedings of the National Water Well Association of Ground Water Scientists and Engineers and the American Petroleum Institute Conference on Petroleum Hydrocarbons and Organic Chemicals in Ground Water: Prevention, Detection and Restoration*, November, 1987, pp. 235–253.

Hall, R., Blake, S. B., and Champlin, S. C., Jr., 1984, Determination of Hydrocarbon Thickness in Sediments Using Borehole Data: In *Proceedings of the National Water Well Association of Ground Water Scientists and Engineers, Fourth National Symposium on Aquifer Restoration and Groundwater Monitoring*, p. 300–304.

Hampton, D. R., 1988, Laboratory and Field Comparisons between Actual and Apparent Product Thickness in Sands: *American Geophysical Union Abstract*, Fall Meeting, December, EOS, Vol. 69, No. 44, pp. 1213.

Hampton, D. R., 1989, Laboratory Investigation of the Relationship between Actual and Apparent Product Thickness in Sands: In *Proceedings of Symposium Conference on Environmental Concerns in the Petroleum Industry* (edited by S. M. Testa), Pacific Section American Association of Petroleum Geologists, pp. 31–55.

Hampton, D. R. and Miller, P. D. R., 1988, Laboratory Investigations of the Relationship between Actual and Apparent Product Thickness in Sands: In *Proceedings of the National Water Well Association of Ground Water Scientists and Engineers and the American Petroleum Institute Conference on Petroleum Hydrocarbons and Organic Chemicals in Ground Water: Prevention, Detection and Restoration*, Vol. I, November, pp. 157–181.

Hampton, D. R., Wagner, R. B., and Heuvelhorst, H. G., 1990, A New Tool to Measure Petroleum Thickness in Shallow Aquifers: In *Proceedings of the National Water Well Association of Groundwater Scientists and Engineers Fourth National Outdoor Action Conference on Aquifer Restoration, Ground Water Monitoring and Geophysical Methods*, May 1990, in press.

Hedgcore, H. R. and Stevens, W. S., 1991, Hydraulic Control of Vertical DNAPL Migration: In *Proceedings of the National Water Well Association of Ground Water Scientists and Engineers Conference on Petroleum Hydrocarbons and Organic Chemicals in Ground Water: Prevention, Detection and Restoration*, pp. 327–338.

Hughes, J. P., Sullivan, C. R., and Zinner, R. E., 1988, Two Techniques for Determining the True Hydrocarbon Thickness in an Unconfined Sandy Aquifer: In *Proceedings of the National Water Well Association of Ground Water Scientists and Engineers and the American Petroleum Institute Conference on Petroleum Hydrocarbons and Organic Chemicals in Ground Water: Prevention, Detection and Restoration*, Vol. I, November, pp. 291–314.

Huling, S. G. and Weaner, J. W., 1991, *Dense Non-Aqueous Phase Liquids*: U.S. Environmental Protection Agency, U.S. EPA/540/4-91/002, 21 pp.

Hunt, J. R., Sitar, N., and Udell, K., 1988, Nonaqueous Phase Liquid Transport and Cleanup, Part I, Analysis of Mechanisms: *Water Resources Research*, Vol. 24, No. 8, pp. 1247–1258.

Hunt, W. T., Wiegand, J. W., and Trompeter, J. D., 1989, Free Gasoline Thickness in Monitoring Wells Related to Ground Water Elevation Change: In *Proceedings of the Auburn University Water Resources Research Institute Conference on New Field Techniques for Quantifying the Physical and Chemical Properties of Heterogeneous Aquifers*, National Water Well Association, March, pp. 671–692.

Huntley, D., 2000, Analytic Determination of Hydrocarbon Transmissivity from Baildown Tests: *Ground Water*, Vol. 38, No. 1, pp. 46–52.

Huntley, D., Hawk, R. N., and Corley, H. P., 1992, Non-Aqueous Phase Hydrocarbon Saturations and Mobility in a Fine-Grained, Poorly Consolidated Sandstone: In *Proceedings of the 1992 Petroleum Hydrocarbons and Organic Chemicals in Groundwater: Prevention, Detection, and Restoration*, National Ground Water Association, Columbus, OH, pp. 223–237.

Kaluarachchi, J. J. and Parker, J. C., 1989, An Efficient Finite Element Method for Modeling Multiphase Flow: *Water Resources Research*, Vol. 25, pp. 43–54.

Kaluarachchi, J. J., Parker, J. C., and Lenhard, R. J., 1989, A Numerical Model for Areal Migration of Water and Light Hydrocarbon in Unconfined Aquifers: *Advanced Water Resources*, in press.

Keech, D. H., 1988, Hydrocarbon Thickness on Groundwater by Dielectric Well Logging: In *Proceedings of the National Water Well Association of Ground Water Scientists and Engineers and the American Petroleum Institute Conference on Petroleum Hydrocarbons and Organic Chemicals in Ground Water: Prevention, Detection and Restoration*, Vol. I, November, pp. 275–289.

Kemblowski, M. W. and Chiang, C. Y., 1988, Analysis of the Measured Free Product Thickness in Dynamic Aquifers: In *Proceedings of the National Water Well Association of Ground Water Scientists and Engineers and the American Petroleum Institute Conference on Petroleum Hydrocarbons and Organic Chemicals in Ground Water: Prevention, Detection and Restoration*, Vol. I, November, 1988, pp. 183–205.

Kemblowski, M. W. and Chiang, C. Y., 1990, Hydrocarbon Thickness Fluctuations in Monitoring Wells: *Ground Water*, Vol. 28, No. 2, pp. 244–252.

Kessler, A. and Rubin, H., 1987, Relationships between Water Infiltration and Oil Spill Migration in Sandy Soils: *Journal of Hydrology*, Vol. 91, pp. 187–204.

Kimberlin, D. K. and Trimmell, M. L., 1988, Utilization of Optoelectronic Sensing to Determine Hydrocarbon Thicknesses within Confined Aquifers: In *Proceedings of the National Water Well Association of Ground Water Scientists and Engineers and the American Petroleum Institute Conference on Petroleum Hydrocarbons and Organic Chemicals in Ground Water: Prevention, Detection and Restoration*, Vol. I, November, pp. 255–274.

Kool, J. B. and Parker, J. C., 1988, Analysis of the Inverse Problem for Unsaturated Transient Flow: *Water Resources Research*, Vol. 24, pp. 814–830.

Kool, J. B., Parker, J. C., and van Genuchten, M. Th., 1987, Parameter Estimation for Unsaturated Flow and Transport Models — A Review: *Journal of Hydrology*, Vol. 91, pp. 255–293.

Kramer, W. H., 1982, Groundwater Pollution from Gasoline: *Ground Water Monitoring Review*, Vol. 2, No. 2, pp. 18–22.

Kueper, B. H. and Frind, E. O., 1988, An Overview of Immiscible Fingering in Porous Media: *Journal of Contaminant Hydrology*, Vol. 2, pp. 95–110.

Kueper, B. H., Abbott, W., and Farquhar, G., 1989, Experimental Observations of Multiphase Flow in Heterogeneous Porous Media: *Journal of Contaminant Hydrology*, Vol. 5, pp. 83–95.

Lenhard, R. J. and Parker, J. C., 1987a, A Model for Hysteretic Constitutive Relations Governing Multiphase Flow, 2. Permeability-Saturation Relations: *Water Resources Research*, Vol. 23, pp. 2197–2206.

Lenhard, R. J. and Parker, J. C., 1987b, Measurement and Prediction of Saturation-Pressure Relationships in Three-Phase Porous Media Systems: *Journal of Contaminant Hydrology*, Vol. 1, pp. 407–424.

Lenhard, R. J. and Parker, J. C., 1988, Experimental Validation of the Theory of Extending Two-Phase Saturation-Pressure Relations to Three-Fluid Phase Systems for Monotonic Drainage Paths: *Water Resources Research*, Vol. 24, pp. 373–380.

Lenhard, R. J., Parker, J. C., and Kaluarachchi, J. J., 1989, A Model for Hysteretic Constitutive Relations Governing Multiphase Flow, 3. Refinements and Numerical Simulations: *Water Resources Research*, Vol. 25, pp. 1727–1736.

Leverett, M. C., 1939, Flow of Oil–Water Mixtures through Unconsolidated Sands: *American Institute of Mining and Metallurgy Engineering, Petroleum Development Technology*, Vol. 132, pp. 149–171.

Leverett, M. C., 1941, Capillary Behavior in Porous Solids: *Transactions AIME, Petroleum Engineering Division*, Vol. 142, pp. 152–169.

Levorsen, A. I., 1967, *Geology of Petroleum*: W. H. Freeman and Company, New York, 724 pp.

Levy, B. S., Riordan, P. J., and Schreiber, R. P., 1990, Estimation of Leak Rates from Underground Storage Tanks: *Ground Water*, Vol. 28, No. 3, pp. 378–384.

Littlefield, K. V., Wehler, N. E., and Heard, R. W., 1984, Identification and Removal of Hydrocarbons from Unconsolidated Sediments Affected by Tidal Fluctuations: In *Proceedings of the National Water Well Association Fourth National Symposium on Aquifer Restoration and Ground Water Monitoring*, pp. 316–322.

Lundegard, P. D. and Mudford, B. S., 1998, LNAPL Volume Calculation: Parameter Estimation by Nonlinear Regression of Saturation Profiles: *Ground Water Monitoring & Remediation*, Vol. 18, No. 3, pp. 88–93.

McCaulou, D. R. and Huling, S. G., 1999, Compatibility of Bentonite and DNAPLs: *Ground Water Monitoring & Remediation*, Spring, pp. 78–86.

Meinardus, H. W., Jackson, R. E., Jin, M., Londergan, J. T., Taffinder, S., and Ginn, J. S., 1998, Characterization of a DNAPL Zone with Partitioning Interwell Tracer Tests: In *Nonaqueous-Phase Liquids, Remediation of Chlorinated and Recalcitrant Compounds* (edited by G. B. Wickramanayake and E. Hinchee), Battelle Press, Columbus, OH, pp. 143–148.

Mercer, J. W. and Cohen, R. M., 1990, A Review of Immiscible Fluids in the Subsurface — Properties, Models, Characterization and Remediation: *Journal of Contaminant Hydrology*, Vol. 6, pp. 107–163.

Mishra, S., and Parker, J. C., 1989, Effects of Parameter Uncertainty on Predictions of Unsaturated Flow: *Journal of Hydrology*, Vol. 108, pp. 19–33.

Mishra, S., Parker, J. C., and Singhal, N., 1989a, Estimation of Soil Hydraulic Properties and Their Uncertainty from Particle Size Distribution Data: *Journal of Hydrology*, Vol. 108, pp. 1–18.

Mishra, S., Parker, J. C., and Kaluarachchi, J. J., 1989b, Analysis of Uncertainty in Predictions of Hydrocarbon Recovery from Spill Sites: *Journal of Contaminate Hydrology*, in review.

Muskat, M., 1937, *Flow of Homogeneous Fluids through Porous Media*: McGraw-Hill, New York, 763 pp.

Parker, J. C. and Kaluarachchi, J. J., 1989, A Numerical Model for Design of Free Product Recovery Systems at Hydrocarbon Spill Sites: In *Proceedings of the National Water Well Association Fourth International Conference on Solving Groundwater Problems with Models*, National Water Well Association, Columbus, OH.

Parker, J. C. and Lenhard, R. J., 1989, A Model for Hysteretic Constitutive Relations Governing Multiphase Flow, 1. Saturation-Pressure Relations: *Water Resources Research*, Vol. 23, pp. 2187–2196.

Parker, J. C., Kaluarachchi, J. J., and Katyal, A. K., 1987, Areal Simulations of Free Product Recovery from a Gasoline Storage Tank Leak: In *Proceedings of the National Water Well Conference on Petroleum Hydrocarbons and Organic Chemicals in Groundwater*, November, 1988, pp. 315–332.

Patrick, G. C. and Anthony, T., 1998, Creosote and Coal-Tar DNAPL Characterization in Fraser River Sands: In *Nonaqueous-Phase Liquids, Remediation of Chlorinated and Recalcitrant Compounds* (edited by G. B. Wickramanayake and E. Hinchee), Battelle Press, Columbus, OH, pp. 149–154.

Rau, B. V., Vass, T., and Stachle, W. J., 1992, DNAPL: Implications to Investigation and Remediation of Groundwater Contamination: In *Proceedings of the Hazardous Materials Control Research Institute, HMC-South 92*, New Orleans, LA, pp. 84–94.

Schiegg, H. O., 1985, Considerations on Water, Oil, and Air in Porous Media: *Water Science and Technology*, Vol. 17 (4–5), pp. 467–476.

Schwille, F., 1984, Migration of Organic Fluids Immiscible with Water in the Unsaturated: In *Pollutants in Porous Media* (edited by B. Yaron, G. Dagan and J. Goldschmid), Vol. 47 of Ecological Studies, Springer-Verlag, New York, pp. 27–48.

Schwille, F., 1985, Migration of Organic Fluids Immiscible with Water in the Unsaturated and Saturated Zones: In *Proceedings of the National Water Well Association Second Canadian/American Conference on Hydrogeology*, pp. 3–35.

Schwille, F., 1988, *Dense Chlorinated Solvents in Porous and Fractured Media*: Lewis Publishers, Chelsea, MI, 146 pp.

Sitar, N., Hunt, J. R., and Udell, K. S., 1987, Movement of Nonaqueous Liquids in Groundwater: In *Proceedings of Geotechnical Practice for Waste Disposal*, ASCE, Ann Arbor, MI, pp. 205–223.

Testa, S. M. and Paczkowski, M. T., 1989, Volume Determination and Recoverability of Free Hydrocarbon: *Ground Water Monitoring Review*, Winter, Vol. 9, No. 1, pp. 120–128.

Todd, D. K., 1959, *Ground Water Hydrology*: John Wiley, New York, 336 pp.

Trimmell, M. L., 1987, Installation of Hydrocarbon Detection Wells and Volumetric Calculations within a Confined Aquifer: In *Proceedings of the National Water Well Association of Ground Water Scientists and Engineers and the American Petroleum Institute Conference on Petroleum Hydrocarbons and Organic Chemicals in Ground Water: Prevention, Detection and Restoration*, November, pp. 255–269.

U.S. Environmental Protection Agency, 1992, *Dense Nonaqueous Phase Liquids — A Workshop Summary*: EPS/600/R-92/030, 81 pp.

Van Dam, J., 1967, The Migration of Hydrocarbons in a Water-Bearing Stratum: In *The Joint Problems of the Oil and Water Industries* (edited by P. Hepple), Institute of Petroleum, London, pp. 55–96.

Wagner, R. B., Hampton, D. R., and Howell, J. A., 1989, A New Tool to Determine the Actual Thickness of Free Product in a Shallow Aquifer: In *Proceedings of the National Water Well Association of Ground Water Scientists and Engineers and the American Petroleum Institute Conference on Petroleum Hydrocarbons and Organic Chemicals in Ground Water: Prevention, Detection and Restoration*, November, pp. 45–59.

Weyer, K. U. (editor), 1990, *Proceedings of the IAH Conference on Subsurface Contamination by Immiscible Fluids*: Balkema Publications, Rotterdam.

White, N. F., 1968, *The Desaturation of Porous Materials*: Ph.D. dissertation, Colorado State University, Fort Collins.

Williams, D. E. and Wilder, D. G., 1971, Gasoline Pollution of a Groundwater Reservoir — A Case History: *Ground Water*, Vol. 9, No. 6, pp. 50–56.

Wilson, J. L., Conrad, S. H., Hagan, E., Mason, W. R., and Peplinski, W., 1988, The Pore Level Spatial Distribution and Saturation of Organic Liquids in Porous Media: In *Proceedings of the National Water Well Association of Ground Water Scientists and Engineers and the American Petroleum Institute Conference on Petroleum Hydrocarbons and Organic Chemicals in Ground Water: Prevention, Detection and Restoration*, Vol. 1, November, pp. 107–133.

Yaniga, P. M., 1982, Alternatives in Decontamination for Hydrocarbon-Contaminated Aquifers: *Ground Water Monitoring Review*, Vol. 2, pp. 40–49.

Yaniga, P. M., 1984, Hydrocarbon Retrieval and Apparent Hydrocarbon Thickness: Interrelationships to Recharging/Discharging Aquifer Conditions: In *Proceedings of the National Water Well Association of Ground Water Scientists and Engineers and the American Petroleum Institute Conference on Petroleum Hydrocarbons and Organic Chemicals in Ground Water: Prevention, Detection and Restoration*, November, pp. 299–329.

Yaniga, P. M. and Demko, D. J., 1983, Hydrocarbon Contamination of Carbonate Aquifers: Assessment and Abatement: In *Proceedings of the National Water Well Association of Ground Water Scientists and Engineers Third National Symposium on Aquifer Restoration*, pp. 60–65.

Yaniga, P. M. and Warburton, J. G., 1984, Discrimination between Real and Apparent Accumulation of Immiscible Hydrocarbons on the Water Table: A Theoretical and Empirical Analysis: In *Proceedings of the National Water Well Association Fourth National Symposium and Exposition on Aquifer Restoration and Ground Water Monitoring*, pp. 311–315.

Young, C. M., Jackson, R. E., Jin, M., Londergan, J. T., Mariner, P. E., Pope, G. A., Anderson, F. J., and Nouk, T., 1999, Characterization of a TCE DNAPL Zone in Alluvium by Partitioning Tracers: *Ground Water Monitoring & Remediation*, Winter, pp. 84–94.

Zilliox, L. and Muntzer, P., 1975, Effects of Hydrodynamic Processes on the Development of Ground-Water Pollution: Study on Physical Models in a Saturated Porous Medium: In *Progress in Water Technology*, Vol. 7, pp. 561–568.

7 Remedial Technologies for NAPLs

"A number of general and site-specific considerations must be made in devising a rational, cost-effective aquifer restoration program tailored to the site's unique hydrogeologic setting."

7.1 INTRODUCTION

Recovery of spilled hydrocarbons has been occurring almost as long as petroleum has been refined. The earliest attempt reported was the use of pitcher pumps attached to shallow posthole depth wells along a breached pipeline. This pre-1900 effort was not driven by environmental concerns, but by its ease in recovery and the perceived economic value of the oil. Most recovery efforts were continued until the labor value exceeded the product value, and then stopped. Primitive equipment, coupled with a lack of understanding of the mechanics of product migration in the subsurface, and the relatively low value placed on the recovered product provided little incentive for the development of remedial technologies.

Oil field production technology continued to develop at a steady pace, since the economic value of large volumes of oil produced justified the expense of continued research and development. Improvements in spilled petroleum recovery technology began to develop when a few very large spills occurred, and recovery proved economically rewarding. Subsequent environmental regulations, as discussed in Chapter 2, provided the mechanism and in most cases the incentive for the recovery of leaked product. Many ingenious approaches have been implemented over the past couple of decades. Most of the equipment designed for these purposes were, and continue to be, based on existing oil field production, refining or water technology, albeit modified in scale and complexity.

Remediation of aquifers contaminated with NAPLs is a multitask endeavor. NAPL can occur in several forms:

1. Free-phase liquid filling the pore spaces of the aquifer matrix and thus free to migrate;
2. Free-phase liquid occurring in pore spaces, but trapped between aquifer grains which are "water-wet," and not able to migrate because the NAPL

cannot displace enough water to pass through the narrow necks of the pores; and

3. NAPL that is attenuated to the surface of soil grains (oil-wet grains), and retained too tightly to migrate.

LNAPLs by definition have a density less than water, and thus tend to float as a layer overlying the water table. Several approaches have been implemented for the recovery of LNAPL depending in part on the depth to the water table, soil type, stratigraphy, the ability to handle coproduced water, and site-specific limitations and constraints. Remedial technologies used in recovery of free-phase LNAPL are in general similar to those employed in petroleum engineering, under conditions of a low-head water drive reservoir. The general form of petroleum equations can be used to define the migration and retention characteristics of LNAPL.

The approach for the recovery of DNAPL presents special problems, since it is very difficult to assess its lateral and vertical distribution in the subsurface. DNAPLs are heavier than water and, if present in sufficient volume, will descend downward through the water. DNAPLs migrate downward until (1) they reach a retarding layer where they cannot penetrate additional pore spaces, (2) they are completely attenuated on aquifer grains, or (3) they are dissolved in groundwater. Delineation of the aerial distribution and reclamation of free-phase DNAPLs from the bottom of an aquifer requires extensive fieldwork since the free-phase form of these compounds collects in topographic depressions, stratigraphic traps (pore size and capillary forces), or as lense(s) within the saturated zone. Free-phase DNAPL recovery strategies are in many ways similar to those for LNAPLs, although recovery commonly consists of a series of horizontally and vertically spaced pumping intakes at the lower portion of the impacted saturated zone–lower confining layer interface, and possibly at select intervals within the saturated zone. Due to the low aqueous solubility of DNAPLs, limitations associated with the handling of coproduced water are also increased. Pumping of large amounts of water relative to DNAPL is typically required to dissolve DNAPLs. Other factors such as hydrogeologic conditions and site-specific contraints, etc., can plague overall system performance regardless of whether LNAPLs or DNAPLs are being recovered.

This chapter focuses on conventional methods to remove (or immobilize) NAPLs. Recovery of LNAPL can be generally divided into two groups: passive and active systems. A summary of the more conventional approaches to the recovery of LNAPL is presented in Table 7.1. Discussion of the LNAPL recovery approach, along with a general discussion of the equipment and applicability, is presented. As in all design decisions, no single approach or piece of equipment may be totally adequate to render a solution to the problem. Innovative and experimental strategies to the recovery of DNAPLs are also discussed. The designer is encouraged to combine and match to suit the particular needs of the site. With new equipment and approaches continually being introduced, manufacturer's catalogs and professional conferences provide excellent resources to keep abreast of emerging technology. A brief discussion of coproduced water handling is also presented, with a more in-depth discussion presented in Chapter 8.

TABLE 7.1
Summary of Conventional LNAPL Hydrocarbon Recovery Alternatives

LNAPL Recovery Approach	System Type	Hydrogeologic Conditions[a] (Peremeability)	Depth to Water (ft)	LNAPL Presence	Dissolved Hydrocarbon Presence	Limitations
Linear interception	Passive	Low to high	0–10	Yes	No	Loosely consolidated formations; underground structures
	Active	Low to high	0–10	Yes	No	
One-pump systems	Submersible turbine	Moderate to high	Unlimited	Yes	Yes	
	Mechanical	Moderate to high	Unlimited	Yes	Yes	
	Positive displacement	Moderate to high	Unlimited	Yes	Yes	
Two-pump systems	Submersible turbine	Moderate to high	Unlimited	Yes	Yes	
Skimming units	Floating	Irrelevant	Unlimited	Yes	No	
	Suspended	Irrelevant	Unlimited	Yes	No	
Other systems	Timed bailers	Irrelevant	0–100	Yes	No	
	Rope skimmers	Irrelevant	0–15	Yes	No	Surface water use only
	Belt skimmers	Irrelevant	0–15	Yes	No	
	Vacuum-assisted	Low- to moderate	0–30	Yes	Yes	
	Vapor extraction	Moderate to high	0–100	No	No	
	Air sparging	Moderate to high	0–100	No	No	Product volatility
	Biodegradation		0–50	Yes	Yes	Availability of oxygen and nutrients

[a] Low = $<10^{-5}$ cm/s; moderate = 10^{-5} to 10^{-3} cm/s; and high = $>10^{-3}$ cm/s.

7.2 PASSIVE SYSTEMS

The term *passive interception* is used to describe recovery systems that rely upon natural groundwater flow to deliver free-phase NAPLs to the collection facility without the addition of external energy (such as pumping). These systems often include linear interception-type systems such as trenches (or French drains), subsurface dams ("funnel-and-gate" structures), combined hydraulic underflow with skimming, and density skimming units.

7.2.1 LINEAR INTERCEPTION

At favorable locations, linear interception procedures have been effective for both LNAPL and dissolved organic chemicals. The concept behind ditches, trenches, and similar structures is to create an area of lowered hydraulic head that directs subsurface flow to a recovery location.

Site conditions favorable for the construction of these systems are generally those that are suitable for agricultural drainage. Most functioning passive recovery systems are based on the principles of drainage design. Typically, passive systems are constructed where:

- The area has unrestricted access by excavating equipment;
- Air quality concerns from vapors are minimal;
- Water–LNAPL depth is less than 25 ft;
- Few or no subsurface utilities or excavation restrictions are present.

All passive systems rely on the natural hydraulic gradient to transport LNAPL to the recovery location. Under most circumstances, the flow of LNAPL into this type of system is very slow. At open surface recovery sites (trenches and ponds) constructed in low-permeability soils, the LNAPL migrates in so slowly that free volatile product often evaporates before it accumulates sufficiently to be collected. High-permeability soils typically are subject to a low hydraulic gradient, which limits the rate of flow into the system. Conditions that are more favorable to passive recovery, shown schematically in Figure 7.1, include:

- Moderate- to high-permeability soils;
- Favorable hydraulic gradient;
- Small seasonal water table fluctuation.

7.2.1.1 Trenches

A passive recovery trench is often considered suitable as an interim procedure prior to initiation of active remediation. Its advantages include the relative ease, cost of construction (with conventional construction equipment), and low maintenance effort.

Major disadvantages include the slow rate of recovery, exposure of large surface areas of flammable fluids (in open trenches), minimal containment, and odor or air quality concerns. Additionally, this recovery method does not address the treatment of dissolved hydrocarbon constituents in groundwater.

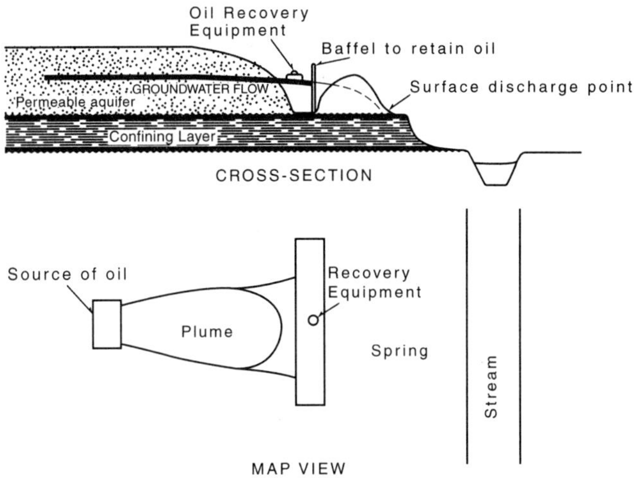

FIGURE 7.1 Idealized subsurface hydrogeologic conditions for use of a passive trench system.

7.2.1.2 Funnel and Gate Technology

This relatively new application of subsurface dams has proved useful to focus groundwater flow to a recovery area. A typical application, referred to as a funnel-and-gate system (Figure 7.2), involves construction of a sheet pile, or slurry wall,

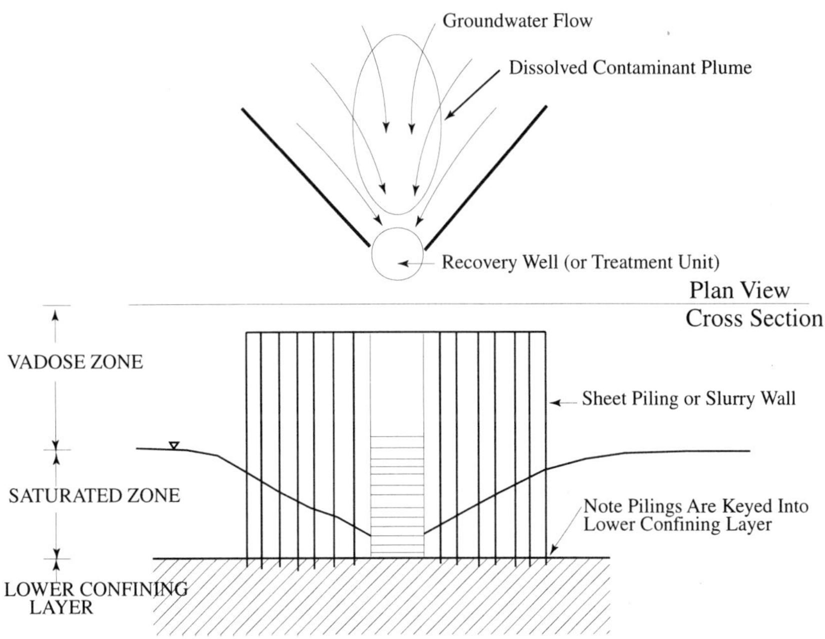

FIGURE 7.2 Schematic of a funnel-and-gate system.

in a manner to direct the groundwater in a desired direction. The recovery unit is installed at the throat, where it can collect concentrated LNAPL. Variations of this design include installation of a flow-control pump to accelerate the flow toward the recovery unit.

Design of funnel-and-gate recovery systems requires a thorough understanding of groundwater hydraulics. A poorly designed system only retards the groundwater flow, resulting in development of a water table mound upgradient of the funnel. The affected groundwater then tends to find alternate pathways around the ends of the funnel walls. Another consideration is what becomes of the subsurface structures after remediation is complete. Removal of the structures to allow the original groundwater flow configuration to be re-established may be difficult and expensive.

7.2.1.3 Hydraulic Underflow and Skimmer

Construction of a subsurface structure that penetrates the water table only a short distance can be an effective LNAPL retention technique. Most underflow structures function in the same way as a surface baffle in an oil water separator or a septic tank. The structure must be carefully installed perpendicular to groundwater flow, and have some arrangement to collect the free-phase LNAPL. Often, a simple French drain (constructed parallel to the retaining wall) leading to a recovery well (with skimmer) is effective. A schematic diagram of a hydraulic underflow structure with a skimming unit is presented in Figure 7.3.

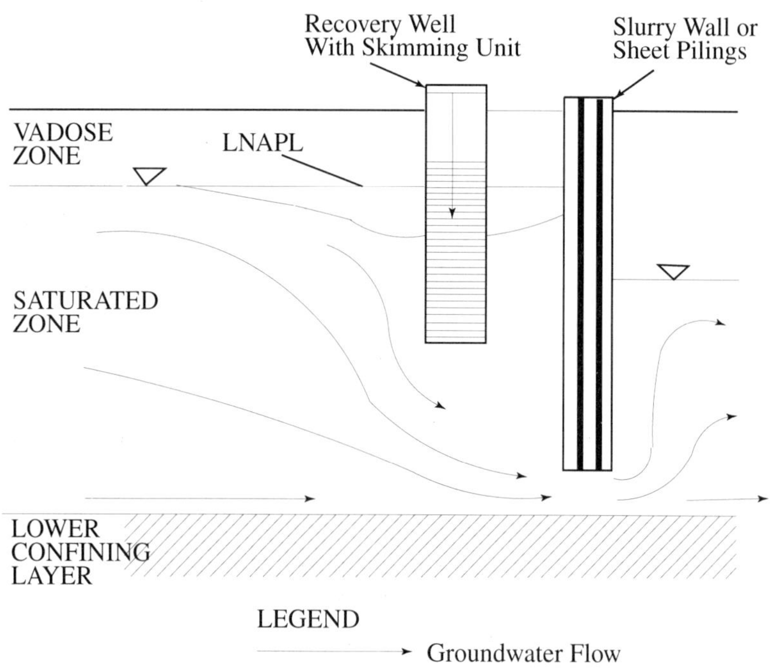

FIGURE 7.3 Schematic of hydraulic underflow with a skimming unit.

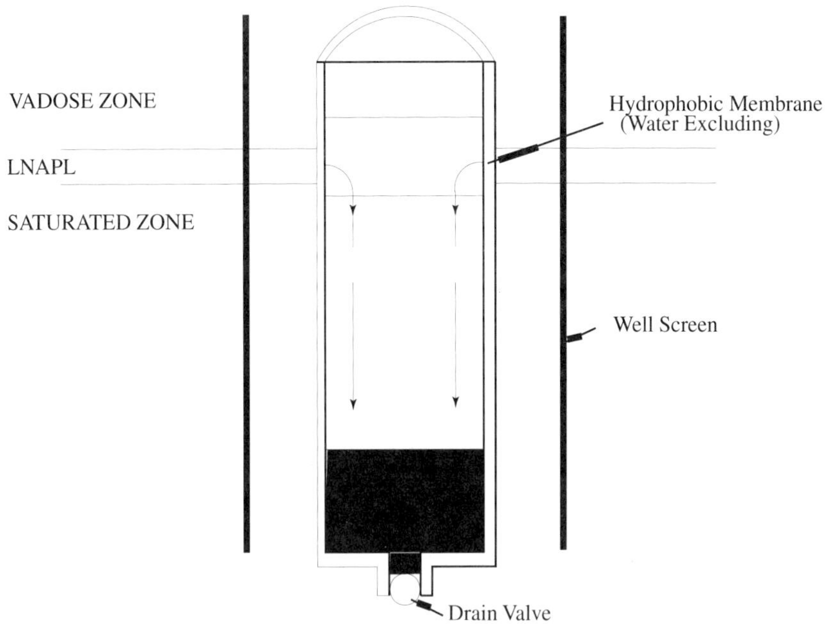

FIGURE 7.4 Schematic of a density skimmer.

Design of an underflow structure requires the knowledge that the water table will not drop below the bottom of the baffle system. If the seasonal water table fluctuation is not known, use another approach.

7.2.2 Density Skimmers

Density skimmers are a special type of bailer that are designed to remove LNAPL from the water surface in wells (Figure 7.4). Most of these circular skimmers are constructed with a hydrophobic (water-repelling) membrane around the intake area. When installed in a well (or other structure), such that the membrane is at or slightly below the water surface, the LNAPL seeps into the storage compartment. Periodically, the bailer is manually retrieved and LNAPL is recovered. Density skimmers are very useful at locations that have slow recovery and are in remote locations. A properly designed skimmer unit can remove LNAPL down to a thickness of $^1/_8$ in. (0.01 ft or 3 mm).

7.3 ACTIVE SYSTEMS

A method frequently used to improve performance of trenches is to incorporate flow enhancement. Typically, this involves the use of a pump to lower the fluid level in the trench, and thus increase the hydraulic gradient. An additional benefit is that the continued flow toward the pump also tends to collect the LNAPL in a smaller area, where it is easier to recover. Water recovered from the trench may be treated for off-site disposal or reinjected upgradient to enhance the flow further. Increased

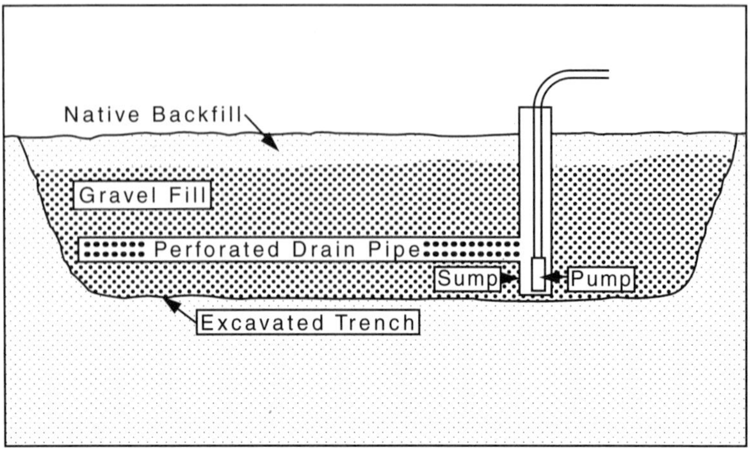

FIGURE 7.5 Schematic showing a pumpage-enhanced recovery trench.

hydraulic gradient reduces travel time in direct proportion to the slope of the water table. Downgradient migration is also reduced or eliminated because water flows into the trench from both sides, thus recovering LNAPL product that had migrated immediately downgradient of the trench.

A refinement of the trench design is to shorten fluid travel distance by extension of the trench directly into the plume. The use of a dendritic pattern, as shown in Figure 7.5, can assure that flow time is reduced to a practical minimum. Recovery systems of this design are well suited for large, open areas with shallow groundwater and flexible air quality regulations. A rough approximation of the water discharge expected from a pumped trench can be made simply by applying Darcy's equation:

$$Q = PIA \tag{7.1}$$

where Q = discharge, I = hydraulic gradient on either side of the trench, P = permeability of soil, and A = area of the trench sides. For initial calculations, a hydraulic gradient of 0.3 is often used for fine sandy soils. When the hydraulic gradient adjacent to the trench is near this value, the flow velocity in the trench is usually sufficient to bring the LNAPL to the recovery unit.

Variations of linear interceptors include closed trenches that are backfilled with highly permeable material or contain a conduit pipe (classic French drain). When set slightly below the yearly low water table elevation, a French drain can serve as an effective interception device (Figure 7.6).

7.3.1 Well-Point Systems

Where LNAPL is situated at shallow depth (<20 ft or 6 m), in moderate to high permeability soil (>10^{-5} cm/s), a well-point system may be the appropriate recovery technology. The use of well points for LNAPL recovery is similar to construction dewatering. A major advantage to this type of system is that it is possible to lower

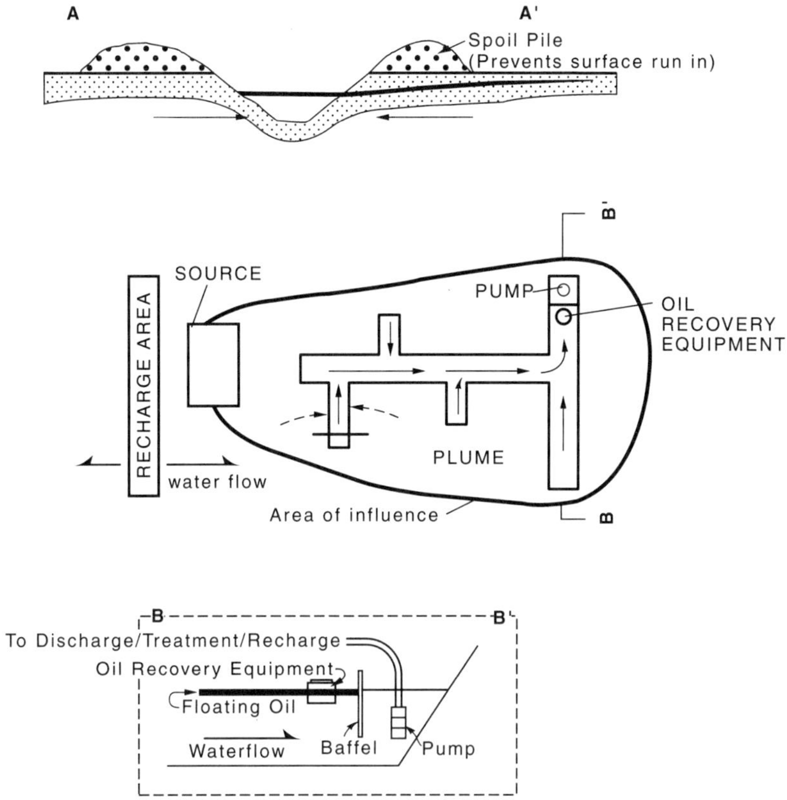

FIGURE 7.6 Schematic of a trench drain system.

a shallow fluid surface over a predetermined area and to encourage inflow from adjacent areas at a reasonable cost.

The well-point system is used to create a flow barrier, as opposed to an open trench collection system, for three primary reasons:

1. Open trenches with free-phase LNAPL can be a fire hazard.
2. Site facilities such as underground piping preclude practical trench excavation.
3. Well construction minimizes the generation of significant quantities of potentially hazardous soils in contrast to significant volumes potentially generated by trench excavation.

Well points are commonly small diameter pipes (2 to 4 in., 5 to 10 cm) attached to a short length of well screen (2 to 3 ft, 60 to 90 cm), which are installed in sets and are connected to a common suction pump. This type of system is best suited where the water table is under unconfined conditions. In field practice the maximum height that the fluid can be lifted is a function of the vacuum maintained by the central pump. Although some vacuum pumps can lift water 20 to 22 ft (6 to 7 m), most systems have sufficient leakage to limit the lift to 15 to 17 ft (4.5 to 5 m).

The design of a well-point system for remediation is similar to that of a dewatering system, with a few notable exceptions relating to the nature of the product to be recovered. Prior to the design effort, a hydrogeologic investigation is necessary to assess:

- Geologic and hydrogeologic conditions (grain size distribution, packing density, hydraulic conductivity, etc.);
- Depth to top of fluids;
- NAPL characteristics (density, viscosity, solubility, vapor pressure, etc.);
- The range of fluid level fluctuation during the recovery operation;
- Aerial extent of floating and dissolved NAPL;
- Estimated quantity of recoverable NAPL.

Calculation of well-point spacing and expected pumping rate is usually based on procedures using the Dupuit (steady-state) equation, as described in standard tests. The following factors must be considered in addition to standard water pumping considerations:

- The top of the screen must be located at or very near the top of the fluid level during pumping.
- If recovery of contaminated water is not a prime concern (after initial drawdown), the screens should be short (<1 m) with closer well spacing to compensate.
- Flowmeters (or at least site tubes) are helpful to determine if each well is functioning.
- Large quantities of volatile NAPL (i.e., gasoline) will reduce the available lift height due to increased vapor pressure.

Well points are typically installed at calculated spacings to cause a limited drawdown over the area necessary to recover the product and contaminated water, but still maintain containment, notably on the downgradient side. A typical well point is shown in Figure 7.7; a typical well-point installation layout is shown in Figure 7.8.

A common variation of a well-point recovery system is based on the concept of the eductor (jet) pump. The principle of eductors is the Bernoulli principle, the conservation of energy. The system works by pumping pressurized fluid through nozzles in the eductors, creating a partial vacuum on the suction side of the eductors. The design of an eductor unit is shown in Figure 7.9. Fluid or vapor is then pulled up from recovery wells into suction tubes attached to the suction side of each eductor. The recovered fluid (or vapor) is entrained by the pressurized fluid, and the mixture is discharged through the eductor to the discharge piping. One main pump can

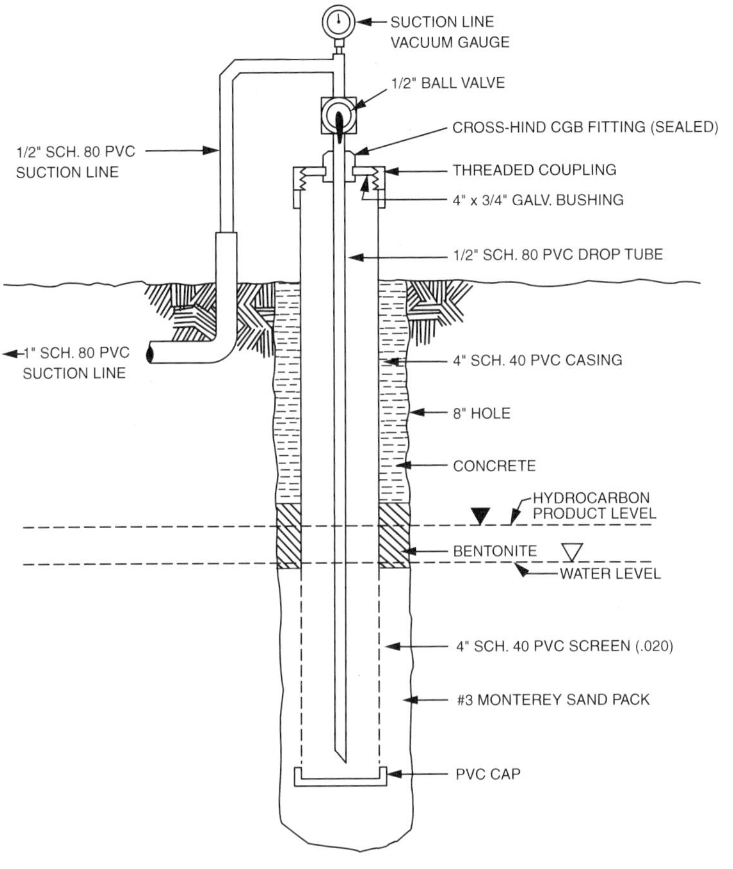

FIGURE 7.7 Typical well-point construction.

provide pressurized fluid to several eductors at multiple wells. Two layouts for well-point systems using eductors are illustrated in Figure 7.10.

The use of a well-point system can be highly effective in the remediation of contaminated shallow aquifers; however, several safety hazards are associated with this or any vacuum-lift process. Vacuum increases vaporization, especially with products similar to gasoline. Discharge of these vapors may not be permitted by regulation. When mixed with air, these vapors can potentially result in explosive mixtures. Air can mix with vapors at the discharge point through leaks in vacuum lines or can be introduced through permeable soil if the fluid level in any well point drops below the top of the screen. Because of the potential explosion hazard, it is important to use only certified explosion-proof equipment for pumping or vapor handling.

7.3.2 Vacuum-Enhanced Suction-Lift Well-Point System

The double-diaphragm suction-lift pump LNAPL recovery system is patterned after the concept of a shallow well-point dewatering system commonly used in the con-

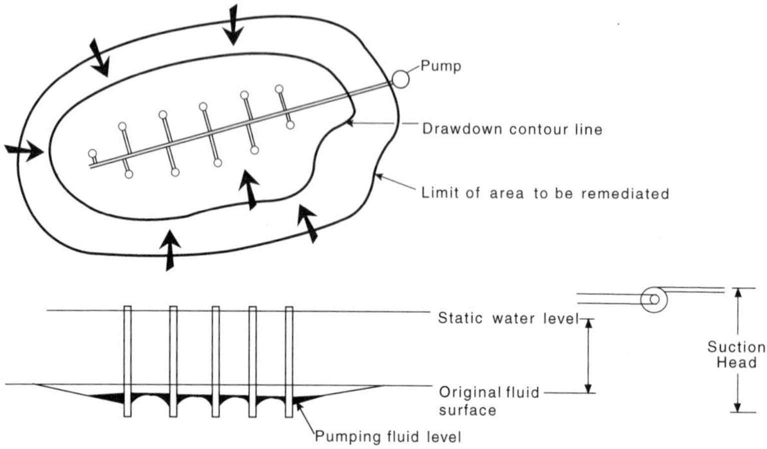

FIGURE 7.8 Typical well-point installation layout shown in plan view (top) and cross-section (bottom).

struction industry. Well-point dewatering systems typically consist of multiple shallow wells or drive points manifolded to a large-capacity suction-lift pump. By pumping at the well points, overlapping cones of depression are imposed on the water table and a barrier to groundwater migration is maintained.

The general recovery comprises primarily the following components:

- Monitoring and recovery wells
- Pumping system

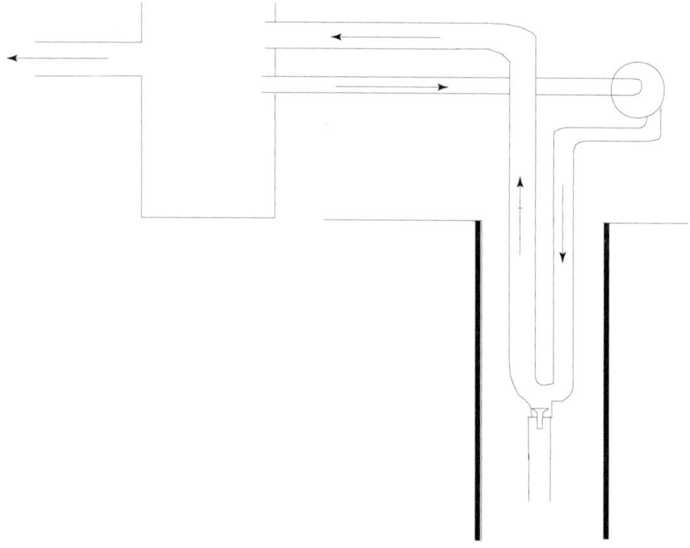

FIGURE 7.9 Design of an eductor unit.

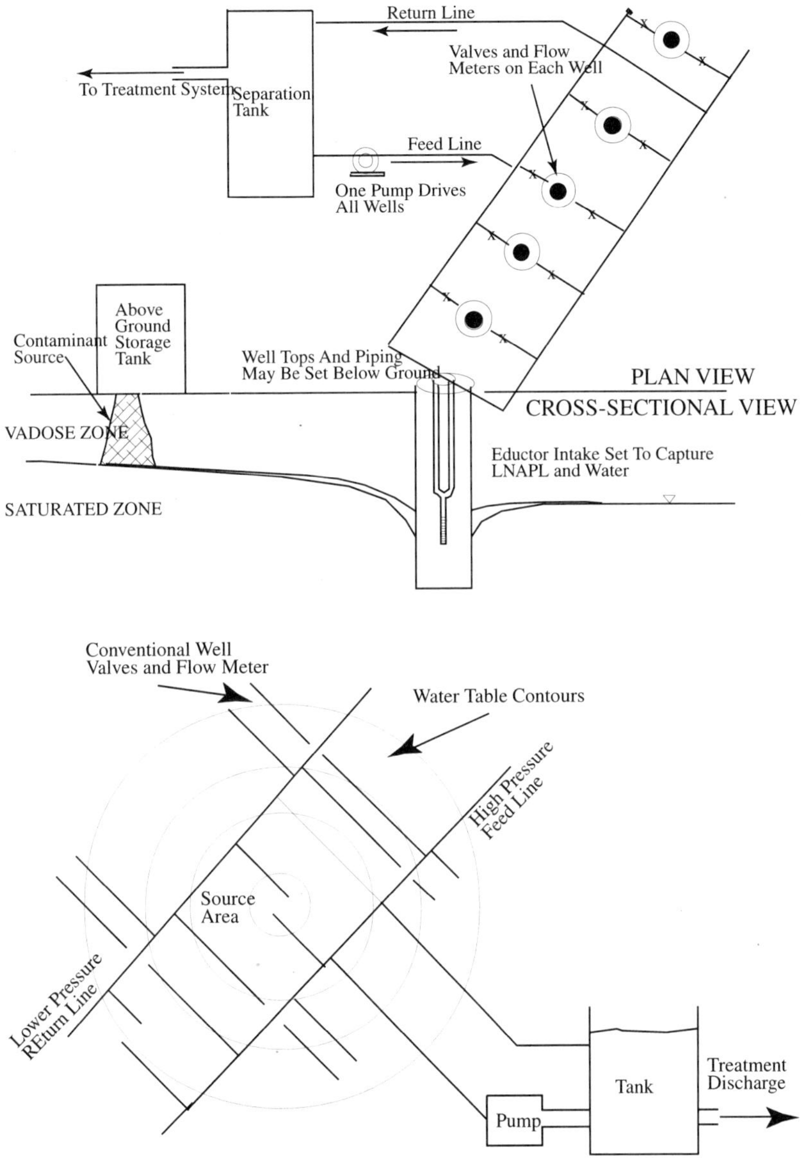

FIGURE 7.10 Two layouts for well-point system using eductors.

- NAPL storage and separation system
- Necessary materials such as air and discharge lines

Well spacing is determined based on aquifer pumping tests, which provide data on aquifer permeability and the radius of influence per well at various pumping rates. Wells are placed to provide overlapping cones of influence within the constraints of site-specific factors such as interference with buildings, above-ground tankage and retaining walls, subsurface piping, etc.

Modification to the well-point system for use in LNAPL recovery includes use of explosion-proof equipment and specific recovery well alterations. The boreholes for recovery well installation can be drilled by either machine-driven, hollow-stem auger or by hand auger in less accessible areas. Two basic alterations from standard monitoring well design are made during the construction of the recovery well: (1) screen length and placement and (2) sealed well heads. To maximize the efficiency of recovery wells, the screened interval is generally placed beginning at the bottom of each recovery well. However, the top of the screen is placed up to 5 ft below the water table (or LNAPL) surface under nonpumping conditions. Under pumping conditions, the water level (and, thus, LNAPL level) is drawn down to the point where LNAPL can enter the well. The deeper screened interval allows for minimization of vacuum loss, since less screened area is in contact with the unsaturated zone as the water level is lowered by pumping.

The well head on each recovery well is sealed to promote additional fluid entry into the well. Sealing the well head allows creation of a vacuum within the well casing while using suction-lift pumping equipment. The generation of a vacuum increases the effective difference in pressure head between the inside of the casing and the adjacent formation, thereby stimulating additional drawdown within the pumped well. As the amount of drawdown (or simulated drawdown) increases, the flow of LNAPL and groundwater into the well increases. With a well that is relatively shallow in depth, the additional simulated drawdown induced by vacuum enhancement results in increased flow to the well in the absence of additional available drawdown.

Components of the pumping system reflect both practical aspects related to the characteristics of the site facilities and the subsurface hydrogeologic conditions encountered. The pumps selected are pneumatic, double-diaphragm suction-lift pumps. The advantages of these pumps include:

- Suction-lift pumps are pneumatically operated and thus are intrinsically safe.
- Air supply typically is readily available from existing on-site, large-capacity compressors.
- Pumps are self-priming to a depth of 22 ft and can be effective to depths of 27 ft, providing suction is not broken.
- Pumps provide pressurized pumping from the well to the storage/separation tank in addition to suction of fluids from the wells.
- In the event a well is pumped dry, the pumps can pump air without damage. Pumping air also induces the vacuum-enhanced fluid entry into the well.

The pumps utilized are commercially available and constructed of cast-iron or stainless steel with Buna-N rubber diaphragms. Each pump is conventionally manifolded to withdraw groundwater from up to four recovery wells. Discharge is pumped into a piping system leading to a storage tank for NAPL/water separation and NAPL storage. Pump controls include intake and discharge valves, a compressed air dryer, a compressed air supply regulator, and a pump/oiler. A sample discharge

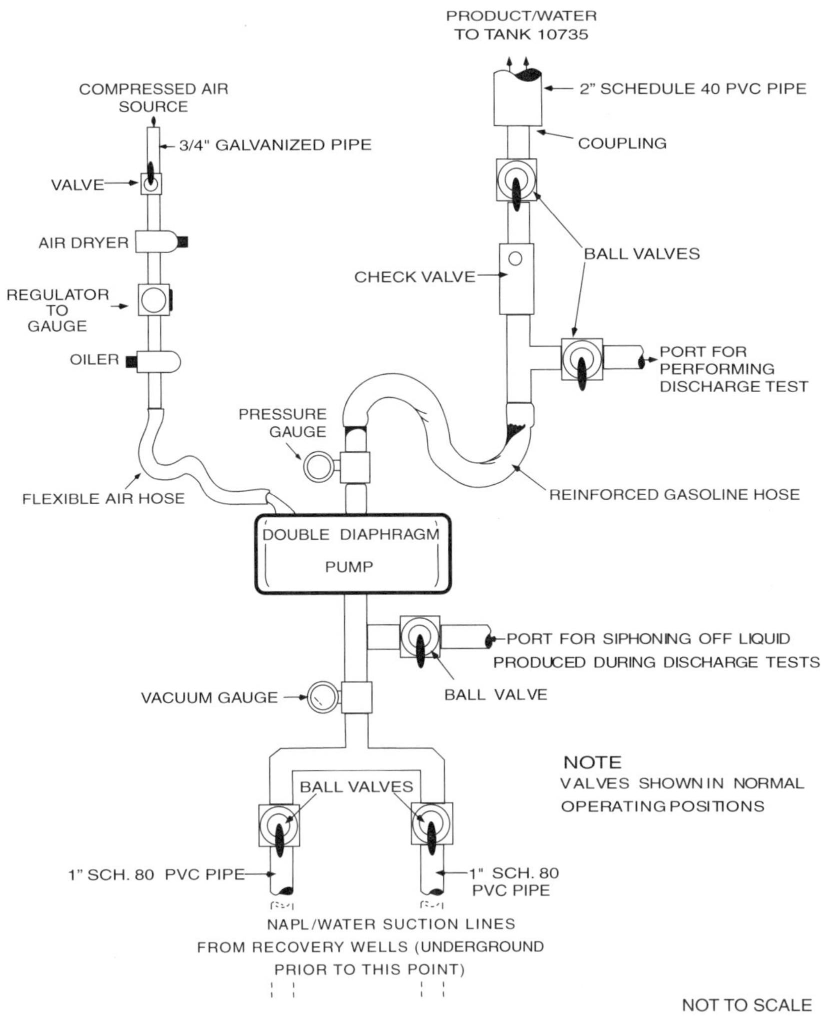

FIGURE 7.11 Schematic of double-diaphragm pump controls.

port is usually installed for monitoring the pumping rate and assessment of NAPL volume discharge by each pump. A schematic of controls for a double-diaphragm pump is presented in Figure 7.11.

The number of wells manifolded to an individual pump primarily reflects the depth to groundwater, but is also in part dependent on the pump size. In low-permeability formations, wells are typically pumped dry. As long as the pumps sustain sufficient vacuum to all wells at the intake depth, LNAPL eventually enters the well and is evacuated. In higher-permeability formations, fluid is pumped consistently from each well at a maximum rate. In both low- and high-permeability formations, individual pumps work harder to pump from greater depths. Decreasing the number of wells per pump maximizes the pumping response of individual wells. Although not specifically quantified, this is inferred to be the result of a single pump not being capable of applying the necessary vacuum to the piping and well volume of multiple wells under deeper groundwater conditions. Where lift requirements are at or near the suction lift capability of the pump (about 22 ft.), only one well per pump is used. For shallow groundwater conditions of 10 ft. or less, up to four wells have been successfully manifolded to one pump.

The fluid recovered from the wells is a mixture of groundwater and LNAPL. Since water and LNAPL are immiscible, the two liquids will separate in an open container, such as a storage tank. The LNAPL, having a lesser specific gravity than water, floats to the top, while water descends to the bottom of the tank. Depending on the configuration of the storage tank, the water is pumped or allowed to drain out the bottom of the tank. Most petroleum-handling facilities maintain existing tanks for routine product separation and storage.

Water discharge is typically handled by the existing treatment and discharge facilities, or treated by carbon filtration prior to discharge where the quantity of recovered water has been minimal. LNAPL is periodically pumped out of the tank and transported to a recycling facility. Periodic tank gauging and product transfers avoid the accumulation of LNAPL to the point where it could be inadvertently discharged with the water.

For effective LNAPL recovery, the suction-lift pumping system requires routine monitoring and operation and maintenance of the equipment. Since the pumps have a minimum of moving parts, and control is accomplished by valving of the air supply, the maintenance of the equipment is minimal.

To assess the effectiveness of the system and to maximize recovery, data collection is periodically performed. Data routinely collected include monitoring of pumping rates at each pump and calculating the percent product of the discharge in total volume to estimate the volume of product recovered. Storage tank gauging is performed to monitor the overall product volumes recovered. Periodic gauging of monitoring wells installed throughout the recovery area is also conducted to monitor reductions in the apparent LNAPL thickness over time.

Gauging is performed using an electric petroleum hydrocarbon and water level probe or steel tape with water-and-oil finding paste. For each gauging event, the depth to LNAPL and depth to water is measured for each monitoring well. These measurements are typically entered into a spreadsheet format for subsequent compilation and analysis. The corrected groundwater elevation from depth to fluid measurements is calculated based on wellhead elevations and specific gravity of the LNAPL as determined through laboratory analysis.

7.3.3 One-Pump System

At sites where the fluid level (water or LNAPL) is below the suction-lift depth, the use of a down-hole pumping unit is necessary. When a single-pump system is used, the water and/or LNAPL is delivered to the surface for separation or treatment, as shown in Figure 7.12. Several types of single-pump systems are in common usage.

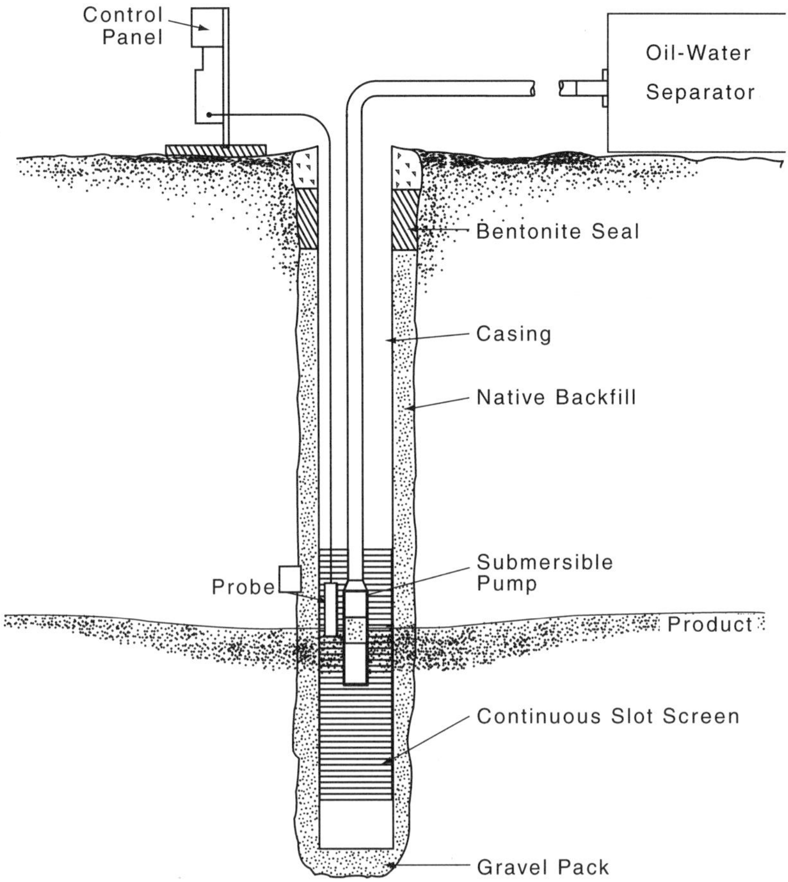

FIGURE 7.12 Typical single-pump LNAPL system with one recovery well.

The only feature shared by all of these is a single intake port. General pumps used are submersible turbine (electric) and mechanical-lift pumps.

7.3.3.1 Submersible Turbine Pumps

Submersible turbine pumps are a variety of vertical turbine pumps with the motor attached below the pumping unit. Water passes in through an intake port located between the motor and bowl assembly, then upward through the bowl stages to the surface through the pump pipe. Electric power is supplied to the motor by specially insulated wires. Submersible turbine pumps are manufactured for water supply and oil well usage by a wide variety of manufactures in sizes ranging from 1/3 to several hundred horsepower, and are constructed of materials suited for many chemicals.

After a recovery–remediation well has been constructed and tested for well capacity, an appropriate submersible pump may be installed. Materials of construction must be compatible with the fluids to be pumped (water, oil, silt/sand). Most pumps intended for domestic water well usage contain internal parts that are not resistant to the chemicals that may be encountered in remediation (either dissolved or LNAPL). Motor seals, impellers, check valves, and electric lead wires are the most common components to fail. When the pumped fluid is primarily a petroleum product, the pitch angle of the impellers must be compatible with the viscosity of the fluid. Also, the pump manufacturer should warrant the pump as explosion-proof if there is any possibility of pumping vapors or LNAPL hydrocarbon.

Installation of submersible pumps in recovery wells requires specific attention to the intake elevation setting. If the intent of the recovery is to reclaim LNAPL product, the intake must be set at the elevation of the air–fluid interface during pumping. Prior to installation of the pump, the well must be thoroughly tested to determine the specific capacity so that the pumping level for each pumping rate can be calculated. The pump should be supported by cable, rigid pipe, or other support attached to a hoisting mechanism, which will allow the pump to be raised or lowered to accommodate fluctuations of the interface surface. Regular adjustments are necessary to maintain optimum performance, maximum LNAPL recovery with minimum water retrieval.

Fluid-level monitoring is a necessary precaution to protect the pump from pumping dry. Level control can be accomplished by the use of a simple float switch, an air-bubbler pressure switch, amperage control, an interface probe, or other device. Usually these controls are attached to an electronic timing unit that is adjusted to stop the pump for a specific period of time, which allows the fluid level in the well to recover and then restarts the pump. Although this type of control can be used to operate the recovery system on a cyclic (on–off) manner, it is not as efficient as continual pumping and is very hard on the pump-starting mechanism (Figure 7.13).

FIGURE 7.13 Photograph showing aboveground wellhead and control box for one-pump LNAPL recovery system.

A recent innovation to improve recovery functions is a variable-speed submersible pump. When used with electronic controls, it is possible to select a fluid surface elevation and adjust the pumping rate periodically to maintain that fluid level at that elevation. Another adaptation of existing control technology uses an electrically adjustable valve to throttle the pump output. As the fluid level declines, the throttle restricts flow to prevent a further decline.

7.3.3.2 Positive Displacement Pumps

The use of traditional rod and piston pumps continues at many LNAPL recovery locations, particularly at refineries and distribution terminals. These units are usually powered by single-speed electric motors and have adjustable stroke lengths to control the pumping rate. When installed with the intakes set at the "optimum" pumping depth, they function fairly well. The primary advantage of rod and piston pumps is that the smooth slow stroke rate can pump mixtures of product and water without creating a significant emulsion.

Several disadvantages should be carefully considered prior to their installation. Standard construction of these units consists of a fixed-length column pipe containing the sucker rods. The intake is attached at the bottom, at a predetermined distance from the pump jack. Raising or lowering of the intake to accommodate fluctuations of the fluid surface requires either adding/subtracting lengths of the pump pipe and sucker rod or changing the height of the pump jack. Both of these options require significant field effort. Another mechanical difficulty is that the reciprocating stroke action is strenuous on the moving parts. Structural components necessary for these units to provide satisfactory service cause them to be uneconomical compared with other recovery options. Finally, the packing seals around the rods at the surface have a great tendency to leak, causing LNAPL to be discharged onto the surface casing. At remediation sites, this does not usually present an appropriate visual image.

7.3.3.3 Pneumatic Skimmer Pumps

Another variety of low-rate NAPL recovery pump is a pneumatic-operated chamber pump, which is suspended with its intake set at the product–water interface to recover mostly product. Most of these pumps have diameters of 1.5 to 4 in. (although larger units are available) with fixed chamber volumes that may be either top-filling or bottom-filling (for DNAPL), and are controlled by a timing control unit. By varying the pulse cycle, pumping rates can be set at 0.01 to 4 gpm for larger units. Many commercial brands can be adapted to this usage (or for groundwater monitoring wells). A schematic of an open-chambered pump is shown in Figure 7.14.

The bladder pump is similar in concept, with a flexible membrane bladder attached to the air line inside the chamber. A bladder prevents operating air from contacting pumped fluids; therefore, when air is released to the atmosphere it does not contain vapors.

Suspended pneumatic skimmers recover LNAPL where sufficient product is accumulated to allow timed recovery. They are installed as single-pump recovery units and also as an NAPL recovery pump for two-pump recovery units. After the cycle time is adjusted for a semi-steady-state pumping situation, they are easily controlled.

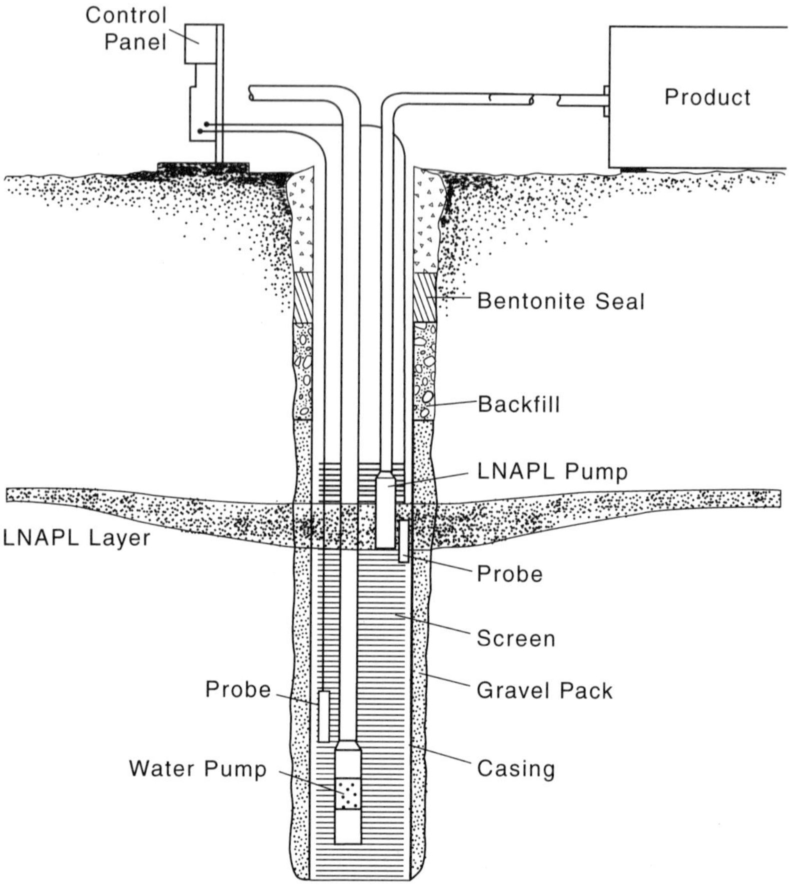

FIGURE 7.14 Typical two-pump LNAPL system with one recovery well.

Raising and lowering the intake to accommodate fluctuating fluid levels or adjusting the time cycle as LNAPL accumulates are the primary adjustments necessary.

7.3.4 Two-Pump System

The recovery most often used at sites with significant quantities of recoverable LNAPL is the two-pump system. A submersible water pump installed below the lowest possible probable interface level is used to create a drawdown cone of depression, while a second pump is suspended with its intake port located at the oil–water interface. A typical two-pump system installation is shown in Figure 7.15.

The pumping rate of the lower pump is adjusted to produce sufficient water to cause a cone of depression extending outward to intercept and retrieve the LNAPL. While the water pump operates continually, the upper product pump cycles on and off as necessary to recover the product as it accumulates.

Automatic interface detection probes attached near the intakes of both pumps provide an operating "logic." The upper probe is adjusted to detect both air–product

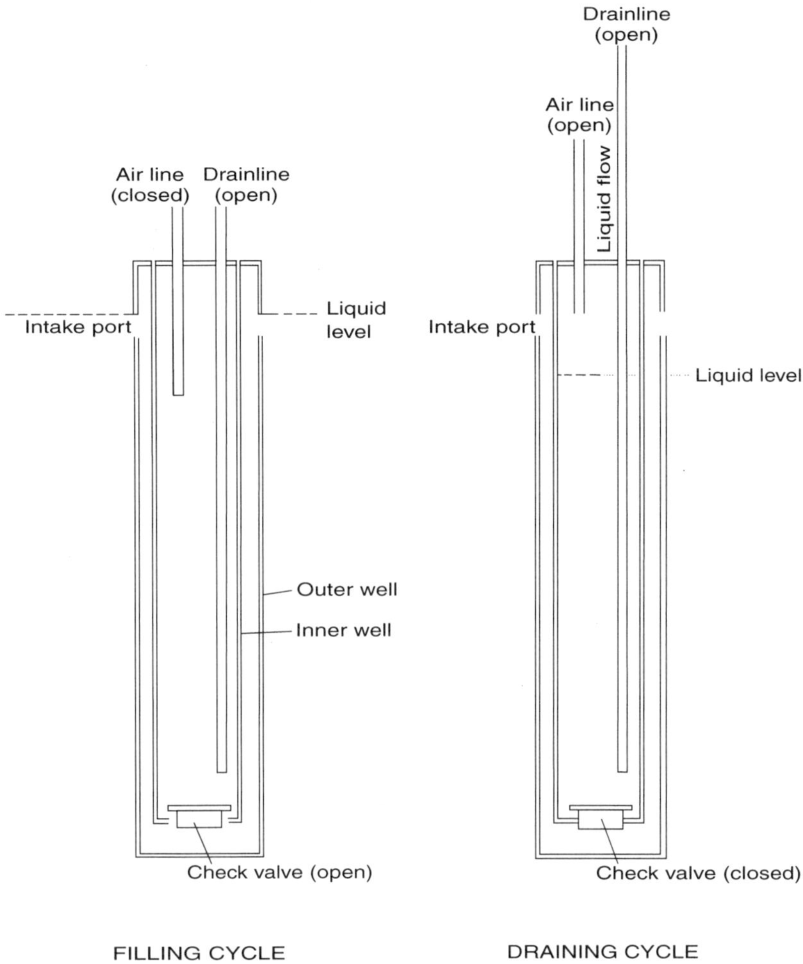

FIGURE 7.15 Schematic of an open-chambered skimmer pump.

and product–water interfaces to assure that the pump only recovers product and does not run dry. The lower probe is set to detect the presence of a product–water interface and stop that pump before it discharges any LNAPL. If for some reason the upper pump should not work and product accumulates, forcing the interface down toward the lower pump intake, the entire system ceases to operate.

There are numerous advantages to a two-pump system. LNAPL is separated in the well, reducing the need for aboveground separators. Occasionally, the LNAPL can be recycled without further treatment. Because the mixing of product and water in the well is minimized, soluble components are not added to the pumped water. The most important advantage is that the system is fully automatic and can be operated continuously with minimal adjustment after start up.

Although this system is the preferred choice for many situations, it is commonly overused, and several disadvantages must be considered. Because all of the operating components are located down the well, when they require regular cleaning or adjustment, they must be brought to the surface. Also, a larger-diameter well is required to contain all of these components. It is common to specify an 8 to 12 in. (minimum) well to assure that there is sufficient space for operation and access. Routine maintenance is essential, as interface probes tend to become coated with oil, which interferes with their function.

The initial start-up adjustment of two-pump systems requires extreme care. The water pumping rate and pump setting must be adjusted to set the LNAPL–water interface at nearly a constant level while maintaining the necessary drawdown to assure the preservation of the capture zone. Regular gauging must be made during the first few days of operation to assure proper settings. Particular attention must be used in nonhomogeneous or low-yield aquifers because stabilization of pumping levels may be difficult. Experienced staff is very important for start-up and maintenance of this type of system.

7.3.5 Other Recovery Systems

7.3.5.1 Timed Bailers

The ancient art of using a bailer bucket to recover fluid from a well has been adapted to recover LNAPL. Commercial units available are constructed to lower a sealed-bottom bailer into a well until the top of the bailer is slightly below the fluid level. When the bailer fills, it is retrieved to the surface where it is discharged into an oil–water separator. After a set period of time, the cycle is repeated. When properly adjusted, and with continued attention, these units can be effective in the removal of accumulated product from wells. Manufactured units are usually available for wells of 2 in. and greater diameter. The advantage of the automatic bailer bucket is that commercial units are self-contained and readily mobile. They can be set up in a short period of time and can be adjusted to be highly effective.

The primary disadvantage is that if these bailers are adjusted to remove only the LNAPL layer, they do not create a drawdown cone. Therefore, unless the product layer is relatively thick and quite fluid, the recovery rate tends to be slow. Also, use of bailers in deeper wells requires significant bailer travel time, which can decrease the effectiveness.

7.3.5.2 Rope and Belt Skimmers

Rope and belt skimmer units are quite efficient at removing thin layers of LNAPL from both open-trench wells and larger surface areas (ponds, or even manholes). Recent developments include development of belts that are sufficiently narrow to use in open wells. Both smooth flexible-tube and belt systems rely upon continuous rotation of a flexible closed-loop tube (or belt) constructed of an oleophilic material (Figure 7.16). The automatic unit draws the oil-covered tube (or belt) through scrapers and returns it to the surface to gather more oil.

Where a tube is used, an excess length is allowed to float on the water surface snaking around the recovery area. The skimmer is usually attached to a cantilever

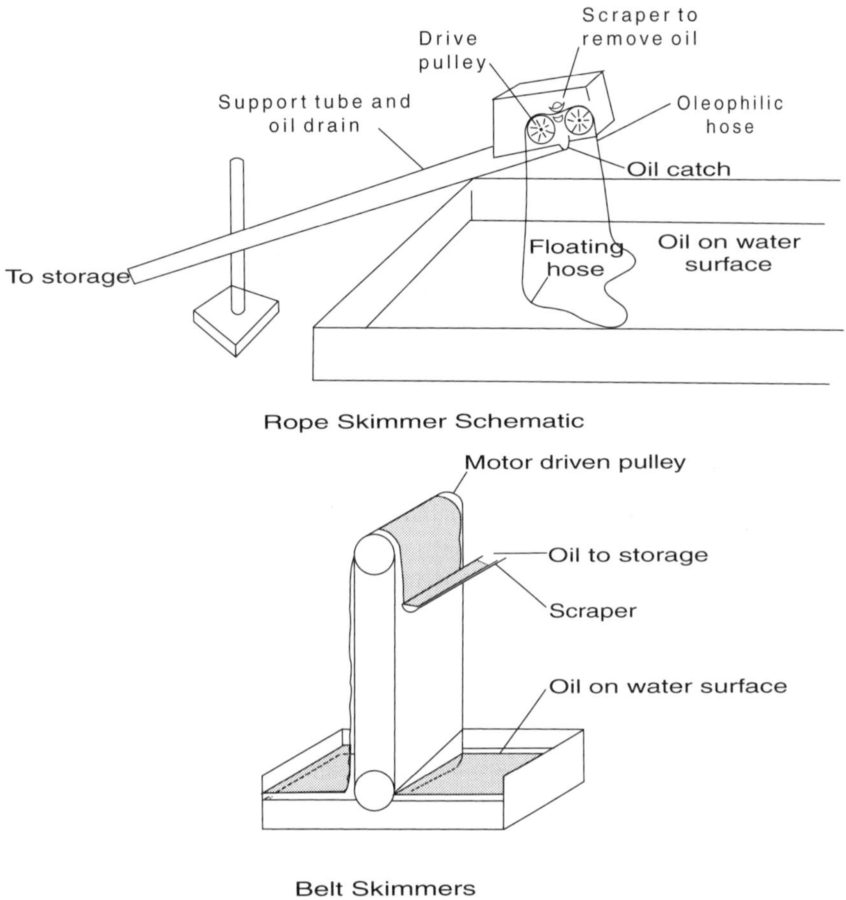

FIGURE 7.16 Schematic of a rope (a) and belt (b) skimmer system.

boom that extends out over the fluid surface. The hollow boom also serves as the conduit to conduct the product to storage containers.

Belt units are constructed such that the continuous belt is suspended between a driven pulley above the fluid to a lower, idling pulley immersed below the fluid surface. As the belt descends into the liquid, the floating oil adheres to both sides of the belt. Wipers are attached to the upper unit to scrape the oil, which drains into a trough that conveys it to storage containers.

The advantage of either rope or belt skimmers is that they do not require significant operational attention. They can be left unattended (except for routine maintenance) for extended periods of time. Also, either unit can recover thin layers of floating product without sophisticated instrumentation. The primary disadvantage is that the recovery capacity is limited to a few hundred gallons per day per unit. Higher-viscosity oils attach more effectively to the oleophilic materials. Less-viscous products are recovered less effectively.

7.3.5.3 Vapor Extraction and Biodegradation

A portion of volatile LNAPL in the subsurface vaporizes into air-filled pore spaces until the vapor–liquid equilibrium concentrations are established. If the soil zone is naturally permeable from the fluid surface to the soil surface, a concentration gradient is established. Eventually, most of the volatile mass can be transferred from the LNAPL to the atmosphere, minus that portion retained by sorption on the soil, or biologically degraded.

Several attempts have been reported by recovery teams to recover LNAPL by the use of vacuum wells set above the fluid surface. While these efforts have attained marginal success, several factors have been seen to interfere. Vacuum wells set in the vadose zone tend to encourage airflow from the surface downward through the soil, as well as to extract vapors. The result is often the enhancement of biological degradation near the wells, which tends to cause the well screens to become plugged with biomass. Routine maintenance of the wells is required to keep them functioning properly.

The major disadvantage to recovery of LNAPL by vapor extraction is the large quantity of energy required to maintain rapid volatilization of the product and airflow to transport the diluted vapors to the surface. The remediation time for this type of system tends to be quite long, often in terms of years.

7.3.5.4 Air Sparging

Air sparging is an effective procedure for removal of volatile organic compounds (VOCs) from saturated soil and groundwater. Compressed air is injected into the subsurface into the groundwater below the deepest contamination as shown on Figure 7.17. As the air begins to rise through the pores between the soil grains, it is able to remove or degrade the contaminant through such mechanisms as volatilization, advection, desorption, and diffusion. As some of the oxygen in the air is dissolved in the water, it stimulates aerobic biodegradation.

When contaminant air rises above the water table into the vadose zone, the VOCs are captured by soil-venting extraction, escape to the atmosphere, or are treated as they encounter indigenous bacteria present in that zone.

Air sparging owes its effectiveness to the mass transfer and transfer mechanisms listed above. Volatilization is the partitioning of the VOC from the liquid phase to the gaseous phase. This is the primary mechanism during the early phase of air sparging. As soon as air sparging is started, an equilibrium is established between the concentration of VOCs in gaseous and liquid phases. VOCs migrate from the liquid to the vapor phase according to Henry's or Raoult's laws. Henry's law is effective for low concentrations of VOCs in groundwater. Raoult's law predicts that in multiple chemical mixtures (i.e., petroleum fuels), the more-volatile constituents will be removed first and less-volatile chemicals will remain. As a result, the VOC content will change with time, and as the constituents with greater volatility are removed, the removal of the remaining chemicals will be more difficult.

When air sparging is halted, and vapor extraction processes are stopped, the subsurface will again approach a state of equilibrium that is consistent with the lower energy effort. The partitioning gradient that had existed under active sparging

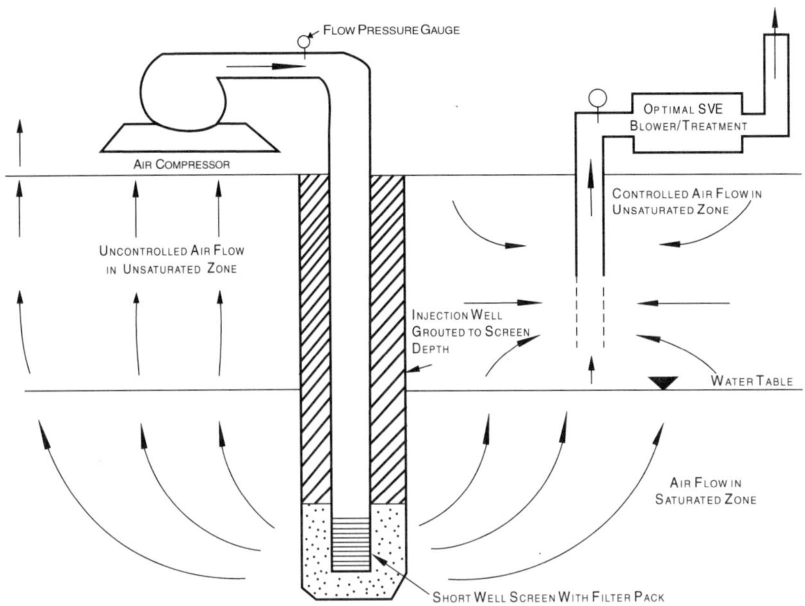

FIGURE 7.17 Conceptual schematic of an air sparging system.

is no longer present. Partitioning that occurs in this semistatic system is the escape of dissolved VOCs from the groundwater surface to the air present in the pore spaces of the unsaturated zone (by Henry's law). Since the air–water contact surface is significantly less than the migrating bubbles during air sparging, the removal rate is dramatically reduced.

During air sparging, injection of air also causes mixing of the groundwater. Additional LNAPL, which is adsorbed on soil particles, trapped in soil pores, or floating as an LNAPL layer on top of the capillary fringe, dissolves in groundwater. Dissolved VOCs near the transition zone or capillary fringe can then be vaporized and removed.

The ultimate mass of contaminant removal depends on the transport mechanisms, advection, diffusion, and dispersion. Advection is the movement of liquids or vapors in response to a pressure gradient. Through advection, the contaminant vapors are removed by the injected airflow. Advection also aids in contaminant dissolution into the groundwater. Molecular diffusion is the migration of contaminants from areas of higher to lower concentrations. Dispersion refers the spreading (dilution) of the contaminant into groundwater or gas phase by mechanical mixing.

Diffusion is the rate-limiting factor for the transport and removal of VOCs. Initially, the high gradient between the more-volatile contaminants and air is greater. Later, as the concentration of less-volatile chemicals is greater, the overall mass removal rate declines significantly. Diffusion is very important in ensuring thorough contaminant removal since it is the means of removing contaminants from the capillary areas, away from clays, and out of soil locations that are not in direct contact with airways.

The effectiveness of air sparging as a remediation tool depends on several variables, including soil type, the method of air injection, and the type and concentration of contaminants. The particle size and gradation of a particular soil determine the flow pattern in which the injected air will travel.

Injected air may migrate either as bubbles or as discrete channels. In soils that have large particle size (i.e., gravel), the air travels in bubble form. Where the soil grains are smaller, the air travels through in the form of discrete channels. In sandy soils low injection pressures will cause the air to flow through the pore spaces; higher pressures result in the air flowing through preferential channels. In clay soils, higher pressures are required. Airflow in clay soils is through developed channels which tend to remain open, thus limiting the contact of air with most of the contaminated soil grains. Air sparging has limited success in clay soils.

Use of pulsed air injection rather than continuous air injection may improve performance at many sites. After a system has been operating for a period of time, the concentration of contaminant in the recovered air declines. Continued injection of the same volume of injected air recovers less VOC. Injection of air in a pulsed pattern allows time for reestablishment of a greater concentration gradient toward air channels and a more uniform concentration throughout the contaminant plume area. When the system is resumed, volatilization is again the predominant removal mechanism.

7.3.5.5 Bioslurping

Despite its humorous name, this technology is a fairly efficient procedure to combine the benefits of vacuum-enhanced recovery and "bioventing" to promote vapor recovery and *in situ* biodegradation. Integration of these technologies into a single step results in LNAPL recovery and remediation of residual soil contamination in the vadose zone.

Bioslurping uses a vacuum to extract NAPL and water from the subsurface soil (Figure 7.18). A small-diameter well is installed so that the well screen is open slightly above and below the water table. A smaller-diameter dip tube is inserted into the well and sealed against the outer well casing. The dip tube does not extend to the bottom of the well, but ends below the water table (or oil–water interface). The maximum depth of the bottom of the slurp tube is the suction-lift capacity of the vacuum pump.

A vacuum pump connected to the dip tube enables the unit to function similarly to a drinking straw in a soda glass, which draws both soda and air from the bottom of the glass (thus slurping). The vacuum pump draws liquid (including LNAPL) and vapor into the vacuum tank where oil, water, and vapor are separated. Bioslurping differs from well-point systems because the bioslurp pump is capable of maintaining a vacuum while removing the vapors along with the liquid. Oil may be recovered for recycling; water and vapor are sent to treatment units prior to discharge.

Effective operation of the bioslurp system manages the problem of diminishing returns. When liquid recovery declines due to decreased saturation, the percentage of the recovered flow is vapor. The airflow to the well can be continued until water and/or LNAPL again enters the well. Or the dip tube can be raised or lowered to

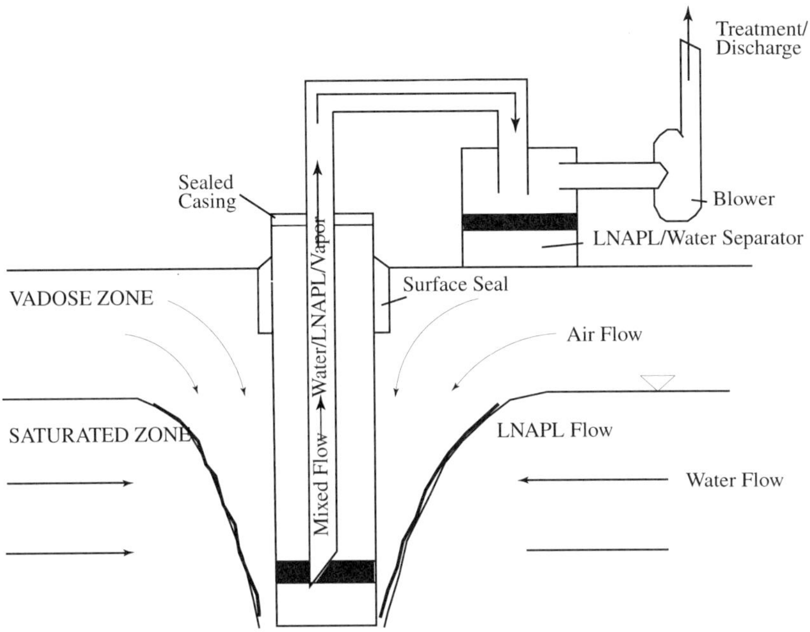

FIGURE 7.18 Conceptual schematic of a bioslurping system.

adjust the percentage of air or liquid flow as required. Several factors may limit the effectiveness of bioslurping. If the vacuum applied is too great, solid particles may be drawn through the gravel pack and well screen, or cause them to have reduced flow. The increased flow of air into the previously saturated zone may result in accelerated bacterial growth or cause precipitation of minerals (particularly iron), either of which may cause plugging of the formation.

Bioslurping is most effective in fine- to medium-textured soils, where there is a significant quantity of LNAPL and associated soil contamination at the water table, with minimum drawdown and groundwater extraction. The practical maximum depth of liquid recovery is suction lift (27 ft of water column).

7.4 COPRODUCED WATER-HANDLING CONSIDERATIONS

NAPL recovery can produce a significant quantity of coproduced water, especially if the system employs water table depression for collection of LNAPL. Discharge of water is a necessary part of the system. The general options available for disposal for recovered groundwater are:

- Transportation to a commercial disposal facility;
- Discharge to a treatment plant or surface water body;
- Recharge to a water-bearing geologic formation.

Selection of the best alternative depends on the quantity of water produced, type of product, degree of contamination, air quality consideration, site-specific considerations, and local regulations. For sites that will be operational for only a short period of time, and producing a small quantity of water, hauling can sometimes be an economical alternative. The costs associated with initial cost of equipment, acquisition of permits, and continued operation must be compared with a per-unit cost of employing a commercial disposal firm.

The cost of pretreating contaminated groundwater on site, for discharge to a publicly owned treatment works is often the preferred alternative (provided the facility has the capacity and local regulations allow acceptance). Pretreatment is usually required to prevent explosive vapors in the sewers and disruption of the biological treatment at the plant. The most common pretreatment includes phase separation and reduction of dissolved contaminants to an assured safe concentration. At small sites, it is not unusual to use phase separation, air stripping, and activated charcoal filtration prior to discharge to a sanitary sewer.

Where the system is located at an industrial facility with its own treatment plant or where large quantities of water are produced, complete on-site treatment is often the most economical procedure. Phase separation followed by air stripping and contact biological (or activated sludge) treatments are common procedures. A more in-depth treatment of coproduced water handling is presented in Chapter 8.

Surface water discharge of coproduced water usually requires a discharge permit, such as a National Pollutant Discharge Elimination System (NPDES) permit and, thus, has greater treatment demands and costs. Discharge under an NPDES permit requires that the water be treated to a degree that it does not degrade the receiving water body. Each regulatory authority has established standards and monitoring frequency. Typical standards are listed below:

Benzene	5 μg/l
Sum of BTEX	100 μg/l
Polynuclear aromatic hydrocarbons (PAH)	10 μg/l
Phenols	0.5 mg/l
Oil and grease	10 mg/l
pH	6–9 standard units
Total lead	50 μg/l
Total organic carbon (TOC)	50 mg/l

Recharge to the aquifer for water table stabilization, plume management, or enhanced recovery must be considered carefully, as it may directly affect contaminant capture. The location of injection in relation to the LNAPL can be used to direct LNAPL to recovery wells and prevent or minimize off-site migration of contamination. The chemical quality of the recovered water must be considered to assure compatibility with water and product in the injection zone. Injection of highly oxygenated water into an oxygen-depleted aquifer may, however, cause chemical or biological reactions to plug soil pores and thus redirect flow paths and increase injection pressures.

7.5 DNAPL RECOVERY STRATEGIES

Conventional pump-and-treat techniques are not very effective in restoring aquifers impacted by DNAPLs. This ineffectiveness is a result of the relatively low solubility of the DNAPL and the large capillary forces that immobilize the nonaqueous phase. Over the past decade, several innovative and experimental strategies have been tested for more effective recovery of DNAPLs. These strategies include the more conventional use of surfactants, and thermally enhanced extraction or steam injection. Other more experimental approaches include cosolvent flooding and density manipulations. Each of these approaches is discussed below.

7.5.1 Surfactants

Surfactants have been used to lower interfacial tensions between the water and the DNAPL, thus increasing the solubility, and thus mobility, of the DNAPL. This is important when dealing with chlorinated solvents such as carbon tetrachloride, methylene chloride, perchloroethylene (PCE), and trichloroethylene (TCE). With these types of organic compounds present, the pumping of contaminated groundwater extracts only a small amount of the contaminant in a dissolved phase. This situation is exacerbated with the passing of uncontaminated groundwater through the impacted zone via pumping, resulting in immediate contamination of the formerly unaffected groundwater. Conventional pumping under these conditions can go on for decades.

Surfactants are selected based primarily on the degree of solubilization. Other factors to be considered include toxicity, biodegradability, surfactant sorption, and surfactant solubility and compatibility with the separation process. Surfactants have the ability to lower the interfacial tension between water and the contaminant by as little as a factor of three to four orders of magnitude. Combined with a sufficient reduction in capillary forces, this allows pumped groundwater theoretically to move the DNAPL toward the recovery or extraction well. This is accomplished by injecting surfactant solution into the contaminated zone. Impacted groundwater characterized by an increase in the concentration of the contaminant is then recovered and treated.

7.5.2 Thermally Enhanced Extraction (Steam Injection)

Thermally enhanced extraction is another experimental approach for DNAPL source removal. Commonly know as steam injection, this technique for the recovery of fluids from porous media is not new in that it has been used for enhanced oil recovery in the petroleum industry for decades, but its use in aquifer restoration goes back to the early 1980s. Steam injection heats the solid-phase porous media and causes displacement of the pore water below the water table. As a result of pore water displacement, DNAPL and aqueous-phase chlorinated solvent compounds are dissolved and volatilized. The heat front developed during steam injection is controlled by temperature gradients and heat capacity of the porous media. Pressure gradients and permeability play a less important role.

DNAPL recovery is accomplished essentially in two concurrent phases: (1) viscosity reduction of the NAPL with the extraction of water and DNAPL via recovery wells and (2) distillation or volatilization by the injection of steam using steam injection wells with recovery of the vapor phase via vapor extraction wells. Thermally enhanced strategies have been used for the recovery of chlorinated solvents, mixed solvents, gasoline, and JP-5 fuel. Other thermal techniques under development include radio-frequency heating, electrical resistance heating, thermal blankets, and *in situ* vitrification. As with the use of surfactants, increasing the mobilization of DNAPLs also raises the concern that such mobilization may result in downward movement into noncontaminated areas.

7.5.3 Cosolvent Flooding

Cosolvent flooding is an experimental method for removing DNAPLs trapped below the water table. It involves injecting a highly concentrated aqueous mixture of solvents, such as alcohols, a chemical that is miscible with either phase in the aquifer. This process has the tendency to increase or enhance DNAPL (or LNAPL) solubility greatly, and to reduce the NAPL–water interfacial tension. Depending upon the phase behavior between the cosolvent and NAPL, a cosolvent flood can be developed to emphasize either enhanced dissolution (i.e., use of methane flooding for the dissolution of TCE) or NAPL mobilization.

Cosolvent flooding is accomplished by the introduction of a cosolvent solution, with subsequent extraction of contaminated groundwater and NAPL. In one reported field test study that focused on enhanced dissolution, the use of about nine pore volumes of a 70% ethanol, 12% pentanol solution injected into a test cell resulted in about 81% bulk NAPL removal, with a higher removal efficiency for several other individual compounds. In another field test study, where mobilization removal was emphasized, injection of about four pore volumes of a mixture of tert-butanol and *n*-hexanol into a test cell resulted in the removal of about 80% of the bulk NAPL, and higher removal efficiency of the more-soluble NAPL compounds.

7.5.4 Density Manipulations

Density manipulation is a process by which chlorinated solvents such as TCE are functionally converted to an LNAPL by passing a concentrated salt solution such as potassium iodide (KI) through the impacted zone. This experimental process increases the density of the aqueous environment making it easier to recover the NAPL.

REFERENCES

Abdul, A. S., Gibson, C. C., Ang, C. C., Smith, J. C., and Sobczynski, R. E., 1992, *In Situ* Surfactant Washing of Polychlorinated Biphenyls and Oils from a Contaminated Site: *Ground Water*, Vol. 30, pp. 219–231.

Adams, T. V. and Smith, G. J., 1998, DNAPL Remediation in Clay Till Using Steam-Enhanced Extraction: In *Physical, Chemical, and Thermal Technologies — Remediation of Chlorinated and Recalcitrant Compounds* (edited by G. B. Wickramanayake and R. E. Hinchee), Battelle Press, Columbus, OH, pp. 103–108.

American Petroleum Institute (API), 1980, *The Migration of Petroleum Products in Soil and Groundwater — Principles and Countermeasures*: API Publication No. 1628.

American Petroleum Institute (API), 1988, *Phase Separated Hydrocarbon Contaminant Modeling for Corrective Action*: American Petroleum Institute, Health Environmental Sciences Department, Publication No. 4474, 60 pp.

Bellandi, R., 1988, *Hazardous Waste Site Remediation, The Engineer's Perspective*: Van Nostrand Reinhold, New York, 422 pp.

Blake, S. B. and Gates, M. N., 1986, Vacuum-Enhanced Hydrocarbon Recovery: A Case Study: In *Proceedings of the National Water Well Association Conference on Petroleum Hydrocarbons and Organic Chemicals in Groundwater: Prevention, Detection and Restoration*, November, pp. 709–721.

Brandes, D. and Farley, K., 1993, Importance of Phase Behavior on the Removal of Residual DNAPLs from Porous Media by Alcohol Flooding: *Water Environmental Research*, Vol. 65, pp. 869–878.

Brennan, T. P., 1996, Practical Considerations for NAPL Recovery: *Environmental Geosciences*, Vol. 3, No. 2, pp. 83–89.

Brown, S. G., 1998, Physical Containment of a DNAPL Source: In *Designing and Applying Treatment Technologies, Remediation of Chlorinated and Recalcitrant Compounds* (edited by G. B. Wickramanayake and R. E. Hinchee), Battelle Press, Columbus, OH, pp. 115–120.

Burke, M. R. and Buzea, D. C., 1955, Unique Clean-up of Hydrocarbon Product from Low Permeability Formation Progressing in St. Paul, Minnesota: *Hydrological Science and Technology*, Vol. 1, No. 1, pp. 53–58.

Chen, L. and Knox, R. C., 1997, Using Vertical Circulating Wells for Partitioning Tracer Test and Remediation of DNAPLs: *Ground Water Monitoring & Remediation*, Summer, pp. 161–168.

Conner, J. A., Newell, C. J., and Wilson, D. K., 1989, Assessment, Field Testing and Conceptual Design for Managing Dense Non-Aqueous Phase Liquids (DNAPL) at a Superfund Site: In *Proceedings of the National Water Well Association and American Petroleum Institute Conference on Petroleum Hydrocarbons and Organic Chemicals in Groundwater: Prevention, Detection, and Restoration*, November, pp. 519–533.

EPA, Office of Underground Storage Tanks, OSWER, 1996, *How to Recover Free Product at Leaking Underground Storage Tank Sites: A Guide for State Regulators,* 148 pp.

Falta, R. W., 1998, Using Phase Diagrams to Predict the Performance of Cosolvent Floods for NAPL Remediation: *Ground Water Monitoring & Remediation*, Vol. 18, pp. 94–102.

Falta, R. W., Lee, C. M., Brame, S. E., Roeder, E., Wright, C. W., and Coates, J. T., 1998, Field Study of LNAPL Remediation by In-Situ Cosolvent Flooding: In *Nonaqueous-Phase Liquids — Remediation of Chlorinated and Recalcitrant Compounds* (edited by G. B. Wickramanayake and R. E. Hinchee), Battelle Press, Columbus, OH, pp. 205–210.

Farmer, V. E., 1983, Behavior of Petroleum Contaminants in an Underground Environment: In *Proceedings of Groundwater and Petroleum Hydrocarbons — Prevention, Detection, Restoration,* PACE, Toronto, Ontario.

Fountain, J. C., 1997, Removal of Nonaqueous Phase Liquids Using Surfactants: In *Subsurface Restoration* (edited by C. H. Ward, J. A. Cherry, and M. R. Scalf), Ann Arbor Press, Chelsea, MI, pp. 199–207.

Fountain, J. C., Starr, R. S., Middleton, T., Beikirch, M., Taylor, C., and Hodge, D. S., 1996, A Controlled Field Test of Surfactant Enhanced Aquifer Remediation: *Ground Water,* Vol. 34, pp. 910–916.

Green, S. R. and Dorrler, R. C., 1996, Four Bullets Shoot Down Costs: *Soil and Groundwater Cleanup*, March, pp. 35–39.

Heron, G. and Christensen, T. H., 1998, Thermally Enhanced Remediation at DNAPL Sites: The Competition between Downward Mobilization and Upward Volatilization: In *Nonaqueous-Phase Liquids — Remediation of Chlorinated and Recalcitrant Compounds* (edited by G. B. Wickramanayake and R. E. Hinchee), Battelle Press, Columbus, OH, pp. 193–198.

Kaluarachichi, J. J. and Elliot, R. T., 1995, Design Factors for Improving the Efficiency of Free Product Recovery Systems in Unconfined Aquifers: *Groundwater,* Vol. 33, No. 6, Nov.–Dec., p. 909.

Lindsly, V. E. and Berwald, F. B., 1930, Effect of Vacuum on Oil Wells: *U.S. Bureau of Mines Bull.*, No. 322, 133 pp.

Lipe, K. M., Sabatini, D. A., Hasegawa, M. A., and Harwell, J. H., Micellar Enhanced Ultrafiltration and Air Stripping for Surfactant-Contaminant Separation and Surfactant Reuse: *Ground Water Monitoring & Remediation*, Winter, pp. 85–92.

Martel, K. E., Martel, R., Lefebure, R., and Gelinas, P. J., 1998, Laboratory Study of Polymer Solutions Used by Mobility Control During *in Situ* NAPL Recovery: *Ground Water Monitoring & Remediation*, Vol. 18, No. 3, pp. 103–113.

McKee, J. E., Laverty, F. B., and Hertel, R. M., 1972, Gasoline in Groundwater: *Journal of Water Pollution Control Federation*, Vol. 44, pp. 293–302.

Mysona, E. S. and Hughes, W. D., 1999, Remediation of BTEX in Groundwater with LNAPL Using Oxygen Releasing Materials (ORM): In *In Situ Bioremediation of Petroleum Hydrocarbons and Other Organic Compounds* (edited by B. C. Alleman and A. Leeson), Battelle Press, Columbus, OH, Vol. 5, No. 3, pp. 283–288.

Parker, J., 1995, Bioslurping Enhances Free Product Recovery: *Soil and Groundwater Cleanup*, October, pp. 53–56.

Powers, J. P., 1981, *Construction Dewatering*: John Wiley and Sons, New York, 484 pp.

Reddy, K. R. and Jeffery, A. A., 1988, Systems Effects on Benzene Removal from Saturated Soils and Ground Water Using Air Sparging: *Journal of Environmental Engineering*, March, 1998, Vol. 124, No. 3, pp. 288–299.

Sanchez, M. and Ely, R., 1998, Recovery of Trichloroethylene from a Bench-Scale Aquifer by Density Manipulations: In *Nonaqueous-Phase Liquids — Remediation of Chlorinated and Recalcitrant Compounds* (edited by G. B. Wickramanayake and R. E. Hinchee), Battelle Press, Columbus, OH, pp. 181–185.

Schroth, M. H., 1998, Field Method for the Feasibility Assessment of Surfactant-Enhanced DNAPL Remediation: In *Nonaqueous-Phase Liquids — Remediation of Chlorinated and Recalcitrant Compounds* (edited by G. B. Wickramanayake and R. E. Hinchee), Battelle Press, Columbus, OH, pp. 199–200.

Smith, G., Adams, T. V., and Jurka, V., 1998, Closing a DNAPL Site through Source Removal and Natural Attenuation: In *Physical, Chemical, and Thermal Technologies — Remediation of Chlorinated and Recalcitrant Compounds* (edited by G. B. Wickramanayake and R. E. Hinchee), Battelle Press, Columbus, OH, pp. 97–102.

Stewart, L. D., 1998, Field Demonstration of Thermally Enhanced Extraction for DNAPL Source Removal: In *Nonaqueous-Phase Liquids — Remediation of Chlorinated and Recalcitrant Compounds* (edited by G. B. Wickramanayake and R. E. Hinchee), Battelle Press, Columbus, OH, pp. 211–216.

Vogel, K. D. and Peramaki, M. P., 1996, Innovative Remediation Technologies Applicable to Soil and Groundwater Impacted by Polycyclic Aromatic Hydrocarbons: *Environmental Geosciences*, Vol. 3, No. 2, pp. 98–106.

West, C. C. and Harwell, J. H., 1992, Surfactants and Subsurface Remediation: *Environmental Science Technology*, Vol. 26, pp. 23–24.

Wickramanayake, G. B. and Hinchee, R. E. (editors), 1998, *Nonaqueous-Phase Liquids — Remediation of Chlorinated and Recalcitrant Compounds*: Battelle Press, Columbus, OH, 255 pp.

8 Handling of Coproduced Water

"Effectiveness of most recovery operations is limited not by available technology, but rather by the inability to handle coproduced water."

8.1 INTRODUCTION

The practice of free-phase NAPL recovery, soil vapor extraction, and hydraulic containment at remediation sites almost always generates some volume of water contaminated with dissolved fractions. Depending on the size of the facility and the scale of the recovery and restoration project, the amount of groundwater coproduced can possibly exceed 1000 gal/min. Handling of these volumes of contaminated water can be very expensive if treatment is required prior to disposal or reinjection. Treatment of water derived from sites contaminated by other organic chemicals will involve adaptations of these procedures to the specific situation.

The degree and type of treatment required prior to disposal of coproduced water will depend primarily on the type of specific contaminants, the intended water use, and the treatment levels required. Treatment systems may be relatively simple in the rare case of a single chemical contaminant, or extremely complex for cases involving numerous contaminants. Disposal options range from direct discharge or reinjection without treatment, to discharge and reinjection after treatment, or to reinjection via recharge basins and allowing percolation.

Normally, treatment of coproduced groundwater during hydrocarbon recovery operations will include, as a minimum, oil–water separation and the removal of dissolved volatile hydrocarbon fractions (i.e., benzene, toluene, and total xylenes). In addition, removal of inorganic compounds and heavy metals (i.e., iron) is often required. Dissolved iron, a common dissolved constituent in groundwater, for example, may require treatment prior to downstream treatment processes to prevent fouling problems in air-stripping systems. Heavy metals removal is normally accomplished by chemical precipitation.

Technologies applicable for dissolved hydrocarbon removal include air stripping, activated carbon adsorption, biological treatment, and various combinations of these technologies. In some cases, especially under low groundwater pumping flow rate requirements, existing refinery or industrial wastewater treatment facilities may be used as part of the groundwater treatment system.

An overview of treatment technologies, along with site-specific considerations, and cost comparisons for various approaches that have been utilized for coproduced groundwater treatment are discussed below. These treatment approaches are:

- Oil–Water Separation
  - Gravity separation
  - Dissolved air floatation
  - Chemical coagulation — flocculation sedimentation
  - Coalescers
  - Membrane processes
  - Biological processes
  - Carbon adsorption
- Inorganics
- Organics
  - Air stripping
  - Carbon adsorption
  - Biological treatment

In addition, disposal options are presented with a discussion on the reinjection of untreated groundwater during free-phase LNAPL recovery.

8.2 OIL–WATER SEPARATION

Removal of NAPL can present problems, particularly if emulsion is involved. Emulsion is an intimate mixture of two liquids not miscible with each other, as oil and water. Water-in-oil emulsions have water as an internal phase and oil as the external phase, whereas oil-in-water emulsions reverse the order. Oil–water separation is required prior to downstream treatment processes. Several specific oil removal technologies are presented below.

8.2.1 Gravity Separation

Gravity oil separation equipment, which includes API separators, tanks with skimming, and various skimming clarifier designs, is efficient in removing large quantities of free oil. While gravity separation is effective in removing free and unstable oil emulsions, soluble oil fractions and many emulsions are not removed by this type of separation equipment. Primary gravity separation is the most economical and efficient way to remove large quantities of LNAPL hydrocarbon. The effluent oil concentrations from gravity separators in refineries generally range from 30 to 150 mg/l, although deviations on either side of this range may occur. The removal efficiency for all gravity separation equipment is a function of temperature and density differences between the oil and water.

8.2.2 Dissolved Air Flotation

Dissolved air flotation (DAF) is another process used to remove “oil and grease” constituents. In this process, groundwater (or some fraction thereof) is saturated under pressure with a gas (usually air). Upon release of this pressure, the air in excess of the atmospheric saturation concentration is released from solution, forming bubbles of approximately 30 to 120 μm in diameter. The bubbles form on the surfaces of the suspended or oily materials or are attached to the particles by interfacial tension. Consequently, an aggregate is formed with an average density substantially less than that of water, causing the aggregate to rise. DAF can be used by itself but, for maximum effectiveness, it is used with chemical coagulation and flocculation. The chemical treatment aspects of DAF operation are extremely important, particularly when colloidal or emulsified oil components are present. Coagulants, such as lime, alum, ferric salts, or polyelectrolites, are used to improve floc formation and to provide good separation. They can be injected at several points depending on the flotation process and the chemical used. Effluent oil concentrations averaging less than 20 mg/l are typical from well-operated DAF units.

8.2.3 Chemical Coagulation–Flocculation and Sedimentation

The important design factors that must be determined for a particular water during treatability studies include:

- Best chemical addition system
- Optimum chemical dose
- Optimum pH conditions
- Rapid mix requirements
- Flocculation requirements
- Sludge production
- Sludge flocculation, settling, and dewatering characteristics

Laboratory-scale test procedures consisting of jar test studies have been used for years, and the test methodology developed is such that full-scale designs can be developed from these studies with a high degree of confidence. A jar test is a series of bench-scale laboratory procedures made on 1- or 2-l water samples to determine the most effective water treatment method. Tests are performed to identify the most effective coagulants, optimum dosage, optimum pH, and most effective order in which to add various chemicals.

Chemical precipitation has traditionally been a popular technique for the removal of heavy metals and other inorganics from wastewater streams. However, a wide variety of other techniques also exist. For example, ion-exchange, reverse osmosis, evaporation, freeze crystallization, electrodialysis, cementation, catalysis, distillation, and activated carbon have all been used for removal of inorganics.

8.2.4 Coalescers

Fibrous bed coalescers generally have a fixed filter element constructed of fiberglass or other material that acts to coalesce (bring together) the oil droplets and to break emulsions. The coalesced oil droplets released from the filter are readily separated downstream by gravity. Coalescence in a fibrous bed coalescer involves three steps:

1. Interception of fine droplets by fibers;
2. Attachment of droplets to the fibers or to retained droplets;
3. Release of enlarged droplets from the fibers.

Demulsification by induced coalescence requires the rupture of the protecting film as the emulsion flows through the small passages in the fibrous media. Coalescence of the dispersed phase is then possible because of the preferential oil-wetting characteristics of the media surface.

8.2.5 Membrane Processes

Membrane processes such as ultrafiltration or reverse osmosis have been proposed as oil removal processes. Laboratory tests have indicated favorable oil removal, although relatively low flux rates, membrane fouling, and membrane life problems have presented concerns for the practical application of membrane processes to oil removal.

8.2.6 Biological Processes

Biological treatment processes have limitations in their applicability to NAPL removal. Microorganisms are efficient in oxidizing most soluble organic compounds, including some dispersed or emulsified oils. Large amounts of NAPL must be avoided, as they coat the biological floc, interfere with efficient oxygen transfer within the biomass, and produce oily sludge scums.

8.2.7 Carbon Adsorption

Activated carbon adsorption has very limited use in the removal of LNAPL. Adsorption is primarily effective for removal of low levels of soluble hydrocarbons. Groundwater applied to activated carbon adsorption units must be pretreated to prevent clogging and coating of the activated carbon with free oil. If the activated carbon adsorption units are not adequately protected, the units will have to be backwashed frequently and the activated carbon will have to be replaced with unacceptable frequency.

8.3 REMOVAL OF INORGANICS

Chemical addition for the removal of inorganic compounds is a well-established technology. There are three common types of chemical addition systems that depend upon the low solubility of inorganics at a specific pH. These include the carbonate system, the hydroxide system, and the sulfide system.

In reviewing the basic solubility products for these systems, the sulfide system removes the most inorganics, with the exception of arsenic, because of the low solubility of sulfide compounds. This increased removal capability is offset by the difficulty in handling the chemicals and the fact that sulfide sludges are susceptible to oxidation to sulfate when exposed to air, resulting in resolubilization of the metals. The carbonate system is a method that relies on the use of soda ash (sodium carbonate) and pH adjustment between 8.2 and 8.5. The carbonate system, although workable in theory, is difficult to control. The hydroxide system is the most widely used inorganic/metals removal system. The system responds directly to pH adjustment, and usually uses either lime (CaOH) or sodium hydroxide (NaOH) as the chemical to adjust the pH upward. Sodium hydroxide has the advantage of ease in chemical handling and in producing a low volume of sludge. However, the hydroxide sludge is often gelatinous and difficult to dewater.

Chemical precipitation can be accomplished by either batch- or continuous-flow operations. If the flow is less than 30,000 gpd (21 gpm), a batch treatment system may be the most economical. In the batch system, two tanks are provided, each with a capacity of 1 day's flow. One tank undergoes treatment while the other tank is being filled. When the daily flow exceeds 30,000 gpd, batch treatment is usually not feasible because of the large tankage required. Continuous treatment may require a tank for acidification and reduction, then a mixing tank for chemical addition, and a settling tank.

8.4 REMOVAL OF ORGANICS

The primary concern with coproduced water during NAPL recovery operations will normally be removal of the dissolved fractions of hydrocarbons. As previously indicated, there are many treatment technologies available for the removal of dissolved hydrocarbons. The commonly used processes are discussed in the following sections.

8.4.1 Air Stripping

Countercurrent packed towers (in their various forms) appear to be the most appropriate equipment configuration for treating contaminated groundwaters for the following reasons:

- They provide the most liquid interfacial area.
- High air-to-water volume ratios are possible due to the low air pressure drop through the tower.

Emission of stripped organics to the atmosphere may be environmentally unacceptable; however, a countercurrent tower is relatively small and can be readily connected to vapor recovery equipment.

The design of an air-stripping process for stripping volatile organics from contaminated groundwater is accomplished in two steps. The cross-sectional area of the column is determined and then the height of the packing is determined. The cross-sectional area of the column is determined by using the physical properties of the air flowing through the column, the characteristics of the packing, and the air-to-water flow ratio. A key factor is the establishment of an acceptable air velocity. A general rule of thumb used for establishing the air velocity is that an acceptable air velocity is 60% of the air velocity at flooding. Flooding is the condition in which the air velocity is so high that it holds up the water in the column to the point where the water becomes the continuous phase rather than the air. If the air-to-water ratio is held constant, the air velocity determines the flooding condition. For a selected

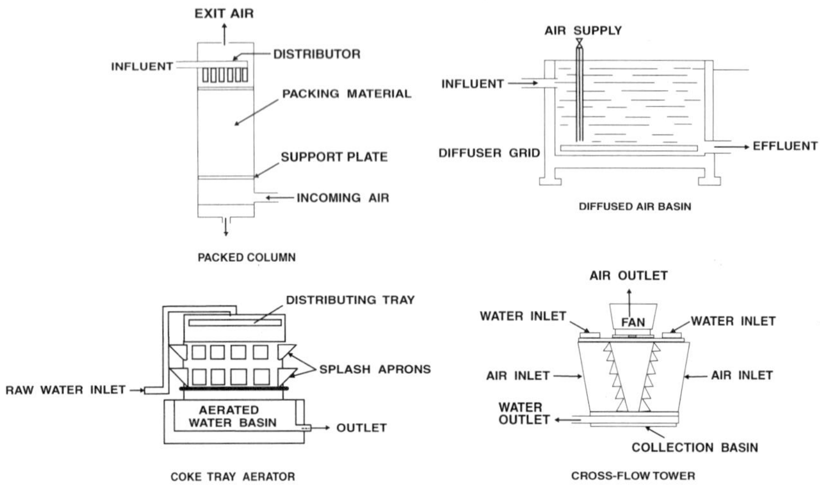

FIGURE 8.1 Schematic showing conventional air stripping equipment configuration. (After Knox et al., 1986.)

air-to-water ratio, the cross-sectional area is determined by dividing the airflow rate by the air velocity. The selection of the design air-to-water ratio must be based upon experience or pilot-scale treatability studies. Treatability studies are particularly important for developing design information for contaminated groundwater.

Air stripping and various combinations, such as steam and high-temperature air stripping, have been successfully used for removing volatile organic hydrocarbons from coproduced groundwaters. There are four basic equipment configurations used for air stripping. These include diffused aeration, countercurrent packed columns, cross-flow towers, and coke tray aerators, as shown in Figure 8.1. Diffused aeration stripping uses aeration basins similar to standard wastewater treatment aeration basins. Water flows through the basin from top to bottom with the air dispersed through diffusers at the bottom of the basin. The air-to-water ratio is significantly lower than in either the packed column or the cross-flow tower.

In the countercurrent packed column, water containing one or more impurities is allowed to flow down through a column containing packing material with airflow countercurrent up through the column. In this way the contaminated water comes into intimate contact with clean air. Packing materials are used that provide high void volumes and high surface areas. In the cross-flow tower, water flows down through the packing as in the countercurrent packed column; however, the air is pulled across the water flow path by a fan. The coke tray aerator is a simple, low-maintenance process. The water being treated is allowed to trickle through several layers of trays. This produces a large surface area for gas transfer.

8.4.2 Carbon Adsorption

Adsorption occurs when an organic molecule is brought to the activated carbon surface and held there by physical and/or chemical forces. The quantity of a compound or group of compounds that can be adsorbed by activated carbon is determined by a balance between the forces that keep the compound in solution and the forces that attract the compound to the carbon surface. Factors that affect this balance include:

- Adsorptivity increases with decreasing solubility.
- The pH of the water can affect the adsorptive capacity. Organic acids adsorb better under acidic conditions, whereas amino compounds favor alkaline conditions.
- Aromatic and halogenated compounds adsorb better than aliphatic compounds.
- Adsorption capacity decreases with increasing temperature, although the rate of adsorption may increase.
- The character of the adsorbent surface has a major effect on the adsorption capacity and rate. The raw materials and the process used to activate the carbon determine its capacity.

When activated carbon particles are placed in water containing organic chemicals and mixed to give adequate contact, adsorption of the organic chemicals occurs. The organic chemical concentration will decrease from an initial concentration of C_0 to an equilibrium concentration of C_1. By conducting a series of adsorption tests, it is usually possible to obtain a relationship between the equilibrium concentration and the amount of organics adsorbed per unit mass of activated carbon. The Freundlich isotherm and the Langmuir isotherm are most often used to represent the adsorption equilibrium.

From an isotherm test it can be determined whether a particular organic material can be removed effectively. It will also show the approximate capacity of the carbon for the application and provide a rough estimate of the carbon dosage required. Isotherm tests also afford a convenient means of studying the effects of pH and temperature on adsorption. Isotherms put a large amount of data into concise form for ready evaluation and interpretation. Isotherms obtained under identical conditions using the same contaminated groundwater for two or more carbons can be quickly and conveniently compared to determine the relative merits of the carbons.

Activated carbon adsorption may be accomplished by batch, column, or fluidized-bed operations. The usual contacting systems are fixed-bed or countercurrent moving beds, as shown in Figure 8.2. The fixed beds may employ downflow or upflow of water. The countercurrent moving beds employ upflow of the water and downflow of the carbon, since the carbon can be moved by the force of gravity. Both fixed beds and moving beds may use gravity or pressure flow.

In a typical fixed-bed carbon column, the column is similar to a pressure filter and has an inlet distributor, an underdrain system, and a surface wash. During the adsorption cycle, the influent flow enters through the inlet distributor at the top of the column, and the groundwater flows downward through the bed and exits through the underdrain system. The unit hydraulic flow rate is usually 2 to 5 gpm/ft^2. When the head loss becomes excessive due to the accumulated suspended solids, the column is taken off-line and backwashed.

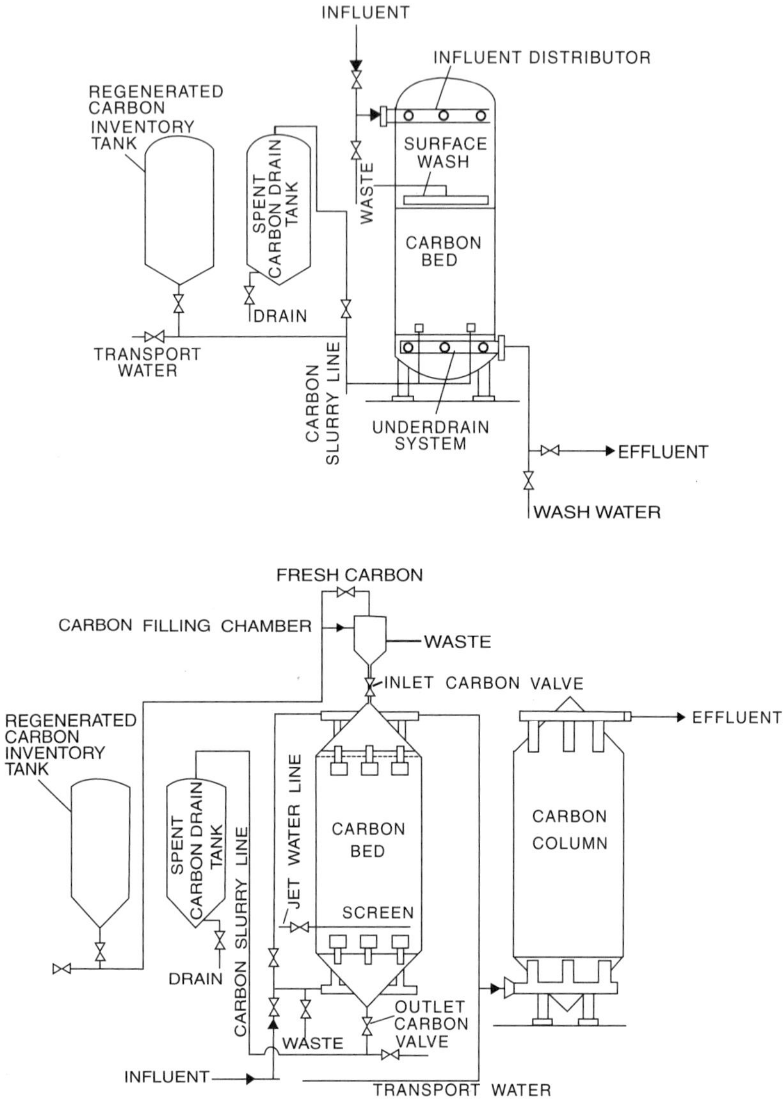

FIGURE 8.2 Schematic showing fixed-bed and moving-bed adsorption system. (After Knox et al., 1986.)

In a typical countercurrent moving-bed carbon column employing upflow of the water, two or more columns are usually provided and are operated in series. The influent contaminated groundwater enters the bottom of the first column by means of a manifold system that uniformly distributes the flow across the bottom. The groundwater flows upward through the column. The unit hydraulic flow rate is usually 2 to 10 gpm/ft^2. The effluent is collected by a screen and manifold system at the top of the column and flows to the bottom manifold of the second column. The carbon flow is not continuous, but instead is pulsed.

The fluidized bed consists of a bed of activated carbon. The water flows upward through the bed in the vertical direction. The upward liquid velocity is sufficient to suspend the activated carbon so that the carbon does not have constant interparticle contact. At the top of the carbon there is a distinct interface between the carbon and the effluent water. The principal advantage of the fluidized bed is that waters with appreciable suspended solids content may be given adsorption treatment without clogging the bed, since the suspended solids pass through the bed and leave with the effluent.

Although the treatability of a particular groundwater by carbon and the relative capacity of different types of carbon for treatment may be estimated from adsorption isotherms, carbon performance and design criteria are best determined by pilot column tests. Design-related information that can be obtained from pilot tests includes:

- Contact time
- Bed depth
- Pretreatment requirements
- Breakthrough characteristics
- Head loss characteristics
- Carbon dosage in pounds of pollutants removed per pound of carbon

The design of an activated carbon adsorption column can be accomplished using kinetic equations that require data obtained from the development of breakthrough curves.

8.4.3 Biological Treatment

In biological treatment of contaminated groundwater, the objective is to remove or reduce the concentration of organic and inorganic compounds. Because many of the compounds that may be present in contaminated groundwater can be toxic to microorganisms, pretreatment of the groundwater may be required. When a groundwater containing organic compounds is contacted with microorganisms, the organic material is removed by the microorganisms through metabolic processes. The organic compounds may be used by the microorganisms to form new cellular material or to produce energy that is required by the microorganisms for their life systems.

Heterotrophic microorganisms are the most common group of microorganisms providing the metabolic process for removing organic compounds from contaminated groundwater. Heterotrophs use the same substances (organic compounds) as sources of both carbon and energy. A portion of the organic material is oxidized to provide energy, while the remaining portion is used as building blocks for cellular synthesis. Three general methods exist by which heterotrophic microorganisms can obtain energy. These are fermentation, aerobic respiration, and anaerobic respiration.

In the case of fermentation, the carbon and energy source is broken down by a series of enzyme-mediated reactions that do not involve an electron transport chain. In aerobic respiration, the carbon and energy source is broken down by a series of enzyme-mediated reactions in which oxygen serves as an external electron acceptor. In anaerobic respiration, the carbon and energy source is broken down by a series of enzyme-mediated reactions in which sulfates, nitrates, and carbon dioxide serve as the external electron acceptors. The three processes of obtaining energy form the basis for the various biological wastewater treatment processes.

Biological treatment processes are typically divided into two categories: suspended growth systems and fixed-film systems. Suspended growth systems are more commonly referred to as activated sludge processes, of which several variations and modifications exist. The basic system consists of a large basin into which the contaminated water is introduced, and air or oxygen is introduced by either diffused aeration or mechanical aeration devices. The microorganisms are present in the aeration basin as suspended material. After the microorganisms remove the organic material from the contaminated water, they must be separated from the liquid stream. This is accomplished by gravity settling. After separating the biomass from the liquid, the biomass increase resulting from synthesis is wasted and the remainder is returned to the aeration tank. Thus, a relatively constant mass of microorganisms is maintained in the system. The performance of the process depends on the recycle of sufficient biomass. If biomass separation and concentration fails, the entire process fails. The process requires the skills of well-trained operators.

Fixed-film biological processes differ from suspended growth systems in that microorganisms attach themselves to a medium that provides an inert support. Biological towers (trickling filters) and rotating biological contactors are the most common forms of fixed-film processes. Biological towers are a modification of the trickling filter process. The media, which normally comprises polyvinyl chloride (PVC), polyethylene, polystyrene, or redwood, are stacked into towers that typically reach 16 to 20 ft in height. The contaminated water is sprayed across the top, and, as it moves downward, air is pulled upward through the tower. A slime layer of microorganisms forms on the media and removes the organic contaminants as the water flows over the slime layer.

A rotating biological contactor (RBC) consists of a series of rotating disks, connected by a shaft, set in a basin or trough. The contaminated water passes through the basin where the microorganisms, attached to the disks, metabolize the organics present in the water. Approximately 40% of the disk surface area is submerged. This allows the slime layer to come into contact alternately with the contaminated water and the air where oxygen is provided to the microorganisms.

Removal efficiencies are generally the same for fixed-film and suspended growth processes. However, fixed-film processes have the potential to be lower in cost, due to the absence of aeration equipment, and they are easier to operate. Both systems may be operated under anaerobic conditions, which may offer advantages for certain contaminated waters.

The addition of powdered-activated carbon (PAC) to the activated sludge process has received considerable attention, particularly with respect to the removal of specific organics. The applicability of activated carbon in removing specific substrates depends on the molecular weight, solubility, polarity, location of functional groups, and overall molecular configuration. Investigations of PAC systems have centered around process enhancement factors. These include:

- Enhancement of a system to receive shock loadings or temperature changes;

- Improved nonbiodegradable organics removal;
- Improved removal of specific organics;
- Improved color removal;
- Enhanced resistance to biologically toxic or bacteriostatic substances;
- Improved hydraulic capacity of existing facilities;
- Improved nitrification;
- Suppressed foaming in the aeration system;
- Improved sludge handling; and
- Reduced sludge bulking.

In recent years research has been primarily directed toward removal of priority pollutant organics, removal of other residual organic compounds, enhancement of nitrification, and improvement in the settling of the sludge.

8.5 TREATMENT TRAINS

Because of the complex composition of most groundwaters, no one unit operation is capable of removing all of the contaminants present. It may be necessary to combine several unit operations into one treatment process to remove effectively the contaminants required. To simplify and make visible the selection of the applicable treatment trains, a number of unit operations and the waste types for which they are effective are presented in Table 8.1.

TABLE 8.1
Summary of Treatment Alternatives Suitability

Remediation Alternatives	Volatile Organics	Nonvolatile Organics	Inorganics
Air Stripping	Suitable for most cases	Not suitable	Not suitable
High-temperature air stripping	Effective removal technique	May be suitable	Not suitable
Steam stripping	Effective concentrated technique	May be suitable	Not suitable
Carbon adsorption	Effective removal technique	Effective removal technique	Not suitable
Biological	Effective removal technique	Effective removal technique	Not suitable, metals toxic
pH adjustment precipitation	Not applicable	Not applicable	Effective removal technique
Membrane processes	May not be applicable	Effective removal technique	Effective removal technique
Electrodialysis	Not applicable	Not applicable	Inefficient operation; inadequate removal
Ion exchange	Not applicable	Not applicable	Inappropriate technology; difficult operation

8.6 COST COMPARISONS

At one refinery, the following groundwater flow and water quality conditions were determined to be applicable during hydrocarbon recovery operations:

Flow	500 gpm
BTEX	20 mg/l (total)
Iron	5–10 mg/l
Manganese	<1.0 mg/l

These water quality conditions are fairly typical for coproduced groundwater during hydrocarbon recovery operations at refineries. The principal treatment processes evaluated consisted of primary treatment for inorganics removal followed by air stripping and activated carbon adsorption. One alternative, also evaluated, consisted of primary treatment for inorganics removal followed by use as cooling tower makeup water and then biological treatment in existing refinery processes. However, this alternative may not be possible in most refineries because of the limited hydraulic capacities in their biological treatment systems.

8.6.1 ALTERNATIVE 1

Alternative 1 consists of preliminary treatment for heavy metals removal with the primary concern being iron removal (Figure 8.3). The levels of iron observed in the groundwater at this site would be very detrimental to the downstream treatment processes. This pretreated water would then be used for cooling tower makeup water followed by biological treatment. This approach would be the easiest and cheapest alternative. This combined process should provide effective removal of BTEX.

Two advantages of this approach would be the minimized capital and operating costs, along with minimal space requirements. The best process (for iron removal) would probably be chemical precipitation with lime through a rapid mix, flocculation, and DAF step. DAF size requirements would be in the range of a 300- to 350-ft^2 unit. Total costs for operation and maintenance of the preliminary treatment facility using an interest rate of 10% over 10 years are estimated at \$0.44/1000 gal. A summary of the capital and operating cost estimates is presented in Table 8.2.

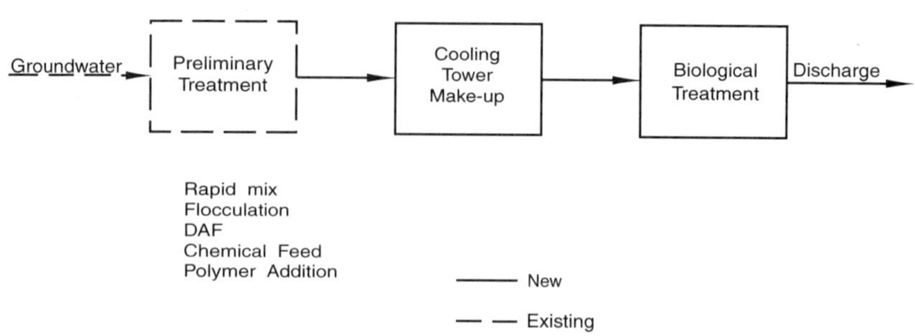

FIGURE 8.3 Preliminary treatment, cooling towers, and biological treatment (Alternative 1).

TABLE 8.2
Summary of Total Annual Estimated Cost (in Dollars) for Groundwater Treatment Alternatives

Expenditures	Alternative 1	Alternative 2	Alternative 3	
			Without Regeneration[a]	With Regeneration[b]
Capital cost	216,000	800,000	634,000	634,000
Annual projected cost at 10% interest for 10 years	35,000	130,000	103,000	103,000
Annual operation & maintenance cost	81,500	205,500	638,000	372,500
Total annual estimated	116,500	335,500	741,000	475,500
Actual water treated cost ($1000 gal)	0.44	1.28	2.82	1.81

[a] Costs do not include sludge dewatering and disposal costs.
[b] Existing carbon regeneration furnace on site.

8.6.2 Alternative 2

Alternative 2 consists of preliminary treatment followed by dual-media pressure filtration, and two-stage air stripping (Figure 8.4). The preliminary treatment step for iron removal would be exactly the same as specified under Alternative 1. The filters would be recommended to remove suspended matter and particulate iron prior to the air strippers. The required filtration capacity could be provided with either a duplex system of two 60-in.-diameter filters or a triplex system of three 42-in.-diameter filters.

A two-stage air-stripping system would be required to get the BTEX levels down to nondetectable levels. The first stage would be designed to achieve effluent levels of 100 to 200 ppb, with the latter stage getting down to the level of detection. Preliminary sizing indicated four 4-ft-diameter columns with two columns in the first stage and two columns in the second stage. The off-air emission from the first stage would require treatment to prevent atmospheric discharges of BTEX. The emission control system was costed-out for activated carbon treatment of the air emissions with steam regeneration; however, both the capital and operating costs could be reduced significantly if there were incinerator(s) on site that could use the off-air for combustion air. The total cost with an emission control system was estimated to be $1.28/1000 gal.

8.6.3 Alternative 3

Alternative 3 consists of preliminary treatment followed by filtration and activated carbon adsorption (Figure 8.5). The preliminary treatment step and filtration requirements and costs would be the same as specified for Alternative 2 (air stripping).

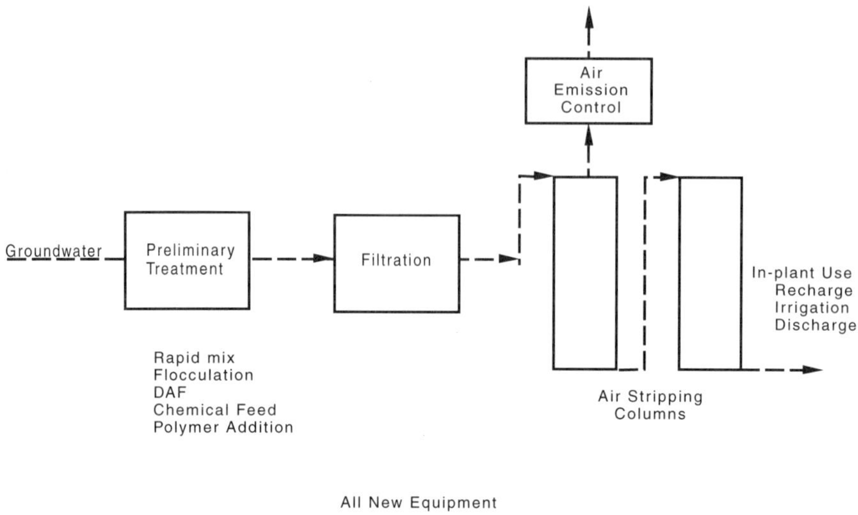

FIGURE 8.4 Preliminary treatment, filtration, and two-stage air stripping (Alternative 2).

A two-stage activated carbon system would also be required to get the BTEX levels down to nondetectable levels. Two 10-ft-diameter columns should provide the required treatment capacity. Each column would be charged with 20,000 lb of carbon. An estimated carbon usage rate of 2 lb/1000 gal of water treated was used for cost estimates. Costs were estimated both with and without on-site regeneration of the carbon. As can be observed in Table 8.2, it would be cheaper to regenerate on-site. The difference in costs is estimated at \$1.80 vs. \$2.82/1000 gal of water treated. Costs for carbon regeneration were based on using an existing multiple-hearth carbon regeneration furnace on site at the refinery. A total breakdown of costs (capital and operations and maintenance) is presented in Table 8.2. These costs do not include sludge dewatering and disposal facilities.

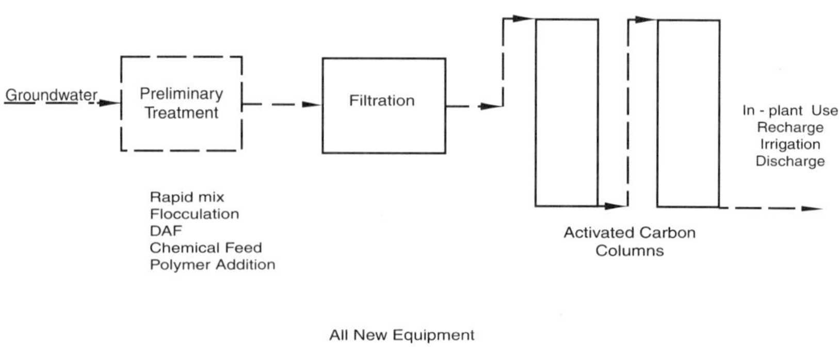

FIGURE 8.5 Preliminary treatment, filtration, and granular-activated carbon adsorption (Alternative 3).

8.7 DISPOSAL OPTIONS

8.7.1 Surface Discharge

Discharge of the coproduced water without treatment is dependent upon the level of contamination, the volume produced, and limitations set forth by existing discharge permits. If the concentration levels are low enough and a discharge permit (with regular sampling) is in place, generally only a volume limit adjustment to the permit is required; however, discharge permits are difficult to obtain in certain areas. In terms of operation and maintenance costs, surface discharge is a cost-effective option. At most large facilities, refinery and bulk terminals, retention basins (for storm water runoff) and/or oil–water separators are available. In-line placement of one or both of these features prior to surface discharge may be desirable depending upon volume and concentration limits imposed by the permit. The oil–water separator will be utilized as a safety precaution in the event that the NAPL is discharged. The retention ponds can accept cascading coproduced water, which may act as a passive air-stripping method. Concentration levels could be reduced through offgassing.

Surface discharge without treatment will ultimately depend on the total mass of contaminants discharged and the acceptable mass-loading rate for the particular discharge point. Therefore, recovery programs producing groundwater at low volumes with high BTEX concentrations or high volumes with moderate BTEX concentrations will likely require some level of pretreatment prior to direct discharge.

8.7.2 Site Reuse

Typically at NAPL recovery sites (i.e., refineries and bulk terminals), some quantity of groundwater is coproduced. A major concern arises from this process in that the coproduced water must be treated, disposed of, or both. An evaluation process then follows on how to handle the coproduced water. A number of factors that control the ultimate fate of the water include the average volume produced on a regular basis, the level of contamination, and site-specific physical constraints.

It is one thing to design, test, and maintain a recovery system that effectively recovers NAPL, but the handling of the coproduced groundwater is another matter. The effectiveness of a recovery system can suffer if the disposal capacity is insufficient or treatment costs are high. Additionally, a cost-effective, low operation and maintenance recovery system can quickly be turned into a high-cost system if the coproduced water fate is not coordinated with the recovery system. A key factor will be the level of treatment required, if any, prior to disposal. Three alternatives are explored: surface discharge; refinery reuse; and reinjection.

Refiner reuse is a valid disposal option. Again, depending upon the level of contamination and the ability of the recovery system to provide a constant volume, this option can be a desirable one if the facility is currently purchasing large quantities of water imported from local water utilities. With this option, costs for disposal design, construction, and operations and maintenance are eliminated. In addition, a reduction in the refinery's water bill can be realized. Problems generally encountered with this option include the natural quality of the groundwater, including iron and manganese content, and alkalinity. Typically, minor or passive treatment methods, such as oil–water separators and retention pond settling, can eliminate more costly treatment methods depending upon the use of the water. One possible option is to use the water for cooling tower makeup water. Depending on the volume, this makeup water could then be routed through the existing wastewater treatment facilities prior to direct discharge.

8.7.3 Reinjection

Reinjection of coproduced groundwater through the use of wells is commonly used to return the water to the same aquifer and to set up hydraulic barriers in an effort to contain the plume. Injection wells are commonly used in conjunction with withdrawal systems to enhance the recovery of hydrocarbons. Injecting water at appropriate locations will create a pressure ridge to increase the hydraulic gradient effectively toward the withdrawal point. Normally, the water pumped from the recovery wells is used as the injection water and is injected, without treatment. This method provides an economical way of handling the produced water, as well as being beneficial to the recovery effort.

8.7.3.1 Regulatory Aspects

In the United States, the EPA has considered (but to date has not implemented) a ban on this type of injection by reclassifying these injection wells as Class IV wells or wells that inject hazardous substances into or above a potential underground source of drinking water. Class IV injection wells are banned by law in the United States. There is technical justification to argue against this potential ban at hydrocarbon recovery sites. As long as free and residual hydrocarbons are present above the water table in the area of reinjection, there will be continuous contamination by dissolved fractions. The argument for reinjection without treatment would be that the injection water is the same quality of water already present within the aquifer, as long as it is reinjected into the same geologic horizon from which it was originally pumped.

The issue of dissolved fractions, in terms of aquifer restoration objectives, would be more appropriately addressed at a phase of the remediation program after the source of the dissolved contamination is controlled. Of course, containment of the dissolved plume remains a priority throughout the program. For large-scale recovery programs, requiring treatment prior to reinjection would place an excessive economic burden on the overall remediation effort without technical justification, at least during the course of NAPL recovery.

8.7.3.2 Zones of Reinjection

Coproduced groundwater can be reinjected within, above, or below the LNAPL zone depending on subsurface conditions, LNAPL occurrence and removal strategies, and site-specific constraints. These options and their respective advantages and limitations are discussed below.

8.7.3.2.1 Within LNAPL Zone

Probably the most common procedure is to utilize reinjection wells which are completed in the same aquifer as the LNAPL production wells. If the geologic setting of the formation is homogeneous and isotropic (without zones of preferential hydraulic conductivity), this strategy is effective. The rate of water injection can be controlled to maintain most of the water–LNAPL interface at a reasonably stable elevation and optimize the horizontal flow velocity for efficient product flushing. However, the effectiveness of this strategy is limited if a zone of significantly greater (or lesser) hydraulic conductivity is present near, but not within, the production zone, the water table is highly variable on a seasonal basis, or the LNAPL pool is in direct vertical communication with the source of the spill.

8.7.3.2.2 Below LNAPL Zone

At specific locations where the aquifer below the water–LNAPL interface has not been contaminated, it may be advantageous to inject clean water beneath the production zone. Injection at this level creates a hydraulic mound, which enhances recovery while commingling only a small amount of clean water with the contaminated zone. A second advantage can be gained by injecting oxygen-deficient water to inhibit biological growth. Limitations exist, however, when the aquifer underlying the water–LNAPL interface has a hydraulic conductivity that is significantly less than the interface zone, or the vertical component of hydraulic conductivity is significantly less than the horizontal. Care must be taken, however, to prevent the LNAPL layer from spreading upward in a reverse "smear."

8.7.3.2.3 Above LNAPL Zone

At locations where the LNAPL pool is near or directly under the source of the spill, hydrocarbon-affected soil may exist from the surface to the saturated zone. Residual LNAPL within the vadose zone can be flushed using recharge water in a leach field, surface ponding, and spray irrigation techniques. Use of surface recharge to flush residual hydrocarbon involves at least two separate processes. The added pressure caused by the continual source of downward moving water can mechanically displace trapped LNAPL from soil pores. Also, the introduction of oxygen-saturated water will often enhance the growth of petroleum-consuming microbes. Many varieties of these microbes produce a surfactant-like chemical which reduces surface tension of the hydrocarbon to enable the microbe to adsorb and digest it. However, under the dynamic conditions of flushing, some of the LNAPL is released from its retention and migrates ahead of the approaching water front. Recharge, however, is effective only if the vertical hydraulic conductivity of the formation is sufficient to accept the water. Other limitations include biological growth and mechanical pore plugging, which may require frequent service (Figure 8.6). The vadose zone must also be sufficiently thick to prevent surface mounding.

8.7.3.3 Injection Well Construction

Reinjection wells are usually constructed of either 2-, 4-, or 6-in. PVC. The screen portion of the well usually has slotted openings. Oversized screen openings and

FIGURE 8.6 Photographs showing microbial activity and scale buildup on injection well screen (a) and within casing (b).

larger gravel packs are normally designed for reinjection wells as opposed to withdrawal well design. The reason for the overdesign is to gain the most open area possible. Also, groundwater is being injected into the well or placing a positive pressure on the formation. Therefore, silting of the well from this overdesign is not of major concern. However, gross overdesign should be avoided, because eventually the well will have to be redeveloped and the most effective way to develop the reinjection well is to reverse the flow by pumping or air lifting. Well diameter is also important; however, doubling screen diameter only increases the injection rate by 17 to 25%. Furthermore, screen type can affect injection well performance more than screen diameter. For a long service life, injection wells with high screen open area and injection rates are desirable.

8.7.3.4 Well Design

Appropriate injection well design and construction techniques vary according to the specific conditions present at a given site. In all cases, the greatest care in design and construction must be taken due to the consequences of water-chemistry problems, air entrainment, thermal interferences, sand or silt pumping (i.e., from the recovery wells), and microbial reactions; injection wells are more likely to fail than recovery wells. Clogging of screens is the most serious problem in injection well operation. Thus, screen open area and screen length must be optimal. The typical injection well should consist of sand-free pump water, be efficient (i.e., maximum

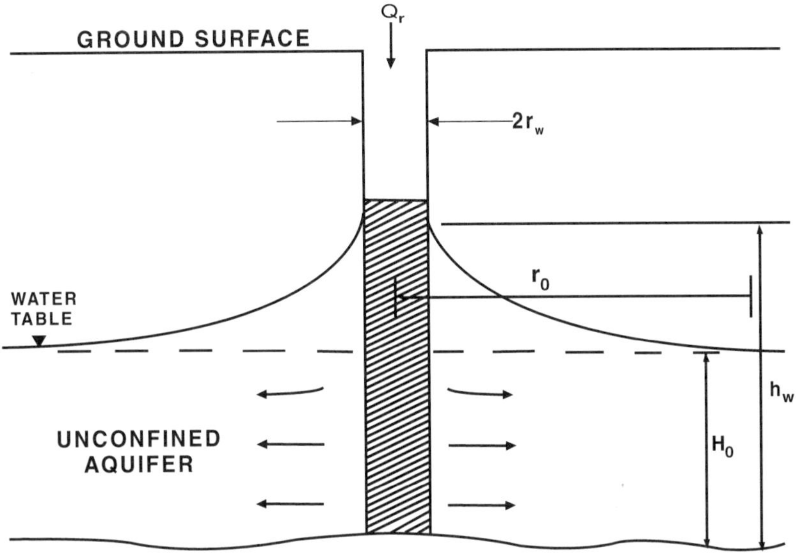

FIGURE 8.7 Schematic showing radial flow from a recharge well penetrating an unconfined aquifer.

recharge at minimum pressure buildup), and be cost-effective in terms of initial investment and operation and maintenance costs.

Injection tubing forms an important part of the design for recharge wells. The injection tube should terminate below the static water level (so the injected water is exposed to air), and must be designed so that positive pressure exists along its entire length. Backpressure valves may be installed to eliminate negative pressures in the injection tube. Another important criterion is that the injection tube should provide for full flow to eliminate the possibility of air entrapment or degassing.

When water is discharged into an injection well, a cone of impression will be formed which is similar in shape (but in reverse) to a cone of depression surrounding a pumping well (Figure 8.7). The equation describing the cone for various discharges can be derived using the same assumptions applied to a pumping well. For an unconfined aquifer with water being recharged into a well completely open to the aquifer at a constant rate (Q in gallons per minute) the following equation is applicable:

$$Q_r = \frac{K(b_w^{\ 2} - H_o^{\ 2})}{1055 \log_{10}(r_o / r_w)} \tag{8.1}$$

where Q_r = rate of injection (in gpm), K = hydraulic conductivity (in gpd/ft^2), b_w = head above bottom of aquifer while recharging (in ft), H_o = head above the bottom of aquifer when no pumping is occurring, r_o = radius of influence (in ft), and r_w = radius of injection well (in ft).

In summary, when undertaking a project such as the recovery of LNAPL, treatment of the coproduced water, prior to reinjection, may not be beneficial or technically necessary. A large percentage of the spilled or leaked petroleum hydrocarbon (40 to 60%) will be retained in the unsaturated zone as residual saturation. This residual hydrocarbon cannot be recovered by conventional withdrawal techniques. Without removing this continual source of contamination to the groundwater system, dissolved contamination will continue. Therefore, in most cases, it may be pointless and extremely costly to treat the coproduced groundwater prior to reinjection while the free- and residual-phase hydrocarbon contamination exists.

8.7.3.4.1 Matrix Fracturing

In weakly consolidated, stratified sediments, the injection pressure must be controlled so that the surrounding formation is not fractured. If fracturing occurs, there is usually a severe loss in hydraulic conductivity because the bedding planes are disturbed. Pressures that will cause fracturing range from a low of 0.5 psi/ft of depth for poorly consolidated coastal plain sediments, to 1.2 psi/ft depth for crystalline rock. For most recharge wells in unconsolidated sediments, the injection pressure should be carefully controlled so that the positive head (in psi at the surface) does not exceed $0.2 \times h$, where h is the depth (in ft) from the ground surface to the top of the screen or filter pack.

8.7.3.4.2 Design Alternatives

Injection well design for unconsolidated sediments often includes screening the well through some portion of the vadose zone (Figure 8.8a). Most of the wells commonly constructed in this manner inject low-iron-content (i.e., <0.03 mg/l) water, either as treated water (i.e., treated wastewater) or surface water. Injecting high-iron-content water into the vadose zone may result in rapid precipitation due to the presence of air in the voids and substantial plugging of that zone. Since the allowable injection pressure is $0.5(h)$ psi, where h is the distance from the surface to the top of the screen, there is a trade-off between gained injection capacity from the additional open area, and shortened service life due to the plugging of the unsaturated zone and lower allowable infection pressures.

Injection into unsaturated deposits must also consider that when the injected water leaves the injection tubing, it is at atmospheric pressure, and the driving head is lost which lowers injection efficiency. In addition, the capillary and surface tension forces in the void spaces provide resistance to water movement, thus limiting injection efficiency. Injection into only the saturated zone (Figure 8.8b) maintains a positive head pressure on the water until it exits the well screen. This limits the chemical reaction that might occur in the well bore. Since the void spaces are saturated, there are no capillary or surface tension forces to overcome.

8.7.3.5 Injection Well Operations

Groundwater quality at most refinery and terminal facilities is not pristine. Elevated concentrations of iron, manganese, BTEX, and petroleum hydrocarbons exist. Upon exposure to the atmosphere or air mixing, changes begin to occur in the makeup of the coproduced groundwater. Iron will precipitate out and form a scale on the

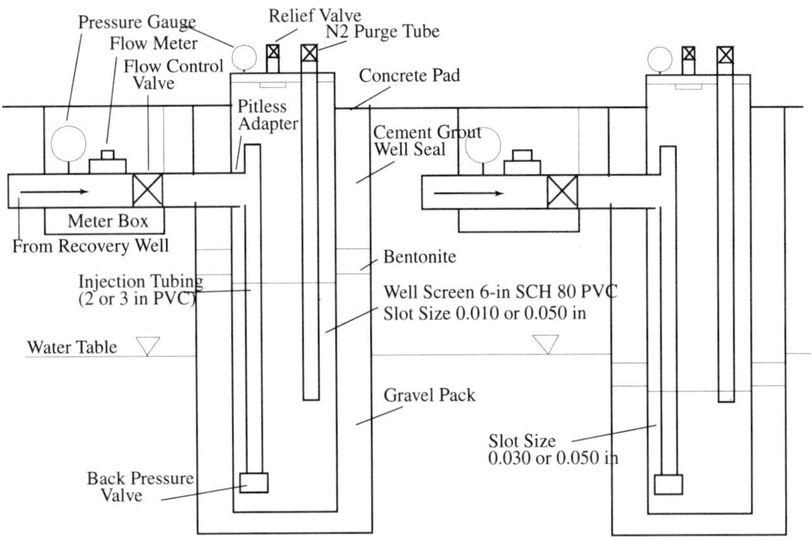

FIGURE 8.8 Schematic showing recommended injection well design screening the vadose (a) and saturated (b) zone.

reinjection well openings. This will reduce the efficiency of the reinjection well quickly. Iron bacteria will also form when three important components are present. These are iron, air, and dissolved petroleum hydrocarbon. If a reinjection well is allowed to degrade to such a point whereby significant amounts of iron bacteria have formed in the screen openings of the well, the gravel pack, and the surrounding formation, redevelopment of the well may not be feasible. Reinjection systems should, therefore, if possible, employ the logic of a closed system (i.e., eliminate exposure of the water to the atmosphere). To do this, the system must be constructed with a closed pipe from the discharge point to the reinjection well. At the reinjection well, a drop pipe is installed usually with a flow restriction at the bottom to reduce the potential for cascading. The bottom of the drop tube is terminated at or near the bottom of the well, beneath the static fluid level, also to aid in the prevention of cascading water.

An important factor to keep in mind when designing reinjection wells or systems is that eventually they will have to be redeveloped. Reinjection wells and the redevelopment of these wells unfortunately follow the law of diminishing returns. Constant reinjection of coproduced groundwater without regular development results in irreversible clogging and ultimately the abandonment of the well.

The location and spacing of reinjection wells is a design area that is sometimes overlooked. Most of the design effort is placed in constructing the most efficient and strategically located recovery well. Granted, the recovery well design is extremely important, but the location and spacing of reinjection wells can be critical in the total system performance. Proper spacing and location of reinjection wells can be advantageous in the overall scheme. Strategic placement of the injection wells upgradient or downgradient in relationship to the LNAPL hydrocarbon pool and the distances are all realistic considerations. Placing the reinjection wells in a relative upgradient location, in relationship to the LNAPL pool, forms a groundwater mound that creates an increased gradient in the direction of the recovery well. This, of course, is a desirable effect where the ultimate goal is to recover the largest amount of free-phase hydrocarbon possible in a shorter time frame.

Locating reinjection wells downgradient of the LNAPL pool can be advantageous as well. In this instance, a line of downgradient reinjection wells creates the same groundwater high or mound. This mound in essence forms a groundwater wall or barrier. The barrier serves two purposes: first, the barrier can prevent off-site migration; and, second, the barrier blocks the migrating free-phase hydrocarbon, causing it to accumulate in this area for recovery. Obviously, the location of recovery wells and reinjection wells is closely associated.

The location of the reinjection wells with respect to the recovery well is also an important consideration. Locating the wells too closely can create counterproductive effects. Locating the reinjection wells too far away from the recovery well may significantly reduce the desired effect or eliminate the benefits altogether. If the latter is the case, "dead spots" may result and may cause short-circuiting of the LNAPL pool hydrocarbon plume. Once this happens, off-site migration of the plume is possible. In fact, reinjection wells placed and spaced improperly can, in some instances, accelerate off-site migration.

Reinjection wells are not without disadvantages. One disadvantage referred to above is the possibility of accelerating off-site migration or dispersion of the pool. Another disadvantage is the additional costs for operation and maintenance (i.e., redevelopment). Redevelopment costs can be kept to a minimum by initial design considerations, but, as the time frame between redevelopment decreases, redevelopment costs increase. Eventually the cost of constructing a new reinjection well becomes the only cost-effective alternative.

REFERENCES

Bellandi, R., 1988, Hazardous Waste Site Remediation, *The Engineer's Perspective,* O'Brien & Gere Engineers, Inc.: Van Nostrand Reinhold, New York, 422 pp.

Dougherty, P. J. and Paczkowski, M. T., 1988, A Technique to Minimize Contaminated Fluid Produced during Well Development: In *Proceedings of the Association of Ground Water Scientists and Engineers and USEPA Second National Outdoor Action Conference on Aquifer Restoration, Ground Water Monitoring, and Geophysical Methods*, May, Vol. 1, pp. 303–317.

Driscoll, F. G., 1986, *Groundwater and Wells,* Second Edition: Johnson Division, St. Paul, MN, 1089 pp.

Ford, D. L., 1980, *Technology for Removal of Hydrocarbon from Surface and Groundwater Resources*: Engineering Science, Inc. and the University of Texas, Austin, Texas.

Fryberger, J. S. and Shepard, D. C., 1987, Reinjection of Water at Hydrocarbon Recovery Sites: In *Proceedings of the International Symposium on Class V Injection Well Technology*: USEPA and UIPC, September, Washington, D.C.

Havash, J. and Oster, J., 1998, Is Your Wastewater Toxic to the Municipal Treatment Plant? *Pollution Engineering*, March, pp. 52–54.

Kaufmann, H. G., 1982, Granular Carbon Treatment of Contaminated Supplies: In *Proceedings of the National Water Well Association Second National Symposium on Aquifer Restoration and Ground Water Monitoring*, May, pp. 94–104.

Knox, R. C., Canter, L. W., Kincannon, D. F., Stover, E. L., and Ward, C. H., 1986, *Aquifer Restoration — State of the Art*: Noyes Publications, Park Ridge, NJ, 490 pp.

Lamarre, B. L., McGarry, F. J., and Stover, E. L., 1983, Design, Operation and Results of a Pilot Plant for Removal of Contaminants from Ground Water: In *Proceedings of the National Water Well Association Third National Symposium on Aquifer Restoration and Ground Water Monitoring*, May, pp. 113–122.

Olsthoom, I. T. N., 1982, *The Clogging of Recharge Wells, Main Subjects*: The Netherlands Waterworks Testing and Research Institute, KIWA.

Pitre, M. P., Enegess, D. N., and Unterman, R., 1999, Bioreactors: The New Wave in Waste Water Treatment: *Environmental Protection*, September, pp. 30–33.

Stover, E. L., 1982, Removal of Volatile Organics from Contaminated Ground Water: In *Proceedings of the National Water Well Association Second National Symposium on Aquifer Restoration and Ground Water Monitoring*, May, pp. 77–84.

Stover, E. L., 1989, Coproduced Ground Water Treatment and Disposal Options during Hydrocarbon Recovery Operations: *Ground Water Monitoring Review*, Winter, Vol. 9, No. 1, pp. 75–82.

Stover, E. L. and Kincannon, D. F., 1983, Contaminated Groundwater Treatability — A Case Study: *American Water-Works Association*, Vol. 75, No. 6, pp. 292–298.

Stover, E. L., Fazel, A., and Kincannon, D. F., 1985, Powdered Activated Carbon and Ozone Assisted Activated Sludge Treatment for Removal of Toxic Organic Compounds: *Science and Engineering*, Vol. 7, No. 3, pp. 191–203.

Stover, E. L., Gates, M. M., and Gonzalez, R., 1986, Treatment and Removal of Dissolved Organics and Inorganics in a Contaminated Ground Water — A Case Study: In *Proceedings of the National Water Well Association and American Petroleum Institute Conference on Petroleum Hydrocarbons and Organic Chemicals in Ground Water — Prevention, Detection and Restoration*, Houston, TX, pp. 689–708.

Vance, D., 1998, Redox Reaction for Insitu Groundwater Remediation: *Environmental Technology*, September/October, p. 45.

Wilson, J. T., Leach, L. E., Michalowski, J., Vandegrift, S., and Callaway, R., 1989, *In-Situ Bioremediation of Spills from Underground Storage Tanks: New Approaches for Site Characterization, Project Design, and Evaluation of Performance*: U.S. Environmental Protection Agency Report No. EPA/60012-89/042, July, 56 pp.

Zinner, R. E., Hodder, E. A., Cartoll, W. E., and Peck, C. A., 1991, Utilizing Groundwater Reinjection in the Design of a Liquid Hydrocarbon Recovery System: In *Proceedings of the AGWSE and API Conference on Petroleum Hydrocarbons and Organic Chemicals in Groundwater: Prevention, Detection and Restoration*, November 20–22, Houston, TX, pp. 469–483.

9 Remediation Strategies for Dissolved Contaminant Plumes

"Experience demonstrates that, at most sites, 10% of the total remediation cost can clean up 90% of the contamination. Is it better to have one site 100% clean or 10 sites 90% clean? What is the risk?"

9.1 INTRODUCTION

Restoration of an aquifer that has been contaminated by petroleum hydrocarbons and organic compounds is a multifaceted endeavor that requires the remediation specialist first to assess the risk and potential use of the aquifer. If the aquifer is a primary source of drinking water, then every reasonable effort is set forth to develop a remedial strategy to achieve promulgated cleanup or drinking water standards. Where the aquifer is either not capable of producing a reasonable quantity of water or if the water contains other compounds (i.e., elevated dissolved solids, chlorides, salts, etc.) at concentrations that restrict its use, cleanup standards may not be as significant and can be negotiated with the lead agency depending upon the future use of the site, among other factors. This assumes that the source has been mitigated and that what recoverable free-phase NAPL exists has been recovered, regardless of the presence of residual hydrocarbons adsorbed to soil and in the dissolved phase.

Aquifer restoration planning should be based on the degree of restoration required, and a strategy developed based on an evaluation of the risk presented to public health and the environment. This is important since complete aquifer restoration is rarely accomplished despite over two decades of intense groundwater remediation research and activities. Aquifer restoration requires several objectives to be accomplished: contaminant plume containment, source containment, contaminant plume removal, and source zone removal.

Plume containment involves stabilization of the dissolved plume hydraulically downgradient. Containment can be achieved by several means, including natural-gradient renovation with minimal pumping and treatment directly within the source area; pump-and-treat scenarios, which rely on aquifer hydraulics and solute transport mechanisms and consist of the select placement of extraction wells; or natural flushing with capture.

Source containment involves the elimination of future contribution of contaminants to the groundwater by the recovery of recoverable NAPLs, removal or treatment of impacted soil within the vadose zone, and source elimination (i.e., removal of leaking underground storage tank, etc.). Source zone containment can be accomplished with pump-and-treat scenarios, *in situ* treatment, the use of cutoff walls, or a combination thereof.

Contaminant plume removal is achievable through a variety of strategies and involves the removal or treatment of the dissolved contaminant plume. Plume recovery involves a pump-and-treat strategy, which can be enhanced with bioremediation (i.e., microorganisms), chemical flushing, and/or steam or thermal extraction. During this phase, the quantity of groundwater generated is significantly greater than the quantity of the contaminant plume.

Source zone removal is by far the most difficult to achieve and involves adsorbed immiscible hydrocarbons and organic compounds that can be extremely difficult to address, except for those cases where the impacted zone is shallow and can easily be excavated. With exception to source zone removal, these objectives can be achieved with some level of success with conventional pump-and-treat technology. Source zone removal requires much more than simple hydraulics to remove the mass of dissolved and adsorbed phases effectively. This is especially true in low-permeability subsurface environments or within complex geologic settings. Experimental approaches currently being attempted in small-scale field tests involve the use of surfactants or alcohol flushing. In reality, complete aquifer restoration is seldom accomplished under the best of conditions. Although complete aquifer restoration is rarely accomplished, the restoration effort is not deemed necessarily unsuccessful, and site closure can be achieved with a well-designed and developed strategy, and realistic and reasonable objectives.

Over the past two decades, several strategies (including some exotic experimental approaches) have been employed to remove petroleum hydrocarbons and organic compounds from aquifers or to treat them in place. These remediation approaches include physical systems such as removal through pump-and-treat and air sparging, biological systems utilizing microorganisms, or chemical systems with chemical or fixation processes by either mechanical entrapment or chemical reaction (Table 9.1). The majority of successful organic chemical cleanup projects have been based on physical and biological (bioremediation) strategies. This chapter focuses on the more conventional strategies including pump-and-treat technology, air sparging (and complementary modifications and enhancements), and bioremediation. Overall, *in situ* bioremediation appears to offer the most promise for cost-effective remediation, albeit not necessarily timely remediation, and thus is given the most attention.

9.2 PUMP-AND-TREAT TECHNOLOGY

A traditional approach to aquifer remediation is to remove the contaminated water by pumping, treating the water at the surface, and then either discharging it or reinjecting it back into the aquifer. When the recovery wells are properly located, this approach has the advantage of creating a capture area which contains and prevents the contamination from migrating.

Setting up an effective pumping recovery system requires a thorough understanding of subsurface conditions, especially the hydraulic conductivity, storativity, variations in vertical and horizontal geologic conditions, regional hydraulic gradient,

TABLE 9.1
Summary of Conventional Aquifer Restoration Strategies for Dissolved Phases

Strategy	Description	Remarks
Physical Strategies		
Pump-and-treat	Pumps water via wells for aboveground treatment and subsequent discharge	Cleanup to promulgated standards difficult to accomplish once asymptotic conditions are reached Generates large volumes of water relative to contaminant mass
Pulsed or variable pump-and-treat	Varied pumping rate allowing for contaminants to dissolve, desorb, and/or diffuse from stagnant areas	Same as pump-and-treat May increase cleanup time
Air sparging	Injects air below the water table and captures it above the water table to extract volatile contaminants and promote biodegradation	Can be inefficient in low-permeability zones and complex geologic settings Typically limited to depths less than 30 ft Multicomponent mixtures can adversely affect extractability
Steam-enhanced extraction	Injects steam above and/or below water table to promote volatilization of contaminants	Can be inefficient in low permeability zones or complex geologic settings
In situ thermal	Injects heat above the water table via Joule heating, radio frequency heating, or means to promote volatilization of contaminants	Difficulty in attaining uniform heat distribution
Natural attenuation	Allows contaminants to biodegrade naturally without human intervention other than monitoring	Requires demonstration that biodegradation is occurring Periodic monitoring required Source elimination required Plume stable or reducing Nonbeneficial water or naturally poor water quality preferred
Physical containment	Physically contains contaminant plume with use of cutoff walls, caps, liners, etc.	Site specific Limited in depth
In situ reactive barriers	Treat contaminated water as it passes through a physical barrier containing reactive chemicals, organisms, or activated carbon	Site specific Slow process

TABLE 9.1 *(continued)*
Summary of Conventional Aquifer Restoration Strategies for Dissolved Phases

Strategy	Description	Remarks
Biological Strategies		
In situ bioremediation	Pumps nutrients through subsurface to promote growth of microorganisms that biodegrade contaminants	Can be inefficient in low-permeability zones and complex geologic settings NAPLs can impede progress and growth of microorganisms Compound specific when chlorinated solvents are present Accumulation of intermediate compounds considered hazardous may result with chlorinated solvents
Chemical Strategies		
Soil flushing	Flushes surfactants or cosolvents below water table to promote recovery of contaminants with low water solubility	Can be inefficient in low-permeability zones or complex geologic settings Slow reaction rates Adverse chemical reactions a concern
In situ chemical treatment	Injects chemicals to transform contaminants in place	Inefficient in low-permeability zones and complex geologic settings Slow reaction rates Adverse chemical reactions a concern

and natural sources of recharge. When these factors have been defined, it is possible to calculate the necessary well spacing, pumping rate, and possibly reinjection locations required to establish hydraulic control. Recovery wells can be constructed as conventional vertical wells, or as horizontal wells. A number of public-domain and commercial computer models, with varying degrees of sophistication, are readily available to assist the environmental professional with system design.

Pumping is an important aspect for recovery of chemicals that are not easily degraded or attenuated in the subsurface. Oil field saltwater releases are often remediated via pumping. Most oil field salt water is primarily sodium chloride with a solubility in the range of several thousand ppm. While some ion exchange occurs between soil minerals and the groundwater such as sodium substituting for calcium, magnesium (or other cation), there is always a cation in solution. Since anions are not readily exchanged, the chloride ion is not easily attenuated and remains in solution. The only practical procedure to remove the salt dissolved in the aquifer is to pump it out. Figure 9.1 illustrates a capture zone map at an oil field brine spill area. The wells were located to capture the migrating chloride solution plume.

Experiments in relatively permeable aquifers (i.e., sand) indicate that between 6 and 10 pore volume flushes with clean water will reduce the highly soluble, nonattenuated salt contamination to a relatively acceptable concentration. This is usually an effective approach because the secondary drinking water standard for chloride is a relatively high value of 250 ppm. This concentration was established because it is the threshold, where one can taste the salt, not because the salt is a hazardous substance.

When an aquifer is contaminated by a toxic, sparingly soluble, organic chemical that is readily attenuated, the acceptable concentration is usually determined to be in the low ppm to ppb range. The concentration of contaminant dissolved in groundwater is always related to the equilibrium distribution between the chemical retained on the

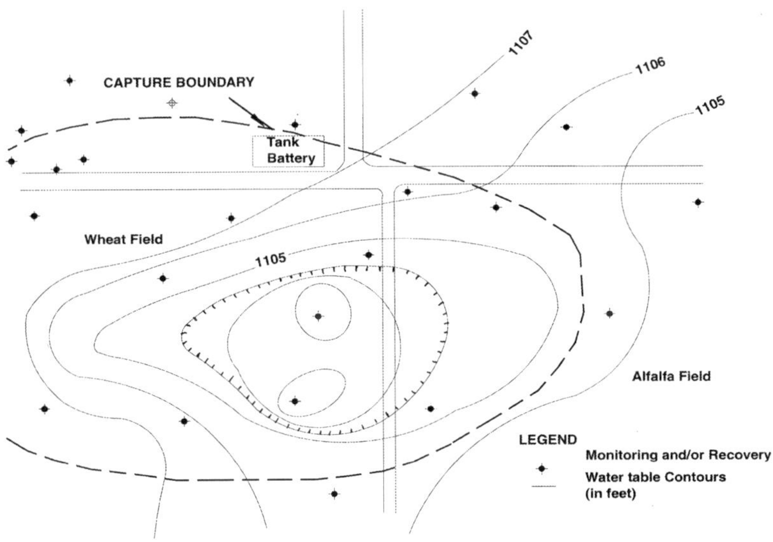

FIGURE 9.1 Capture zone at a brine spill site.

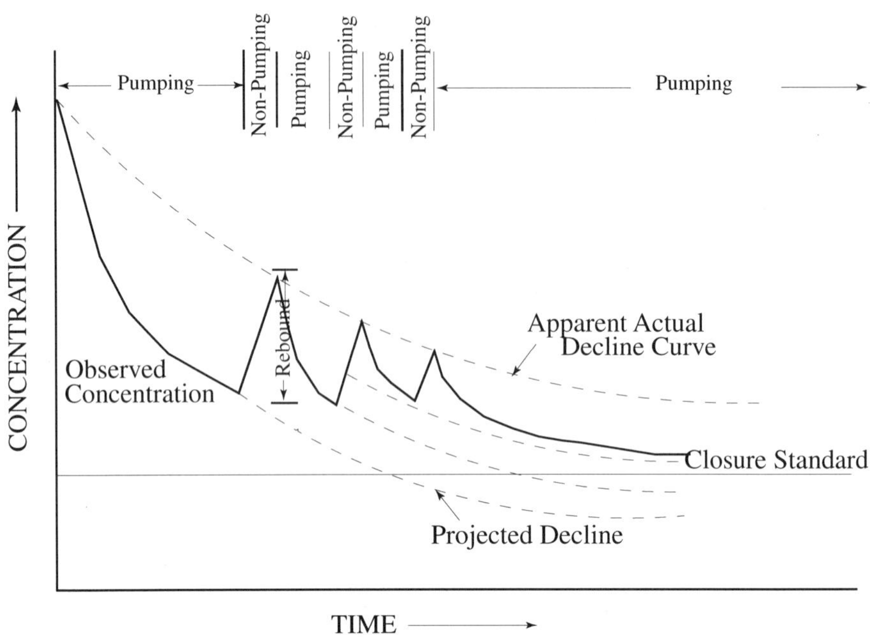

FIGURE 9.2 Typical concentration decline curve.

soil particles and that dissolved in the groundwater. This distribution is usually expressed as a distribution coefficient (K_d), as discussed in Chapter 5 (Fate and Transport).

When a recovery well is located within a contaminant plume and the pump is started, the initial concentration of contaminant removed is close to the maximum level during preliminary testing. As the pump continues to operate, cleaner water is drawn from the plume perimeter through the aquifer pores toward the recovery well. Some of the contaminant is released from the soil into the water in proportion to the equilibrium coefficient. For example, if the K_d is 1000, at equilibrium, 1 part is in the water and 1000 parts are retained in the soil. If the water–soil contact time is sufficient, complete equilibrium will be established. After the first pore volume flush (theoretically), the concentration in the water will be 0.9 and that on the soil will be 999. With each succeeding flush, the 1000:1 ratio will remain the same. If the time of water–soil contact is not sufficient to establish equilibrium, the recovered water will contain a lesser concentration. A typical decline curve is shown on Figure 9.2. Note the asymptotic shape of the curve where the decline rate is significantly reduced.

After a period of time, analysis of the pumped water will indicate that the contaminant concentration has declined toward the target level. When pumping is halted for a few days, then restarted, the concentration of contaminant will be significantly greater. With continued pumping time, the concentration will again approach the target level. After stopping, the concentration will again increase. Each time the pump is halted, the concentration will rebound upward. This rebound phenomenon is the result of the equilibrium balance. During pumping, the water–soil contact time is almost always insufficient for true equilibrium to develop.

Examination of the cyclic start–stop curve indicates two decline curves: the baseline resulting from each prolonged pumping period; and the decline curve formed by the peaks of the restart data. For long-term projections of time to cleanup, the curve formed by the peaks should be used. The shape of the curve almost always indicates a very long remediation period (often many years) would be necessary if pump-and-treat were used as the sole remediation method. Although some improvement can be effected by injecting clean water into the margins of the plume, the time of pumping required and the quantity of water to be treated are almost always considered excessive.

For slightly soluble organic chemicals, aquifer remediation by pump-and-treat technology as the sole approach is considered by most professionals as an obsolete practice. Pump-and-treat serves as a highly effective means of creating a capture zone to prevent plume migration. As the containment phase of a combined system, such as air sparging, soil vapor extraction, or bioremediation, the traditional pumping system is invaluable.

Concurrent treatment of contaminants within the aquifer, with possibly some pumping for hydraulic control, is most often a more efficient procedure than pumping alone. When the contaminant is removed, by air sparging or soil vapor extraction (SVE), produced vapors can be recovered and treated at the surface. If dissolved contaminants are biotreated within the aquifer, the resulting degradation products (daughter compounds) are usually themselves degraded into chemicals of lesser concern. The only notable common exception where the degradation product has a greater environmental concern is the anaerobic degradation of chlorinated hydrocarbons (i.e., trichloroethylene and similar compounds), which sometimes produce vinyl chloride, a known carcinogen. However, when toxic degradation products are anticipated, additional treatment can be arranged. Vinyl chloride can be also be biologically treated or intercepted in vapor form by SVE and treated at the surface.

9.3 AIR SPARGING

The term *air sparging* is used to describe the introduction of compressed air below the water table to promote aquifer restoration. Treatment of dissolved chemicals by air sparging relies upon two mechanisms, volatilization and biodegradation, working together or independently. Volatilization occurs when the contaminant dissolved (in groundwater) or attenuated (on the soil matrix) evaporates to the vapor phase and into passing air by the process of mass transfer. Once present in the vapor phase, it is transported into the vadose zone and ultimately into the atmosphere. Aerobic biological activity within the aquifer depends on the continuing presence of oxygen (20% oxygen in air). The oxygen content of groundwater is typically less than 1 to 2 ppm. Since the solubility of oxygen in groundwater is only approximately 8 ppm, continued replenishment and diffusion of oxygen from the area of sparging results in significantly faster biochemical reactions. In practice, the two processes are mutually supportive (Figure 9.3). There are very few cases where either bioreactions or physical processes occur independently.

The relative ratio of volatilization to bioremediation depends on Henry's law constant, solubility, and biodegradability of the chemicals of concern. Many hydro-

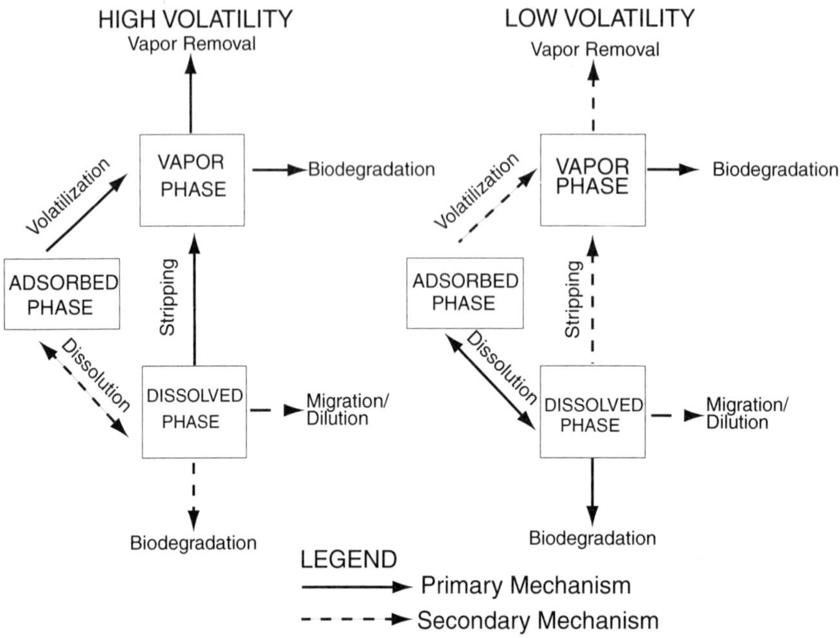

FIGURE 9.3 Air sparging partitioning and removal mechanisms as a function of volatility.

carbon compounds have greater solubility and higher Henry's law constants than does oxygen. When this is the case, volatilization becomes the dominant process. For example, oxygen has an approximate solubility of 8 ppm in groundwater, with a Henry's law constant of 1.5×10^{-3} atm·l/mmol, while benzene has a solubility of 1780 mg/l and a Henry's law constant of 4.4×10^{-3} atm·l/mmol. Thus, volatilization is the predominant remediation process.

9.3.1 Applications

The most common applications of air sparging are in-well aeration and injection well aeration. Where air sparging releases a significant quantity of volatile organic chemicals, an SVE system is often installed in the vadose zone to intercept the vapors for aboveground treatment.

In-well aeration is the process of injecting air into the lower portion of a dual-screened well with perforations at the bottom and above the water table. As the bubbles rise, they expand, which causes the mixed mass of air and water to have less density. The result is an air-lift pump effect. When the water rises and exits the upper perforations, replacement water enters the bottom of the well. The result is a circulation cycle. Free air does not enter the aquifer, but dissolved air (and oxygen) travels with the circulating water. Figure 9.4 is a schematic diagram of in-well aeration.

Injection well aeration is the process of introducing air directly into the aquifer through either a vertical well, which has an isolated short well screen at its base, or a horizontal well installed with the perforations below the contaminated zone (Figure 9.5). The goal is to force compressed air outward through the perforations into the

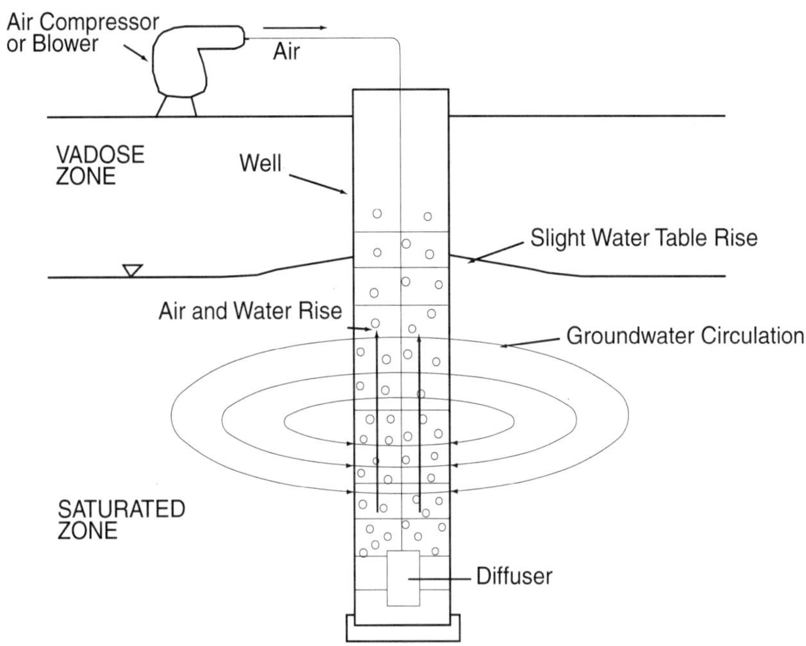

FIGURE 9.4 Typical in-well aeration application.

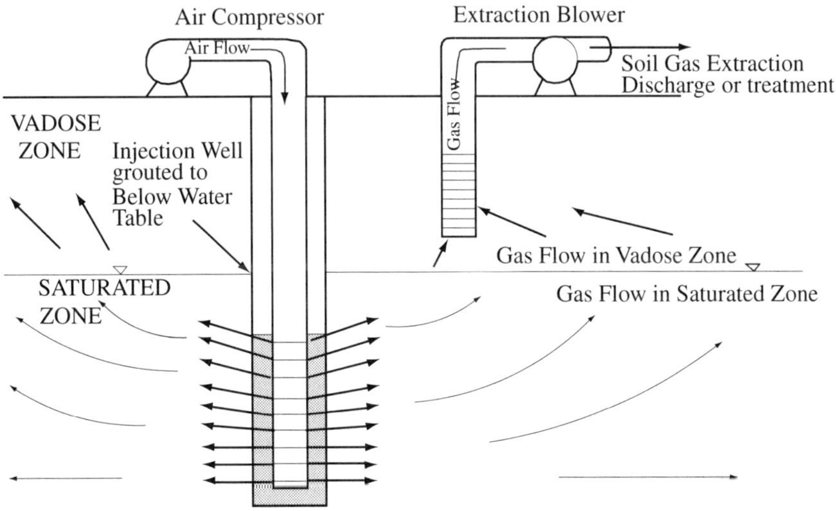

FIGURE 9.5 Schematic *in situ* air sparging.

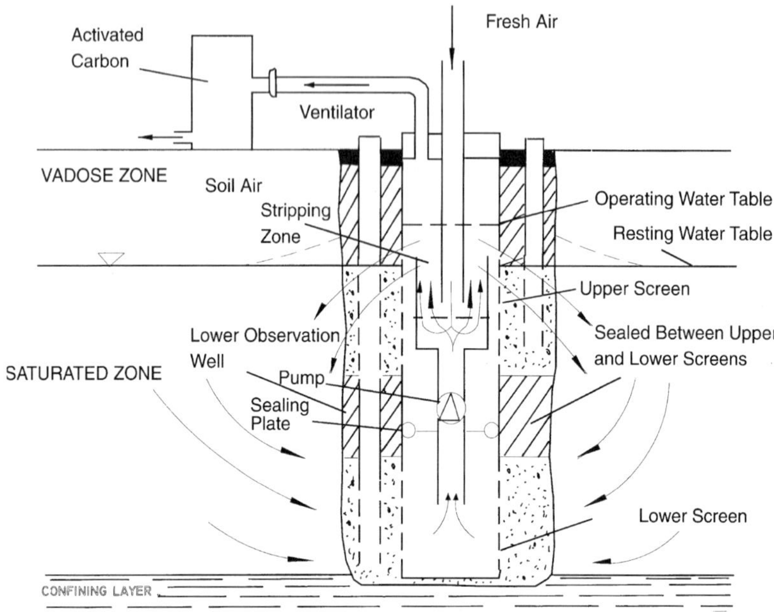

FIGURE 9.6 Vacuum vaporized well application.

aquifer. As the air migrates upward through soil pores (or fractures) it is in intimate contact with both dissolved and attenuated contaminant. Some of the oxygen dissolves to support bioactivity and the remainder of the air functions as an in-place air stripper.

Several variations of air sparge procedures are currently in common use, with each designed for site-specific situations. Vacuum vaporization is accomplished by attaching a vacuum pump to the top of an in-well aeration system. The vacuum removes the vapors and also draws soil vapors from the vadose zone. The combination of air sparging and SVE can provide effective capture of vapors with a minimum of wells.

Air sparging designs for specific applications include addition of several pieces of equipment in the same well. In the German-designed vacuum-vaporized well (German initials UVB), a vacuum is drawn to the top of the casing, and an additional down-hole pump and separation plate are added to enhance the stripping effect. A schematic diagram of a UVB is shown in Figure 9.6.

In some applications, it is necessary to inject nutrients or other chemicals into the aquifer to effect a more efficient restoration. Most of the time, additives are injected into separate wells. These additives may include surfactants, nutrients, pH adjustment chemicals, or additional carbon sources. Some success has been achieved with injected heated air to improve volatility of the chemicals. Where a *small* quantity of methane (as a primary substrate) is required, it can be added with the injection air. The lower explosive limit (LEL) of methane in air is 5%; thus, extreme care must be used to control the mixture and the methane content of the vapor that reaches the surface.

TABLE 9.2
Air Sparging Pilot Test Parameters

Parameter	Significance
Vacuum/pressure vs. distance	Provides an estimate of radius of influence
VOC concentrations in soil and groundwater under static and dynamic conditions	Indicates area being affected and rates of removal
CO_2 and O_2 in soil vapor	Indicates biological activity
Increase in dissolved oxygen (DO) levels	Indicates radius of influence; air travel increases DO
Water level before, during, and after	Indicates airflow as air sparging causes water table mounding; water table "collapses" after sparging

Source: Modified from Brown et al. (1994).

9.3.2 Field Testing

Design of an air sparging system requires a balanced airflow to maintain effectiveness and control of the migration of fluids and vapors. Because of the potential for loss of control, an air sparge system should not be designed or installed without a field pilot test. Design data from pilot tests include:

- The radius of influence conducted at different pressures and injection flow rates;
- The radius of influence of a vacuum extraction system, if necessary, to recover sparged vapors; and
- Pressure and vacuum requirements for effective treatment and capture of volatilized materials.

Pilot tests consist of testing for sparge radius of influence, vacuum radius, and a combined sparge vent operation. These tests are necessary where vapors are of concern. The primary parameters that should be determined during the pilot test, and their significance, are summarized in Table 9.2. Reliable data from the pilot test are used to complete the full-scale remediation design. Site pilot and test data required for design of an air sparging system is presented in Table 9.3.

9.3.3 Limitations

Air sparging in its various forms is a valuable tool; however, every remediation strategy has its strengths and weaknesses. No one procedure is universally suited for all chemicals, geologic settings, or cleanup goals. When designing an air sparging system, the following limitations should be considered:

- Air injection systems, by their very nature, start with no initial hydraulic control because flow is away from the injection point. Contaminants are sometimes spread laterally by water displacement until control is established.

TABLE 9.3
Site and Pilot Test Data Required for Air Sparging System Design

Data	Impact on Design
Vertical extent of contamination	Sparging depth
Lithologic barriers	Feasibility/sparging depth
Horizontal extent of contamination	Number of sparging wells
Volatility of contaminant	Vapor control (venting)
Radius of sparge influence	Well spacing/flow requirements
Optimal flow rates	Compressor size
Vent/pressure balance	Blower size/well placement
Vapor levels	Vapor treatment

Source: Modified after Brown et al. (1994).

- There is always some risk of off-site vapor migration.
- Some uncertainty is associated with the efficiency of oxygen transfer to the groundwater; some designs estimate only a 2% transfer rate.
- Airflow may be impeded by lithological barriers or travel preferentially through high-permeability channels (or fractures).
- Overpressurization will occur when the injection pressure exceeds three times the minimum pressure and overcomes the water column pressure.

Designing an effective air sparging system is as much art as science. In its best form, it has the potential to provide an effective and efficient remedial alternative. If it is poorly designed and operated, air sparging can lead to adverse environmental consequences.

9.4 *IN SITU* GROUNDWATER BIOREMEDIATION

In situ treatment of organic chemicals in the saturated zone by biological means is a technology that encourages growth and reproduction of microorganisms which are capable of degrading the chemicals. *In situ* bioremediation can effectively degrade chemicals that are either dissolved or attenuated to the aquifer matrix. Both aerobic and anaerobic degradation are common. *In situ* groundwater bioremediation is often combined with other procedures (such as air sparging) and vadose zone remediation (i.e., bioventing or soil vapor extraction).

Bioremediation usually requires a procedure for stimulation of and maintaining the activity of microorganisms. For biodegradation to be successful, it is necessary to provide a continuous supply of a suitable electron acceptor (such as oxygen or nitrate), nutrients (nitrogen, phosphorus), and a carbon source for energy and cell material. The most commonly deficient components in the subsurface are electron acceptors and nutrients.

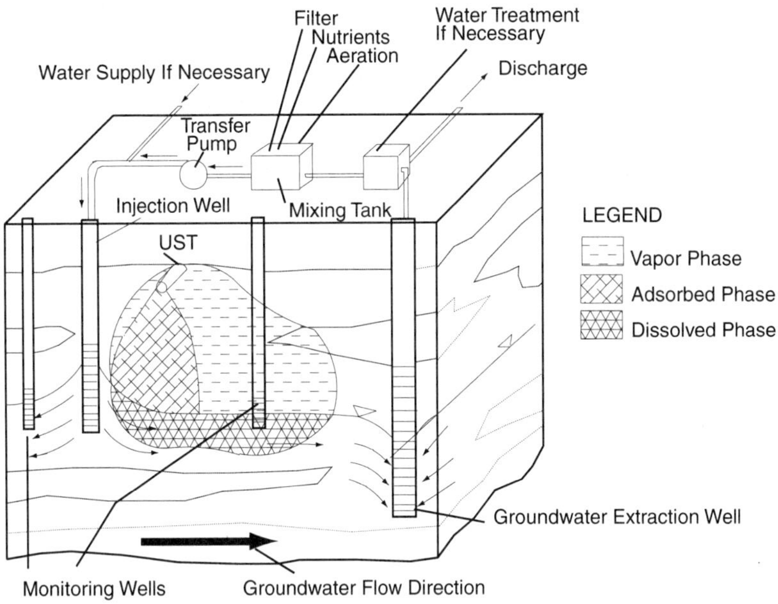

FIGURE 9.7 Schematic of *in situ* groundwater bioremediation using injection wells.

In a typical *in situ* bioremediation system, necessary components are introduced into a hydraulically controlled system to assure the nutrients and electron acceptors are in contact with the contaminants long enough to be utilized. In some cases, groundwater is pumped from recovery wells, treated to remove excess contaminants (if necessary), and inoculated with nutrients, electron acceptors, and other additives (i.e., eH adjusting agents), then reintroduced into an upgradient area of the aquifer. Reinjection may be through wells or infiltration galleries. Selection of the process of a reintroduction procedure is dependent on physical site conditions and local regulations.

An ideal system would have a balance of the rate of recovered water equal to the reinjected water, in a closed-loop system as shown on Figure 9.7. With a continual recycle system, the pumping would continue until cleanup was achieved. Some regulatory authorities do not allow reinjection of untreated or partially treated water. In that case, fresh makeup water must be continually provided from a nearby source and pumped water must be discharged to a suitable location such as a publicly owned treatment works or surface water body.

In situ bioremediation can be conducted in several treatment modes, including aerobic (oxygen respiration) or anaerobic (nonoxygen respiration), utilizing alternate electron acceptors such as nitrate, iron, or sulfate reduction. Some recalcitrant compounds are more effectively degraded by the process of cometabolism that occurs when the compound of interest is degraded along with another, often innocuous, compound. The mode of biodegradation is highly site specific depending on the chemicals, nutrients, and microbiological species present.

Aerobic degradation has been shown to be most effective in reducing the concentration of aliphatic and aromatic hydrocarbon fuel compounds. Anaerobic reactions are more effective in degrading chlorinated hydrocarbons and long-chain animal fats and oils. Detailed study is necessary to determine the most-effective procedure for a specific site.

Aerobic treatment of aquifer oxygen is usually supplied by one of three methods: direct air sparging of air or oxygen into wells screened below the contaminated zone, saturation of water with air or oxygen prior to reinjection, or addition of an oxidant (typically a peroxide compound) directly into an injection well or injection water. Regardless of the mechanism of introduction, the important factor is that the oxidant is distributed throughout the contaminated zone at a concentration and rate such that it can be utilized by the microorganisms.

In situ aerobic remediation of groundwater by bioremediation can be effective for most fuel and nonchlorinated compounds. In general, short-chain, low-molecular-weight, more water soluble compounds are degraded more rapidly to lower concentrations than compounds than are longer-chain, high-molecular-weight, less soluble compounds. Some compounds are exceptions to these generalities; an example is MTBE, a common additive to gasoline. MTBE is highly soluble but can only be degraded by a limited number of microbial species. A thorough understanding of the chemicals is thus necessary prior to attempting any biorestoration project.

The key factors that determine the effectiveness of bioremediation in aquifers are:

- Hydraulic conductivity, which controls the distribution of electron acceptors and nutrients in the aquifer matrix.
- Biodegradability of the organic constituents, which determines the rate, degree of degradation, rate of degradation, and reaction products (sometimes called daughter products).
- Location of the contaminants in the subsurface. Degradable compounds must be dissolved in groundwater or adsorbed onto more permeable matrix materials within the saturated zone.

In general, the aquifer medium will determine hydraulic conductivity. Fine-grained media (e.g., silts and clay) have lower intrinsic permeability than coarser media (e.g., sands and gravels). Bioremediation is generally effective in more permeable materials (sand and gravel). However, depending on the extent of contamination, bioremediation also can be effective in lesser permeable silty or clayey media. In general, lower-permeability materials will require longer time to remediate than more-permeable material.

Biodegradability of an organic compound is a measure of its ability to be metabolized (or cometabolized) by bacteria or other microorganisms. Chemical characteristics dictate the degradation ability of an organic compound. The ease of biodegradation depends upon chemical structure and physical chemical properties (solubility, water chemical partition coefficient). Highly soluble organic compounds with low molecular weight will tend to be more rapidly degraded than less soluble compounds with higher molecular weights. The low water solubility of more complex compounds causes them to be more difficult to degrade. More complex chemical compounds may be slow to degrade or may even be recalcitrant to biological degradation (e.g, No. 6 fuel oil).

The location, distribution, and disposition of chemical contaminants in the aquifer can strongly influence the likelihood of success for bioremediation. This technology generally works well for dissolved contaminants and contaminants adsorbed onto higher-permeability sediments. However, if the majority of the contamination is trapped in lower-permeability sediments or outside the "flow path," where it is in contact with nutrients and electrons acceptors, this technology will have reduced impact, or none at all.

9.4.1 Site Characteristics That Control Aquifer Bioremediation

Several important parameters control aquifer bioremediation projects. These include hydraulic conductivity, soil structure and stratification, groundwater mineral content, groundwater pH, temperature, microbial presence, and bench-scale testing, as further discussed below.

9.4.1.1 Hydraulic Conductivity

Hydraulic conductivity, as discussed in Chapter 3, is a parameter that describes the ability of water to move through the aquifer matrix. As such, hydraulic conductivity controls the rate and distribution of electron acceptors and nutrients delivered to the indigenous bacteria present in the aquifer. Hydraulic conductivity of aquifer media varies over a wide range depending on the constituent materials and their stratigraphic orientation and their in-place density (compaction). Fine-grained soils such as silts and clays typically have low hydraulic conductivity; however, if the soils have good connectivity, a layered structure, or are fractured, the hydraulic conductivity may be significantly greater than estimated. Generally, field hydraulic conductivity is somewhat higher than that determined by laboratory tests on "undisturbed" samples or estimated based on particle size distribution.

For general purposes, for aquifers with hydraulic conductivity greater than 10^{-4} cm/s, bioremediation can be expected to be successful. Where the sites have lower hydraulic conductivities in the range of 10^{-4} to 10^{-6}, bioremediation can proceed, but at a slower rate and more management is required. Where the hydraulic conductivity is less than 10^{-6} cm/s, the success of biodegradation is less likely, unless procedures such as hydrofracking are used.

9.4.1.2 Soil Structure and Stratification

Structural characteristics such as layering and microfracturing can result in preferential flow paths. Stratification of soils with different hydraulic conductivity can dramatically increase the lateral flow of groundwater in the more permeable strata while reducing the flow through less permeable strata. Preferential flow paths can result in reduced effectiveness and extended remedial times for less permeable strata. Fluctuations of the groundwater table are also a consideration. Significant fluctuations will submerge some of the soil in the vadose zone or isolate a part that was previously saturated, and thus result in a "smear zone."

9.4.1.3 Groundwater Mineral Content

Minerals naturally occurring in groundwater are always available to participate in any biochemical reaction. Calcium, magnesium, or iron in groundwater can react with phosphate (either naturally occurring or added), and make the phosphate unavailable. Calcium and magnesium may precipitate in the presence of carbon dioxide produced by aerobic bioreactions. The resulting carbonate minerals can plug pore spaces.

During the biodegradation process, one of the most common first-stage breakdown products of organic compounds is formation of an organic acid. Acids tend to dissolve minerals such as calcium and magnesium. Where soluble iron is present, it can be utilized by iron bacteria or react with injected air to produce an insoluble slime that plugs well screens and pore spaces. The subsurface environment involves complex interactions among water chemistry, dissolved minerals, dissolved and sorbed organic compounds, and the aquifer matrix or formation media. Biologically mitigated reactions can alter established or natural conditions and significantly affect the effectiveness of the restoration endeavor. Thus, a thorough understanding of initial water quality conditions and key parameters is an important aspect of any biodegradation project. The general water quality parameters: cation–anion balance, hardness, alkalinity, pH, Eh, electrical conductivity, and temperature should be defined and monitored throughout the project.

Once a bioremediation effort is started, the bioreactions that occur in the presence of added electron acceptors will result in significant variations of water chemistry across the three-dimensional area of the aquifer. Careful monitoring of these variations is an important indicator of the effectiveness of the remediation process.

9.4.1.4 Groundwater pH

Values of pH ranging from 6.5 to 8 are generally within the optimal range for normal biological activity in soils. Extreme pH values (<5 or >10) are unfavorable for remediation purposes. These ranges are "typical," however, since indigenous microbes are adapted to their site-specific environment and some sites do maintain healthy microculture in the extreme ranges.

If the pH is adjusted as part of an engineered remediation, the microbial balance may be upset and bioreactions slowed until the microbe cultures adjust to the new conditions. Alternatively, if the release of organic chemicals has altered the pH outside the natural range, it may be necessary to add certain chemicals (i.e., aluminum sulfate, carbon dioxide, sodium hydroxide, etc.) to return the pH to preexisting conditions. Changes of pH should be monitored since rapid changes of more than 1 to 2 pH units over a short period can inhibit microbial activity and may extend the acclimation period before the microbes adapt and renew activity.

9.4.1.5 Groundwater Temperature

The rate of microbial activity in the subsurface is a direct function of temperature. Subsurface microbial activity decreases significantly below 10°C and almost ceases below 5°C. Above 45°C, the microbes responsible for degradation of most organic chemicals cease. Within the range of 10 to 45°C, the rate of bioactivity typically doubles for every 10°C. Since groundwater naturally has a relatively stable temperature, usually near the annual local average temperature, the rate of biological activity is not easily altered. If an engineered remediation system involves injection of heated air, or water, the rate of microbial activity can be increased in the local area.

9.4.1.6 Microbial Presence

Natural soil normally contains large numbers of diverse organisms, including bacteria, algae and fungi, and actinomycetes. Bacteria are responsible for most biodegradation because they are the most numerous and biochemically active organisms, particularly at low oxygen levels.

When biological degradation is initiated at a site, the indigenous microbial population adjusts in a stepwise manner. Initially, the microbes undergo an acclimation period, during which they adapts to their new environment and food source. Second, the variety of microbes that adapts most quickly tends to multiply the fastest and can consume nutrients that other microbes would need. Third, as the environmental conditions change and the nature of the food supply changes, the microbial populations change as well. Organisms capable of accommodating to the stress of their changing environment will usually be those that will contribute most to bioremediation of the site.

Laboratory analysis of site soils should be made to determine the presence and population density of naturally occurring microbes that are capable of degrading the contaminant. At a minimum, these analyses should include plate counts to determine the relative number of microbes of several types, the substrate (food) type that reflects the type of chemicals they are likely to consume, and if toxic substances are present.

9.4.2 Bench-Scale Testing

Basic biodegradability can be determined by making a few relatively inexpensive tests at the bench scale. To evaluate treatability characteristics, the following tests are commonly required:

- Confirm presence of a responsive microbial population.
- Determine biodegradability of the chemicals of concern.
- Establish relative degradation rate.
- Determine optimum concentrations of oxygen, nutrients, and micronutrients required.
- Indicate likely interactions among the substrate, microbes, and natural dissolved minerals in groundwater.
- Evaluate cleanup goals achievable by the indigenous microbial population.
- Determine if specialized microbes must be introduced to reach cleanup goals

Two types of bench-scale studies are common: a flask (respirometer) study and a column study. For flask studies, samples of the aquifer matrix and contaminated groundwater are analyzed to determine the presence of organic, inorganic, and heavy metal compounds, and to estimate the relative number of microbes present. A minimum of three treatment options are tested, with usually 12 to 16 separate tests performed:

1. Flask tests involve the testing of native soil and groundwater without additives, in which the oxygen uptake (or carbon dioxide produced) is continually monitored. After about 1 to 5 days, the concentration of contaminant is analyzed and compared with the initial concentration. Results indicate the rate of degradation with only oxygen addition, and the quantity of oxygen required per unit of contaminant.
2. Control tests involve flasks of native soil and groundwater that are chemically sterilized to halt microbial activity. Oxygen uptake (or carbon dioxide produced) is continually monitored for about 1 to 5 days. The sample is then analyzed and compared with the initial starting concentration. Results indicate the quantity of nonbiological degradation occurring. Almost all remediation activities involve some level of nonbiological reactions.
3. Amended testing involves adding nutrients to flasks of native soil and groundwater, and proceeding in the same manner as the previous task descriptions. If several nutrient ranges are tested, the optimum concentration of nutrients can be determined. After evaluation of several test sets, the relative concentrations of nutrients and pH can also be determined. Alternative testing may include addition of specialized microbes to evaluate their performance.

While the above discussion describes testing of aerobic microbial activity, the same scenario is applicable for anaerobic bioreactions. The primary difference is the analytical parameter. The uptake of carbon dioxide, nitrate degradation, sulfate reduction, or iron reduction may be monitored instead of oxygen utilization.

Column studies employ the same general techniques as flask studies, except glass tubes are packed with aquifer matrix, and groundwater, with or without amendments, is percolated through the column. Sterile and nutrient-amended columns are used as controls. Leachate is tested to determine the rate and type of chemical change that occurs in the column. While column studies do not accurately reflect the actual subsurface environment, they do provide an indication of the likely effects of sorption and precipitation within the aquifer.

9.4.3 Pilot Studies

Pilot-scale treatability studies are enlargements of bench-scale tests, which more closely approach the full-scale project. The objective of pilot-scale tests is to verify treatability outside of a laboratory setting and verify the actual field conditions. Many pilot-scale studies are performed at the actual project site. Data from pilot-scale studies are used to design full-scale field operations. A pilot test may include the following:

- A pumping test may be performed to evaluate hydrogeologic characteristics of the aquifer. Derived data can be applied to computer models or at least included in the conceptual site model.

- Dyes or tracers may be introduced into the aquifer to measure the actual groundwater flow rate, or indicate the dispersivity that can be expected.
- Microbial response to injection of electron acceptors and nutrients may be conducted.
- Long-term operation monitoring may be performed to evaluate the potential for injection well fouling, pH changes, and temperature variations and to identify unexpected variables.

Data derived from pilot-scale testing can be "scaled-up" for design of the full-scale site remediation plan or used as input to computer models for further evaluation of design options.

9.4.4 Groundwater Modeling

For design of large aquifer remediation projects or complex smaller ones, the use of groundwater modeling can be a valuable tool to develop a more accurate conceptualization of the site and evaluation of design options. Once the model has been set up and calibrated (i.e., determined to approximate actual site conditions), it can be used to assess the optimum spacing of extraction and injection locations for efficient remediation and hydraulic containment of the contaminant plume, or to minimize water recovery, or any of a number of design variables. Some new models, including the USGS MODFLOW (and commercial variations), have the ability to simulate hydraulic parameters and biological activity within an integrated package. Data generated from site characterization and bench-scale and pilot testing can be incorporated into a model that provides projections and predictions of aquifer conditions with time. Typical model results can be used to evaluate the following:

- Hydrogeologic conditions such as flow rates and direction, water levels, number and location of injection and extraction wells under pumping conditions;
- Injection and extraction points and aquifer sensitivity to various operating scenarios;
- Number, location, and configuration of injection, extraction, and monitoring wells that will optimize system efficiency; and
- Fate and transport of contaminants including concentration, distribution, migration, and degradation with time.

9.4.5 System Design

After the feasibility of *in situ* groundwater bioremediation has been established, the engineering design can proceed. Detailed design documents should be based on:

- The volume and area of aquifer to be treated, based on site characterization and identified action levels;
- Initial concentration of contaminants of concern encountered during conduct of preliminary investigations. This information is used to predict likely toxic effects on the indigenous microbes and to estimate electron

acceptor and nutrient requirements and the extent of treatment required; and
- Required final constituent concentration, as defined by remediation action goals, or those determined on a site-specific basis using transport models along with risk-based calculations.

9.4.5.1 Estimates of Electron Acceptor and Nutrient Requirements

The most important criterion to assure that hydraulic control of the contaminated area is maintained during the remediation program is the proper layout of injection and extraction wells. This is important obviously to minimize and exclude the significant spreading of contaminants into clean areas and to ensure the focus of bioremediation efforts in the areas of highest concentration of contaminants. Important parameters to be considered are as follows:

- Design area of influence, which is the volume and area of the aquifer to which an adequate amount of electron acceptor and nutrients can be supplied to sustain microbial activity;
- Location and design of monitoring wells or points essential to providing operational data;
- System controls and alarms necessary to assure the safety of site operators and the environment in case of a malfunction; and
- Cleanup goals required for final closure.

Extraction wells are usually necessary to maintain hydraulic control of the plume and to ensure that the plume does not migrate into clean areas or accelerate migration toward sensitive receptors. Placement of extraction wells is especially important with systems that use nutrient injection wells or infiltration galleries. These sources of fluids can alter natural groundwater flow patterns, which may cause contaminant migration in an unintended direction or rate. If the natural groundwater system has a sufficient concentration of electron acceptors and nutrients, to achieve remediation at an acceptable rate, it may not be necessary to add any additional materials.

9.4.5.2 Well Placement

The placement of injection (or injection galleries), extraction, and monitoring wells is very site specific depending on the local site and subsurface conditions. The essential criteria for locating these units are as follows:

- Extraction wells should be located where hydraulic control is achieved at the boundary of the contaminant plume. Cones of depression created by recovery wells should intersect so that the overall hydraulic gradients ensure capture and recovery.
- Injection wells or infiltration galleries should be located to provide distribution of the electron acceptor and nutrients throughout the area targeted for cleanup. The impacts on water table gradients caused by injection well

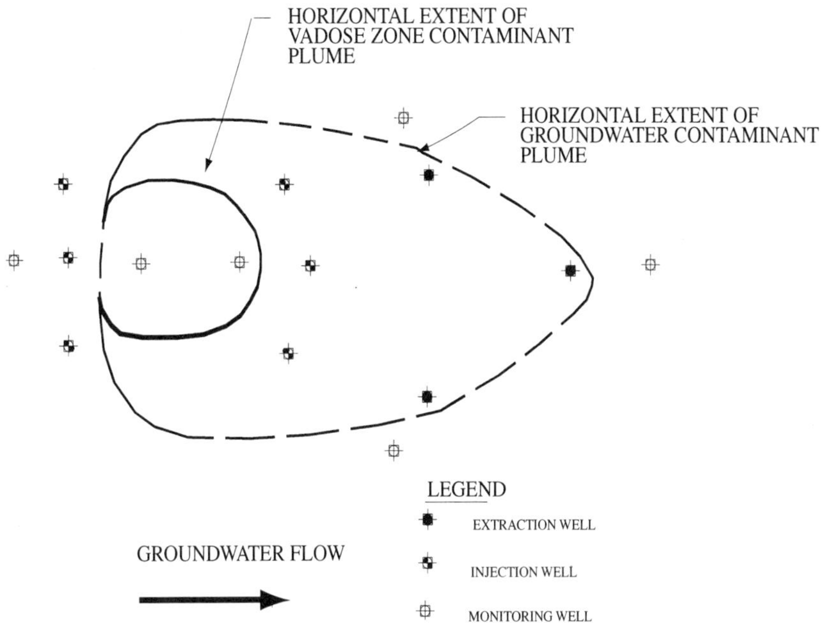

FIGURE 9.8 Idealized layout of extraction, injection, and monitoring wells for *in situ* groundwater remediation.

location and rate of liquid injection should be considered carefully. Excessive water table mounding can induce migration of contaminants in unintended directions or alter the effectiveness of the extraction well in achieving hydraulic control.

- Monitoring wells should be located within the plume at horizontal and vertical locations to monitor progress. Perimeter wells should be placed to ensure that extraction wells achieve the desired hydraulic control and prevent further migration.

An idealized configuration of extraction, injection, and monitoring wells is shown in Figure 9.8. The design area of influence for extraction and injection points will determine the number of wells required. The area of influence of neighboring extraction wells should overlap sufficiently to achieve hydraulic control.

9.4.5.3 Electron Acceptor and Nutrient Addition

At any given site, selection of electron acceptor and nutrient addition will be determined by results of previous laboratory, plate counts, and treatability studies. The most widely used electron acceptor is oxygen, which promotes aerobic degradation. Oxygen can be delivered by several systems. Injection water can be aerated to oxygen saturation with atmospheric air, typically 8 to 10 ppm at groundwater temperature. Pure oxygen can be bubbled into the aquifer below the water table, resulting in dissolved oxygen ranging up to 40 to 50 ppm.. Air can also be bubbled directly into the aquifer below the water table via air sparging.

Higher dissolved oxygen concentrations can be achieved by the use of hydrogen peroxide (H_2O_2). However, oxygen concentrations of greater than about 400 ppm act as a biocide. An initial concentration of dissolved oxygen at a dissolved oxygen level of 50 to 60 ppm is usually safe. After the microbial activity is well established, the dissolved oxygen concentration can be slowly increased to 200 to 250 ppm without risk of killing the microbes.

Without an inhibitor, hydrogen peroxide degrades rather quickly to water and oxygen, and is sometimes difficult to distribute throughout the cleanup zone. Another factor to consider is that oxygen from hydrogen peroxide will react with any oxidizable material. Dissolved iron, manganese, organic carbon, and sulfide compounds are easily oxidized, usually causing a precipitate which may plug pore spaces.

An electron acceptor addition system includes the following basic elements:

- Oxygen enrichment equipment including an air blower, or pure oxygen source, and a contacting chamber.
- Injection well or sparging system. Injection wells with long screens are useful to create horizontal flow in the aquifer. Air sparge wells have short well screens and allow gas (air or oxygen) injection below the water table. The rising gas travels as bubbles or ganglions between the soil grains where it is dissolved into the water.
- Hydrogen peroxide system including a hydrogen peroxide source, storage, metering, and injection pump.

A nutrient addition system includes the following:

- Reagent (nutrient) storage facilities;
- Mixing tanks for nutrient solutions;
- Meters for measuring the rate of introduction of nutrient solutions into carrier streams; and
- Control system for metering systems.

9.4.5.4 System Controls and Alarms

Control of site equipment is always important for safety and operational concerns. Many sites are remote or not staffed on a daily basis. Remediation equipment should be fitted with failsafe systems to shut down the system in the event of failure, fire, or unusual conditions (such as injection well plugging). Alarm systems should be included, which may be as simple as illuminating a warning light, or as complex as a teleconnection to a remote computer station or telephone alert to the operator's residence. Many commercial companies offer remote monitoring equipment.

9.5 START-UP OPERATIONS

Almost all new systems have need of adjustment of flow balances, level switch adjustment, or injection rates. A normal system start-up operation should consist of daily monitoring until the system stabilizes at a relatively constant rate. At the end of the start-up when all appears to be operating, a round of sampling should be collected for constituents of concern and indicator parameters (e.g., dissolved oxygen, pH, Eh). This round of sampling will represent the initial baseline data to compare with later data.

9.6 OPERATIONAL MONITORING

After start-up, the system should be checked at least weekly, with some observations, notably in the early phases, requiring daily monitoring. Information such as groundwater levels, extraction and injection flow rates, groundwater electron acceptor concentrations, nutrient concentrations, pH, and conductivity should be recorded at least on a weekly basis. Complete records of rates, concentrations, electrical usage, and other operational data can be invaluable when evaluating operational efficiency or documenting unit costs.

9.7 REMEDIAL PROGRESS MONITORING

Monitoring remedial progress requires monitoring of both groundwater and aquifer media samples for compounds of concern and other indicator parameters. Typically, groundwater samples should be collected and analyzed monthly, or quarterly at a minimum. More frequent sampling is not usually justified because groundwater flow is relatively slow and biochemical reactions in the subsurface are typically slow compared with laboratory reactions. Confirmatory soil samples are routinely collected, prior to site closure, to demonstrate that cleanup objectives have been achieved.

Periodic monitoring points should include extraction, injection, and monitoring points. When sampling injection wells, the injection system should be shut down for approximately 24 h to allow ambient conditions to be reestablished. Monitoring wells should be purged and sampled according to standard procedures. Extracted fluid material (pumped water) can be sampled by the same procedures as collecting a surface water sample.

9.8 LONG-TERM OBSERVATIONS

During remediation, contaminant levels decrease until they achieve an asymptotic level. Once asymptotic conditions are reached for several successive sampling periods, continuing remediation activities generally result in little further decrease in contaminant reduction. However, frequently when active remediation is halted, levels of dissolved contaminants abruptly increase (rebound). This increase is the result of the diffusion into solution of contaminants that were previously adsorbed onto the surface of the aquifer media. Sometimes more efficient cleanup is achieved by operating the remediation system on a cycle of several days on and several days off. Cyclic operation allows the operator to time the remediation to treatment of the higher rebound concentrations.

If asymptotic conditions occur before cleanup standards are reached, the operator should evaluate alternative procedures that will complete the remediation, or consider renegotiation with the lead agency regarding more reasonable cleanup goals. Possible procedural options include adjusting the rate and concentration of electron acceptors or nutrients, examining the possibility of another remedial procedure, or considering changing from an active to a passive (monitored natural attenuation) remediation system.

REFERENCES

Ahlfeld, D. P. and Sawyer, C. S., 1990, Well Location in Capture Zone Design Using Simulation and Optimization Techniques: *Ground Water*, Vol. 28, No. 4, pp. 507–512.

Alleman, B. C. and Leeson, A. (editors), 1999a, *In Situ Bioremediation of Petroleum Hydrocarbon and Other Organic Compounds*: Battelle Press, Columbus, OH, Vol. 5, No. 3, 588 pp.

Alleman, B. C. and Leeson, A. (editors), 1999b, *Bioreactor and Ex Situ Biological Treatment Technologies*: Battelle Press, Columbus, OH, Vol. 5, No. 5, 221 pp.

Alleman, B. C. and Leeson, A. (editors), 1999c, *Bioremediation of Nitroaromatic and Haloaromatic Compounds*: Battelle Press, Columbus, OH, Vol. 5, No. 7, 302 pp.

Bajer, K. H. and Herson, D. S. (editors), 1994, *Bioremediation*: McGraw-Hill, New York, 375 pp.

Batu, V., 1998, *Aquifer Hydraulics — A Comprehensive Guide to Hydrogeologic Data Analysis*: John Wiley & Sons, New York, 727 pp.

Brown, R. A., Hicks, R. J., and Hicks, P. M., 1994, Use of Air Sparging for Insitu Bioremediation: In *Air Sparging for Site Remediation* (edited by R. E. Hinchee), Lewis Publishers, Boca Raton, FL, pp. 38–55.

Brubaker, G. R., 1993, In-Situ Bioremediation of Groundwater: In *Geotechnical Practice for Waste Disposal* (edited by D. E. Daniel), Chapman & Hall, New York.

Cherry, J. A., Feenstra, S., and MacKay, D. W., 1997, Developing Rational Goals for in Situ Remedial Technologies: In *Subsurface Restoration* (edited by C. H. Ward, J. A. Cherry, and M. R. Scalf), Ann Arbor Press, Chelsea, MI, pp. 75–98.

Conger, R. M. and Trichel, K., 1993, A Ground-Water Pumping Application for Remediation of a Chlorinated Hydrocarbon Plume with Horizontal Well Technology: In *Proceedings of the Seventh National Outdoor Action Conference and Exposition*, Las Vegas, NV, National Ground Water Association, Dublin, OH, pp. 47–61.

Copty, N. D. and Findikakis, A. N., 2000, Quantitative Estimates of the Uncertainty in the Evaluation of Ground Water Remediation Schemes: *Ground Water*, Vol. 38, No. 1, pp. 29–37.

de Marsily, G., 1997, Hydraulic Control for Contaminant Containment: In *Subsurface Restoration* (edited by C. H. Ward, J. A. Cherry, and M. R. Scalf), Ann Arbor Press, Chelsea, MI, pp. 161–171.

Devlin, J. F. and Barker, J. F., 1999, Field Demonstration of Permeable Wall Flushing for Biostimulation of a Shallow Sandy Aquifer: *Ground Water Monitoring & Remediation*, Vol. 19, No. 1, pp. 75–83.

Domenico, P. A. and Schwartz, F. W., 1990, *Physical and Chemical Hydrogeology*: John Wiley & Sons, New York, 824 pp.

Gaily, R. M. and Gorelick, S. M., 1993, Design of Optimal, Reliable Plume Capture Schemes; Application to the Gloucester Landfill Groundwater Contamination Problem: *Ground Water*, Vol. 31, No. 1, pp. 107–114.

Gorelick, S. M., Freeze, R. A., Donohue, D., and Keely, J. F., 1993, *Groundwater Contamination — Optimal Capture and Containment*: Lewis Publishers, Chelsea, MI, 385 pp.

Green, S. R. and Dorrler, R. C., 1996, Four Bullets Shoot Down Costs: Combined Technology Approach Cleans Up Faster and Cheaper than Pump-and-Treat: *Soil and Groundwater Cleanup*, March.

Havash, J. and Oster, J., 1998, Is Your Wastewater Toxic to the Municipal Treatment Plant? *Pollution Engineering*, March, pp. 52–54.

Hinchee, R. E. (editor), 1994, *Air Sparging for Site Remediation*: Lewis Publishers/CRC Press, Boca Raton, FL, 142 pp.

Javandel, I. and Tsang, C. F., 1986, Capture-Zone Type Curves: A Tool for Aquifer Cleanup: *Ground Water*, Vol. 24, No. 5, pp. 616–625.

Keely, J. F., 1984, Optimizing Pumping Strategies for Contamination Studies and Remedial Actions: *Ground Water*, Vol. 4, No. 3, pp. 63–67.

Kinsella, J. V. and Nelson, M. J. K., 1993, In Situ Bioremediation: Site Characterization, System Design, and Full Scale Field Remediation of Petroleum Hydrocarbon and Trichloroethylene Contaminated Groundwater: In *Bioremediation Field Experience* (edited by P. E. Flathman and D. E. Jerger), CRC Press, Boca Raton, FL.

Leeson, A. and Alleman, B. C. (editors), 1999a, *Engineered Approaches for In Situ Bioremediation of Chlorinated Solvent Contamination*: Battelle Press, Columbus, OH, Vol. 5, No. 2, 336 pp.

Leeson, A. and Alleman, B. C. (editors), 1999b, *Bioremediation Technologies for Polycyclic Aromatic Hydrocarbon Compounds*: Battelle Press, Columbus, OH, Vol. 5, No. 8, 358 pp.

Lesage, S., Brown, S., and Millar, K., 1996, Vitamin B_{12} — Catalized Dechlorination of Perchloroethylene Present as Residual DNAPL: *Ground Water Monitoring & Remediation*, Vol. 16, No. 4, pp. 76–85.

Maccagno, M. D., Ross, S. D., and Hardisty, P. E., 1996, Containment of a Dissolved-Phase Plume Using a Horizontal Well: In *Proceedings of the Tenth National Outdoor Action Conference and Exposition,* Las Vegas, NV, May 13–15, National Ground Water Association, pp. 221–236.

MacDonald, J. A. and Kavanaugh, M. C., 1994, Restoring Contaminated Groundwater: An Achievable Goal? *Environmental Science and Technology*, Vol. 28, No. 8, pp. 363 A–368 A.

MacKay, D. M. and Cherry, J. A., 1989, Limitations of Pump-and-Treat Remediation: *Environmental Science and Technology*, Vol. 23, No. 6, pp. 630–637.

Mercer, J. W., Skipp, D. C., and Giffin, D., 1990, *Basics of Pump-and-Treat Ground-Water Remediation Technology*: U.S. Environmental Protection Agency, R. S. Kerr Laboratory, Ada, OK, EPA/600/8-90/003.

Murdoch, L. C., Losonsky, G., Cluxton, P., Patterson, B., Klich, I., and Braswell, B., 1991, *Feasibility of Hydraulic Fracturing of Soil to Improve Remedial Actions. Project Summary*: U.S. Environmental Protection Agency Office of Research and Development, Cincinnati, OH, EPA/600/S2-91/012.

Norris, R. D., 1994, In-Situ Bioremediation of Soils and Groundwater Contaminated with Petroleum Hydrocarbons: In *Handbook of Bioremediation* (edited by R. D. Norris, R. E. Hinchee, et al.), CRC Press, Boca Raton, FL.

Norris, R. D., Hinchee, R. E., et al., 1993, *In-Situ Bioremediation of Groundwater and Geologic Material: A Review of Technologies*: U.S. Environmental Protection Agency, Washington, D.C., EPA/600/R-93/124 (NTIS:PB93-215564/XAB).

Norris, R. D. and Dowd, K. D., 1993, In-Situ Bioremediation of Petroleum Hydrocarbon-Contaminated Soil and Groundwater in a Low-Permeability Aquifer: In *Bioremediation Field Experience* (edited by P. E. Flathman and D. E. Jerger), CRC Press, Boca Raton, FL.

Sims, J. L., Suflita, J. M., and Russell, H. H., 1992, *In-Situ Bioremediation of Contaminated Groundwater*: U.S. Environmental Protection Agency, Washington, D.C., EPA/540/S-92/003 (NTIS: PB92-224336/XAB).

Staps, S. J. M., 1990, *International Evaluation of In-Situ Biorestoration of Contaminated Soil and Groundwater*: U.S. Environmental Protection Agency, Office of Emergency and Remedial Response, Washington, D.C., EPA/540/2-90/012.

Steinle, R. R., 1993, *An Inventory of Research and Field Demonstrations and Strategies for Improving Ground Water Remediation*: U.S. Environmental Protection Agency, 500/K-93/001.

U.S. Army Corps of Engineers, 1995, *Soil Vapor Extraction and Bioventing*: Engineers Manual EM 1110-1-4001.

Van der Heijde, P. K. M. and Elnawawy, O. A., 1993, *Compilation of Groundwater Models*: U.S. Environmental Protection Agency, Office of Research and Development, Washington, D.C., EPA/600/R-93/118 (NTIS: PB93-209401).

Ward, C. H., Cherry, J. A., and Scalf, M. R. (editors), 1997, *Subsurface Restoration*: Ann Arbor Press, Chelsea, MI, 491 pp.

Wilson, J. L., 1997, Removal of Aqueous Phase Dissolved Contamination: Non-Chemically Enhanced Pump-and-Treat: In *Subsurface Restoration* (edited by C. H. Ward, J. A. Cherry, and M. R. Scalf), Ann Arbor Press, Chelsea, MI, pp. 271–285.

10 Treatment of Impacted Soil in the Vadose Zone

"The microscopic organisms are very inferior in individual energy to lions and elephants, but in their united influence they are far more important than all of these animals."

— Ehrenburg, 1862

10.1 INTRODUCTION

In virtually every situation where the source of a hydrocarbon spill occurs at or near the ground surface, some quantity of hydrocarbon is retained in the soil as residual saturation within the vadose zone above the water table. Most unsaturated soils are capable of retaining hydrocarbons in a quantity equivalent to approximately 30% of their holding capacity (the amount of water that a particular soil can retain at saturation). In most cases, however, by the time remediation of the impacted soil is implemented, the residual hydrocarbon is less than maximum.

Treatment of soil impacted by petroleum hydrocarbons or organic compounds in the vadose (unsaturated) zone is an important step in preventing and minimizing the significance of contaminant migration to underlying water-bearing zones. The objective of soil treatment is to immobilize contaminants or reduce their concentration to a level that is "irreducible," or at least not a continuing threat to human health or the environment. Regardless of the approach to the remediation of impacted soil, four specific criteria always need to be met:

- Technical efficiency;
- Cost-effectiveness;
- Ability to be accomplished within a reasonable time frame; and
- Reduction of potential liability to the lowest of reasonable terms.

These criteria in most cases are resolved through discussions and negotiations among the consultant, the client, and the regulatory agency. After consideration of these basic issues, several strategies can be made to work either independently or sequentially, or in combination, to accommodate the ultimate project goals. Some of the more common and conventional approaches to mitigate hydrocarbon-impacted soil within the vadose zone are discussed here. Conventional and more common strategies for the remediation of hydrocarbon- and organic-impacted soil are summarized in Table 10.1.

The primary focus of this chapter is to introduce approaches that limit (or prevent) migration of hydrocarbon and organic contaminants from the vadose zone into underlying water-bearing zones. The more conventional strategies, including a brief discussion of the process, relative cost, practical constraints, and limitations, are presented. Also presented are two typical soil vapor extraction case histories.

10.2 *IN SITU* SOLIDIFICATION/STABILIZATION

Solidification describes those techniques that encapsulate the waste material into a monolithic solid that has some degree of structural integrity. Encapsulation may be based on adsorption of waste onto fine matrix particles, or agglomeration into a large solid block, or solidification of a container of wastes. Solidification does not necessarily involve a chemical reaction between the waste and the solidifying reagents, but may mechanically bind the waste into the mass. Contaminant migration is restricted by vastly decreasing the surface area exposed to leaching or by isolating the waste within an impervious capsule.

Stabilization refers to techniques that immobilize the contaminant to reduce the hazard (or mobility) potential of a substance by converting the contaminants into their least migratory (or leachable) form. Processes of stabilization interact with the contaminant to solidify by chemical means, insolubilize, destroy, or sorb it into a relatively stable form. The following factors may limit the applicability and effectiveness of the process:

- Depth of contaminants;
- Local environmental conditions that may affect immobilization of contaminants;
- Some processes that result in a significant increase in total waste volume;
- Certain wastes that are incompatible with variations of this process; treatability studies may be required.

General processes used to prevent or minimize continued downward migration or leaching of contaminants into the groundwater can be generally categorized as follows:

- Enhanced attenuation of contaminants to the surface of the soil particles;
- Reduced potential for diffusion from soil grains into infiltrating water;
- Chemical procedures that "fix" the contaminant into a solidified form;
- Solidification of the upper soil to form a "roof," which substantially reduces or eliminates infiltration.

Solidification of the upper layers can be accomplished by blending pozzolanic additives, modified clay, or stabilization reagents into moist soil and compacting the mass. Pozzolanic additives include such fixatives as portland cement, quick lime,

TABLE 10.1
Summary of Conventional Soil Remediation Strategies to Minimize, Contain, or Remediate Petroleum Hydrocarbons- and Organics-Impacted Soil

Remedial Strategies	Process	Practical Constraints	Remarks
		***In Situ* Strategies**	
SVE	Evaluate hydrogeologic conditions Assess extent of impacted soil Conduct pilot study Design and install system Monitor effectiveness Confirm effectiveness	Fine-grained soils and low volatile hydrocarbons limit effectiveness	Not technically viable for clayey soils Requires disposal of air medium
Air sparging	Evaluate hydrogeologic conditions Assess extent of impacted soil Conduct pilot study Design and install system Monitor effectiveness Confirm effectiveness	Fine-grained soils and low volatile hydrocarbons limit effectiveness	Not technically viable for clayey soils Requires disposal of air filtration medium
Steam injection and stripping	Evaluate hydrogeologic conditions Assess extent of impacted soil Conduct pilot study Design and install system Monitor effectiveness Confirm effectiveness	Fine-grained soils and permeability constrasts limit ability to inject steam and recover fluids from subsurface	Overall effectiveness difficult to assure pending pilot study Relatively high operation and maintenance costs
Soil washing/extraction (in-place leaching)	Evaluate hydrogeologic conditions Assess extent of impacted soil Excavate and crush Mix with washing fluids	Limited to granular soils and moderate to high solubility hydrocarbons	High costs; limited applicability Often used in biotreatment practices Permit approval difficult

TABLE 10.1 *(continued)*
Summary of Conventional Soil Remediation Strategies to Minimize, Contain, or Remediate Petroleum Hydrocarbons- and Organics-Impacted Soil

Remedial Strategies	Process	Practical Constraints	Remarks
	***In Situ* Strategies**		
	Treat wash water Construct infiltration and recovery system Irrigate washing fluids Retrieve and treat fluids		
Bioventing	Evaluate hydrogeologic conditions Assess extent of impacted soil Conduct pilot study Design and install venting system Monitor effectiveness Confirm effectiveness	Fine-grained soils and low volatile hydrocarbons limit effectiveness	Not viable for clayey soil Requires disposal of air filtration medium
Bioremediation (chemical degradation)	Evaluate hydrogeologic conditions Assess extent of impacted soil Conduct pilot study Design and install pumping and injection system Monitor effectiveness Confirm effectiveness	Fine-grained soils or permeability contrasts limit effectiveness Requires ongoing operation and maintenance	Overall effectiveness difficult to assure Extensive on-site monitoring required
Natural attenuation	Evaluate hydrogeologic conditions Assess extent of impacted soil Document source elimination Confirm plume stability Conduct feasibility study Perform periodic monitoring	Requires regulatory approval	Not usually viable for sensitive and/or beneficial-use groundwater

	Non-*In Situ* Strategies		
Excavation/disposal as waste	Evaluate hydrogeologic conditions Assess extent of impacted soil Excavate and transport Maintain documentation	Cradle-to-grave liability as generator	High cost; easy to overexcavate
Excavation/aeration and disposal	Evaluate hydrogeologic conditions Assess extent of impacted soil Excavate, spread, and turn Import and compact clean fill Transport and dispose of aerated soil Maintain documentation Confirm effectiveness	Cradle-to-grave liability as generator Emissions considerations Chemical testing can be extensive	Permitting can be difficult
Excavation/landfarming and replacement	Evaluate hydrogeologic conditions Assess extent of impacted soil Excavate, aerate, and add nutrients and water Replace and compact Maintain documentation Confirm effectiveness	Emissions considerations	Permitting can be difficult
Mechanically enhanced volatilization	Evaluate hydrogeologic conditions Assess extent of impacted soil Excavate, crush, aerate, and replace Maintain documentation Confirm effectiveness	Requires dust control Requires vapor treatment	High costs but suitable under certain circumstances
Soil washing (aboveground leaching)	Evaluate hydrogeologic conditions Assess extent of impacted soil Excavate and place over collector bed Flush with wash fluid Replace and treat fluids Confirm effectiveness	Requires total fluid collection Requires temperature and odor control Requires sufficient open area	May be used in association with biotreatment Permitting not very difficult

TABLE 10.1 *(continued)*
Summary of Conventional Soil Remediation Strategies to Minimize, Contain, or Remediate Petroleum Hydrocarbons- and Organics-Impacted Soil

Remedial Strategies	Process	Practical Constraints	Remarks
	Others		
Reuse/recycled (asphalt incorporation)	Evaluate hydrogeologic conditions Assess extent of impacted soil Conduct pilot test Develop mix design Determine end use Excavate and produce product Confirm effectiveness Construct end use (i.e., road base, pavement, etc.) Maintain documentation	Soil must pass flash test	Requires comprehensive understanding of reuse/recycling regulations On-site end use easier to deal with
No action	Evaluate hydrogeologic conditions Assess extent of impacted soil Demonstrate low or minimal risk	Requires regulatory acceptance	Site specific

lime, cement kiln dust, and coal fly ash mixtures. These materials are typically highly alkaline and react with water and soil particles to "set up" like concrete.

Modified clays are also effective solidification agents. These clay materials are natural products that have been treated by organic chemical reactions to alter them from hydrophilic (water-liking) to hydrophobic (water-hating) substances. Thus modified, the clays have a greater capacity for sorption of organic constituents. These products have the potential of immobilizing organics so that they will not desorb under usual site conditions; some appear to catalyze reactions that may detoxify the organic compounds. Most modified clays are proprietary commercial compounds. Once in place and compacted, the solid paved surface severely reduces rainfall infiltration.

An almost infinite variety of chemical reactions is possible among soil, additives, and organic contaminate. However, at the moisture, temperature, and pressure conditions present at most sites, only a few reactions are responsible for most stabilization processes. Aside from such processes as absorption, volatilization, and biodegradation, chemical reactions include processes such as hydrolysis, oxidation, reduction, compound formation, and fixation on an insoluble substrate.

The decision of whether to blend stabilization agents *in situ* or in aboveground mixing equipment is determined based on the quantity of material to be blended, the depth of mixing required, and the quantity of soil requiring handling. When remediation is limited to surface materials (<2-ft depth) which are to be stabilized with cement, fly ash, or asphalt and used as a site cap or liner, only thorough mixing is required. The most practical equipment for aboveground blending is concrete mixers, pug mills (Figure 10.1), and auger mixers. These implements allow the user to control carefully the quantities of additives, water, and soil to develop the consistency required for replacement. Once the desired blend is achieved, the material may be placed and compacted as necessary to complete covering of the site.

For blending of soil in the range of 2 to 5 ft deep, such as may occur in shallow ponds or soil piles, a tractor-mounted rotary tiller can be used. Tillers are designed for thorough mixing, provided the blending agent is properly dispersed over the surface.

When mixing at moderate depths is required, mixing can be accomplished by use of backhoe, vertical auger, or rotary-tilling device. Typically, the blending agent is spread on the ground ahead of the mixing equipment and the soil is mixed as it is excavated. Water can be added to enhance blending. If the operator is diligent and mixes the blending uniformly, this procedure is satisfactory. However, the procedure is more easily described than accomplished. Quality control inspection and testing are vital to assure that project goals are achieved.

Where deeper soils are involved, the mixing process is often accomplished by use of large multiple-auger units, which work somewhat like a kitchen mixer. Soil auguring may use a single 1- to 3-m-diameter paddle auger or several smaller-diameter augers. Usually, the augers are attached to a crane capable of lifting the auger equipment from the ground surface to the bottom of the contaminated zone. Uniform mixing of soil is accomplished by repeated upward and downward movement of the rotating auger. Treatment zones are typically horizontally overlapped to assure complete aerial coverage. This process is effective both above and below the water table. Soil auguring is very flexible because it is readily adaptable to deliver

FIGURE 10.1 Photograph showing use of a pugmill for the incorporation of hydrocarbon-impacted soil into cold-mix asphalt.

hot air, steam, chemical stabilizers, nutrients, cement grout, or other additives directly to specific areas or depths at controlled rates and volumes.

Soil auguring can be applied to a wide variety of sites, including wood treatment facilities, oil and gas production units, bulk storage sites, pipeline, manufactured gas plants and other sites contaminated with residual oils, petroleum products, VOCs, semi-VOCs, and other contaminants. This technology is also applicable to sites with complicated, heterogeneous subsurface geology that can render more traditional approaches ineffective, or at sites where excavation is not cost-effective or practical.

At locations where aboveground blending takes place, the resulting soil material can be placed back in the original excavation (or selected location) and compacted to the desired density. If the solidified material is to have a desired structural strength (i.e, subbase, pavement, controlled fill, etc.) it can be compacted by conventional construction equipment (vibrating or sheeps-foot rollers).

A less rigid pavement can be created by blending an asphalt–water–surfactant emulsion with the upper soil layers. During the curing process, the asphalt–water emulsion deteriorates, leaving the asphalt to bind the hydrocarbon contaminants and soil to create a low-permeability pavement. Asphalt–soil pavements tend to be less rigid than pozzolanic–soil structures.

10.3 SOIL VAPOR EXTRACTION

In the vadose (unsaturated) zone, liquid organic compounds partition into four distinct but related phases: NAPL adsorbed onto soil particles, free-phase NAPL, soluble constituents in soil water, and vapor-phase components as part of the soil

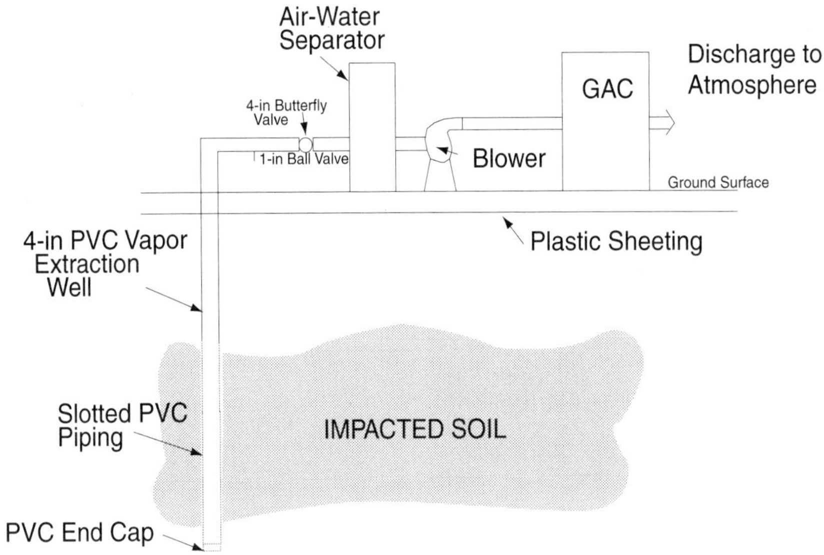

FIGURE 10.2 Schematic of a typical SVE system configuration.

atmosphere. Soil vapor extraction (SVE) can serve to control the contaminant source by removing volatile constituents in the vadose zone before they migrate to groundwater or to reduce the overall concentration of hydrocarbon- and organic-impacted soil at depths where excavation is not feasible or cost-effective. In its conceptual form, SVE is a simple technology involving a well screen open to the vadose zone connected to a vacuum pump. A schematic sketch of a typical setup is shown in Figure 10.2. The vacuum created at the well creates airflow through the soil, and extraction of volatile and semivolatile contaminants.

Sites suitable for conventional SVE have certain typical characteristics. The contaminating chemicals are volatile or semivolatile (vapor pressure of 0.5 mm Hg or greater). Removal of metals, most pesticides, and PCBs by vacuum is not possible because their vapor pressures are too low. The chemicals must be slightly soluble in water, or the soil moisture content must be relatively low. Soluble chemicals such as acetone or alcohols are not readily strippable because their vapor pressure in moist soils is too low. Chemicals to be removed must be sorbed on the soils above the water table or floating on it (LNAPL). Volatile dense nonaqueous liquids (DNAPLs) trapped between the soil grains can also be readily removed. The soil must also have sufficiently high effective porosity (permeability) to allow free flow of air through the impacted zone.

SVE functions by the establishment of a concentration equilibrium between the contaminant adsorbed on the soil and the contaminant vapor concentration in the passing air. Significant factors involved in SVE are shown in Figure 10.3. Since the equilibrium concentrations are essentially linear, a higher concentration in the soil causes a similarly high concentration in the soil air. Before SVE is started, stagnant air in the soil pores contains VOC at the established equilibrium concentration. When SVE is initiated, two processes occur. Fresh VOC free air is brought into contact

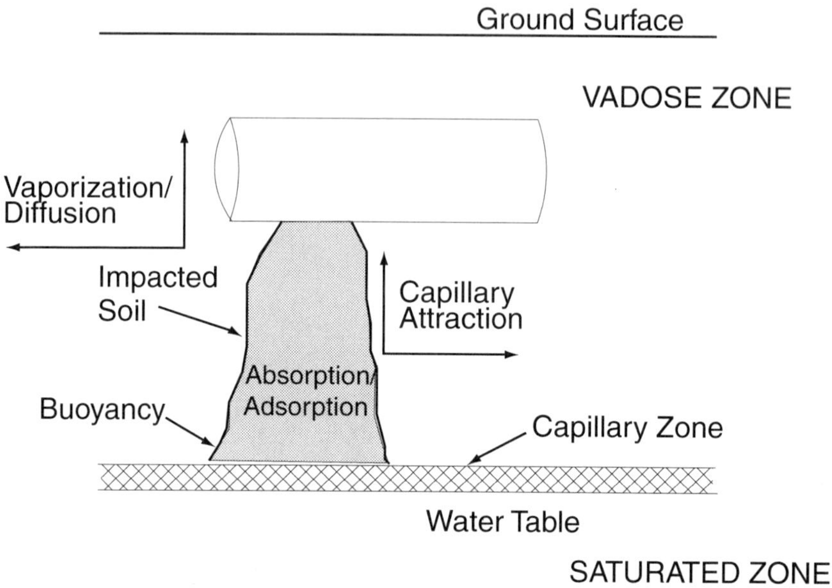

FIGURE 10.3 Factors involved with SVE.

with the soil, resulting in a transfer of compounds from the soil surface into the air. Also, the lower pressure (vacuum) caused by the air pump increases the rate of VOC and water volatilization (evaporation). As the water content of the soil is reduced, the rate of airflow increases. This relationship between soil water content and relative air permeability is shown in Figure 10.4. Dryer soils overall have greater permeability and more effective SVE.

High concentrations of low-solubility compounds with higher vapor pressures will result in greater quantities that will be removed with each flush volume of air. The vacuum developed by the pump, the airflow rate, and the temperature are other factors that also affect removal rates.

Where the volatile compounds are present in soil with low permeability, use of high-vacuum (i.e., up to 25 in Hg) extraction has been successful. Such low pressures enhance evaporation of both contaminants and water attached to the soil particles.

Transfer of VOC from the adsorbed to volatile phase is not instantaneous. If air is passed through the soil too rapidly, it may not have time to reach complete VOC equilibrium. Alternately, if the airflow is too slow, the VOC in the air will be at equilibrium, but the remediation process will be very slow. At most sites, the initial concentration gradient (driving force) is relatively great, and rapid removal of VOC is accomplished. As the VOC content of the soil is decreased, the concentration in the passing air is decreased proportionately. Eventually, this process leads to an asymptotic relationship, where very little VOC is removed with each air flush volume. However, if the system is halted for a period of time (usually days) and then restarted, the VOC removal rate is again increased. This is due to the time-of-release factor. Regular monitoring is required to determine the most effective airflow rate.

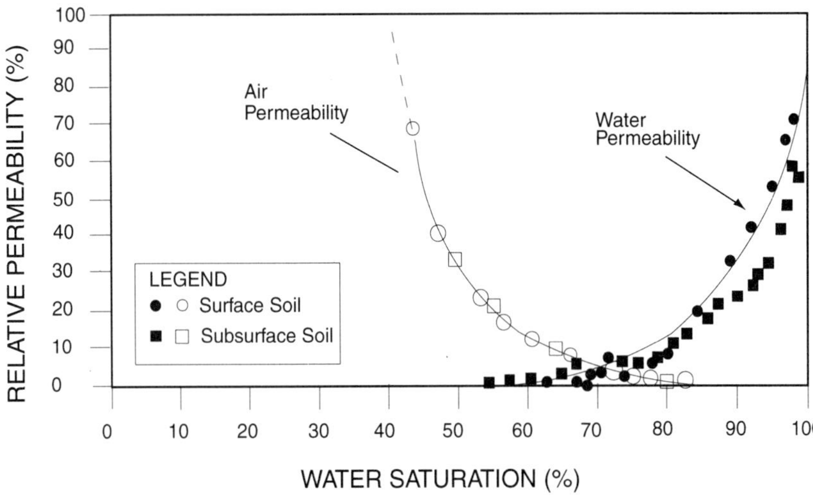

FIGURE 10.4 Relationship between soil water content and relative air permeability.

When the subsurface materials are uniform and isotropic (no gradational changes or confining layers) the airflow pathways are also uniform. The airflow paths developed for an open system and a covered system are shown in Figure 10.5. Selection of covered or uncovered is determined by the air paths necessary to contact the contamination. At some sites, inlet vent wells are installed to ensure air entry at specific locations.

Most sites, however, are not uniform or isotropic. Where zones of soils with differing permeability are present, airflow will follow preferential pathways. These can occur either in vertical settings as sedimentary sequences or in horizonal settings, such as alluvial settings. When preferential flow paths are present, SVE can still operate, but it relies upon both advective and diffusive flow paths. Advective flow paths are the direct airflow routes. Diffusive pathways are the result of concentration gradients caused by the removal of contaminants by air flowing through preferential flow channels. Contaminants then migrate by diffusion and dispersion from areas of higher concentration toward lower concentration. SVE will still work, but the rate of removal is diminished. The relationship between advective flow and diffusive recovery is shown in Figure 10.6.

An additional benefit of SVE is that the flushing air introduces oxygen into the soil pores. Under most circumstances, indigenous soil bacteria in a VOC-contaminated vadose zone have adapted to recognize the VOC as a source of carbon or energy. If the soil bacteria are aerobic, the addition of oxygen will stimulate their growth. Where the soil system is anaerobic, the increased presence of oxygen may interfere with degradation processes. An example SVE extraction pilot test is presented at the conclusion of this chapter.

12.4 AIR SPARGING

Air sparging is the process of introducing compressed air below the VOC- or semi-VOC-contaminated water table, and allowing it to migrate upward into the vadose

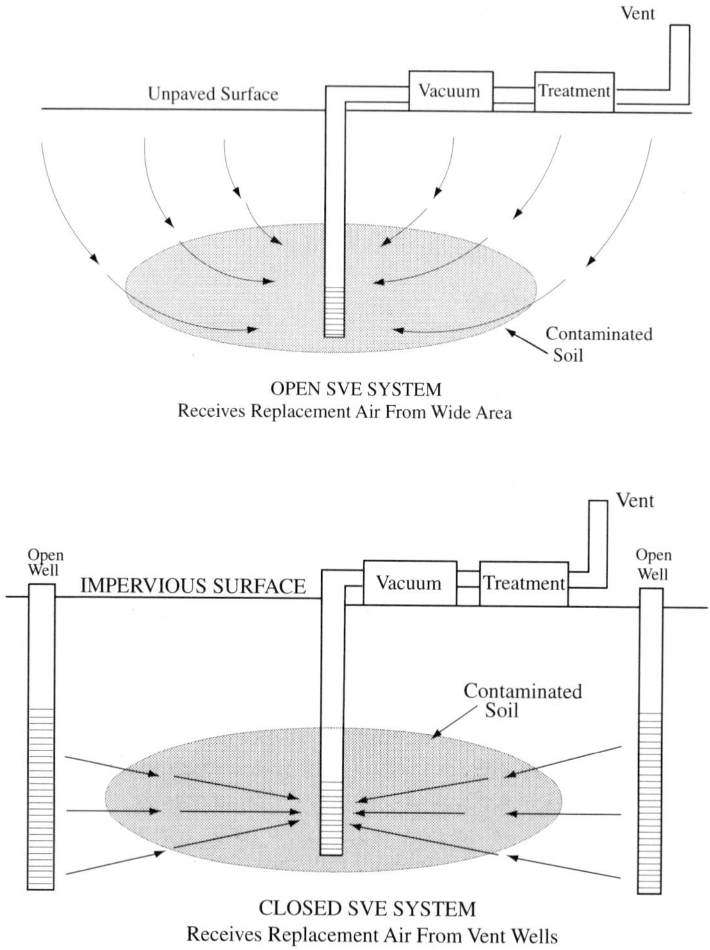

FIGURE 10.5 Airflow paths developed for an open and closed system.

zone. As the air passes through the tortuous pathways between the soil grains, VOCs are "air stripped"; that is, they are transferred by concentration gradient from the water and soil into the air. After rising into the vadose zone, the VOC-containing air may be collected by an SVE system for treatment or, if the biological activity in the vadose zone is adequate, the vapors may be degraded before they are captured or they escape to the atmosphere.

Traditionally air sparging has been used as a groundwater remediation tool. Occasionally, however, it has been successfully used to remediate the vadose zone. In this application, the compressed air is injected through a well screen that is open to the VOC-contaminated area. The injection wells may be either vertical or horizontal (Figure 10.7). In this setting, the injected air is usually captured by a corresponding set of SVE wells (Figure 10.8). Properly spaced patterns of injection and recovery wells are necessary for efficient operation.

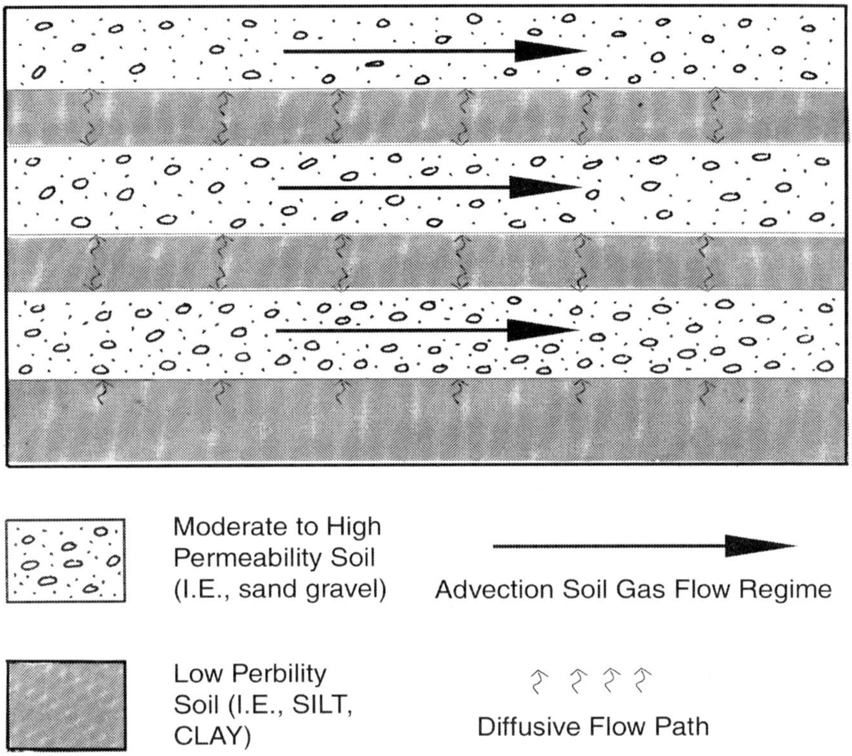

FIGURE 10.6 Relationship between advection flow and diffusive recovery.

10.5 STEAM INJECTION AND HOT AIR STRIPPING

In situ injection of steam has been a common practice in the oil fields of southern California for several decades. The addition of heat reduces the viscosity of the crude oil, allowing it to migrate more easily. The increased temperature also beneficially increases the vapor pressure of the oil, which allows volatile compounds to be more easily recovered. For most VOCs, vapor pressure doubles for every 20°C increase in soil temperature. The use of injected heat, steam, hot water, or air exemplifies the effective transfer of technology from other fields.

Injection of steam or heated air into the subsurface provides large amounts of thermal energy, which speeds the mobilization of adsorbed organic contaminants and results in their removal as either a vapor or liquid phase. Elevated temperature increases the vapor pressure of the chemicals involved and promotes transfer of constituents across the air–water interface, which results in the increased removal of contaminants in high-humidity or nearly saturated soil systems. Additionally, the presence of high-temperature water sometimes results in oxidation or hydration of organic contaminants.

Steam (or hot air) injection requires the placement of injection wells outside the contaminated area and extractions near the center of the targeted treatment area.

FIGURE 10.7 Air sparging pilot test showing portable compressor, airflow regulator meter/oil filter, and injection well layout (A), injection well with quick connect air line and pressure gauge (B), and observation well with magnihelic pressure gauge sensitive to 0.01-in. water column (C).

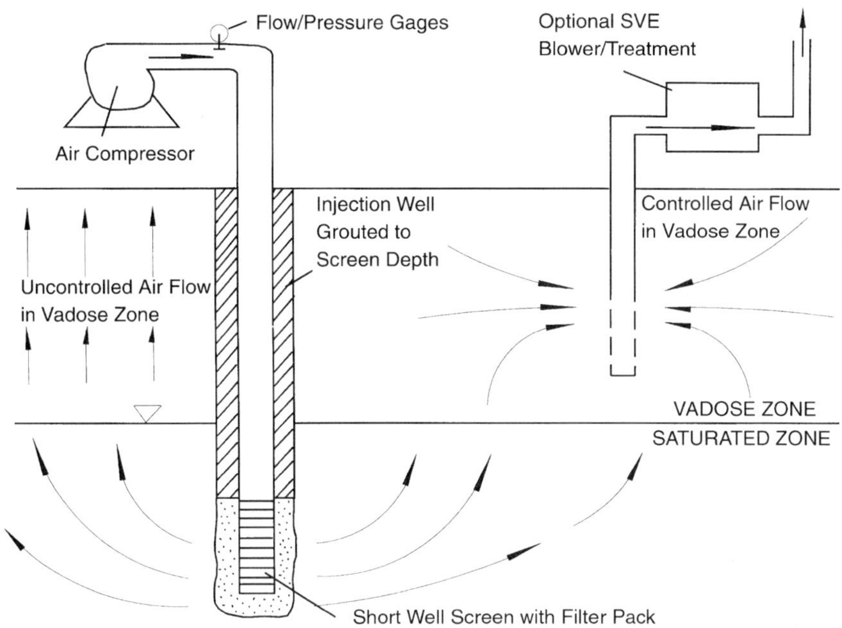

FIGURE 10.8 Schematic showing use of injected steam for the remediation of hydrocarbon-impacted soil.

Key parameters include temperature, pounds of steam injected (or similar factors for air), and duration and depth of treatment. Steam at a pressure 3.5 to 4.2 kg/cm^2 (50 to 60 psi) can heat contaminated soil to 155°C (310°F). The recovery process involves the use of wells to depress the water table and ensure capture of released free-phase NAPLs and vapor-phase hydrocarbons at or near the surface. A conceptual schematic is shown in Figure 10.8.

The system must be operated properly so the vadose zone does not become completely saturated with water, thus reducing the effective permeability to the point that gases and vapors cannot be recovered. When the soil-heating process has progressed to the extent that gaseous steam reaches the recovery wells, all the water within the soil zone has been vaporized. At this juncture, SVE becomes the primary removal mechanism. A blower or vacuum pump can be used to induce airflow in the subsurface, which will facilitate the removal of any remaining residual liquids.

Hot vapors recovered from steam/air stripping are composed of water vapor, volatilized contaminants, and hot air, which must be treated before release to storage or the atmosphere. Treatment trains generally include cooling, demisting (scrubbing), condensing, and dewatering stages, followed by contaminant removal similar to conventional vapor stripping. Activated carbon adsorption is a common treatment option prior to discharge to the atmosphere.

For relative comparisons, tests at a controlled field site resulted in almost complete recovery of kerosene from sand after 126 pore volumes of steam flushing. An EPA SITE program reported 85 to 99% removal of chlorinated VOCs from clay soils. Limitations to the application of steam/heated air injection include subsurface heterogeneities and significant permeability contrasts, which can interfere with the uniform free flow of vapors. A great advantage of steam stripping is that remediation occurs in a relatively short time compared with other technologies. The primary disadvantage is the volume of steam required and the cost of generating it.

10.6 SOIL WASHING

Soil washing involves the leaching of contaminants by the flushing of water through the contaminated soil. Soil washing can be accomplished either in place or aboveground as discussed below.

10.6.1 Leaching in Place

Use of leaching processes to remove adsorbed hydrocarbon from unsaturated soil requires the addition of energy sufficient to overcome the attractive forces that hold the petroleum to the soil grains. Once released, the hydrocarbon is free to migrate as a liquid, emulsion, or vapor. The forces that attract the hydrocarbon to the soil are determined by several factors. Where the surface of the soil grains is initially water-wet (i.e., the individual soil grains are coated primarily by water), the contaminant is loosely held and is often easily flushed by migrating water. Where the soil and hydrocarbon have been in intimate contact for an extended time, the soil particles may have become oil-wet (i.e., individual soil grains are coated primarily by petroleum a few molecules thick). Separation of hydrocarbons from oil-wet soil is a much more difficult task. Use of water as a flushing agent usually requires the addition of surfactants to reduce the surface tension between the water. This allows the water to enter the smaller pore spaces to displace a greater portion of the hydrocarbon, allowing it to migrate.

In natural unsaturated soil environments, oil-degrading biological activity is relatively common. Many soil bacteria are physically attached to the soil grains. These bacteria absorb their nutrients by passing solubilized organic material through their cell walls. To enhance the mobility of the "food," many of the bacteria types produce and secrete enzymes that act as surfactants releasing oil (and other food sources). Since solubilized materials tend to migrate from areas of higher concentration to areas of lower concentration, the solubilized material spreads aerially to be consumed. When the hydrocarbon is solubilized at a more rapid rate than it is consumed, free-phase NAPL migration occurs. Stimulation of indigenous microorganisms can result in migration of a significant quantity of free-phase NAPL. Flushing of the unsaturated zone to release hydrocarbons is a task that should be undertaken only if provisions have been made to intercept and control the migration of free-phase NAPL.

10.6.2 Leaching Aboveground

When soil contaminated by organic chemicals has been excavated for aboveground "washing," the procedure involves both physical and chemical processes to achieve solid and liquid separation. The first stage involves separation of large particles by passing the soil through a bar screen and into a pugmill, where the soil clumps are pulverized to a size sufficiently small to be intimately contacted with a water–surfactant solution. After mixing with the water–surfactant, the soil slurry is then introduced into a mixing chamber where it is thoroughly agitated. The remainder of the process is a series of particle size separations that progressively segregate the washed soil into smaller and smaller grain sizes. Since the smallest soil grains contain the bulk of the surface area, they also contain the majority of the contaminant. After several sortings, the remaining fines are passed through a hydraulic centrifuge for final separation.

Throughout the process, water, surfactant, and NAPL are recovered (as much as possible) for reuse and/or recycling. Many of the aboveground leaching facilities are essentially closed-loop systems. In most cases, the resulting clean soil can be used for backfill, landscaping, or similar purposes.

10.7 BIOVENTING

Organic compounds released into the vadose zone exist in four closely interrelated forms: free-phase NAPL, attenuated to surface of soil grains, dissolved in water retained on and between the soil particles, and present as a gaseous phase. The mass distribution of each of these phases is controlled by such factors as concentration gradients, distribution coefficients, and Henry's law constants.

The soil atmosphere within the pore spaces of the vadose zone contains vapors that result from the chemical and biological processes active at that location. While normal air contains about 79% nitrogen, 20% oxygen, and 0.03% carbon dioxide (by weight), plus minor other gases and a variable saturation of water, the soil atmosphere is usually significantly different. Elemental nitrogen (N_2) is relatively inert, oxygen is very active, and carbon dioxide and methane are typical reaction products of biological reactions. Additionally, since natural ventilation through the pore spaces is very slow, water saturation in the soil atmosphere is almost always near 100%. Where hydrocarbons or other volatile chemicals are present, their vapor phase is also present.

At most release sites, the rate of organic chemical degradation by bacteria is limited by terminal electron acceptors (oxygen, nitrates, sulfates, or iron). The typical distribution pattern for oxygen and carbon dioxide in and around a hydrocarbon plume in porous unsaturated soil is shown in Figure 10.9. At sites where the unsaturated zone is reasonably permeable, the introduction of air will stimulate bacterial activity. A steady supply of air drawn (or injected) through the interconnected soil pores will supply the oxygen the bacteria require. Approximately 10% of the indigenous soil bacteria population can generally use hydrocarbons either as a source of energy or as carbon for cell building.

Wells constructed with perforations and an open unsaturated zone are an efficient procedure to introduce air into the subsurface. A vacuum pump attached to a central well(s) will draw air through the soil to purge the existing soil atmosphere with fresh air. Three commonly used scenarios for bioventing are shown in Figure 10.10.

Consideration of the air permeability of the soil, rate of biological degradation, and rate of volatilization of contaminants is important in the design of a bioventilation system. Reactions that take place in the soil are concentration, temperature, and time

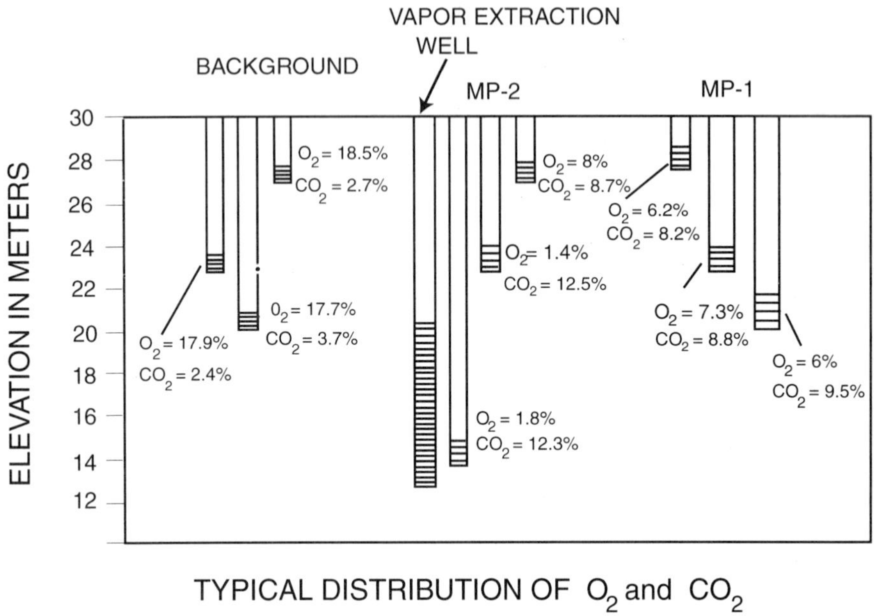

FIGURE 10.9 Typical distribution pattern for oxygen and carbon dioxide in and around a hydrocarbon plume in porous unsaturated soil.

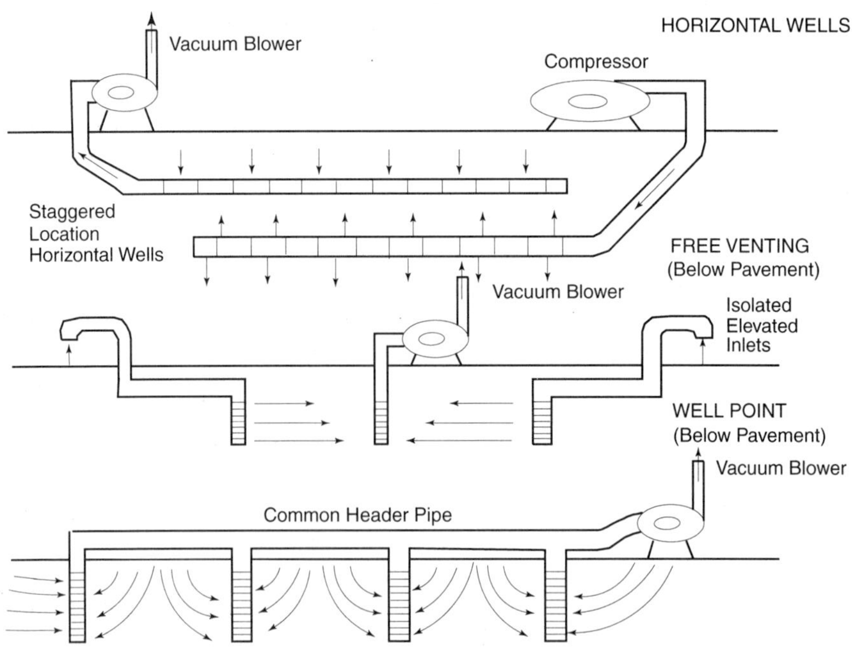

FIGURE 10.10 Schematic showing bioventing scenarios.

related. The airflow must be adjusted to optimize the process. If airflow is too rapid, the mass of vapors removed is less than equilibrium. Where a large quantity of relatively dry air (less than 100% humidity) is drawn through the soil, water may be removed by evaporation to the extent that it interferes with the ability of the bacteria to function. While each site is an entity onto itself, general experience indicates a slow, steady flow of air is more likely to result in a relatively more effective and efficient remediation effort. An advantage to the replacement of air in the soil is the extraction of organic vapors. With volatile compounds, the quantity removed by evaporation can become a significant part of the remediation process.

10.8 BIOREMEDIATION

Bioremediation as a remediation procedure within the vadose zone can take several forms including bioventing as previously discussed. When conditions of low or inadequate permeability exist, the site may not be suitable for ventilation, and the contaminants may be best degraded by anaerobic reactions.

In its basic form, bioremediation of the vadose zone involves introduction of nutrients and electron acceptors necessary to stimulate the indigenous bacteria and provide for removal of waste products generated by the reactions. This sometimes takes the form of a series of injections of a "soup" of nutrients and electron acceptors into the vadose zone through wells, or infiltration galleries. Other sites may require pressure fracturing of the soil before the stimulant blend can be injected.

Prior to the implementation of *in situ* bioremediation as a remedial option, biological testing in addition to routine hydrogeologic characterization is required. Initial biological testing typically includes evaluation of samples by plate count (or respirometer) to determine which types of microorganisms are present and to identify any antibacterial agents (poisons) that may inhibit the growth of the microbes. Biological testing is also used to determine which nutrients may be required to stimulate bioactivity. Tests for available (soluble) mineral content (nitrates, sulfates, and iron) are relatively inexpensive and will assist the designer with selection of remedial options. Upon subsurface characterization, and if the necessary conditions for biological action have been determined, the design phase needs only to provide the necessary nutrients, electron acceptors, and proper moisture content for the remediation process to proceed.

In some situations, indigenous bacteria capable of degrading a specific contaminant are not present. When this situation occurs, addition of imported adapted microbes may be necessary. Added microbes for a variety of chemicals can be purchased from remediation suppliers, or can be cultured from sites where they are known to be active. As an example, a spill of butyl acrylate (a plasticizer compound) along a railroad right-of-way required remediation without disrupting rail traffic. Indigenous soil bacteria at the site seemed incapable of adapting to the chemical within a reasonable time. A sample of waste treatment sludge from the chemical plant that manufactured the butyl acrylate was cultured with the appropriate nutrients in a large tank and delivered to the site. Within a few weeks after the "soup" (culture and nutrients) was spread along the right-of-way, significant decreases in butyl acrylate were observed.

10.9 NATURAL ATTENUATION

At certain locations the best approach to remediation of the unsaturated zone is to allow nature to take its course. Natural attenuation (sometimes called intrinsic remediation) is not a "do-nothing" solution; instead, it relies upon a two-phase approach. The first phase is to complete a thorough analysis of the site to account fully for all the biological and geochemical processes that are capable of containing the spread of contaminants and degrading them by unaided means. The second phase may include the need to establish institutional controls, such as new zoning regulations for the affected sites, to reduce the risk of public exposure to the contaminants until they are degraded.

Processes that allow for the degradation of hydrocarbons and other constituents include advection, dispersion, sorption, dissolution, redox buffering, and abiotic (chemical) and biological transformations. Natural attenuation is considered a "passive" remedial approach because it does not require the application of outside energy. Only recently has natural attenuation of petroleum hydrocarbon and other organic chemicals been considered an acceptable remedial strategy by regulatory agencies.

Sites impacted by chlorinated solvents and other complex chemicals may not be well suited for natural attenuation, since the processes of degradation involved become more site specific dependent. Halocarbons can be metabolized by certain organisms, and will ultimately dechlorinate without bioassistance; however, the intermediate products that develop include numerous other compounds that may be considered toxic. To demonstrate the appropriateness of natural attenuation at these sites, comprehensive understanding of subsurface soil conditions, concurrent with a program incorporating long-term soil and groundwater monitoring and rigorous fate-and-transport modeling, may be necessary.

Confirmation of certain conditions usually needs to be demonstrated for natural attenuation to be deemed acceptable as a remedial option. This demonstration may involve showing the contaminant plume to be stable and contained, shrinking, or to be intrinsically remediated at the same rate as its spreading (Figure 10.11). When the plume is essentially being remediated at the same rate that it is spreading, dynamic equilibrium conditions exist.

10.10 OTHER TECHNOLOGIES

Other technologies that have been used under certain conditions include vacuum vaporization, hydrofracturing enhancement, electrochemical, and vitrification and electric heating techniques.

10.10.1 VACUUM-VAPORIZED WELL

The vacuum-vaporized well technology is a proprietary process that combines vacuum extraction, bioremediation, and soil flushing to mobilize contaminants in the vadose, capillary, and saturated zones. A schematic of the double-screened well with the equipment contained within is shown in Figure 10.12.

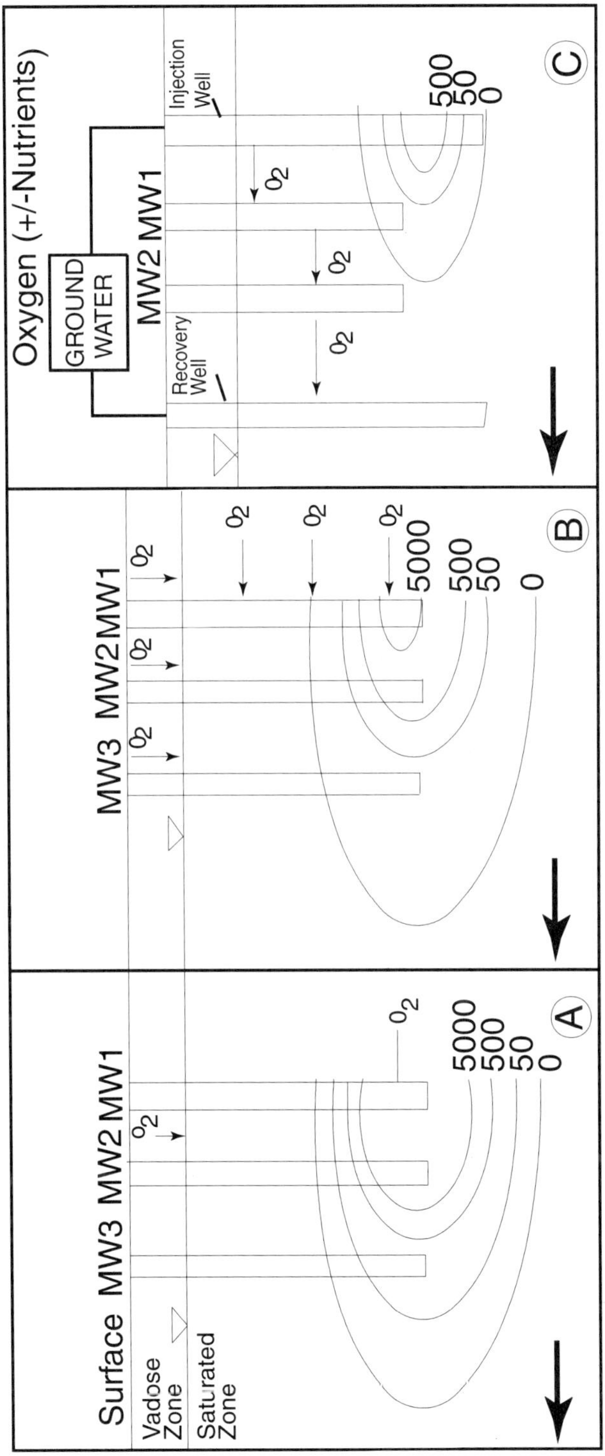

FIGURE 10.11 Schematic showing natural attenuation scenarios for a contaminant plume when the plume is stable under low-permeability conditions (A), shrinking under high-permeability conditions (B), and reducing via enhanced biodegradation (C).

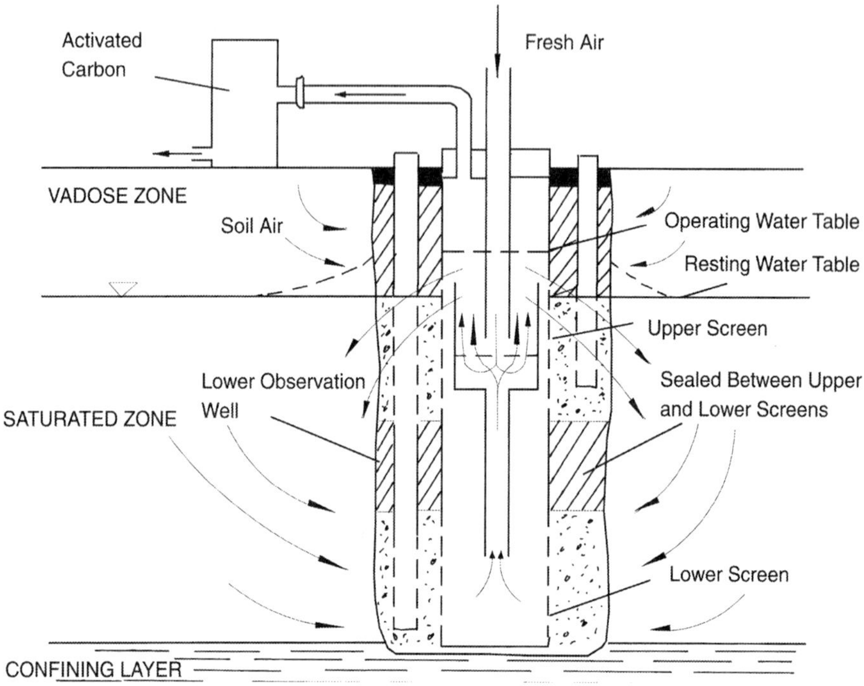

FIGURE 10.12 Schematic of a double-screened vacuum-vaporized well.

Treatment in the vadose zone is achieved by vacuum extraction through a partially exposed upper screen, whereas treatment in the saturated zone is achieved by a combination of soil flushing, air stripping, and bioremediation. Vertical groundwater circulation is established between the well screens by pumping water from the lower part of the well upward to the top of the stripper fixture, where it overflows and exits the upper screen. As water travels through the remediation well, groundwater passes through one or more treatment systems, which may include an air stripper/aerator or *in situ* bioreactor.

Groundwater leaving the upper screen contains oxygen as a result of the stripping reactor. Other additives, such as inorganic nutrients, may be added to facilitate biodegradation of contaminants in the aquifer. The water circulating between the upper and lower well screens distributes the nutrients into the aquifer media surrounding the well.

Application of a vacuum to the well casing creates a lower-pressure condition that assists with vapor removal from the air stripper/aerator as well as increasing airflow through the surrounding unsaturated soil. The radius of treatment by vapor-vaporized wells depends on site-specific subsurface conditions. Factors such as hydraulic conductivity, hydraulic gradient, saturated thickness, and groundwater circulation rates are important considerations. Published results of field tests indicate that this technology can achieve significant reductions of most volatile chemicals including fuel hydrocarbons, paint solvents, and chlorinated solvents.

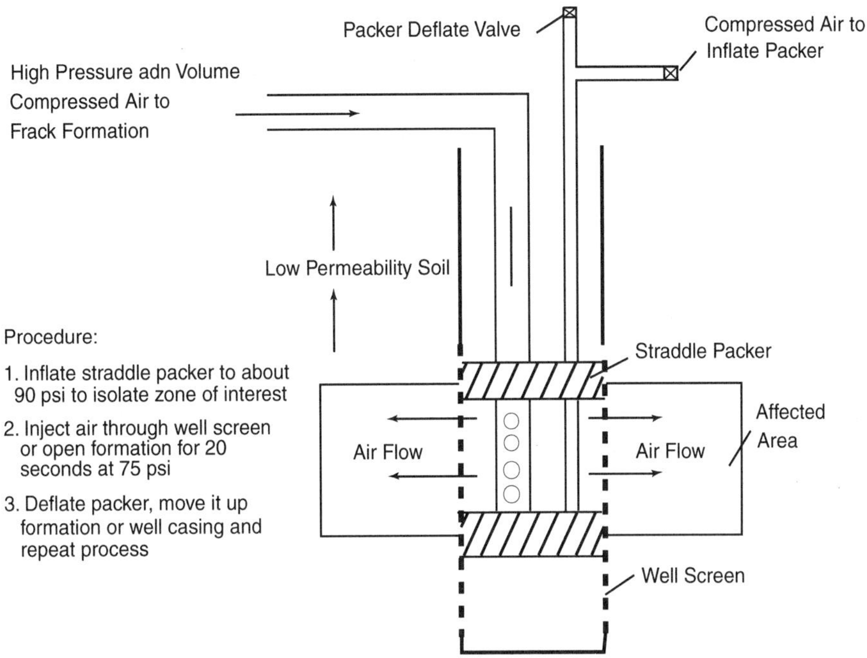

FIGURE 10.13 Schematic showing subsurface conditions favorable for hydrofracturing.

10.10.2 Hydrofracturing Enhancement

When remediation wells are constructed within the unsaturated zone for vapor extraction or enhanced bioremediation, it is common that the wells do not have the necessary permeability. This situation may result from the natural low effective porosity of the formation, "polishing" of the borehole wall by the turning of the hollow stem auger, or a residual mudcake if air or water rotary-drilling equipment was used. A commonly used technique to improve the effective porosity is to fracture the soil formation by injecting either pressurized water or injecting a surge of high-pressure air. The zone to be fractured is identified and isolated by packers set across the zone as shown on Figure 10.13. If the formation has sufficient structural integrity to support packers, they may be set in an uncased borehole. Most often, however, the formation will require a casing and filter pack. After the packers have been set and inflated, a surge of either water or air is injected into the isolated interval and into the formation. The injection fluid then causes the soil to fracture (or heave), making room for the fluid. After the formation has opened, the pressurizing fluid is removed by pumping (or air bleeding). The resulting fractured soil tends to have greater permeability (effective porosity), which enhances the remedial possibilities.

When unsaturated soil is fractured, the stress of the injected fluid is adsorbed by the soil in either of two scenarios. Either the surrounding soil is compressed or the surrounding soil maintains its integrity and conveys the energy to the surface where heaves develop. In unconsolidated soil with relatively low shear strength, the fracturing force (and heaving force) is roughly equal to the weight of the overlying soil. A maximum sustained pressure commonly used for fracturing is 0.5 psi/ft of depth. When fracturing is used on slightly plastic clay soil, some collapse of the soil should be expected. Some sophisticated fracturing efforts involve injection of coarse sand or glass beads into the developed fractures. This is difficult, however, to accomplish if a filter pack has been placed between the screen and the well bore.

Pneumatic or hydraulic fracturing can occasionally be used to enhance the effective porosity (permeability) of unsaturated zones for SVE or air sparging. Successful soil fracturing is as much an art as a science. Pilot testing by experienced professionals is strongly recommended before performance guarantees are granted.

10.10.3 Electrochemical

Certain basic chemical reactions can sometimes be used to remediate organic chemical contaminants in the unsaturated zone. Introduction of reacting chemicals to alter the Eh (oxidation state/electron availability), the pH (hydrogen ion availability), or a contaminant to an immobile state occasionally presents a viable remedial option. This procedure can be rather expensive in dealing with organic chemicals, and typically is only used in specific situations.

10.10.4 Vitrification and Electrical Heating

Recalcitrant organic contaminants in the unsaturated zone, those that are difficult to remediate by vapor extraction, biodegradation, or other biological–chemical processes, can be removed via vitrification and electrical heating. This process involves heating the soil by electrical resistance or applied electrical fields to vaporize semi-volatile and volatile compounds, thermally to break down larger organic molecules, and (if heated sufficiently) to vitrify the mineral matter in the soil.

Application of electrical heating requires installation of electrodes in the soil in a set pattern and at spacings that will uniformly heat the contaminated soil mass. An aboveground vapor capture system is necessary to intercept the vapors produced. Treatment of the vapor phase is always necessary because of the wide variety of original and degradation chemicals produced by heating. A schematic diagram of an electrical heating system is shown on Figure 10.14. While highly effective, electrical heating is suitable for only certain sites. The intense heat, cost of electricity, and resulting altered soil mass limit the applications and overall benefit.

10.11 SVE Case Histories

10.11.1 Case History Example 1

An SVE pilot test was conducted at a service station site that had experienced a release of gasoline. Some LNAPL had been removed, but high concentrations of vapor persisted across the site. The site was underlain by silty clay, which overlies weathered and highly fractured shale at a depth of approximately 7 ft. The water table occurs within the upper portion of the shale at a depth of approximately 11 ft.

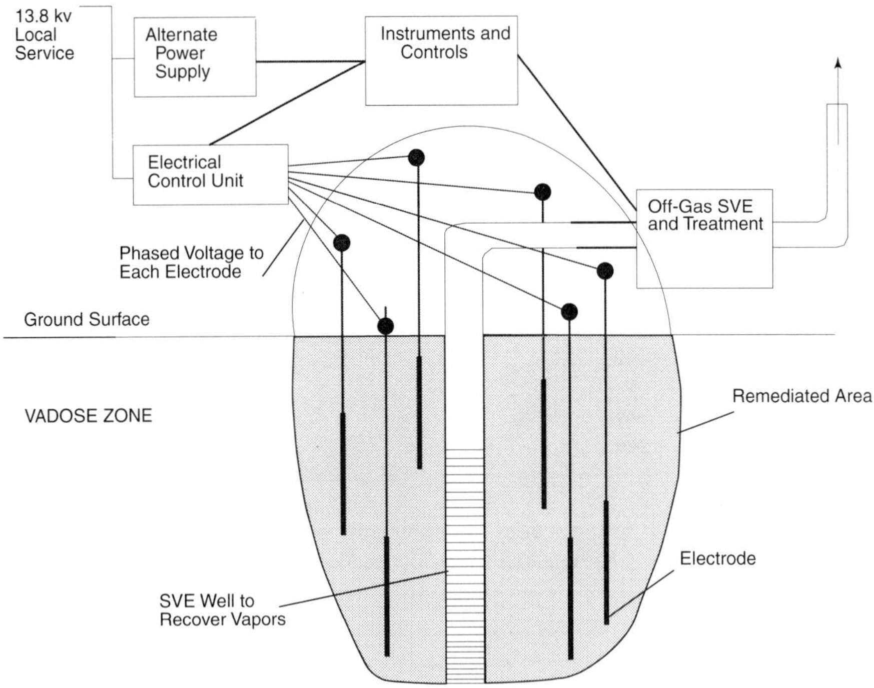

FIGURE 10.14 Schematic of an electrical heating system.

The underground tanks at this site were located in a tank hold that extended into the shale. When the release occurred, gasoline migrated downward into the shale and groundwater. Groundwater in the shallow aquifer in this area is not used, due to limited quantity and poor quality, although a significant concern existed due to elevated hydrocarbon vapors. A local basement fire had been attributed to the presence of the vapors.

Remediation commenced with the interception of the recoverable gasoline. One 4-in.-diameter vapor extraction well and three 2-in. vapor monitor wells were installed. Each well was constructed with 4 ft of well screen placed and filter packed across the unsaturated portion of the shale. The annulus of the wells was sealed to the surface with bentonite and cement grout according to local standards. The monitor wells were located in a triangular pattern around the vapor extraction well at distances of 5.3, 10.6, and 25.4 ft.

A regenerative vacuum blower was connected to the extraction well and operated at a vacuum range of 46 to 49 in. of water column vacuum at a controlled flow of approximately 17 standard ft^3/min (scfm). Vacuum gauges were attached to each of the monitoring wells. During the 8-h testing period, periodic samples were collected for Draeger colorometric analysis, soil vapor monitoring, and confirmatory laboratory testing. Pressure, flow, and vacuum measurements were recorded every few minutes.

As this is an demonstrative example, only a portion of the data will be discussed here. When the SVE unit was started, the vacuum in the recovery well rapidly approached a water column vacuum of 46 to 49 in., where it remained throughout the test. The total petroleum hydrocarbon concentration of the exhaust was recorded at regular intervals as follows:

Soil Vapor Extraction TPH Recovery Results

Time, min	Concentration (ppm-weight)	Recovery Rate (lb/day)
10	Nondetectable	0
30	150	0.28
60	2000	3.8
90	2000	3.8
360	3500	6.6
480	5300	10

At the end of the 8-h test, vacuum measurements in the three monitor wells had stabilized (for at least 2 h) with the following recorded readings:

Well	Distance (ft)	Vacuum (in.)
MW-1	5.3	0.37
MW-2	10.6	0.65
MW-3	25.4	0.06

With only data from three monitor wells, the radius of influence was intuitively determined to be 25 ft. A water column vacuum of 0.06 in. was only slightly detectable. Results of calculations based on this value was considered satisfactory to determine if SVE was a viable remediation option.

Soil permeability was determined by the equation:

$$K = \frac{Q\mu\left(\ln\frac{R_w}{R_i}\right)}{H\Pi P_w\left[1-\left(\frac{P_{atm}}{P_w}\right)^2\right]} \tag{10.1}$$

where K = intrinsic permeability (cm^2), Q = airflow = 7731 cm^3/s (17 scfs), μ = air viscosity = 1.8×10^{-4} g/cm-s, R_w = radius of pumped well = 5.08 cm (2 in.), R_i = radius of influence = 762 cm (25 ft), H = screen length = 121.9 cm (4 ft), P_{atm} = 1.016×10^6 g/cm-s^2 (1 atm), and P_w = pressure in pumped well = 8.88×10^5 g/cm-s^2 (0.88 atm). The solution to Equation 9.1 describes the intrinsic permeability calculated as follows:

$$K = \frac{7731\,\text{cm}^3/\text{s}(1.8\times10^{-4}\ \text{g}-\text{s}^2)\left(\ln\dfrac{5.08}{762}\right)}{(121.9\,\text{cm})\Pi(8.88\times10^5\ \text{g/cm}-\text{s}^2)\left[1-\left(\dfrac{1.016\times10^6\ \text{g}-\text{cm}^2}{8.88\times10^5\ \text{g}-\text{s}^2}\right)^2\right]}$$

$$K = 6.6\times10^{-8}\ \text{cm}^2\ (\approx 6.7\,\text{D})$$

A site with an intrinsic permeability value of $>1 \times 10^{-10}$ or $>1 \times 10^{-12}$ cm^2 (depending on the process) is a good candidate for SVE. An element of caution should be observed in relying on one set of test results for design of an SVE system for an entire site. Air permeability can (and will) vary by one to three orders of magnitude across a site, and can be estimated from boring log data only within an order of magnitude. It was recommended that the system be designed with sufficient operational flexibility to accommodate these extremes.

The actual mass of fuel that could be anticipated to be removed is dependent on several factors including volatility of the hydrocarbon compounds, rate of flow through the soil, moisture condition, applied vacuum, biological and chemical reactions, and partition equilibrium processes. Fresh gasoline is a complex mixture of mostly high-vapor pressure components, weathered gasoline has lost some of the volatile fractions, and diesel fuel is much less volatile. If airflow is too fast, the vapor will not be at saturation in the air, whereas if airflow is too slow, the system is not efficient. Infiltration of rainwater or evaporation by infiltration of dry air can significantly change the air permeability. Applied vacuum can have a direct effect in the recovery rate. Small vacuum leaks are difficult to locate and decrease the recovery. Introduction of fresh air into an environment that was previously relatively devoid of oxygen may result in biological growth or chemical precipitation to change the rate of airflow. Since the SVE relies on partition equilibrium processes, the rate of recovery tends to decline n a semiexponential rate.

While the 8-h test results indicated a total petroleum hydrocarbons (TPH) recovery rate of 10 lb/day, this rate probably will not continue. Estimation of the expected long-term recovery rate of TPH by this short-term test data is not a reliable exercise. The data demonstrate the applicability of SVE as a recovery system, but further testing or detailed computer modeling was recommended prior to final design.

10.11.2 Case History Example 2

The SVE case history discussed is included as a "typical situation," similar to many projects encountered in the field that diverge from the ideal case. The site is located adjacent to a vehicle maintenance building in southern California. A 3000-gal and a 5000-gal UST were used to store unleaded gasoline, and a 550-gal UST was used to store petroleum-based paint thinner (Figure 10.15). To add to the situation, different phases of the project were completed by different consultants, and, due to administrative issues, almost 4 years passed between removal of the USTs and completion of the remediation activities.

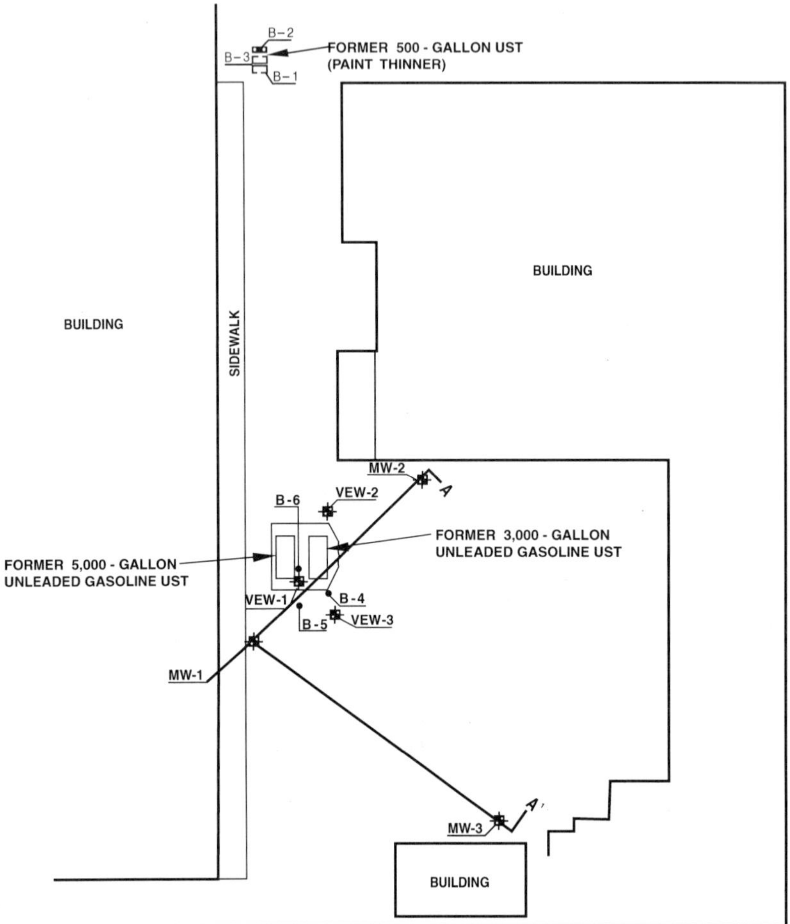

FIGURE 10.15 Site layout for an SVE system.

The site was underlain by stratified but laterally and vertically discontinuous mixtures of sand, silt, and clay (Figure 10.16). Groundwater was encountered under unconfined water table conditions at 68 to 75 ft below the ground surface, and isolated from the impacted area by low-permeability clay and silt lenses. Soil samples retrieved from beneath each former UST during removal had reported TPH (modified for gasoline) concentrations ranging from 7800 to 14,000 ppm immediately below the 3000-gal UST, and nondetectable below the 5000-gal UST. TPH (modified for paint thinner) was reported at a concentration of 35,000 ppm immediately below the 550-gal UST. Soil borings were drilled to assess further subsurface hydrogeologic conditions, and the lateral and vertical extent of the hydrocarbon-impacted area. Certain soil samples retrieved from between and beneath two larger USTs had reported TPH concentrations ranging up to 3590 and 3570 mg/kg at depths of 40 and 45 ft below the ground surface.

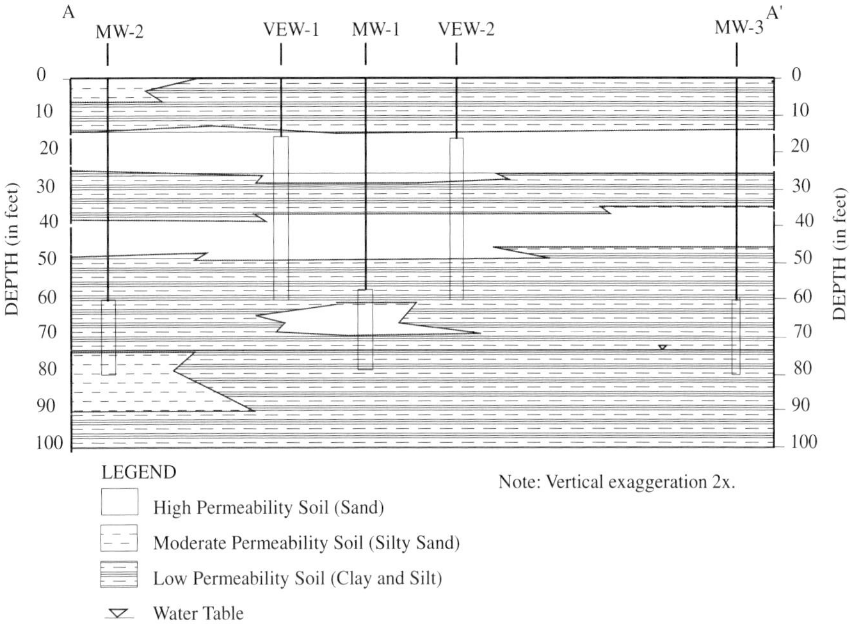

FIGURE 10.16 Hydrogeologic cross section A–A.

A pilot soil vapor extraction test was performed to determine whether this method would be effective to mitigate the concern. Such a test provided information used to design a full-scale recovery and vapor treatment system. With air permeabilities on the order of 1 to 3 D and TPH vapor concentration on the order of 330 ppm (weight), the pilot test supported SVE as a viable remediation approach.

The key features of the SVE, as shown in Figure 10.17, include:

- 170 linear ft of below-grade, 3-in.-diameter Schedule 80 polyvinyl chloride (PVC) vapor-header pipe to connect the vapor extraction wells to the SVE unit.
- Vapor-flow sensor and vapor-sampling ports located near the vapor extraction well heads.
- A trailer-mounted, 200-cfm SVE unit equipped with temperature and LEL sensor controllers, automatic vapor diluter, and water knockout tank (Figure 10.18).
- Vapor treatment equipment consisting of two serial filters containing 1000 lb of granulated activated charcoal (GAC) situated upstream of the SVE blower; and
- Chain-link security fencing enclosing the system.

The operations and maintenance program consisted of routine maintenance of the SVE unit and vapor conveyance lines. Data recording units (disk recorders) required maintenance, as well as emptying the recovery drum of accumulated knock-

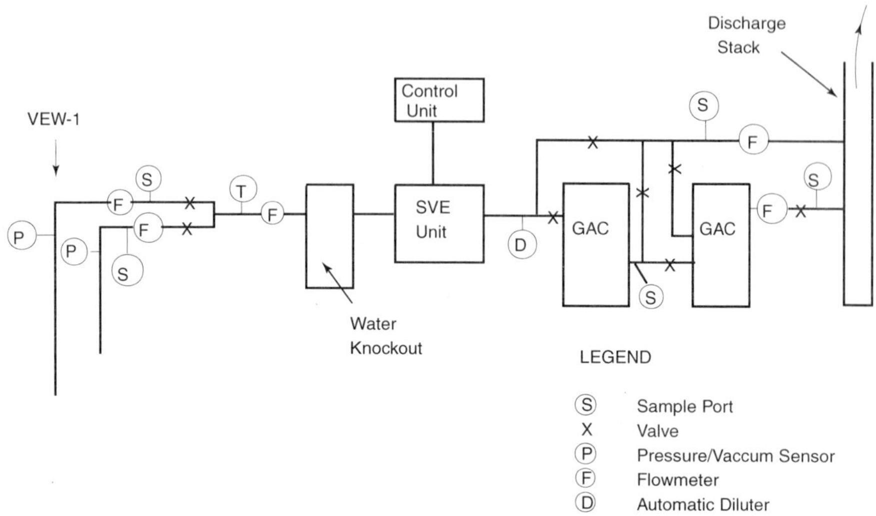

FIGURE 10.17 Schematic showing SVE equipment and configuration.

out water, cleaning the air filter, and servicing the blower motor. The project management program consisted of periodic measurement of airflow rates and vapor concentrations at the vapor extraction wells and the SVE unit. Static and differential vacuum pressures were periodically measured at flow sensors using magnehelic gauges. Vapor samples were collected in 1-l Tedlar bags from sampling ports located near the vapor extraction well and GAS influent/effluent lines. The vacuum pressures were converted into flow rates from the vapor extraction wells, and vapor samples were field monitored for VOCs, with select samples submitted to a certified laboratory for confirmatory analysis.

The SVE system operated almost continuously (92% of the time) throughout the 72 days of operation. During this period, vapor flow remained within the range of 56 to 70 scfm (1615 to 1986 l/min) with an almost constant well head pressure (absolute) of 0.8 atm (vacuum equal to 81 in. H_2O column or 6 in. Hg). Vapor samples were collected approximately every 5 days from each well head, GAC influent and effluent, and exhaust stack effluent.

Analytical results for BTEX and TPH-gasoline for the GAC influent and vapor extraction well (VEW-1) are shown in Table 10.2. Total volatile petroleum hydrocarbons concentrations entering the GAC unit represent the gross vapor removal from the system (Figure 10.19). The GAC filter proved highly effective in recovery of the hydrocarbon compounds. Concentrations of BTEX in the GAC exhaust were reduced to nondetectable levels, whereas TPH-gasoline was maintained at low levels (typically 5 to 15 ppm). Graphs of the total volatile petroleum hydrocarbon and BTEX concentrations measured in one of the vapor extraction wells (VEW-1) is shown in Figure 10.20; the graph for vapor extraction well VEW-2 is not shown since it resembles that of vapor extraction well VEW-1, and BTEX concentrations were reduced to nondetectable levels after 8 days of operation. After 72 days, low

FIGURE 10.18 Graphs of total volatile petroleum hydrocarbons concentration with time at GAC inlet.

asymptotic levels were reached, and the system was shut down for several days, restarted to confirm asymptotic conditions, then shut down permanently.

Confirmatory soil sampling was subsequently performed. Three soil borings were drilled and samples retrieved from the impacted area. Soil from a depth of 30 ft and below were reported as nondetectable; however, samples from a depth of 21 ft still contained significant gasoline components with TPH-gasoline ranging up to 3600 ppm. This zone of elevated hydrocarbons was anticipated due to the presence of clays and silt at this depth. Ventilation of these low-permeability soils was not deemed cost-effective, and significant reduction of the residual hydrocarbon concentrations unlikely.

Since the remaining residual hydrocarbons were localized, and had no impact on groundwater quality, a risk-based closure approach was taken, and subsequently approved for site closure. This case history is typical of many UST sites in that full removal of hydrocarbons below regulatory levels is not achievable for a variety of reasons. This will be further discussed in Chapter 13 (Site Closure).

TABLE 10.2
Analytical Results for Soil Vapor for Vapor Extraction Well VEW-1

	Vapor Extraction Well VEW-1					GAC Influent				
		Volatile Aromatic Hydrocarbons					Volatile Aromatic Hydrocarbons			
Day	TPH[a,b]	Benzene	Toluene	Ethylbenzene	Xylenes	TPH[a,b]	Benzene	Toluene	Ethylbenzene	Xylene
1	270	5.3	26	2.5	7.6	350	4.7	37	3.9	17.2
3	270	4.1	36	4.1	30.4	220	2.9	27	4.6	33.2
11	180	NA[c]	NA	NA	NA	190	NA	NA	NA	NA
17	220	NA	NA	NA	NA	170	NA	NA	NA	NA
25	130	NA	NA	NA	NA	100	NA	NA	NA	NA
30	54	NA	NA	NA	NA	70	NA	NA	NA	NA
40	96	NA	NA	NA	NA	68	NA	NA	NA	NA
44	130	2.4	11	4.0	4.9	72	NA	NA	NA	NA
47	41	NA	NA	NA	NA	46	NA	NA	NA	NA
50	42	NA	NA	NA	NA	46	NA	NA	NA	NA
54	44	NA	NA	NA	NA	46	NA	NA	NA	NA
62	390	ND[d]	7.1	14	57	28	NA	NA	NA	NA
65	24	NA	NA	NA	NA	25	NA	NA	NA	NA
67	153	0.25	5.3	0.51	1.4	NA	NA	NA	NA	NA
69	95	NA	NA	NA	NA	70	NA	NA	NA	NA
72	100	NA	NA	NA	NA	63	NA	NA	NA	NA

[a] Data analyzed using EPA Method 8015 modified.
[b] Data shown presented in parts per million per volume (ppmv).
[c] NA = not analyzed.
[d] ND = not detected at or above its respective analytical detection limit.

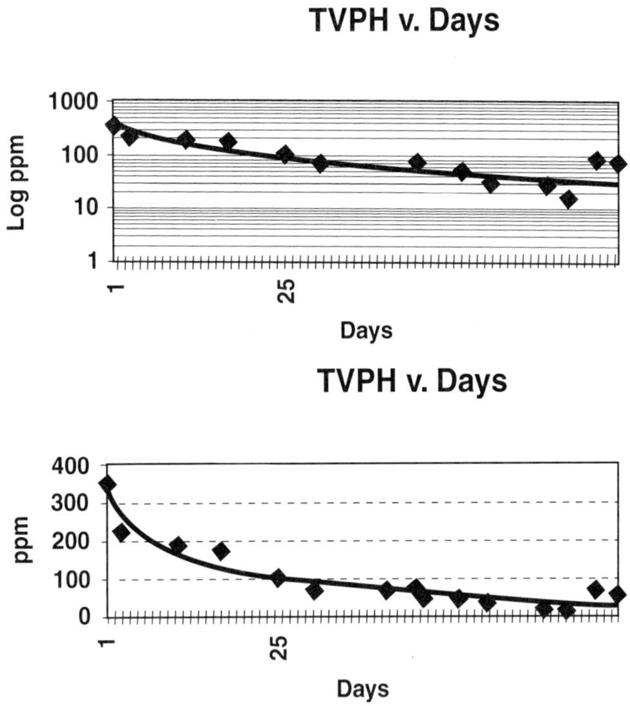

FIGURE 10.19 Graphs of vapor analytical results at vapor extraction well.

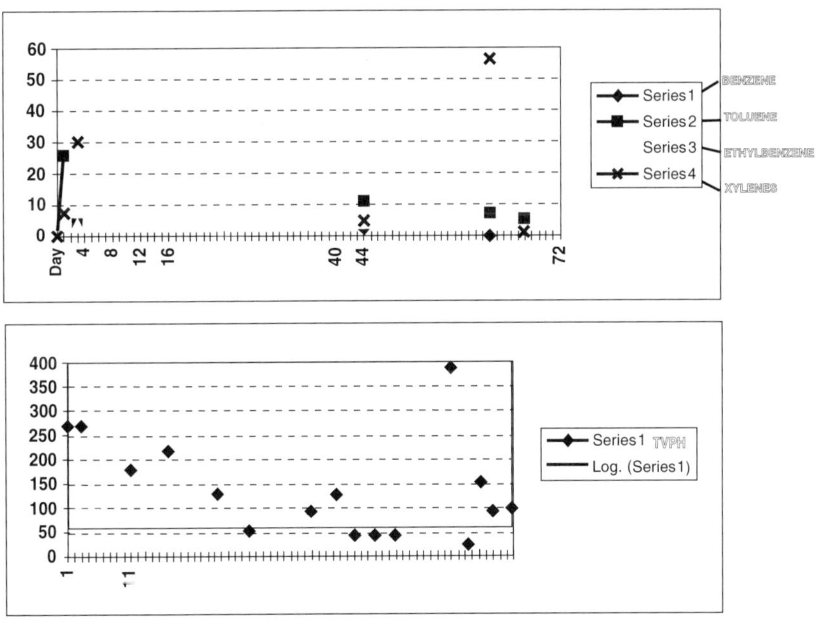

FIGURE 10.20 Time vs. concentration for VEW-1.

REFERENCES

Ahlfeld, D. P., Dahmani, A., and Ji, W., 1994, A Conceptual Model of Field Behavior of Air Sparging and Its Implications for Application: *Ground Water Monitoring Review*, Fall, Vol. 14, No. 4, pp. 132–139.

American Petroleum Institute, 1985, *Subsurface Venting of Hydrocarbon Vapors from an Underground Aquifer*: American Petroleum Institute, Publication No. 4410.

Anastos, G., Corbin, M. H., and Coia, M. F., 1986, *In Situ* Air Stripping: A New Technique for Removing Volatile Organic Contaminations from Soils: In *Proceedings of Superfund, 86*: Hazardous Materials Control Research Institute, Silver Spring, MD.

Bennedsen, M. D., 1987, Vacuum VOC's from Soil: *Pollution Engineering*, Vol. 19, No. 1, pp. 66–68.

Bergsman, T. and Trowbridge, B., 1997, Soil Vapor Extraction to the Sixth Degree: *Soil and Groundwater Cleanup*, July, pp. 7–11.

Borden, R. C. and Chih-Ming, K., 1989, Water Flushing of Trapped Residual Hydrocarbons: Mathematical Model Development and Laboratory Validation: In *Proceedings of the National Water Well Association Conferences on Petroleum Hydrocarbons and Organic Chemicals in Groundwater: Prevention, Detection and Restoration*, November, pp. 473–486.

Brookman, G. T., Flanagan, M., and Kebe, J. O., 1985a, *Literature Survey: Hydrocarbon Solubilities and Attenuation Mechanisms*: American Petroleum Institute of Health and Environmental Sciences Department, No. 4414, 101 pp.

Brookman, G. T., Flanagan, M., and Kebe, J. O., 1985b, *Literature Survey: Unassisted Natural Mechanisms to Reduce Concentrations of Soluble Gasoline Components*: American Petroleum Institute of Health and Environmental Sciences Department, No. 4415, August, 73 pp.

Brubaker, G. R., 1999, Considering Advanced Techniques for Enhanced Recovery of NAPL: *Soil and Groundwater Cleanup*, June/July, pp. 10–11.

Bruce, D. A., Ellen, M., Bruce, C., and DiMillio, A. F., 1998, Deep Mixing Method: A Global Perspective: *Civil Engineering*, December, pp. 39–41.

Calabrese, E. J. and Kostecki, P. T. (editors), 1989, *Petroleum Contaminated Soils,* Vol. 2: Lewis Publishers, Chelsea, MI, 515 pp.

Camp, Dresser, and McKee, 1986, *Superfund Treatment Technologies: A Vendor Inventory*: EPA 540/2-86-004.

Carpenter, M. D., Gierke, J. S., et al., 1993, Vapor Phase Transport in Physically-Mixed Clay Soils: In *Proceedings of the 1993 Petroleum Hydrocarbons and Organic Chemicals in Ground Water: Prevention, Detection and Restoration*: American Petroleum Institute and National Ground Water Association, November 10–12, pp. 479–489.

Castle, C., Bruck, J., Sappington, D., and Erbaugh, M., 1985, Research and Development of a Soil Washing System for Use at Superfund Sites: In *Proceedings of the USEPA Sixth National Conference on Management of Uncontrolled Hazardous Waste Sites*: Washington, D.C., pp. 452–455.

Chaperon, M., 1995, Guaranteeing the Success of Bioremediation: The Biofeasibility Study: *Remediation Management*, September/October, Vol. 1, No. 1, pp. 16–19.

Chen, M. R., Hinkley, R. E., and Killough, J. E., 1996, Computed Tomography Imaging of Air Sparging in Porous Media: *Water Resources Research*, Vol. 32, No. 10, pp. 3013–3024.

Clayton, W. S., 1998, A Field and Laboratory Investigation of Air Fingering during Air Sparging: *Ground Water Monitoring & Remediation*, Summer, Vol. 18, No. 3, pp. 134–145.

Conklin, A. R., 1997, Bacteria 101: *Soil and Groundwater Cleanup*, May, pp. 15–16.

Conner, J. R., 1988, Case Study of Soil Venting: *Pollution Engineering*, Vol. 20, No. 7, pp. 74–78.

Crow, W. L., Anderson, E. P., and Minugh, E. M., 1987, Subsurface Venting of Vapors Emanating from Hydrocarbon Product on Ground Water: *Ground Water Monitoring Review*, Vol. 7, pp. 51–57.

Dragun, J., 1988, *The Soil Chemistry of Hazardous Materials*: Hazardous Materials Control Research Institute, Silver Springs, MD, 458 pp.

Dunlap, L. E., 1984, Abatement of Hydrocarbon Vapors in Buildings: In *Proceedings of the National Water Well Association and American Petroleum Institute Conference on Petroleum Hydrocarbons and Organic Chemicals in Groundwater: Prevention, Detection and Restoration*, November, pp. 504–518.

Fam, S. A., Messmer, M. F., Nautiyal, D., and Hansen, M. A., 1995, Critical State of the Art Review of Vapor Extraction: In *Proceedings of 50th Purdue Industrial Waste Conference*, Ann Arbor Press, Chelsea, MI, pp. 7–22.

Fan, C. Y. and Krishnamurthy, S., 1995, Enzymes for Enhancing Bioremediation of Petroleum Contaminated Soils: A Brief Review: *Air and Water Management Association*, Vol. 45, June, pp. 453–460.

Frankenberger, W. T., Emerson, K. D., and Turner, D. W., 1989, *In Situ* Bioremediation of an Underground Diesel Fuel Spill: A Case History: *Environmental Management*, Vol. 13, No. 3, pp. 325–332.

Fustos, V. and Lieberman, P., 1996, Integrated In-Situ Remediation: *Environmental Protection*, January, pp. 44–49.

Grasso, D., 1993, *Hazardous Waste Site Remediation Source Control*: CRC Press, Boca Raton, FL, 545 pp.

Haddad, J., Geddes, T., and Kurzanski, P., 1996, Rail Site Crosses Excavation, Bioremediation Strategies: *Soil and Groundwater Cleanup*, August/September, pp. 6–11.

Hansen, M. A., Gates, M., and Sittler, S. P., 1998, Using High-Vacuum Technology: *Environmental Technology*, March/April, pp. 16–21.

Harper, B. M., Stiver, W. H., and Zytner, R. G., 1998, Influence of Water Content on SVE in a Silt Loam Soil: *Journal of Environmental Engineering*, Vol. 124, No. 11, November, pp. 1047–1053.

Hazaga, D., Fields, S., and Clemens, G. P., 1984, Thermal Treatment of Solvent Contaminated Soils: In *Proceedings of the USEPA Fifth National Conference on Management of Uncontrolled Hazardous Waste Sites*, Washington, D.C., pp. 404–406.

Hewitt, A. D., 1999, Measurement for Trichloroethylene: Relationship between Soil Vapor and Soil Matrix: *Environmental Testing and Analysis*, May/June, Vol. 8, No. 3, pp. 25–31.

Hinchee, R. E. (editor), 1994, *Air Sparging for Site Remediation*: Lewis Publishers/CRC Press, Boca Raton, FL, 142 pp.

Hinchee, R. E., Downey, D. C., and Coleman, E. J., 1987, Enhancing Bioreclamation, Soil Venting, and Groundwater Extraction: A Cost Effectiveness and Feasibility Comparison: In *Proceedings of the National Water Well Association and American Petroleum Institute Conference on Petroleum Hydrocarbons and Organic Chemicals in Groundwater: Prevention, Detection and Restoration*, November, pp. 147–163.

Hinchee, R. E., Downey, D. C., and Beard, T., 1989, Enhancing Biodegradation of Petroleum Hydrocarbon Fuels through Soil Venting: In *Proceedings of the National Water Well Association Conference on Petroleum Hydrocarbons and Organic Chemicals in Groundwater: Prevention, Detection and Restoration*, November, pp. 235–248.

Hoag, G. E. and Marley, M. C., 1986, Gasoline Residual Saturation in Unsaturated Uniform Aquifer Materials: *Journal of Environmental Engineering*, Vol. 112, No. 3, pp. 586–604.

Jafek, B., 1986, VOC Air Stripping Cuts Costs: *Waste Age*, Vol. 17, No. 10, October, pp. 66–67.

Ji, W., Dahmani, A., Ahlfeld, D. P., Lin, J. D., and Hill, E., 1993, Laboratory Study of Air Sparging: Air Flow Visualization: *Ground Water Monitoring & Remediation*, Vol. 13, No. 4, pp. 115–126.

Johnson, P. C., Stanley, C. C., et al., 1990, A Practical Approach to the Design, Operation, and Monitoring of in Situ Soil-Venting Systems: *Ground Water Monitoring Review*, Spring, pp. 159–177.

Kerfoot, W. B., 1988, The Soil-Scrub™ Process for Rapid Decontamination of Gasoline-Impregnated Soil: In *Proceedings of the Second National Outdoor Action Conference on Aquifer Restoration, Groundwater Monitoring and Geophysical Methods*, Vol. I, Las Vegas, NV, May, pp. 123–134.

Koenigsberg, S., 1997, Enhancing Bioremediation: *Environmental Protection*, February, pp. 19–22.

Kostecki, P. T. and Calabrese, E. J. (editors), 1989, *Petroleum Contaminated Soils,* Vol. I: Lewis Publishers, Chelsea, MI, 357 pp.

Laney, D. F., 1988, Hydrocarbon Recovery as Remediation of Vadose Zone Soil/Gas Contamination: In *Proceedings of the National Water Well Association Second National Outdoor Conference on Aquifer Restoration, Groundwater Monitoring and Geophysical Methods,* Vol. III, Las Vegas, NV, May, pp. 1147–1171.

Lindhult, E. C. and Kwiecinski, D. A., 1996, Cleaning Up Clay: *Civil Engineering*, May, pp. 49–51.

Lundegard, P. D. and Anderson, G., 1996, Multiphase Numerical Simulation of Air Sparging: *Ground Water*, Vol. 34, No. 2, pp. 451–460.

Magee, R. and Gotlieb, I., 1995, Washing Process Cleans Up Multiple Contaminants: *Soils,* April, pp. 14–16.

Malachosky, E. and Hartke, F. J., 1995, Micro-Encapsulation in Action: *Soils*, April, pp. 6–12.

Marley, M. C. and Hoag, G. E., 1984, Induced Soil Venting for Recovery/Restoration of Gasoline Hydrocarbons in the Vadose Zone: In *Proceedings of the National Water Well Association and American Petroleum Institute Conference on Petroleum Hydrocarbons and Organic Chemicals in Groundwater: Prevention, Detection and Restoration*, November, pp. 473–503.

McCray, J. E. and Falta, R. W., 1997, Numerical Simulation of Air Sparging for NAPL Contamination: *Ground Water*, Vol. 35, No. 1, pp. 99–110.

Miller, M. E., 1995, Bioremediation on a Big Scale: *Environmental Protection*, July, pp. 15–16.

Morris, C. E., Thomson, B. M., and Stormont, J. C., 1999, Design of Dry Barriers for Containment of Contaminants in Unsaturated Soils: *Ground Water Monitoring and Remediation*, Summer, Vol. 19, No. 3, pp. 145–155.

Morrow, M. T. and VanDerpool, G., 1988, The Use of a High Efficiency Blower to Remove Volatile Chlorinated Organic Contaminants from the Vadose Zone — A Case Study: In *Proceedings of the National Water Well Association Second Outdoor Action Conference on Aquifer Restoration, Groundwater Monitoring and Geophysical Methods,* Vol. III, Las Vegas, NV, pp. 1111–1135.

Noffsinger, D., 1995, Steam Injection Melts Cleanup: *Soils*, April, pp. 31–36.

O'Conner, M. J., Agor, J. G., and King, R. D., 1984, Practical Experience in the Management of Hydrocarbon Vapors in the Subsurface: In *Proceedings of the National Water Well Association and American Petroleum Institute Conference on Petroleum Hydrocarbons and Organic Chemicals in Groundwater: Protection, Detection and Restoration*, November, pp. 519–533.

Pal, D., Karr, L., et al, 1996, Have Burners, Will Clean: A Study of Hot Air Vapor Extraction: *Soil and Groundwater Cleanup*, October, pp. 24–31.

Parker, B. L., Gillham, R. W., and Cherry, J. A., 1994, Diffusive Disappearance of Immiscible-Phase Organic Liquids in Fractured Geologic Media: *Ground Water*, Vol. 32, No. 5, September/October, pp. 805–820.

Piotrowski, M. and Cunningham, J., 1996, Factors to Consider before Adding Microbes and Nutrients: *Soil and Groundwater Cleanup*, May, pp. 44–51.

Plummer, C. R., Nelson, J. D., and Zumwalt, G. S., 1997, Horizontal and Vertical Well Comparison for In-Situ Air Sparging: *Ground Water Monitoring & Remediation*, Vol. 17, No. 1, pp. 91–96.

Poulsen, T. G. et al., 1999, Predicting Soil-Water and Soil-Air Transport Properties and Their Effects on Soil-Vapor Extraction Efficiency: *Ground Water Monitoring and Remediation*, Vol. 119, No. 3, pp. 61–70.

Rodfes, J. R., 1998, In-Situ Thermal Destruction: *Environmental Protection*, February, pp. 20–21.

Sale, T. and Pitts, M., 1989, Chemically Enhanced in Situ Soil Washing: In *Proceedings of the National Water Well Association and American Petroleum Institute Conference on Petroleum Hydrocarbons and Organic Chemicals in Groundwater: Prevention, Detection and Restoration*, November, pp. 487–503.

Stephens, D. B., 1996, *Vadose Zone Hydrology*: Lewis Publishers/CRC Press: Boca Raton, FL, 339 pp.

Testa, S. M., 1994, *Geological Aspects of Hazardous Waste Management*: CRC Press/Lewis Publishers, Boca Raton, FL, 537 pp.

Testa, S. M., 1995, Natural Attenuation Requires Attention: *Soils*, April, pp. 19–23.

Testa, S. M., 1997, *The Reuse and Recycling of Contaminated Soil*: CRC Press/Lewis Publishers, Boca Raton, FL, 268 pp.

Thomas, J. M., Lee, M. D., Bedient, P. B., Borden, R. C., Canter, L. W., and Ward, C. H., 1987, *Leaking Underground Storage Tanks: Remediation with Emphasis on in Situ Biorestoration*: EPA/600/2-87/008, 144 pp.

Thornton, J. S. and Wootan, W. L., Jr., 1982, Venting for the Removal of Hydrocarbon Vapors from Gasoline Contaminated Soil: *Journal of Environmental Scientific Health*, A17, Vol. I, pp. 31–44.

Towers, D. S., Van Arnam, D. G., and Dent, M. I., 1995, Evaluation of In-Situ Technologies for VHOs-Contaminated Soil: *Remediation Management*, September/October, Vol. 1, No. 1, pp. 21–27.

Treweek, G. P. and Wogee, J., 1988, Soil Remediation by Air/Stream Stripping: In *Proceedings of the Hazardous Materials Control Research Institute Fifth National Conference on Hazardous Wastes and Hazardous Materials*, April, pp. 147–153.

U.S. Environmental Protection Agency, 1984, *Review of Inplace Treatment Techniques for Contaminated Surface Soils*, Vol. I: Technical Evaluation, EPA-540/2-84-003a.

U.S. Environmental Protection Agency, 1990, *Guide for Conducting Treatability Studies under CERCLA: Soil Vapor Extraction*: U.S. Environmental Protection Agency, Risk Reduction Engineering Laboratory and Office of Emergency and Remedial Response, Cincinnati, OH, August.

U.S. Environmental Protection Agency, 1993, *Behavior and Determination of Volatile Organic Compounds in Soil: A Literature Review*: Office of Research and Development, EPA/600/R-93/140.

Vance, D., 1998, Redox Reactions for Insitu Groundwater Remediation: *Environmental Technology*, September/October, p. 45.

Weinstein, W., 1996, Mixing Up a Microbial Cocktail: *Environmental Protection*, June, pp. 28–29.

Weyer, K. U. (editor), 1992, *Subsurface Contamination by Immiscible Fluids*: International Association of Hydrogeologists, Canadian Chapter, A.A. Balkema, Rotterdam, the Netherlands, 576 pp.

Whitley, G., et al., 1999, Contaminated Vadose Zone Characterization Using Partitioning Gas Tracers: *Journal of Environmental Engineering*, June, Vol. 125, No. 6, pp. 574–582.

Wilson, D. J. and Clarke, A. N. (editors), 1994, *Hazardous Waste Soil Remediation; Theory and Application of Innovative Technologies*: Marcel Dekker, Inc., New York, 367 pp.

Wisconsin Department of Natural Resources, 1994, *Naturally Occurring Biodegradation as a Remedial Action Option for Soil Contamination*: Interim Guidance (Revised): Madison, WI, 16 pp.

Yuan, D., Schuring, J. R., and Chan, P. C., 1999, Volatile Contaminant Extraction Enhanced by Pneumatic Fracturing: *Practice Periodical of Hazardous, Toxic, and Radioactive Waste Management*, April, pp. 69–76.

11 Economic Considerations for Aquifer Restoration

"If you fail to plan, you plan to fail."

11.1 INTRODUCTION

Project managers are tasked with the responsibility of identifying the most cost-effective, technologically feasible remedial procedures to restore the site to a usable status, within monetary, time, and regulatory constraints. Starting with the preliminary site investigation phase and continuing throughout the life of the project, the manager continually evaluates procedures to control ultimate (bottom-line) costs. Maintaining a focus on the final goal of the project (end-point approach) allows the project manager to accomplish the task. This approach involves looking at all of the major project tasks simultaneously:

- Description of the physical setting and evaluation of contamination;
- Identification and evaluation of risks to human health and the environment;
- Identification of regulatory restrictions;
- Establishment of site-specific remediation criteria;
- Determination of a closure schedule;
- Evaluation the total project costs of each remediation option.

Preparation of the investigation work plan and overall remediation program should be tailored to the site needs, with use of good engineering practices and consideration of the ultimate use of the site and total project cost. The use of the final goal approach is applicable regardless of the regulatory or administrative forces initiating the remediation.

Up-front planning is a major factor in the determination of the ultimate cost of every remediation project. Definition of remedial goals at an early project stage and development of a strategy for achieving these goals are essential parts of project planning. Collection of the "right data the first time" enhances evaluation of design options and enables defendable remedial design which are cost-effective overall. The principle of the six Ps applies: *P*roper *P*lanning *P*revents *P*itiful *P*oor *P*erformance.

Use of sound engineering judgment is the best procedure to ensure a cost-effective project. A poorly executed minimal initial feasibility study may be costly in the long run. Often basic feasibility studies are not sufficiently broad to determine the full range of acceptable remedial options. The goal of an effective feasibility study is to develop a remedial solution that efficiently meets cleanup goals at the lowest cost, within the time allotted. Depending on the project, it may be beneficial to incur the bulk of the expense up-front with relatively low operation and maintenance. Or at another site, the best choice may be low initial cost with higher operation and maintenance expenses. The optimum solution should be based on sound engineering principles, regulatory approval, and best business practice.

Careful planning of the initial investigation activities is very important because this task sets the stage for future decisions and actions. Physical site characteristics, extent and intensity of contamination, cultural features, and historical background are confirmed in this project phase. Cleanup levels, potential remedial schemes, and probable costs are established with the data derived from the initial work. Regulators typically set conservative closure standards to assure public safety. Once these are established, it is difficult to modify them. Therefore, a truly representative site investigation must be completed before any remediation levels are set.

A complete and accurate site characterization provides the data to identify and design an appropriate remedial program. Poor or incomplete site characterization can lead to a problem project.

Minimizing expense is always important, but deciding where to reduce costs during the investigation phase is difficult. Reducing the number of samples or using less definitive analytical procedures and not applying adequate modeling can be costly. Completion of a thorough site characterization, basic risk assessment, and transport modeling can substantially reduce the overall remediation cost.

Fate-and-transport studies using computer modeling and mathematical calculations can effectively evaluate the potential for contaminant migration and degradation through the natural and anthropogenic processes active at the site. Often a brief fate-and-transport study early in the project can lead to optimal placement of monitor and recovery wells, and selection of sample analytical procedures.

At many sites, the investigation phase is performed in two tasks (Phase 1 and Phase 2). Phase 1 focuses on the large picture, a definition of overall site conditions. Data developed are used to assess potential risks and identify potential remedial options. Phase 2 work fills in data gaps and defines the level of detail necessary to establish site-specific remediation goals and to evaluate probable remediation procedures or to indicate the need for pilot cleanup studies.

Field screening can be used to improve the efficiency of collection and laboratory sample analysis. Inexpensive field tests can be used to screen samples to approximate the concentration and extent of contamination and to determine which samples should be subjected to laboratory analysis. Typical field screening involves chemical testing, such as portable gas chromatograph analysis, colorometric chemical testing, photoionization detector vapor screening, ultraviolet hydrocarbon screening, pH analysis, Eh probe testing, and conductivity testing. Other common field screening tools include soil gas analysis and geophysical methods.

Geostatistical mathematical models can be used to assist investigators to target sampling locations and to avoid sampling in irrelevant areas. These models can be used to generate maps of soil, sediment, water, and groundwater contaminants.

Caution must be used, however, because inexperienced investigators often try to use estimated data instead of actual analyses. While estimates reduce sampling and analysis costs, safe use of estimated data requires that it be conservative. Experienced investigators strike a balance between estimates and field sampling to save costs without sacrificing cleanup levels.

Following completion of the site work, the remediation process moves into evaluation and selection of remediation methods. A feasibility study involves the systematic evaluation of closure methods. The most efficient process is to focus as quickly as possible on the minimum number (usually three or four) of realistic remediation alternatives that are appropriate for the site. Evaluation of a larger number of possible procedures for the sake of formality is a waste of project resources. Typically, feasibility studies have been paper studies without bench or pilot exercises. For straightforward simple sites where cookbook remediation is applicable, this approach may be acceptable. However, where the site is complex or where a recently developed technology appears to be suitable, real data from bench- or pilot-scale demonstrations may be appropriate to determine cost and to show conclusively that the technology offers an effective alternative. Many vendors of new technology offer demonstration studies at reduced cost, and regulators are less likely to challenge the validity of well-documented real data and cost estimates derived from these tests.

Each project has its own unique characteristic physical, environmental, cultural, and regulatory setting. No single approach (technology, process, or cleanup standard) can be universally applied. The following sections discuss the key elements that impact the economic decisions of remediation projects and result in the most efficient, cost-effective restoration.

11.2 IMPACTED SOIL CONSIDERATIONS

Definitions of the extent, type, and concentration of soil contamination are the primary factors that control the cost of soil remediation. The economic effectiveness of a remediation strategy depends on the following:

- Physical and chemical properties of the contaminant;
- Soil type;
- Moisture content;
- Concentration of contaminant;
- Degradation or attenuation rate of contaminant;
- Site-specific cleanup standards;
- Cultural considerations;
- Level of technology available.

11.2.1 Lateral and Vertical Distribution

In the vadose zone, liquid or dissolved contaminants exist in a complex environment that involves interaction among the chemicals, soil grains, water attached to soil grains (or between the soil pores), the atmosphere in void spaces, and numerous biological reactions. Selection of the most economic remediation process must consider all these factors.

Ideally, the site characterization study has defined the vertical and horizontal extent of the contamination. Contoured site maps showing the (three-dimensional) distribution of the contaminants allow identification of areas that require extensive restoration, or may be allowed to be monitored to closure under natural attenuation. Knowledge of "how much contamination exists and its location" is the important first step in the remediation process. Evaluation of these data will permit consideration of the various remediation remedies available. Where the contaminant is contained within the shallow (<6 m) unsaturated zone and is recalcitrant (not readily biodegradable), excavation for off-site treatment or disposal may be the most expeditious procedure. Alternatively, depending on the contaminant, a variety of *in situ* procedures, including bioremediation, air sparging, soil vapor extraction, and fixation, may be applicable.

11.2.2 Contaminant Type

Definition of contaminant type is a critical component of remediation costs. High-molecular-weight hydrocarbons, or those with high attenuation potential, are less likely to migrate and may be better candidates for excavation or fixation in place. Chemicals with small molecular structure or with ionic properties are more likely to be water soluble or volatile, and thus mobile. Where volatile or aerobically degraded chemicals are predominant, soil vapor extraction or air venting may be the most effective system. Many chlorinated hydrocarbons are more readily degraded under anaerobic (reducing) conditions. Careful consideration must be given to the chemicals to be remediated and the possibility of a multiple-phase restoration project.

11.2.3 Time Frame

Time is one of the most important factors of any project. Accurate estimation of the time required for a remediation project requires an understanding of the established management goals as well as a thorough appreciation of the physical, chemical, and biological processes involved. At most contaminated soil sites, excavation for off-site treatment and disposal is mechanically the most rapid cleanup procedure. While "dig and haul" may appear expensive, the majority of cost is incurred within a short time. The property may be returned to income-producing status within a short time. If significant regulatory or air quality concerns are involved, the time required for permit acquisition may be a hindrance.

When all necessary nutrient supply systems are in balance and functioning properly, aerobic biological remediation can be relatively rapid. Gasoline components have been observed to have a half-life of days to months under well-controlled field conditions. Chemicals such as tetrachloroethylene that are best degraded under anaerobic conditions require significantly more time. Published half-lives for similar chlorinated solvents under field conditions are on the order of 300-day half-lives. Several computer programs are available that calculate the probable life expectancy of remedial projects. For best results, these programs require input of real field data.

With each use of (necessarily) conservative assumed data, the precision of the estimate is diminished. Field pilot testing is an effective way to generate valid data for input into the models. When natural attenuation is selected as the best option, continued monitoring for 3 to 10 years should be expected.

Some very helpful computer programs are available to assist the planner estimate the time for remediation. One popular public-domain model from the EPA is Bioscreen: Natural Attenuation Support System. This easy-to-use model accepts data input on hydrogeology, dispersion, adsorption, biodegradation, aerial extent, and source characteristics. The program accepts actual field data, or calculates conservative default values for these data. Output from the model is a distribution of the plume at user-specified times based on no degradation, first-order decay, and instantaneous reactions (all in graphical or digital format). With these data, the project manager can evaluate a series of remediation options to determine which best serves the site. The ultimate choice, however, depends upon the professional judgment of the project manager. After a remedial system is fully functional, the operational data can be used to refine the closure date.

11.2.4 Regulatory Climate

The most cost-effective, technically feasible remediation procedures are not always in agreement with regulatory controls. Environmental regulations, by their very nature, must be applicable for a wide variety of settings and must protect the overall environment. In past years remediation standards were established as general numerical concentrations usually based on drinking water standards or other health-related criteria borrowed from related public health fields. This type of remediation was often generic, not site specific.

Recent regulatory thinking has focused on site-specific criteria derived from assessments of the risks to human health and the environment. Risk assessments are based on concentration of contaminants and exposure pathways.

11.3 LNAPL RECOVERY

Where LNAPL is present, its recovery is an essential and typically immediate part of the cleanup effort. Removal of the floating LNAPL layer is almost always required prior to the initiation of other restoration- and remediation-related activities (enhanced bioactivity, vapor extraction, etc.) and response to other environmental issues (i.e., hydrocarbon-affected soils, vapors, or dissolved hydrocarbon in groundwater).

In principle, the recovery of LNAPL is similar in mechanical operation to production of a low-pressure, water-driven reservoir. Almost all documented petroleum remediations have been characterized by subsurface conditions under water table conditions (i.e., the top surface of the fluids are at atmospheric pressure). Few cases of confined aquifer situations have been reported in the literature, and although the mechanical recovery procedures are slightly different, the economic considerations are similar.

Detailed planning is necessary to ensure that the fieldwork meets predetermined goals with regard to administrative requirements, the development of recovery facilities, and optimized production. At any specific site, the goals may include any or a combination of the following:

1. *Meeting regulatory compliance.* In many situations, recovery efforts are driven by statutory forces, which are result oriented and not particularly oriented toward any specific site. Most projects of this variety have specified quality time constraints, and the economic considerations are secondary.
2. *Recovery of recyclable LNAPL.* Many older petroleum-refining operations have experienced individual spills or continual small-scale leakage over a prolonged period of time. A large percentage of these leaks occur from underground sources and, because of geologic settings, have not been apparent. A loss of 0.5% may not be detected by the operator, but can accumulate significantly over an extended period of time. For example, a refinery with production of 60,000 barrels/day with undetected losses of 300 barrels/day would have accumulated leakage of a total of over 3 million barrels during a 30-year period. Under favorable geologic and hydrogeologic conditions, recovery of this product may be an economic asset. Careful management of recovery operations can often produce a positive cash flow (or at least significantly reduce costs), as well as provide positive environmental benefits.
3. *Maintenance of current commercial status quo.* At a few locations, the decision has been made to continue operation of the facility without disruption, so long as NAPL product and dissolved chemicals do not exit the boundary of the facility or provide a risk to workers or the environment at the site. Hydraulic containment of the aquifer is the procedure that is usually selected for these sites. A system including recovery wells and injection wells can often be operated to balance the subsurface flux so that product loss equals product recovery at a minimal cost.
4. *Accomplishment of "best environmental restoration."* A thorough understanding of the geologic and hydrogeologic setting, including the LNAPL recovery potential, can lead to development of a total remediation plan. Under this type of goal, overall cleanup of the contaminant is the primary goal. The use of variable hydraulic flow and flushing techniques, combined with the utilization of both natural and enhanced biodegradation, will ultimately result in a "best-case" total restoration. Investigative and management effort for this type of project is extensive.

The following discussion is directed toward evaluating the costs of operating LNAPL recovery systems. The principles presented are equally applicable to all types of LNAPL recovery situations.

11.3.1 Preliminary Considerations

Each recovery site has its own individual site-specific characteristics. Subsurface stratigraphy and other geologic considerations, depth to water table, hydraulic conductivity, aquifer(s) thickness, size of product pool, physical characteristics of the product, effects of product weathering, gradient, source of leakage, and groundwater quality are all factors of importance in the design and operation of an LNAPL recovery system.

Initial investigations must be adequate to provide sufficient information to support the design and operation of the recovery system. Progressive and flexible testing procedures during the investigative phases will require sufficient hydraulic testing to characterize the product-producing areas and to evaluate the approximate quantity and rate of recovery. The data procured should also provide guidance in assessing how many recovery and monitoring wells are required and where best to situate them.

Determination of initial recovery well (or trench) locations is an important design parameter. Floating LNAPL product tends to move in the direction of overall groundwater flow, as determined by the water table gradient. As a well or trench is pumped, the fluids (water and/or oil) migrate toward the area of lower pressure to fill the void. A cone of depression develops that extends outward. The fluid surface exhibits a rapid slope near the well, diminishing to a very low gradient at a distance.

Floating product migrates toward the well at a rate proportionate to the hydraulic gradient. As production wells deplete oil reserves in the immediate vicinity, the oil must travel farther, over lesser gradients, to reach the well. The end result is reduced oil production. Construction of new wells at more advantageous intermediate locations will revitalize production. An alternative procedure is to install an injection well between recovery wells. The increase in the hydraulic gradient caused by the injection well would then stimulate oil flow toward the recovery wells.

During the design phase, all of the data derived from the hydraulic characterization are evaluated for use in the selection of recovery pumping equipment and for the determination of the most appropriate subsurface fixtures (whether wells, trenches, or drains, etc.). A variety of generic scenarios may be appropriate to optimize product recovery. If the product thickness is sufficient, the viscosity low, and the formation permeable, a simple pure-product skimming unit may be the best choice. Other combinations of permeability, geology, and product quality will require more active systems, such as one-pump total fluid, or two-pump recovery wells.

The selection of the recovery equipment should be based on functionality, capital cost, operational cost, and ease of movement between individual locations. A well-managed recovery system will include routing relocation of equipment in response to recovery needs. Flexibility of usage is important because the capital cost of the recovery equipment presents a significant percentage of the project cost.

11.3.2 Economics of LNAPL Recovery

All projects involving any significant quantity of LNAPL product recovery require the consideration of economic factors. Careful planning to optimize each project phase can lead to the lowest cost of operation and can occasionally generate positive cash flow, while currently accomplishing aquifer restoration. A basic premise in this discussion is that the recovered LNAPL is suitable for reuse (i.e., as refinery feed stock or fuel for incinerators). Products unsuitable for resale only add to the debit side of the economic equation.

Debits associated with LNAPL recovery projects are the costs for:

- Investigation and testing;
- Evaluation and design;
- Construction of field facilities;
- Operation and maintenance;
- Money invested (time value).

Credits resulting from recycling of recovered LNAPL include cash returns of:

- Value of product recovered;
- Time value of money from sold products.

A summarization of the credits and debits describes the economic status of the project. A carefully planned and executed project results in the greatest possible credit balance (or least expenditure) for the longest period of time.

The following example demonstrates the principles involved in the optimization of a project. A fairly large, older refinery situated on an alluvial floodplain had been in continuous operation for 35 years. Leakage and spillage were common, in almost unnoticed quantities (<0.5% by volume), resulting in a significant accumulation of product on the shallow water table over the course of operation. The intended goal of the recovery program was to accomplish the "best environmental restoration" at the least cost. For convenience, all costs and returns are presented in terms of value per barrel. A graphical presentation of the cumulative costs (in barrels) vs. time and a summation of the volume of recovered product vs. time is shown in Figure 11.1. The following costs (in terms of the value of barrels of LNAPL recovered) were assumed:

- Investigation and testing phase cost 3000 barrels over a 2-month period.
- Evaluation and design cost 1000 barrels over a $1^1/_2$-month period.
- Construction of field equipment cost 8000 barrels over a $1^1/_2$-month period.
- Operational costs remained constant at 1500 barrels per month from initiation to completion of the recovery phase.

Recovery started in the fifth month and continued until completion. Recovery rates associated with Figure 11.1 are summarized in Table 11.1. The example illustrated in Figure 11.1 follows the typical case in which the recovery rate increases to a peak rate, then rapidly declines toward an asymptotic curve over several months. The use of cumulative curves allows a comparison of the relative positions of both cost and production, as illustrated in Figure 11.2. Prior to $7^1/_2$ months, the project operated at a cumulative loss. After that time, the cumulative recovery was greater than the accumulated cost. This positive situation continued until month 45 (projected off the edge of the chart), where the production costs equaled the value of the product recovered. The distance between the two curves was at a maximum near month 18, when the product was being recovered at a rate of 50 barrels/day at a cost of 50 barrels/day. Beyond this time, the return rate is negative. A project planned to focus on positive cash flow would end at this point, although an additional 25

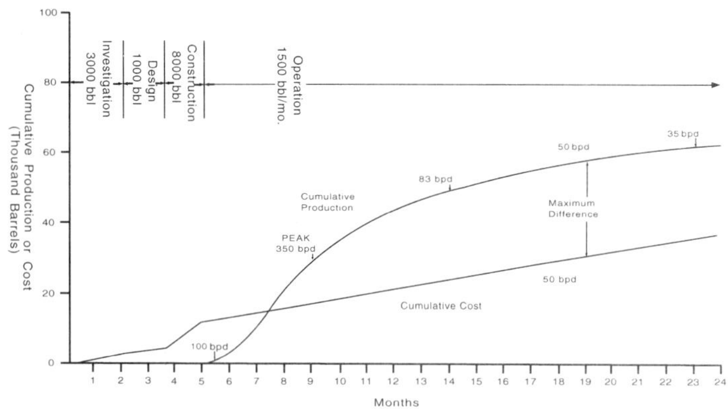

FIGURE 11.1 Example project of LNAPL recovery, cumulative costs, and production vs. time.

months would pass before the total value of recovered product equaled costs. Since this project was planned to accomplish long-term remediation, funds earned during the positive-cash-flow period were retained to finance subsequent longer-term phases of the project. This planning resulted in a greatly reduced overall cost.

Optimization involves careful planning during all phases. The investigation phase is "adequate," with sufficient drilling and testing to characterize the site and to identify optimum locations for recovery wells. Interviews with refinery maintenance staff can prove invaluable in delineating potential source areas. Existing boring logs also provide data on subsurface geologic and hydrogeologic conditions and permeability.

During the design phase, the primary goal is to specify reliable equipment that is easily maintained and can be operated with minimal energy. Downtime, while awaiting a repair specialist or the arrival of unique parts, can cost more than a slightly less efficient system that is easily repaired by on-site staff. Installation and activation of the field equipment should be conducted in a manner that initiates the recovery of LNAPL as soon as practical. As each additional well is installed, piping, energy for pumping, control units, tankage, and other necessary utilities should also

TABLE 11.1
Recovery of LNAPL with Time

Month	Production (barrels/day)
5	100
9	350
14	83
18	58
23	35
45	7 (projected value)

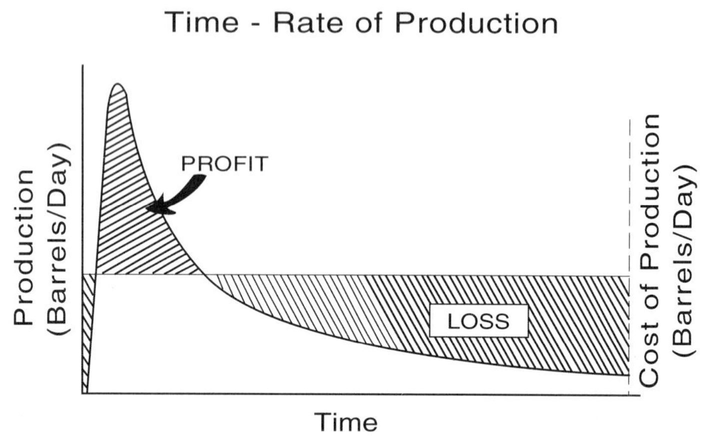

FIGURE 11.2 Time-rate of production vs. operating cost.

be ready for use. Proper prior planning and careful coordination are critical to efficient operations.

Operation of a system on a continuing basis requires regular monitoring and maintenance. This phase represents the largest cost items of a long-term project; however, routine adjustments of pumping units, coupled with preventative maintenance, will result in lower long-term costs.

11.3.3 Project Planning and Management

The preceding paragraphs describe a scenario related to a single-phase project. The initial well system was operated as a single unit throughout the duration of recovery. Although this technique was appropriate for that site, alternative programs may often be better suited for other locations. At sites that have extensive "reserves," it has often been demonstrated that a phased approach can improve recovery efficiency. This procedure involves optimizing the use of recovery equipment by progressive replacement of high-capacity oil-pumping equipment with equipment of lesser capacity as production declines in each well. A typical recovery curve for a well, or system of wells, is shown in Figure 11.3. Because prime areas were selected and exploited first, the peak rate of recovery for each phase tends to be less for each succeeding phase. Also, the break-even line increases as operational costs for each new phase are added. After several phases have been added, the cost of operation will eventually meet or exceed recovery rates.

11.3.4 Estimating Reserves

Preliminary estimates of LNAPL made during the investigative phases of a project are usually based on the results of short-term pumping tests, approximations of actual LNAPL thickness based on gauging data generated from monitoring wells, and other approximate data. The numerical estimates based on this short-term information are often adequate to define the physical parameters sufficiently for prelim-

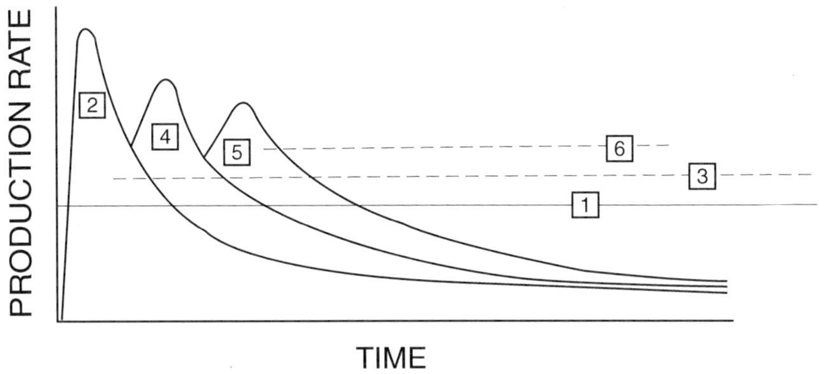

1 1st Phase Break-Even Line Includes Costs Involved With Site Investigation and Equipment and Energy and Labor and Maintenance, etc.

2 Production From Phase 1.

3 Break-Even Line for Phase 2 and Phase 1. Includes Incremental Cost of Installation of Additional Wells.

4 Production From Phase 3 and Phase 1.

5 Production From Phases 3 + 2 + 1.

6 Break-Even Line For Phases 1 + 2 + 3.

FIGURE 11.3 Time-production rate vs. operational cost for a multiphased project.

inary design purposes. However, long-term reserve estimates and the economics of production can only be completed after at least part of the recovery system has been operating long enough to establish a distinctive production pattern.

A simplified analysis of production decline curves was initially intended for evaluation of individual oil well production; however, this type of analysis can also provide reasonable estimates when applied to multiple-well LNAPL recovery systems. This analytical method is applicable to most types of decline curves, whether they tend to follow exponential, hyperbolic, or harmonic forms. The following general differential equation is applicable to all forms of decline curves:

$$D = Kq^n = -\left(\frac{dq}{dt}\right)q \tag{11.1}$$

where D = decline, K = operative constant, dq = differential rate of production, dt = differential time, q = production rate, and n = decline exponent.

Specific equations that are all solutions for the general equation are presented in Table 11.2. Because each of these equations contains two unknowns, solution

TABLE 11.2
Summary of Mathematical Solutions for Production Decline Curves

Decline Exponent	Type of Decline	Rate/Time Relationship	Rate: Cumulative Relationship	$D_i t$ Relationship	$Q_1/q_i t$ Relationship
$n = 0$	Exponential	$q_t = q_i e^{D_i t}$	$Q_t = (q_i - q_t)/D_i$	$D_i t = \ln(q_i / q_t)$	$Q_t / q_i t = \dfrac{1 - (q_i / q_t)^{-1}}{\ln(q_i / q_t)}$
$D < n < 1$	Hyperbolic	$q_t = q_i(1 + nD_i t)^{-1/n}$	$Q_t = \dfrac{q_i^n}{(1-n)D_i} \cdot (q_i^{1-n} - q_t^{1-n})$	$D_i t = \dfrac{(q_i / q_t)^n - 1}{n}$	$Q_t / q_i t = \dfrac{1 - (q_i / q_t)^{n-1}}{(q_i / q_t)^n - 1} \cdot \left(\dfrac{n}{1-n}\right)$
$n = 1$	Harmonic	$q_t = q_i(1 + D_i t)^{-1}$	$Q_t = (q_i / D_i)\ln(q_i / q_t)$	$D_i t = (q_i / q_t) - 1$	$Q_t / q_i t = \dfrac{\ln(q_i / q_t)}{(q_i / q_t) - 1}$

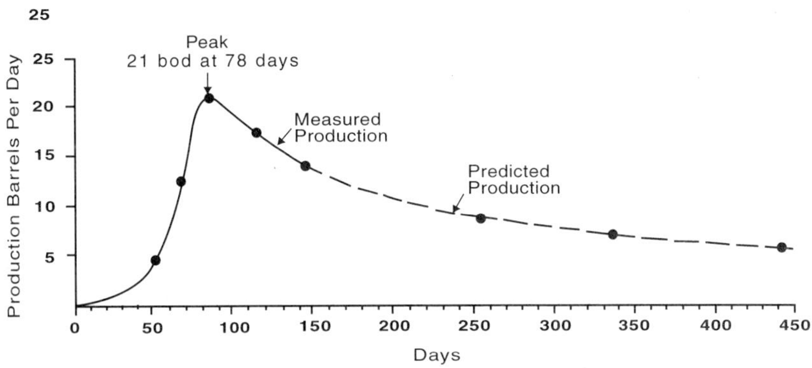

FIGURE 11.4 Example of project time-reduction rate.

requires the input of data from external sources. Necessary values derived from the relationships of field-measured data are applied to the theoretical equation to derive a solution. Any set of consistent measurement units may be used. Field measured data are:

q_i = initial production rate (units per time)
q_t = measured production rate at time t after initial peak production
Q_t = accumulated production rate between q_i and time t

Use of these equations to predict future production from a recovery project is described by the following example. An abandoned refinery property is being dismantled and the underlying aquifer remediated. Substantial LNAPL product accumulations occurred overlying the fine silty sand aquifer. Preliminary investigation indicated that a four-well system would effectively remove most of the product within a reasonable time at a modest cost. The production rate over time is illustrated in Figure 11.4. Peak production occurred on day 78 of operation, then declined. Final measurement occurred on day 141.

The harmonic equation provides the closest answer (within approximately 5%); therefore, that equation is most representative. By use of the rate/time relationship for the harmonic equation, the recovery curve may be projected to predict recovery rates at future dates after peak production.

$$q_t = q_i(1 + D_i t)^{-1} \tag{11.2}$$

where at 173 days, q_t = 8.87 bpd; at 258 days, q_t = 6.91 bpd; and at 3654 days, q_t = 5.40 bpd. As this is an asymptotic curve, it will approach zero at infinity. After more than 10 years of operation, records indicate this site continues to yield approximately 1 barrel per month. In practice, the practical production limit is reached when the actual LNAPL layer is thin enough that it does not migrate freely.

Each combination product type, soil variety, and wetted condition (oil–water saturation) behaves differently and must be considered separately. An estimate of total production from this system at 365 days after peak can be determined by use of the rate/cumulative relationship:

$$Q_t = \frac{q_i}{D_i} \ln\left(\frac{q_i}{q_t}\right) \tag{11.3}$$

$$Q_{t365} = 3590 \text{ barrels}$$

This example demonstrates that it is possible to make reasonably accurate predictions of production after the pattern of production is established. However, caution must be exercised by the professional when developing these data. The example cited occurred during a period when the water table at the facility was within the "normal" ranges. If the water table had risen, or fallen substantially (for any reason), the pattern of production would have changed, and the calculations would not be considered reliable. Estimation of reserves determined by these methods can be fairly reliable if the data used are based on adequate and regular measurements, and are applied with a reasonable measure of professional judgment.

As in any scientific or engineering endeavor, the quantity and validity of input data determine the accuracy of prediction. Frequent gauging of fluid levels in monitoring wells, flow rates, and oil–water ratios, in conjunction with proper quality control, can lead to accurate estimates that support proper project performance.

11.3.5 Other Factors

LNAPL recovery projects require careful planning, operation, and management. Some additional factors become important at unforeseen times. Recovery of free product often produces flammable vapors at concentrations that can violate air quality standards if freely released. Vapor treatment should be considered early in the design process. Storage of flammable liquids in appropriate containers is necessary.

Safe, secure operating conditions are sometimes difficult to maintain, especially at locations where staff is not continually present. Safe operations of the site, and protection of workers and neighbors, may represent a significant but necessary expense to the project.

11.4 Dissolved Hydrocarbons in Groundwater Considerations

11.4.1 Lateral and Vertical Distribution

Organic chemicals dissolved in groundwater originate from a release source, which may be point source (defined leak) or it may emanate from an area source, such as a mass of contaminated soil. Once dissolved, the chemicals disperse into the groundwater by molecular diffusion and advection (combined as dispersion), both of which are influenced by equilibrium distribution relationships with the aquifer materials.

Distribution of individual chemicals within the "plume" is controlled by chromatographic principles. Chemicals that are more soluble and with less tendency to adsorb onto the aquifer matrix migrate outward more rapidly than less soluble chemicals, which have a tendency to adsorb onto aquifer materials. The shape of the resulting plume is almost always three dimensional. In a theoretical aquifer, the shape of a dissolved plume is an elongated tube, expanding as it travels. In reality, the dispersed shape of the plume tends to be affected by zones of different hydraulic conductivity. Monitoring wells screened below or above the plume may miss the plume entirely. Wells with a long screen tend to result in averaged concentration results.

An adequate monitoring program requires a sufficient number of monitoring wells, push points, or other depth-specific sampling equipment to define the plume distribution and the local geologic setting. An efficient recovery or containment system requires careful calculation (or modeling) of the aquifer system to place the recovery (containment) wells at strategic locations to assure that the plume does not bypass recovery, or that contaminated water is not drawn into uncontaminated portions of the aquifer. A number of computer models such as the USGS MODFLOW are very useful in predicting the three-dimensional distribution of a complex plume. Accurate field information is always required for validation of any computer model, with actual field or laboratory-generated data always preferable to calculated or assumed data.

At sites where water-containing dissolved phase hydrocarbon is part of an active NAPL recovery system, the aquifer water has been (or is being) drawn into intimate contact with the NAPL. Where the water and NAPL are in active contact, the water will contain almost saturation concentrations of each of the fuel compounds. A system designed to minimize the quantity of water produced from the water–NAPL contact zone will result in lower treatment costs.

Wherever contaminated groundwater must be recovered or prevented by migrating by hydraulic control, the expense of a valid field investigation is always recouped. Efficient hydraulic control or recovery is based on the principle of handling (and treating) the smallest volume of water containing the highest concentration of dissolved chemicals as possible.

11.4.2 Contaminant Type

Migration of dissolved hydrocarbons in groundwater is related to several factors including solubility of the various components, volatility (based on Henry's law), and the tendency to adsorb on the aquifer materials. Hydrocarbons such as gasoline contain a wide variety of volatile and relatively soluble compounds. The range of solubility of gasoline compounds includes additives such as alcohols (and MTBE) which are almost infinitely soluble, to benzene (1500 ppm), to other components which are much less soluble. Diesel fuel is composed of a large number of chemicals, but these are mostly much less soluble than gasoline compounds. In a plume, the most soluble chemicals will disperse most rapidly and appear at perimeter monitoring points first. Hydraulic oil (or lubricating fluids) is composed of almost insoluble, nonvolatile materials with high viscosity, which are easily adsorbed to soil grains.

The costs of dissolved-phase recovery and treatment involve a series of trade-offs between the quantity of water pumpage necessary to accomplish the task and the concentration of dissolved chemicals that require treatment. Optimization of these factors is often a complex process that requires evaluation of several design options.

11.4.3 Economics and Time Frames

Effective design of a remediation system for dissolved hydrocarbons in groundwater requires consideration of more than only the effectiveness of the technological process involved. At many sites a variety of techniques are capable of completing the cleanup. However, design of a project that is efficient in all aspects — technically, in terms of time, and economically — requires an evaluation of the entire life cycle of the project from inception to closure. Typically, at sites where remediation is expected to continue over a 4-year project life, operation and maintenance account for between 50 and 80% of the total project cost. These percentages increase each year thereafter. The principal components of operation and maintenance are power, labor, and parts. Identification and quantification of these components are critical to the overall cost of a project.

A major failure of estimating "life-cycle costs" of a project is an improper sizing of the treatment system. Conventional initial investigations often do not provide adequate data to size the treatment equipment properly. For example, if several thousand well volumes must be removed from a production well, several hours of pumping are necessary to establish a reasonably stable drawdown and maintainable pumping rate. Short-term pump tests and slug tests are frequently inadequate to define true site conditions. At most sites the rate of contaminant recovery from a soil vapor extraction or air sparge system is significantly greater during the initial period of operation and declines exponentially thereafter. Design of a remedial treatment system based on the initial data results in a highly overdesigned system over the life of the project.

Treatment systems should be demonstrated to be effective, and expandable (or reducible), in a stepwise modular fashion to ensure efficient design and operation. At many sites, it is cost-effective to install several temporary treatment systems in parallel configuration, and remove them as the need is diminished until the stable rate is achieved where a single unit can handle the load. After careful analysis, it may be determined that the "best" procedure is to install and operate the system at the projected long-term rate from the onset. Although this plan may extend the long-term remediation time, cost savings on equipment purchase (or lease) or initial operation labor may be justified.

Most technology used for groundwater treatment was developed for wastewater facilities. Wastewater equipment is usually designed to last a minimum of 20 to 30 years. With continued maintenance, the equipment should survive this long. For the wastewater market, the design is based on the life expectancy of the equipment, not the waste stream. Groundwater treatment equipment systems will typically be in use for a much shorter time; the controlling factor is the life expectancy of the project, not the equipment.

Beyond equipment size and time considerations, selection of appropriate technological complexity is a major factor of design. Water treatment plants are often overengineered with complicated control systems that are expensive and difficult to operate. Sites at remote locations often do not have full-time, highly trained technicians as operators. The complexity of the equipment should be adjusted to the setting. If highly sophisticated equipment is necessary, technical support and monitoring must be readily available or redundant systems must be included. Where the technical competence of the operator is limited, simple designs with minimal controls are a good way to reduce failures and operation and maintenance costs. Certain design parameters such as flow, concentration, capital cost, operator expense, and continued expense are further discussed below:

Design Parameter: Flow

During the initial phase of a groundwater cleanup, the rate of flow is higher than the long-term average. This is due to the nature of groundwater hydrology with the balance of driving heads and resulting change of storage in a system in the process of establishing dynamic equilibrium. After the system has been in operation for some time, the flow rate stabilizes. The treatment system must have sufficient flexibility to function at a range of flow rates.

Design Parameter: Concentration

The time effect on concentration is an important consideration of process design. In treatment of most manufacturing or municipal wastes, the concentration of contaminants is usually relatively high and reasonably uniform over time (or at least predictable). At remediation sites, the concentrations tend to vary with time and not be as predictable. Certain treatment processes are more efficient at minimum concentrations and others function best at higher concentrations. When influent concentrations vary significantly with time, the effectiveness of the treatment system may vary. One system design may not be adequate for a site from start to finish.

Design Parameter: Capital Cost

Most field equipment used at remediation sites has a lifetime substantially longer than the length of the project where it is used. When the total cost of new equipment is included in the project, the unit treatment cost increases substantially as the time of treatment decreases. Two solutions to this dilemma are common: either lease the equipment or use the equipment on several different projects. Both of these options require the equipment to be of some treatment volume rate such that it can be readily applied to another treatment train, and also that it can be portable. If the designer cannot locate lease equipment, or if there is only one project on which to use the equipment, cost calculations should include a salvage value for the equipment.

Design Parameter: Operator Expense

Any continually operating system will require some operator attention. No system is so reliable that it does not require regular attention. Machinery operating on automatic, even the best, will occasionally go astray. Maintenance, adjustment, and monitoring are important factors for systems that have flexibility designed into them. Since most groundwater treatment systems are relatively small compared with municipal or industrial units, the cost of the operator per unit treated becomes substantial.

Design Parameter: Continued Expense

Power, treatment chemicals, maintenance, periodic monitoring, and regulatory reporting are major items of continuing expense. The realistic cost of each of these items is an integral part of a continuing project.

As long as the equipment is operating, power and treatment chemicals will be consumed. Generally this cost can be calculated on a daily or volume basis. Difficulties can occur when operation is halted on an unplanned occurrence, such as an accidental power disruption or other cause for system stoppage. Restarting and return of the system to a semiequilibrium status can sometimes be as costly as the initial startup. Contingencies should be included for periodic happenings.

Periodic performance monitoring (weekly or monthly) of operating systems is a key factor for performance monitoring. Frequent testing of indicator parameters such as volatile analysis, pH, conductivity, dissolved oxygen, or Eh status can often be used as gauges of continuous performance much as a speedometer and oil pressure and amperage indicators are used to monitor the daily performance of automobiles. A full suite of analytical and physical parameters is necessary to document the remediation progress. Project cost plans must include sufficient funds to provide performance monitoring.

Compliance monitoring is almost universally required by regulatory authorities. Typically, this monitoring will involve a suite of chemical analyses along with physical parameters to document the success of the remediation. Quality control sampling of this variety often involves a significant expense. Mobilization, fieldwork, documentation, analysis, and reporting can sometimes be some of the greatest costs of a remediation project.

11.4.4 Site Closure

At the conclusion of each remediation project, the site must be returned to some alternative use. Closure usually involves removal of surface equipment and underground piping, plugging of wells, and reestablishment of surface features (regrading, paving, or landscaping). Although some of the cost closure may be offset by the salvage value of major equipment, site restoration may amount to several percent of the total project cost and must be recognized at project initiation.

Continued postclosure compliance monitoring is usually required by regulatory agencies. During the later stages of remediation when concentration trends are well established, specific compliance points should be identified and monitoring schedules agreed upon, prior to the start of decommissioning activities. Costs associated with postclosure monitoring (at present or future money value), as well as the cost of the final closure report, can amount to a significant part of the overall project cost.

11.4.5 Air Sparging Pilot Study Case History

Air sparging was used to perform a pilot test at an audio equipment manufacturing facility. The release site was an open nearby area where paint shop wastes had been discharged into the subsurface via a dry well over a 9-year period. Previous initial studies confirmed organic and inorganic contaminants in soil and groundwater.

In all, 11 vertical soil vapor extraction wells and 9 vertical air sparge wells were installed in the treatment area. Sand chimneys (boreholes filled with coarse sand) were also installed to facilitate vertical air circulation. Mechanical remediation equipment and systems control installed for remediation were mostly automated with minimal operator control required.

The preliminary study identified approximately 2300 yd^3 of contaminated soil with a "hot zone" of approximately 800 yd^3, which required the bulk of remediation effort. A pilot air sparge/soil vapor extraction study was made to define remediation costs. The remediation effort was estimated to extend for 3 years.

The cost projections are based on assumptions and costs developed from pilot test data. The 12 categories defined to reflect the cleanup activities are as follows:

- Site preparation;
- Permitting and regulatory requirements;
- Capital equipment (amortized over 10 years);
- Start-up;
- Consumables and supplies;
- Labor;
- Utilities;
- Effluent treatment and disposal cost;
- Residuals and waste shipping, handling, and storage services;
- Analytical services;
- Maintenance and modifications;
- Demobilization.

The itemized costs for each of these categories on a year-by-year basis for an estimated 3-year operating life is shown in Table 11.3. The total cost to remediate 21,300 yd^3 of soil was estimated to be \$220,737 or \$10.36/yd^3. This figure does not include any treatment of the off-gases. If effluent treatment costs are included, it would increase costs to \$385,237 or \$18.09/yd^3.

The relative importance of each category on overall costs is shown in Figure 11.5. It shows the largest cost component without effluent treatment was site preparation (28%), followed by analytical services (27%) and residuals and waste shipping, handling, and storage (13%). Labor accounted for a relatively small percentage (9%), excluding travel, per diem, and car rental expenses. These four categories alone accounted for 77% of the total costs. Utilities and capital equipment accounted for 6 and 8%, respectively, and the remaining cost categories each accounted for 4% or less. Effluent treatment costs would have accounted for 43% of the total cleanup if it had been conducted. This estimate does not include costs for site decommissioning (well plugging, equipment disposal, etc.) or extended postclosure monitoring.

TABLE 11.3
Estimated Cost (in Dollars) for Treatment Using the Subsurface Volitilization Ventilation System Process over a 3-Year Application

Cost Category	First Year	Second Year	Third Year
Site Preparation	32,500	—	—
Well drilling and preparation	10,000	—	—
Building enclosure (10 × 15 ft)	5,000	—	—
Utility connections	15,000	—	—
System installation subtotal	62,500	—	—
Permitting and Regulatory Requirements	10,000	—	—
Capital Equipment (amortized over 10 years)			
Vacuum pump	450	450	450
Blower	450	450	450
Plumbing	3,333	3,333	3,334
Building Heater	333	333	334
Subtotal	4,566	4,566	4,566
Start-up	7,957	—	—
Consumables (Health and Safety Gear)	1,000	1,000	1,000
Labor	6,300	6,300	6,300
Utilities			
Electricity (blower and pump)	3,900	3,900	3,900
Electricity (heater)	660	660	660
Subtotal	4,560	4,560	4,560
Effluent Treatment and Disposal Costs	—	—	—
Residuals and Waste Shipping and Handling			
Contaminated drill cuttings	12,500	—	6,000
Contaminated health and safety gear	6,000	1,000	3,000
Subtotal	18,500	1,000	9,000
Analytical Services	20,000	20,000	20,000
Maintenance and Modifications	—	—	—
Demobilization	—	—	2,500
Total Annual Costs	135,383	37,426	47,928
Total Remediation Costs			220,737

Source: Department of Energy (1996).

Cost figures presented here are "order of magnitude" estimates and are generally accurate to +50% to –30%. A remediation contractor whose compensation will be based on performance will wish to refine this type of estimate to include line items for excluded items, plus a risk factor and profit.

11.5 REGULATORY CLIMATE

Since the advent of RCRA in the 1980s, the traditional approach used to determine aquifer cleanup standards has been to establish compound specific concentrations

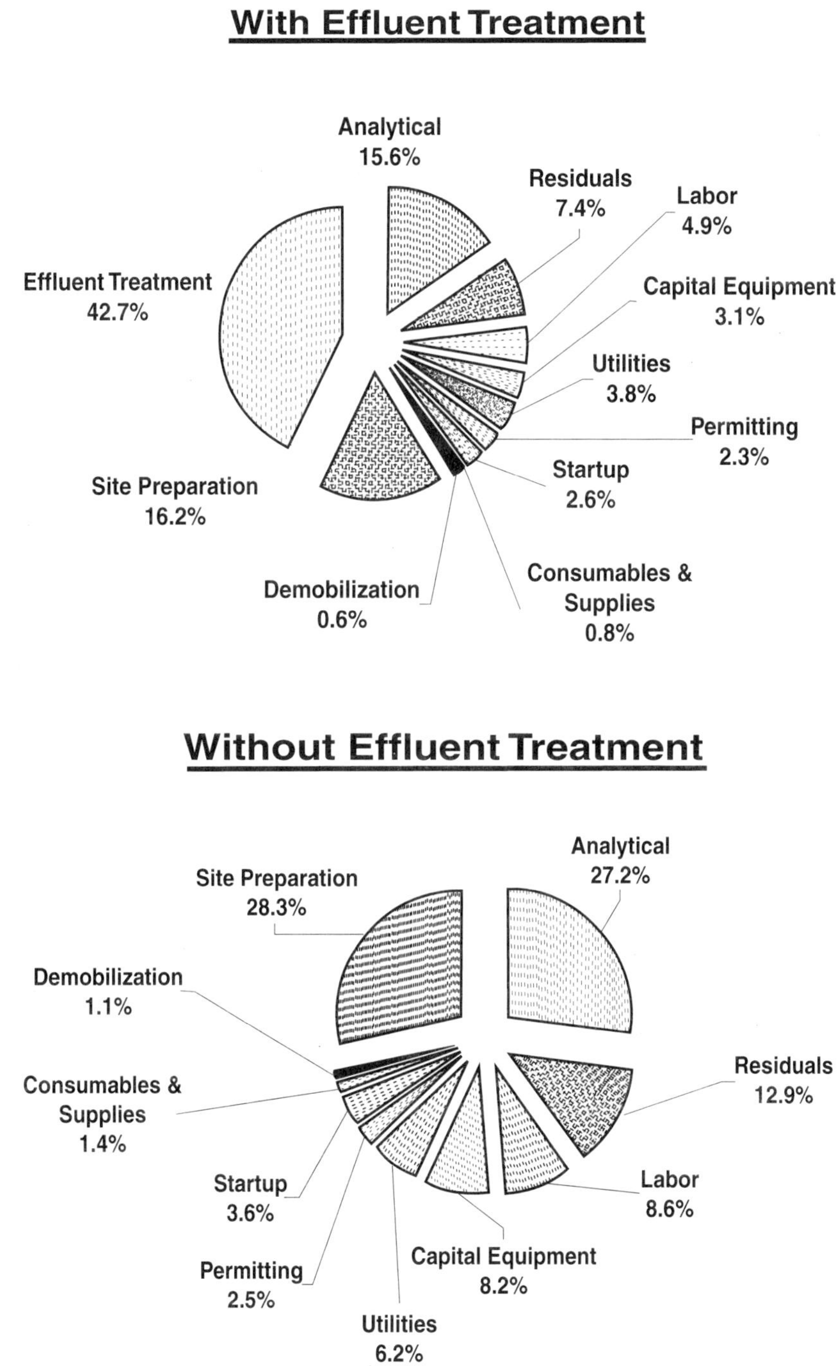

FIGURE 11.5 Cost breakdown for 3-year remediation.

of specific chemicals based on general health risks. This approach generally set concentrations based on some multiplier of the drinking water standard. For example, the drinking water standard for benzene is 0.005 ppm, and a typical industrial area aquifer remediation standard was established at 0.5 ppm (a multiplier of 100). Similar values were often established for other chemicals. The multiplier approach was an attempt to recognize the ratio of soil–water partitioning, probable aquifer usage, and related human health risks.

As the ability to evaluate risks and the understanding of partitioning developed, the multiplier approach was often modified to accommodate specific site conditions such as the presence of a shallow Class 1 (high-quality) aquifer, or the presence of a low-permeability clay layer. Further recognition of site-specific cleanup standards has occurred with the implementation of formal risk assessments. Many state and federal agencies have recently adopted policies that encourage definition of exposure to chemicals present at a site and evaluation of human or environmental health risks to that exposure. Risk-based corrective action (RBCA) was developed by the American Petroleum Institute (API) for sites with petroleum contamination, and other programs were developed by the EPA. Cleanup standards developed by these procedures are highly site specific and reflect the future planned use of the site.

Cost of cleanup is usually directly proportional to the cleanup standard. In practice, the cost of remediation is highest where human or environmental risk is greatest. Established industrial areas that will remain in that use category will have higher acceptable cleanup concentrations than will residential or commercial areas.

REFERENCES

Arps, J. J., 1956, Estimation of Primary Oil Reserves: *American Institute of Mining and Metallurgical Engineers Transactions,* Vol. 207, pp. 182–191.

Bergsman, T. and Trowbridge, F., 1999, Soil Vapor Extraction to the Sixth Degree: *Soil and Groundwater Cleanup*, July, pp. 7–11.

Connor, J. R., 1988, Case Study of Soil Venting: *Pollution Engineering*, July, pp. 74–78.

Diekmann, J. E. and Featherman, D. W., 1988, Assessing Cost Uncertainty: Lessons from Environmental Restoration Projects: *Journal of Construction Engineers and Management*, Vol. 124, No. 6, pp. 445–451.

Downey, D. C., Hinchee, R. E., and Miller, R. N. (editors), 1999, *Cost-Effective Remediation and Closure of Petroleum-Contaminated Sites*: Battelle Press, Columbus, OH, 297 pp.

Gentry, R. W., 1972, Decline-Curve Analysis: *Journal of Petroleum Technology,* pp. 38–41.

Hansen, W. J., Orth, K. D., and Robinson, R. K., 1998, Cost Effectiveness and Incremental Cost Analyses: Alternative to Benefit–Cost Analysis for Environmental Projects: *Practice Periodical of Hazardous, Toxic and Radioactive Waste Management*, January, pp. 8–12.

Kroopnick, P. M., 1998, Selecting the Appropriate Abatement Technology: Estimating Life-Cycle Costs: *Pollution Engineering*, November, pp. 36–40.

Paek, J. H., 1996, Pricing the Risk of Liability Associated with Environmental Clean-up Projects: *Transactions of American Association of Cost Engineers*, cc4.1–cc4.5.

Paleologos, E. K. and Fletcher, C. D., 1999, Assessing Risk Retention Strategies for Environmental Project Management: *Environmental Geosciences*, Vol. 6, No. 3, pp. 130–138.

Sellers, K., 1996, Keep Your Eyes on the Prize: *Soil and Groundwater Cleanup*, January–February, pp. 34–37.

Slider, H. C., 1968, A Simplified Method of Hyperbolic Decline Curve Analysis: *Journal of Petroleum Technology,* pp. 235–236.

Stanforth, M. T., Kulik, E. J., Appel, C., and Hafenmaier, M. F., 1998, Case Study: DNAPL Remediation Alternatives Evaluation Yields Surprising Results: In *Nonaqueous-Phase Liquids-Remediation of Chlorinated and Recalcitrant Compounds* (edited by G. B. Wickramanayake and R. E. Hinchee), Battelle Press, Columbus, OH, pp. 161–168.

Trimmel, M. L., 1987, Installation of Hydrocarbon Detection Wells and Volumetric Calculations within Confined Aquifers: In *Proceedings of the National Water Well Association and American Petroleum Institute Conference on Petroleum Hydrocarbon and Organic Chemicals in Ground Water — Prevention, Detection and Restoration*, Vol. I, November, pp. 225–269.

Trimmel, M. L., Winegardner, D. L., and Testa, S. M., 1989, Cost Optimization of Free Phase Liquid Hydrocarbon Recovery Systems: In *Proceedings of the Hazardous Materials Control Research Institute Conference on Hazardous Waste and Hazardous Materials*, April.

U.S. EPA Office of Solid Waste and Emergency Response, 1997, *Analysis of Selected Enhancements for Soil Vapor Extraction*: EPA-542-R-97-007, September.

12 LNAPL Recovery Case Histories

"Ignorance is no defense."

12.1 INTRODUCTION

Selection of approach or combination of approaches to LNAPL recovery and ultimately aquifer rehabilitation and restoration is dependent on numerous factors as discussed in the previous chapters. The case histories presented below reflect different remediation approaches and strategies in response to varying geologic and hydrogeologic conditions, site-specific constraints, and regulatory environment. A variety of LNAPL recovery systems are first discussed, including:

- Vacuum-enhanced suction-lift well-point system;
- Rope skimming system with bioremediation;
- Vacuum-enhanced eductor system; and
- Combined one- and two-pump system with reinjection (and vacuum enhancement).

A case study is presented that demonstrates the importance of evaluating lithofacies distribution, or lateral and vertical heterogeneities, and depositional environment as a control on LNAPL occurrence and migration, and implementation of an effective and efficient remediation strategy. This is followed by a case history on the development of a long-term remedial strategy for LNAPL recovery and aquifer restoration from a regional perspective.

12.2 VACUUM-ENHANCED SUCTION-LIFT WELL-POINT SYSTEM

Near-shore facilities are typically characterized by shallow groundwater conditions. The occurrence of LNAPL product on the water table presents the need for immediate containment and continued recovery of the product to abate degradation of groundwater quality, hydrocarbon vapor migration, and discharge of hydrocarbon product to surface waters. A pneumatically operated, double-diaphragm, suction-lift pump has been frequently used in such circumstances to contain and recover LNAPL.

Applicable over a wide range of hydrogeologic environments with depths-to-water ranging up to 22 ft, this type of system can also be used to pump from aquifers of varying soil characteristics ranging from low-permeability clay and silt to higher-yielding sand formations. This type of system is also found to be generally compatible at sites such as refineries, terminals, and gasoline stations where they offer certain favorable characteristics, such as pneumatic operation availability and intrinsic safety. Other practical aspects include off-the-shelf availability, and the added benefit that it can be modified to induce additional recovery through application of a vacuum to the recovery wells.

One of several sites that was found to be suitable for a double-diaphragm suction-lift well-point pumping system was a bulk liquid marine terminal located in Los Angeles Harbor (Figure 12.1). This facility had been used as a marine terminal since the early 1920s and is still used for the transfer of petroleum products to and from ships and onshore facilities. Onshore facilities include multiple subsurface petroleum product pipelines to inland facilities and aboveground storage tanks of varying sizes. At this particular site, overfilling of aboveground tanks was a frequent occurrence. Initial remedial response was initiated when LNAPL product seepage was noted discharging into harbor waters.

Subsurface conditions were initially explored by the hand augering of 26 exploratory borings and 61 additional borings to be used for monitoring or recovery well installation. Due to the close proximity of the site to surface water, and the shallow depth to groundwater, the maximum depth drilled was approximately 15 ft. The surficial materials encountered across the site were primarily hydraulically emplaced uncontrolled fill consisting of fine sand and silty sand. The fill materials appeared to be reworked local materials, likely dredgings from the adjacent harbor since shell fragments were observed in many of the borings. The base of the fill was inferred to be coincident with the top of a thin, laterally discontinuous clay layer generally encountered between 2 to 5 ft below the existing ground surface. It appeared that the natural land surface prior to emplacement of fill was fairly irregular. Fill thicknesses varied from 2 to 10 ft across the site, being generally greater along the shoreline behind the riprap bulkhead. Below the fill were natural deposits of marine sand, silt, and clay characteristic of an estuarine depositional environment.

Groundwater was encountered under water table conditions at depths ranging from 6 to 8 ft below the local ground surface. Groundwater flow was perpendicular to the shoreline and generally toward the open water of the harbor with a gradient of approximately 0.02 ft/ft. Shallow groundwater occurred both within the fill and underlying natural sediments. Grain-size distribution analyses performed on soil samples retrieved from the water-bearing strata indicated that the aquifer materials are predominantly fine to medium sand. Pumping test data yielded an estimated aquifer transmissivity of 100 gpd/ft and resultant permeability of approximately 2 ft/day (7.4×10-cm/s).

The approach to containment of LNAPL discharge to the harbor followed the concept of a well-point dewatering system for interception of groundwater discharge. The discharge of groundwater and hydrocarbon product to the adjacent harbor is illustrated in cross section in Figure 12.2. Interception of product discharge to the harbor was accomplished by the pumping of groundwater resulting in a reversal of

FIGURE 12.1 Photographs of a bulk liquid hydrocarbon storage facility with evidence of frequent overspilling of tanks (A), resulting in seepage along the shoreline (B).

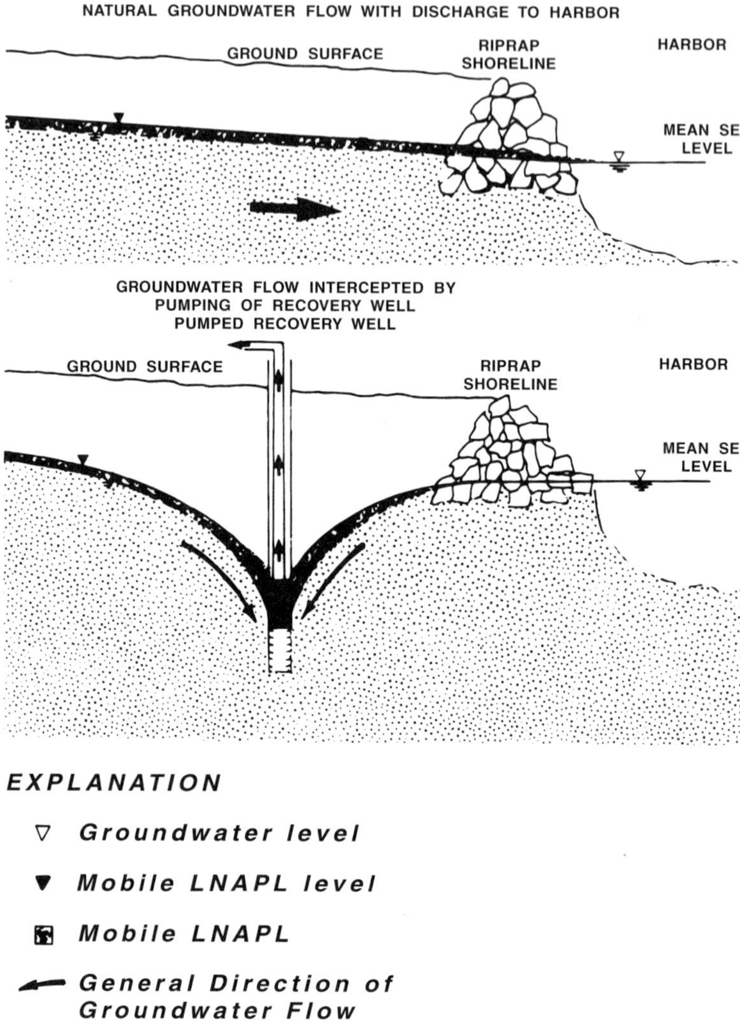

FIGURE 12.2 Cross-sectional view of LNAPL contaminant system under shallow groundwater conditions at a harbor facility.

flow direction. This reversal reflects the lowering of the water table below that of the adjacent harbor water level.

Apparent LNAPL thicknesses measured in monitoring wells ranged up to 7 ft. During the subsurface characterization phase, three distinct petroleum product types were evident, which have apparently coalesced into one LNAPL pool. Product samples retrieved indicated characteristics of gasoline, kerosene, and crude oil. Despite the variation in product types, the pumping system design and construction was similar throughout the site with no significant effect on system efficiency and effectiveness.

Containment of groundwater flow was initially accomplished through installation and pumping of ten recovery wells of which five were oriented parallel and adjacent to the shoreline. Based on pumping test results, recovery wells were installed generally on 60-ft centers. In subsequent phases, additional recovery wells were installed farther inland within the tank farm area to intercept and recover product prior to migration to the shoreline area.

Data generated through a year of operation of the recovery system were analyzed by a variety of means to provide both qualitative and quantitative indication of overall system effectiveness. Seepage of LNAPL from the shoreline to the harbor waters had manifested itself as staining of the riprap and a visible sheen of product on the water surface. No specific quantification was made of the rate of seepage prior to initiation of hydrocarbon recovery. Regardless, after approximately 3 months of operation, no visible seepage was evident. In addition, visible product evident on the riprap within the zone of tidal fluctuation had been eliminated. Estimates of volume of product recovered were developed from pumping rate and percent product data. Average pumping rate ranged from 0.68 to 1.87 gpm. With an average of three wells per pump, estimated pumping rates for individual wells ranged from 0.22 to 0.62 gpm. Average percent product in total fluids pumped for individual pumps ranged from 7.6 to 22% with no observable difference in recovery rates for the different product types present. Overall, an estimated 569,000 gal (13,548 barrels) of product was recovered during the initial year of recovery from 33 recovery wells manifolded to 11 pumps.

Reductions in apparent LNAPL thickness in the subsurface have been effected over most of the site. In general, apparent thickness after 1 year were reduced 30 to 40% to those apparent thicknesses measured prior to recovery. The reduction in apparent thickness in a representative monitoring well with time is shown in Figure 12.3. However, in one area of the site, hydrocarbon thicknesses actually increased

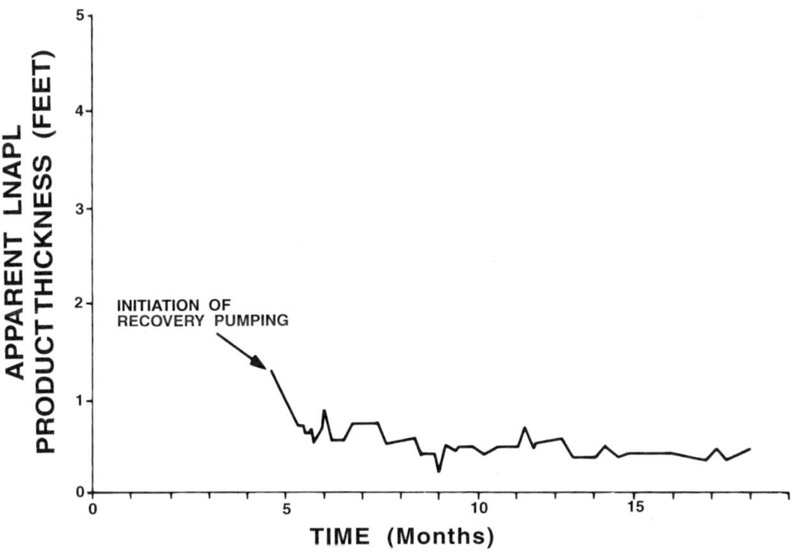

FIGURE 12.3 Apparent LNAPL thickness decreases with time upon initiation of recovery operations.

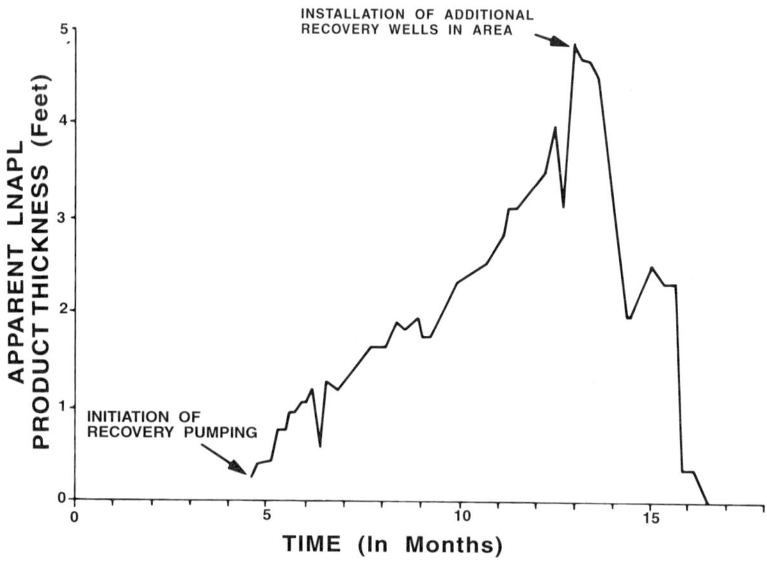

FIGURE 12.4 Apparent LNAPL thickness fluctuations with time reflecting installation of additional recovery wells.

as a result of pumping, reflecting redistribution of the product in response to the creation of a cone of depression in an area of relatively higher permeability. Installation of additional recovery wells in this localized area effected rapid decrease in the apparent LNAPL thickness. Such an increase and subsequent decrease in one monitoring well is graphically illustrated in Figure 12.4.

The effectiveness of a suction-lift well-point pumping system for LNAPL recovery under shallow water table or perched conditions is evident from experience gained in this case history. Primary effectiveness is attributable to the inherent capabilities of suction-lift equipment combined with the practical aspects of using such equipment at petroleum-handling facilities. Due to the flammable and/or explosive nature of hydrocarbon products, pumps that are driven by compressed air are inherently safer than pumps requiring electrical power. In addition, many petroleum-handling facilities, including refineries, terminals, and gasoline stations, have existing compressors that can be used to supply the necessary air volume for pump operation. The pumping equipment and controls (valves) are readily available through many industrial supply distributors. Since double-diaphragm pumps have multiple uses, there are many brands and sizes to select from to suit anticipated pumping requirements.

The use of less-sophisticated electronic controls leads to significantly reduced equipment costs. Double-diaphragm pumps are self-priming at a lift of 22 ft. If suction is not broken, these pumps may recover fluids from as deep as 27 ft. In addition, the pumps will operate against pressures of up to 110 lb/in.^{2}. Double-diaphragm pumps can also handle a wide variety of petroleum products ranging from gasoline through heavy crude oil, and are not significantly affected by the pumping of solids such as sand or silt (up to $^1/_8$-in. diameter).

Double-diaphragm suction-lift pumps also have advantages relative to varying subsurface hydrogeologic characteristics encountered at different sites. These include:

- Wide range in pumping rates;
- Can be pumped dry; and
- Vacuum-assisted recovery.

The pumps can sustain varying pumping rates ranging from virtually no discharge to tens of gallons per minute. Pumping rate is easily controlled by valving the air supply intake or discharge from the pump. In low-permeability formations, where the pumping rate exceeds the yield of the well(s), double-diaphragm pumps are not damaged by pumping without available fluid. A vacuum will be maintained until the fluid level in the well(s) recovers, and recovery of fluids resumes. By sealing the wellhead to air entry, the vacuum that develops within the well enhances fluid flow to the well by simulating additional drawdown, thereby increasing the effective recovery of the well.

Overall, the pumping system has been effective in eliminating surface water discharge of LNAPL product, reducing apparent product thickness 30 to 40% of preexisting conditions while recovering a significant volume of product for recycling.

12.3 ROPE SKIMMING SYSTEM

An abandoned refinery in the Midwest is located near a suburb of a medium-sized city. The site encompasses about 60 acres and has remained unoccupied since surface features at the refinery were removed in the 1960s. During the late 1970s, adjacent residents began to complain that their shallow irrigation wells produced contaminated water that smelled like gasoline.

Subsurface geologic and hydrogeologic conditions were evaluated based on excavations of test pits (on 100-ft centers), installation of monitoring wells, and by conducting *in situ* rising-head slug tests and pumping tests. The site is located within a glaciated region, and is immediately underlain by unconsolidated, stratified, sequence of silt and sand to a depth of approximately 50 ft below the ground surface. These relatively coarse-grain deposits overlie deposits of highly plastic inorganic clay, which acts as a lower confining unit. Geologic contacts between the upper sand and silt and the lower clay deposits are gradational. A site facility layout map showing well locations is illustrated in Figure 12.5. A hydrogeologic cross section is shown in Figure 12.6.

Groundwater occurs under shallow water table conditions at an approximate depth of 5 to 8 ft below the ground surface with a southwest gradient of 0.004 ft/ft. Hydraulic conductivity based on *in situ* rising-head slug tests and pumping tests was on the order of 4.3×10^{-3} cm/s (92 gpd/ft^2). Groundwater supplies for industrial purposes are typically derived from within the upper 75-ft portion of the glacial deposits.

Chemical analysis of water samples collected from monitoring wells and adjacent ditches was performed for BTEX, total organic carbon (TOC), chromium, and lead. Results of analysis presented a complicated chemical distribution in a dynamic

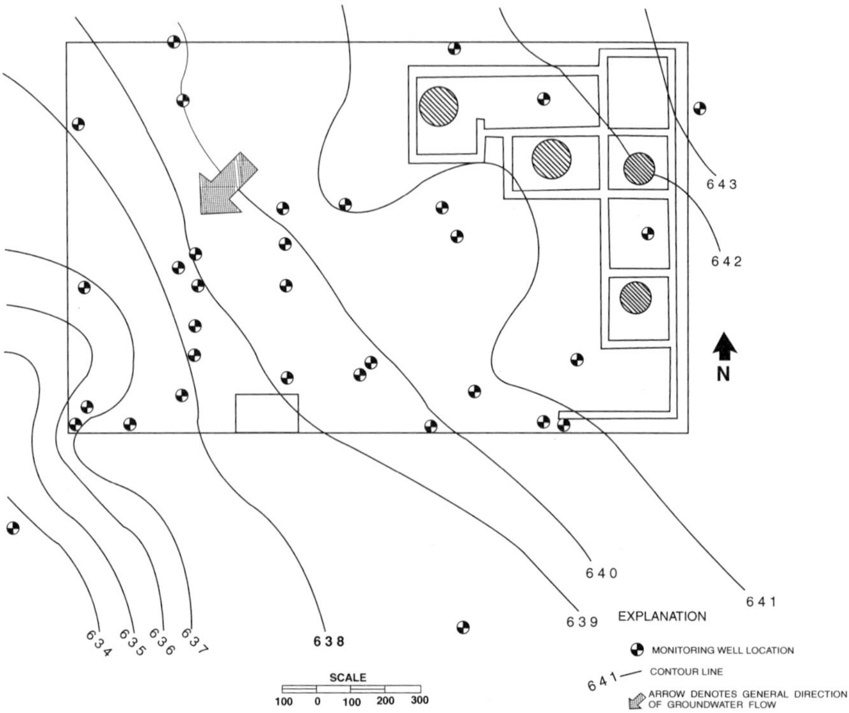

FIGURE 12.5 Refinery site facility layout map showing well locations and configuration of the water table.

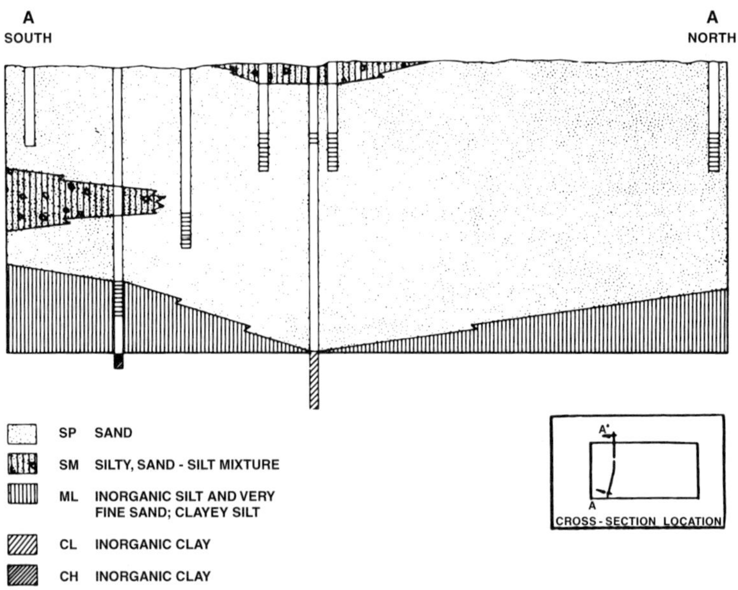

FIGURE 12.6 Hydrogeologic cross section at refinery site.

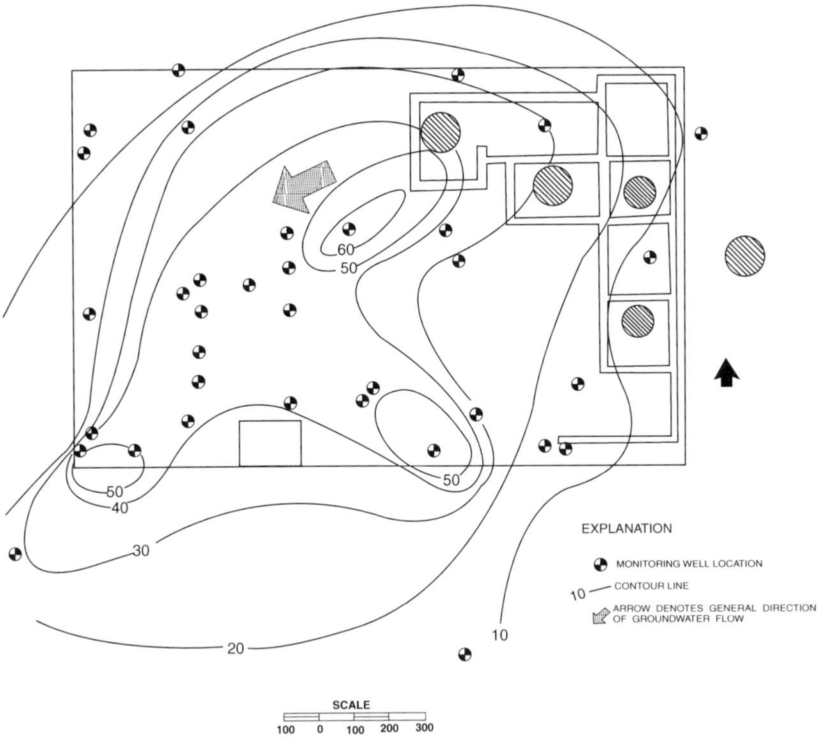

FIGURE 12.7 TOC distribution in shallow groundwater beneath refinery site.

environment. Concentration of each of the analyzed compounds ranged from less than 2 μg/l to nearly 1 mg/l, with a few higher levels reported. The area of primary concern was one well with benzene reported at 1400 μg/l. TOC served as a groundwater contamination indicator, and ranged from 7 mg/l in the upgradient wells to 40 or 60 mg/l in the central portion of the site (Figure 12.7).

LNAPL product was found to be distributed over the central portion of the site with measured apparent thicknesses of approximately 0.2 to 0.5 ft. Concluding that additional work was warranted at the site, two goal-oriented objectives were developed: (1) consideration of on-site remedial alternatives to prepare the site for other industrial uses and (2) determination of the extent of off-site groundwater contamination.

Early in the investigation phase it became apparent that removal of the LNAPL was an essential part of the aquifer restoration program. A pilot recovery well was constructed to determine the feasibility of recovery of LNAPL. The 48-in.-diameter, backhoe-excavated well was equipped with a submersible water pump and an LNAPL recovery skimmer. A cone of depression was developed using the water pump, while the skimmer recovered mobile LNAPL that migrated toward the well. Water produced was discharged onto the ground surface 200 ft from the well. Relatively high soil permeability allowed the water to recharge quickly so that no significant ponding occurred.

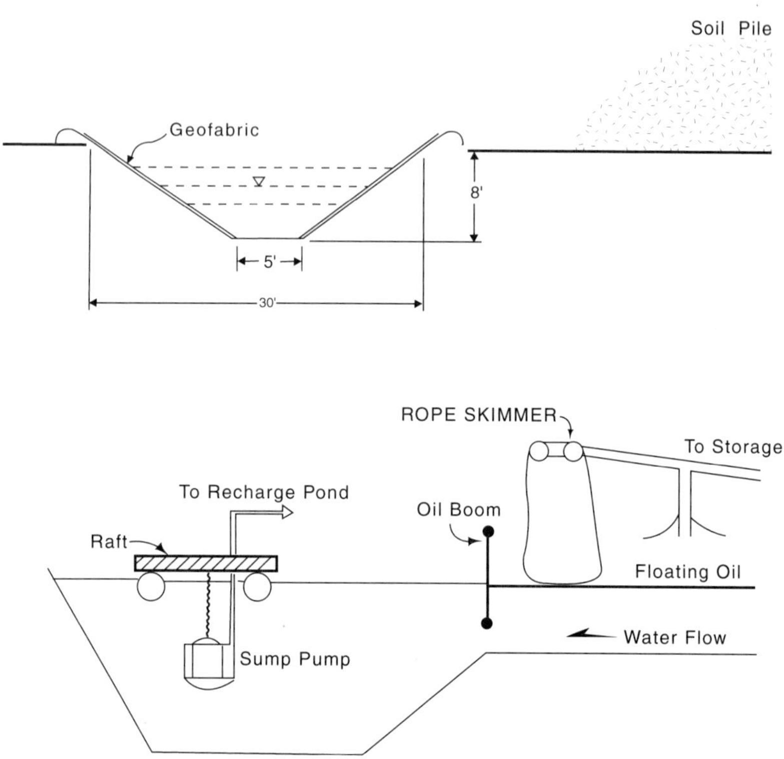

FIGURE 12.8 Schematic of trench recovery system beneath refinery site.

This pilot recovery initially produced significant LNAPL, which decreased rapidly due to a limited area of influence. Based on results of this test, it was concluded that the use of individual wells to recover the mobile LNAPL was not an acceptable option. In addition, the number of wells and pumps and the effort of maintenance were considered restrictive. An alternative to installing individual wells with limited areas of influence was to construct a system of open drainage ditches with a dendritic pattern spread over an area of LNAPL occurrence. A single submersible pump, at the end of the trench, was provided to assure flow toward the recovery area. A single "rope skimmer" recovered LNAPL as it migrated toward the pump. A schematic sketch of the trench recovery system is shown in Figure 12.8. The general layout of the trenches is shown in Figure 12.9. Water from the lower end of the trench was pumped to an upgradient recharge pond. The increased hydraulic gradient with enhanced mobile hydrocarbon recovery rates prevented the system from being dewatered.

Trenches were constructed using conventional earthmoving equipment. During construction, air monitoring was continued to assure that explosive conditions were not present. However, odors persisted from the freshly excavated soils but disappeared within a few days. Operation of this trench system continues throughout frost-free seasons and has proved very successful. Initially, approximately 5 bar-

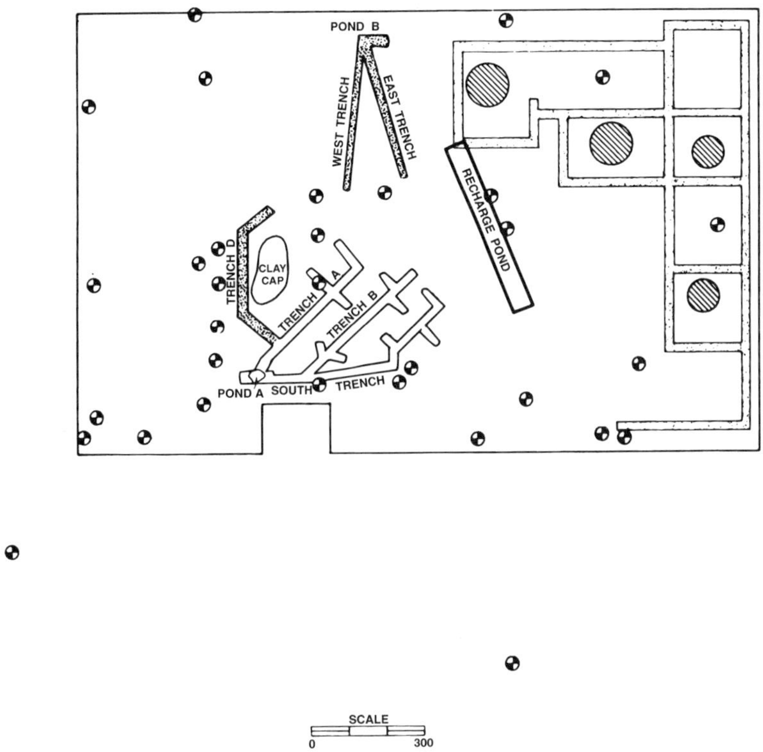

FIGURE 12.9 General layout of monitoring wells and trenches at refinery site.

rels/day, which has since diminished, were recovered. When the product is essentially gone, the trenches will be refilled with the original excavated material.

Cleanup of residual LNAPL in the shallow soils was considered to be a necessary part of the overall remediation program. Enhanced biological activity was initially considered. A laboratory demonstration proved that naturally occurring local oil bacteria could be used to accomplish this task. A pilot laboratory study was conducted to evaluate the scale-up treatment parameters for operation in the field.

A 10-yd^3 soil sample was excavated from the site, blended, and characterized for initial hydrocarbon content and nutrient content. The reactor was filled with soil compacted to field density (Figure 12.10). The tank at the bottom was filled with water nutrients and surfactants. Water from this tank was sprayed over the top of the soil at a rate that maintained aerobic conditions. A significant amount of LNAPL was initially released from the soil, which required additional air to be pumped into the well points to maintain favorable growth conditions. After 105 days of operation, more than 87% of the total aliphatics and 89% of the total aromatics were removed.

Based on the successful reactor study, a 1-acre field demonstration was performed to evaluate the potential for full site remediation of the unsaturated zone. The selected area was surveyed to be exactly 1 acre (circular), carefully sampled to determine the quantity of residual hydrocarbon, and place counts were made to

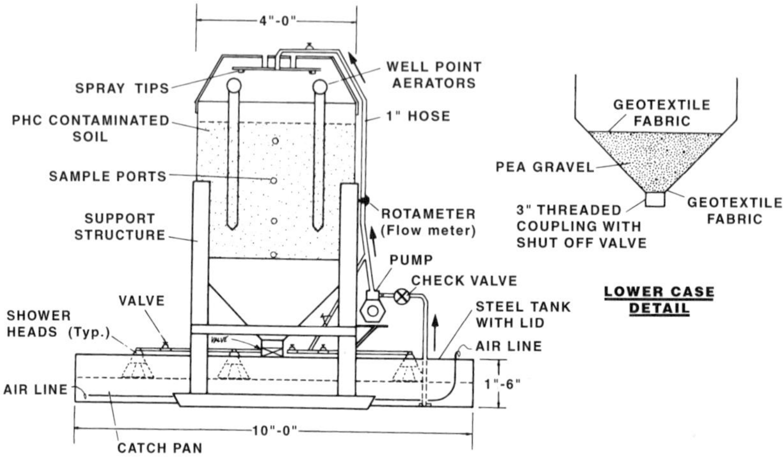

FIGURE 12.10 Schematic of soil-nutrient reactor used for biotreatment pilot study.

identify the species of soil microbes present. A single recovery well was installed and tested in the center of the plot. The purpose of this well was to create a drawdown cone extending to the boundary of the plot to assure collection of all fluids infiltrating during the test.

A 50,000-gal tank was installed adjacent to the test area. This tank was equipped with an aerator, nutrient feed equipment, and a submersible discharge pump. The physical layout of this system was very similar to that shown in Figure 12.11.

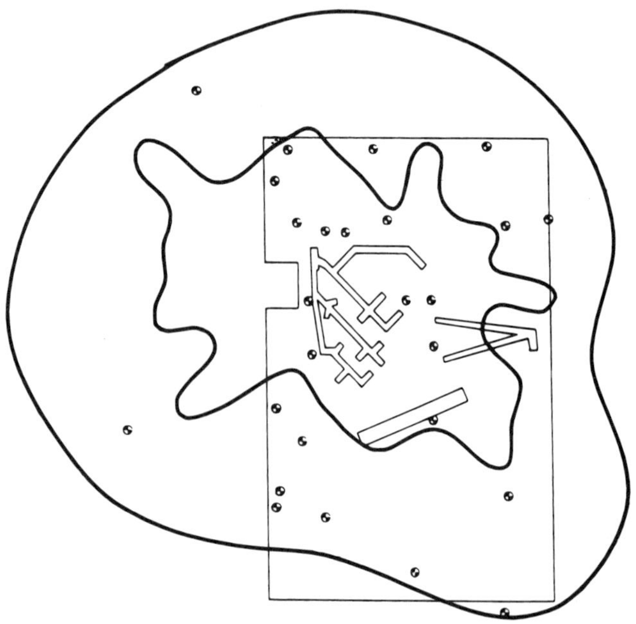

FIGURE 12.11 Initial TOC distribution in shallow groundwater beneath a refinery site.

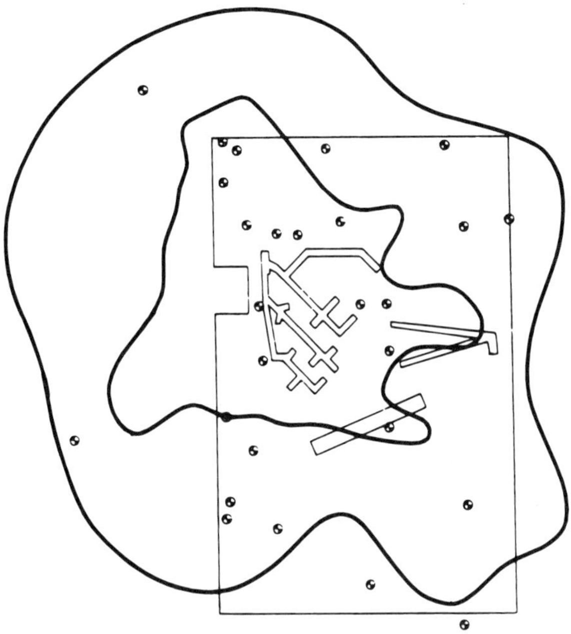

FIGURE 12.12 Predicted TOC distribution in shallow groundwater beneath a refinery site.

Makeup water from the recovery well was pumped into the tank at a rate of approximately 38 gpm. Soluble nitrogen and phosphorus fertilizer compounds were metered into the tank and mixed by the action of the aerator. The submersible pump was used to spray-irrigate the treatment plot with the enriched water. Average application was approximately 2 in./day over the test area (slightly less than the vertical infiltration rate of the shallow soil). During rainstorms, it was necessary to discharge some water to an off-site holding pond; however, this did not appear to affect the long-term test results significantly. After one summer of operation, the field demonstration confirmed that the residual hydrocarbon could be reduced by this procedure. In following years, plans were made to expand the treatment to larger areas.

When off-site contamination was confirmed, computer modeling was used to estimate and predict future dispersion and transport. TOC was the indicator parameter selected. Use of a numerical model allowed projections to be made without causing public concern during fieldwork. The model selected was the Plasm model sequence. Hydraulic parameters were initially based directly on the data derived from fieldwork. These original data were manipulated somewhat to calibrate the model to match observed field conditions. The initial distribution of TOC is shown in Figure 12.11. Further runs were made to project the contaminated spread during the following years. A map of predicted TOC distribution 6 years after the initial study is shown in Figure 12.12. Follow-up sampling indicated that the projections were realistic.

Remediation currently continues at this site. Recovery of LNAPL will continue for the next few years. Further expansion of the biological treatment system is

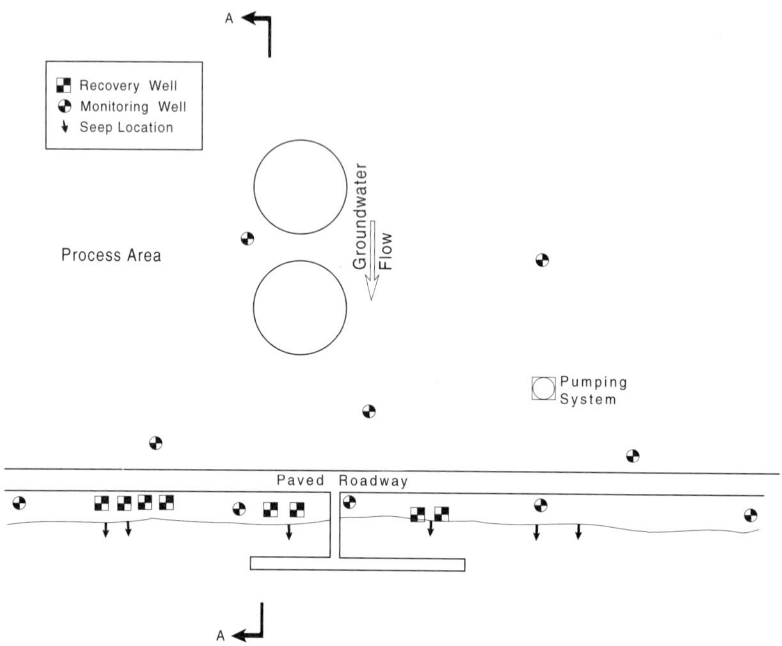

FIGURE 12.13 Site facility layout map.

planned to follow the LNAPL recovery effort. Off-site groundwater remediation has been demonstrated to be unfeasible due to the volume involved and the limited rate of spreading. Municipal water lines have been extended to neighboring residences, providing low-iron-content water, which is of much higher overall quality. The overall consequence is that a reasonable cleanup is being completed that will result in a usable industrial site being available for development without destroying a local industry.

12.4 VACUUM-ENHANCED EDUCTOR SYSTEM

A refinery located along a shipping canal in the southern United States was plagued by continuous oil seepage into the canal (Figure 12.13). Local environmental regulatory authorities were encouraging effective remediation within a limited time period. The general geologic conditions underlying the site are shown in cross section A–A′ (Figure 12.14). During construction of the canal, dredged material was pumped onto the bank behind a thin soil dike. After a number of years, industrial facilities, including a refinery, were constructed on this fill area.

Test borings revealed that the fill material consists predominantly of thinly laminated silt with a relatively high clay content. Occasional discontinuous layers of fine sand were encountered at random locations, both horizontally and vertically. The water table gradient under nonpumping conditions was toward the canal with an approximate gradient of 0.011 ft/ft. Slug tests indicated an effective permeability of 1×10^{-5} cm/s.

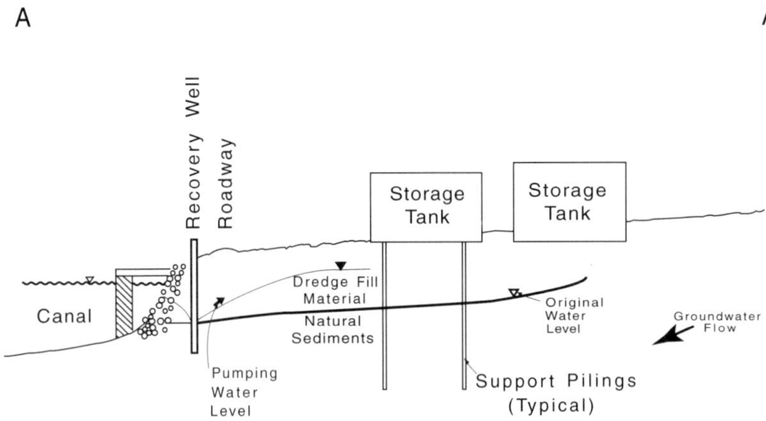

FIGURE 12.14 Hydrogeologic cross section at refinery site.

Apparent LNAPL thicknesses observed in monitoring wells were highly variable, ranging from a sheen up to 3-ft accumulation after several days. The spatial distribution of thickest accumulations suggested a multiple-source origin (within the same refinery). Multiple sources were supported by chemical analysis of the product recovered from certain monitoring wells. Each product sample comprised an admixture, consisting predominantly of gasoline, kerosene, or crude oil.

Several recovery scenarios were considered for remediation. Initially, construction of a narrow, permeable trench parallel to the canal appeared to be an appropriate interception system. The construction technique considered was use of a specially designed deep trenching unit. This type of trench would have included a tile drain leading to a single two-pump recovery well. However, a review of the subsurface site plans and interviews with long-term employees determined that an unknown number of buried pipes traverse the area intended for the trench construction. Disruption of refining operations and safety considerations resulted in rejection of this option.

The second option considered was use of interception wells. One- or two-pump wells could be constructed at calculated spacings to create a hydraulic trough parallel to the canal to intercept the product. This design was considered more acceptable to the safety officer and the facility engineer, but was rejected by the maintenance foreperson because of the relative complexity of the operation system. The number of submersible pumps and sophisticated electronic controls would have required employment (or training) of technical specialists beyond the cost budgeted under normal operations.

After another review of the practical constraints and technical considerations, it was concluded that an eductor vacuum-enhanced recovery system would be feasible and still be cost-effective and efficient. In all, 11 6-in.-diameter wells extending to a depth of 28 ft were installed at strategic locations in the area where product seeps were observed. Each well was serviced by a high-pressure supply and a low-pressure return line. A basic domestic-type deep-well eductor was installed in each well, attached to a drop pipe that extended to 25 ft below the surface. A check valve on the drop pipe prevented backflow into the well during service. The top of the casing

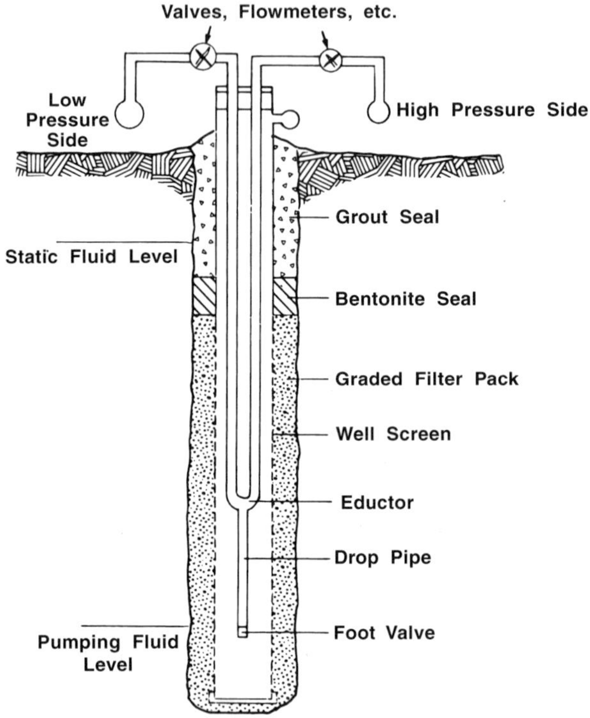

FIGURE 12.15 Schematic of typical recovery well construction detail.

was sealed with a multiport well top seal to maintain the vacuum. A schematic diagram of a typical recovery well is shown in Figure 12.15.

The pumping unit consisted of a submersible pump installed in a 10,000-gal reservoir tank, which also served as a holding tank (Figure 12.16). Fluids pumped from the wells were all returned to the tank for separation. All the oil and some water produced overflowed from the tank into another oil–water separator and then into the slop oil treatment system of the refinery. Clarified water from the bottom of the tank was recycled by the submersible pump through the eductor units to continue operations.

Instrumentation installed on this system was intended to monitor pressure, vacuum, and flow rates and to prevent accidental spills. Each well casing was equipped with a vacuum gauge. Lines leading to and from the eductor have a site glass to observe flow and are equipped with auxiliary piping for temporary attachment of a blow-rate meter. Pressure was monitored on both high- and low-pressure piping by gauges. Overflow from the tank was measured by a continuously recording flowmeter. Oil production is measured in a separate holding tank by periodic withdrawals of oil and water. Level switches in the reservoir tank acted as a safety measure to prevent the levels from becoming too high or too low.

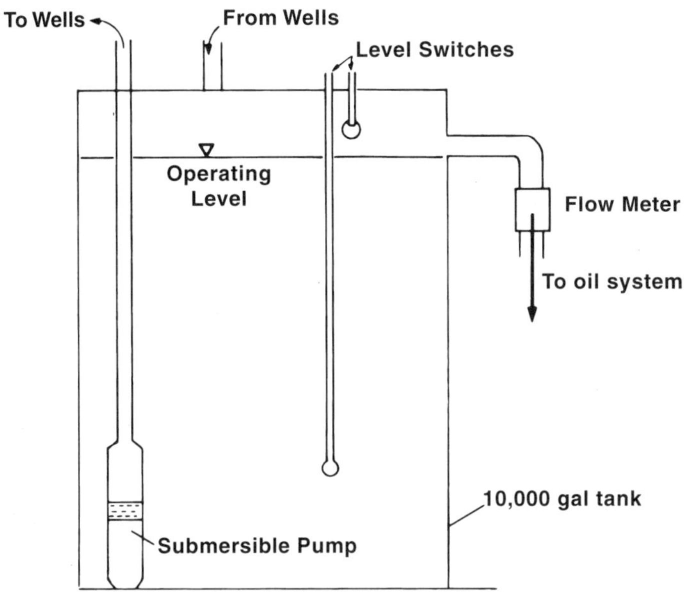

FIGURE 12.16 Schematic of reservoir tank.

Initial production from each of the 11 wells was approximately 1.5 gpm total fluid. Product production from the system was reported to average 15 barrels/day (bpd) for the first 60 days and 12 bpd for the next 60 days. Visible seepage to the canal was achieved and the state regulators dismissed further administrative action in exchange for periodic monitoring data.

12.5 COMBINED ONE- AND TWO-PUMP SYSTEM WITH REINJECTION

For over a decade, LNAPL occurrence has been investigated beneath an active refinery site in southern California. Numerous monitoring wells along with LNAPL samples have been used to evaluate the extent and character of LNAPL occurrence. LNAPL was found to occur as five pools. The main pools each consist of individual accumulations of distinct product types occurring under both perched and water table conditions. Two different recovery and mitigation strategies have been utilized. In relatively high permeability zones, a system of two-pump recovery wells was used to recover fluids; recovered water is reinjected without treatment. In relatively low permeability zones, a system of one-pump recovery wells was used. In the latter case, recovered water is treated prior to disposal.

Over the years, the refinery has produced a range of petroleum products including liquid petroleum gas, gasoline, chemicals, solvents, distillate fuels, gas oils, lubricating oils, greases, asphalt products, and bunker fuels. The primary products of the refinery are currently gasoline, jet fuel, and diesel. Minor products include coke, sulfur, naphtha, and fuel oil. The refinery processes approximately 200,000 barrels of crude oil per day. All crude oil processed is derived from the Alaskan North Slope, although historically some crude was derived from local oil fields via pipelines. This crude is transported to the Port of Los Angeles by tanker ship and delivered to the refinery by pipeline.

Subsurface geologic and hydrogeologic conditions beneath the facility have been investigated by the drilling and sampling of borings at a minimum of 300 locations with subsequent installation of wells. These borings extend to a maximum depth of 140 ft. Four types of wells have been installed: monitoring wells, injection wells, recovery wells, and 39 "weep" wells. Monitoring wells screen either the perched or water table zones. Weep wells, which screen across both perched and water table zones in an attempt to drain perched LNAPL down to the water table for recovery, have since been abandoned

On-site aquifer testing has also been performed in several phases. Typically, pumping tests are performed on newly installed injection and recovery wells to collect data used to determine optimum pumping or maximum injection rates. Monitoring wells are also tested to determine the location or feasibility of injection and recovery wells.

Characterization of LNAPL in pools underlying the site consisted of collecting and analyzing over 120 LNAPL samples. LNAPL product samples were distilled and analyzed to determine API gravity, lead, and sulfur content. Boiling point and percent recovery data were used to generate distillation curves.

The site is underlain by a sequence of unconsolidated, stratified, laterally discontinuous deposits of sand, silty sand, clayey silt, and silty clay of Recent and Upper Pleistocene age. A thin veneer of Recent deposits immediately underlies the eastern portion of the site. These deposits are difficult to distinguish from the underlying Upper Pleistocene deposits, which comprise the lower portion of the Lakewood Formation, due to similarities in lithology.

The subsurface soils that make up the lower portion of the Lakewood Formation can essentially be divided into three informal units, as shown in Figure 12.17. The uppermost unit comprises predominantly sand and silty sand with lenses of finer-grained soils consisting of clay, silt, silty clay, clayey silt, and sandy clay. However, a preponderance of finer-grained soils is evident in the northern and eastern portion of the main refinery adjacent to the flood control channel, and they are, in part, of Recent age. These fine-grained soils reflect a lateral facies change typical of a fluvial depositional environment. Deposits of the uppermost unit extend to depths ranging to about 40 ft below the ground surface.

The middle unit comprises predominantly fine-grained soils consisting primarily of clay and silt with subordinate lenses of sand and silty sand. These deposits are typically laterally discontinuous, as is evident in the hydrogeologic cross section A–A (see Figure 12.17). It is upon these fine-grained layers that perched groundwater occurs beneath the site. These deposits are present at depths ranging from about 35 to 70 ft below the ground surface with a maximum thickness on the order of about 35 ft.

The lower unit comprises predominantly coarser-grained soils similar in bedding and character to those of the uppermost unit described above. These deposits are first encountered at depths of about 50 to 70 ft below the ground surface. All borings

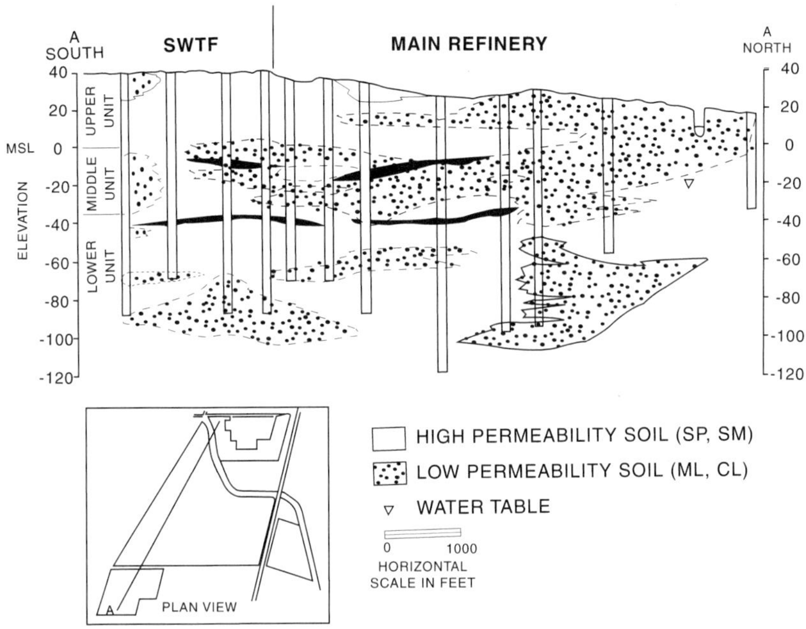

FIGURE 12.17 Generalized hydrogeologic cross section showing occurrence of perched and water table LNAPL. Approximate position of upper and lower informal units of the Lakewood Formation as shown on the left margin. Vertical exaggeration equals 25:1. (After Testa et al., 1989.)

drilled were terminated in this unit. Water table LNAPL occurs predominantly within this unit.

Groundwater beneath the site occurs within the Gage aquifer under perched and water table conditions. Beneath the northern half of the main refinery shallow groundwater appears to occur only under water table conditions. Beneath the southwestern portion and southern half of the main refinery, groundwater occurs under perched conditions upon and within the middle unit. Groundwater is present under water table conditions within the lower unit.

Groundwater encountered under perched conditions beneath the southwestern portion of the site occurs above a laterally discontinuous clay layer, which lies at a depth of about 40 to 50 ft below the ground surface (ranging in elevation from about sea level to 12 ft below sea level). The elevation of perched groundwater beneath the main portion of the refinery ranges from 40 to 70 ft below the ground surface (ranging in elevation from sea level to about 30 ft below sea level). The majority of the wells completed in the perched aquifer are now dry. Perched conditions in this area appear to occur as individual, discontinuous perching layers.

A water table piezometric surface contour map is presented in Figure 12.18. The elevation of this surface varies from 10 to 40 ft below sea level across the facility. Groundwater occurring under water table conditions is first encountered between 30 and 60 ft in depth. The regional direction of flow of shallow groundwater is toward a depression in the piezometric surface of the Gage aquifer located about

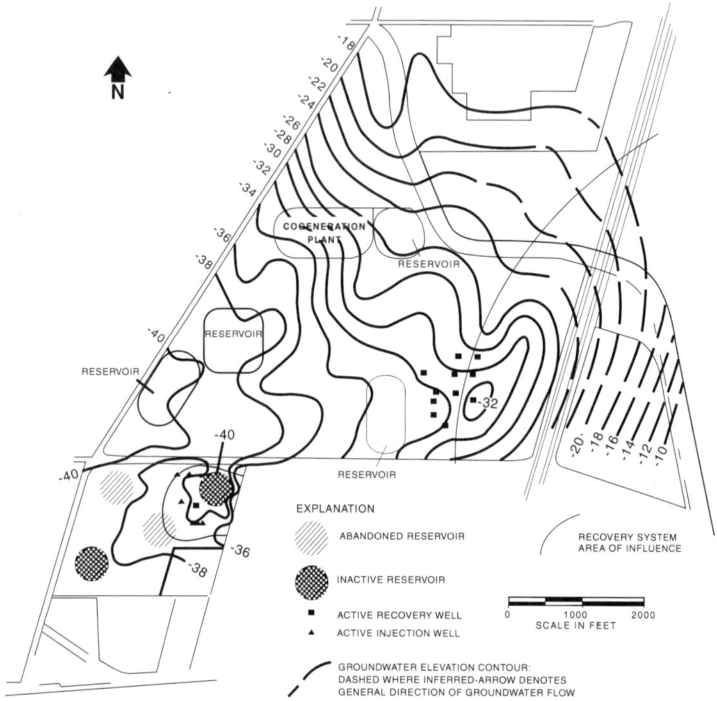

FIGURE 12.18 Corrected water table piezometer surface contour map showing regional hydrogeologic gradient toward the southwest. Note the impact of recovery systems within their area of influence or capture zones. (After Testa et al., 1989.)

2000 ft to the southwest. Water table flow directions are in general agreement with the regional flow direction except within the areas of influence of the two recovery systems. Depressions and mounding in the configuration of the water table reflect continued pumpage and reinjection associated with LNAPL recovery operations.

The series of active recovery wells located in the southeast portion of the main refinery has formed an elongated trough and a closed depression exhibiting about 2 ft of relief. A relatively steep gradient equal to almost 45 ft/mile is present between the flood control channel and the recovery wells beneath the eastern portion of this facility.

A limited number of aquifer pumping tests have been conducted within the perched zone. Two separate tests were conducted on a recovery well located in the southwestern portion of the refinery. Low transmissivity values of 100 and 150 gpd/ft were calculated.

Data collected from aquifer pumping tests conducted on LNAPL recovery wells indicated a wide range of transmissivity values within the water table aquifer. Transmissivity values ranged from 2000 to 74,000 gpd/ft, although typical values were between 2500 and 23,000 gpd/ft. Variation in transmissivity and conductivity values reflects the presence of aquifer heterogeneities, including stratification, varying thickness, and partial penetration of the LNAPL recovery wells. Pumping tests conducted on wells screened within the water table aquifer beneath the main refinery

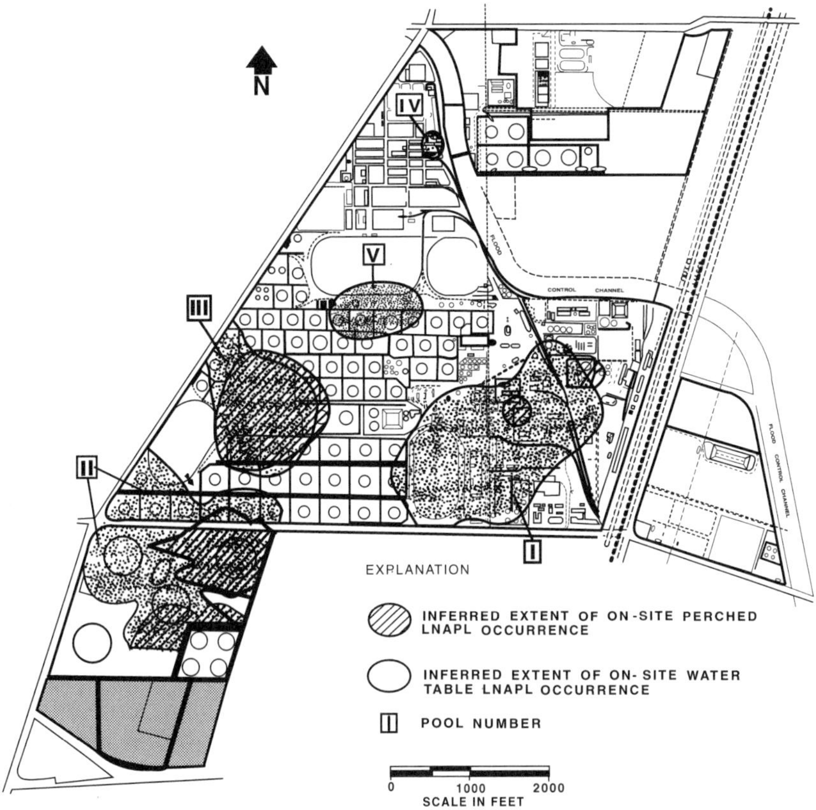

FIGURE 12.19 Generalized LNAPL occurrence map showing individual accumulations that have coalesced to form distinct LNAPL pools.

indicate that the transmissivity of this material is relatively low, ranging from about 30 to 1620 gpd/ft.

Beneath the facility, LNAPL product occurs as three main pools (Pool No. I, II, and III) and two smaller pools (Pool No. IV and V) of localized occurrence. The known extent of Pool No. I, III, IV, and V is entirely beneath the main portion of the refinery. Pool No. II is located beneath the southwestern portion of the site. LNAPL occurrence beneath the facility is shown in Figure 12.19. For purposes of this discussion, a pool is defined as an aerially continuous accumulation of LNAPL. The three main pools consist of individual accumulations of differing product character and are therefore referred to as coalesced pools. Individual product accumulations within these coalesced pools were delineated on the basis of physical and chemical properties characteristic of free hydrocarbon samples retrieved from the wells. Individual accumulations of relatively uniform product are referred to as subpools since it is inferred that they coalesced to form areally continuous occurrences or pools. These data suggest multiple sources over time for coalesced pools. The numbering of individual product accumulations is based on product type and areal extent as summarized in Table 12.1. Data presented include areal extent, groundwater conditions, product types, API gravity range, and lead and sulfur content. Volume determinations for certain pools are presented in Table 12.2. Areal extent is considered a minimum estimate where pools border the site boundaries and/or extend beyond the limits of on-site data.

LNAPL recovery at the facility dates back to 1977 when recovery was initiated at the site by the owner when losses from aboveground storage structures were discovered. The recovery wells, each utilizing a single positive-displacement pump, operated intermittently through 1979 until hydrodynamic conditions caused wells to pump water. About 38,000 barrels were recovered from the perched zone with this method. In late 1982, a more aggressive approach was undertaken to delineate the extent of the LNAPL pool and to design and implement a recovery system. Two two-pump recovery wells were installed and put into operation in early 1983. During the latter part of 1983, three additional two-pump recovery wells were installed. In addition to the two-pump water table recovery system, a single-pump recovery well was installed in the perched zone in 1984. By the end of 1985, 89 monitoring wells had been installed on site to delineate free hydrocarbon occurrence and to monitor the effectiveness of the aquifer rehabilitation program.

Exploration for LNAPL beneath the main portion of the refinery commenced in October 1984; 2-in. monitoring wells were initially installed followed by 4-in. test wells in areas of known occurrence. A one-pump recovery system was subsequently designed and installed in the less permeable material encountered beneath this area. The first well was activated in June 1987.

LNAPL recovery operations in the southwestern portion of the refinery have been conducted using a two-pump recovery system. This system currently includes up to four two-pump hydrocarbon recovery wells. Each two-pump system uses a 16-in.-diameter recovery well that is designed to accommodate two independently operated pumps placed at different levels within the well.

The two pumps within each recovery well are controlled by a series of electrodes that are positioned at predetermined levels within the well. The water pump utilizes a power interrupter probe to detect free hydrocarbon. This probe is positioned above the water intake and is adjusted to turn off the pump automatically when the hydrocarbon interface approaches the pump intake. This prevents the lower pump from accidentally pumping LNAPL to the injection wells.

The upper hydrocarbon pump is switched on and off by a "product" probe, which measures the resistivity of fluids adjacent to the intake of the hydrocarbon pump. When the product probe senses the higher-resistivity hydrocarbon, it switches on the product pump, which removes the accumulated LNAPL. The water level in the well rises in response to the removal of overlying product. As soon as the water reaches the product probe, the probe shuts down the hydrocarbon pump until the probe again senses that a sufficient quantity of free hydrocarbon has accumulated within the well. The well continues to cycle in this manner as long as free hydrocarbon is entering the well.

The groundwater that is produced by the recovery wells is reinjected into the aquifer from which it was originally pumped through a network of 6- and 2-in.-

TABLE 12.1
Summary of LNAPL Hydrocarbon Occurrence at Refinery Site

Pool No.	Estimated Extent (acres)	Groundwater Conditions	Subpool No.	Product Type	API Gravity (g/gal)	Lead Content (wt %)	Sulfur Content (wt %)
I	2	Perched		Two accumulations; light hydrocarbon[a]	No data	No data	No data
I	82[b]	Water table	1-A	Mixture of gasoline, naphtha, and kerosene	53.0–60.6	<0.005–2.00	0.12–0.25
			I-B	Kerosene; small amounts of naphtha and light gas-oil	33.4–44.2	<0.005–0.02	0.13–0.51
			I-C	Mixture of naphtha and kerosene[c]	52.7	<0.005	0.01
II	25[b]	Perched	II-C	Light gas-oil; small amounts of kerosene and heavy gas-oil	23.2–27.9	<0.005–0.02	0.09–0.50
			II-D	Light gas-oil	20.7–22.2	<0.005	0.90–1.13
II	70[b]	Water table	II-A	Mixture of gasoline, naphtha, and kerosene	50.7–54.1	0.18–0.62	0.21–0.26
			II-B	Mixture of gasoline, naphtha, and kerosene[c]	55.9	0.52	0.11
			II-C	Light gas-oil; small amounts of kerosene and heavy gas-oil	22.8–29.6	<0.005–0.08	0.05–0.53
			II-D	Light gas-oil; small amounts of naphtha and kerosene	20.9–22.9	<0.005	0.95–1.14
III	32	Perched	III-A	Mixture of light and heavy gas-oil[c]	19.2–24.0	<0.005	0.92–1.30
III	36[a]	Water table	III-A	Mixture of light and heavy gas-oil[c]	22.1	No data	No data
III			III-B	Mixture of gasoline and naphtha[c]	42.6	<0.005	0.04
III			III-C	Light gas-oil[c]	59.1	No data	No data
III	1	Water table	—	No data	No data	No data	No data
III	17	Water table	—	Gasoline; small amounts of naphtha and kerosene	61.3–69.2	0.12–0.15	0.04–0.51

[a] *Light hydrocarbon* is a field term referring to color and estimated density, and is used where no NAPL samples have been analyzed.
[b] LNAPL pool borders site boundary or lacks sufficient wells for good control, resulting in minimum area estimates.
[c] Only one LNAPL sample was analyzed.

TABLE 12.2
LNAPL Recovery System Effectiveness

LNAPL Pool No.	Media	Estimated Original Volume Recoverable (barrels)	Recovered to Date (barrels)	Estimated Present Volume Recoverable	Estimated Volume Removed (%)
I	Water table	181,000	28,000	151,000	15
II	Perched	110,000	2,000	108,000	1.8
II	Water table	310,000	183,000	127,000	59

diameter injection wells. The water that is produced at each well is metered but not consumed or treated in any manner in accordance with an existing court order.

To evaluate enhanced recovery, a single-pump vacuum-assisted recovery system was employed. This system was temporarily installed to evaluate recovery of LNAPL from the relatively low permeability perched zone beneath the northern portion of the site. The system was installed in 1984 and consists of a single 6-in. well. The well was completed with a 10-ft section of screen across the perched zone. The screen was connected to an underlying 30-ft-long sump. The annular space from the bottom of the sump to the bottom of the screen was sealed, allowing for accumulation of hydrocarbon and water within the sump for recovery without providing a conduit for downward migration of recoverable product. Another advantage of this construction was that it created a vacuum opposite the area of contamination, instead of pulling in air from zones above and below the LNAPL layer, which would reduce the overall efficiency of the recovery effort.

A submersible pump was installed within the sump and used to remove the accumulated LNAPL and water from the sump, and to pump it to a storage tank for separation. A vacuum pump created suction, which increased the recovery rate. The effect of vacuum assistance on product recovery rates is illustrated in Figure 12.20.

The main refinery LNAPL recovery system consists of 11 single-pump 4- and 6-in.-diameter production wells. Recovery wells are constructed of slotted PVC screens and casing. For the recovery of LNAPL, 4-in.-diameter submersible pumps have been installed. The submersible pumps installed within the recovery wells are of stainless steel construction. These pumps were not adversely affected by water, hydrocarbon, or minor amounts of fine sand and silt produced by the recovery wells.

Because of the low transmissivities of the formations being pumped, the pumps frequently broke suction. Current fluid production rates ranged from as low as 0.4 up to 20 gpm. The total fluid production from the system was approximately 50 gpm. Produced fluid was piped via overhead pipelines to a tank that was modified to serve as a "gunbarrel" oil–water separator. The water was constantly drained from the separator tank through a "water leg." The water leg resulted in a fairly constant hydrocarbon–water interface level inside the tank while allowing the water to drain from the tank as it accumulates. The free hydrocarbon was drained off the top of the water, entering a pipe set above the static interface. The hydrocarbon product was transported to a storage tank prior to reprocessing.

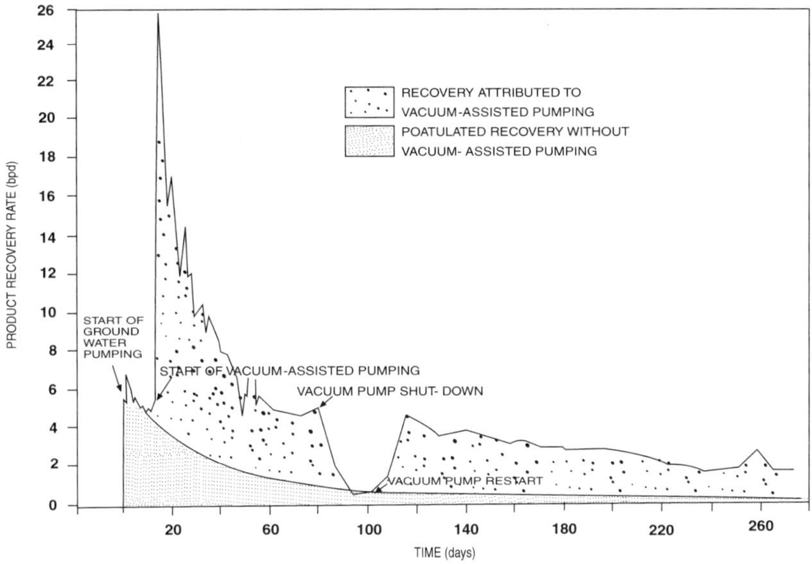

FIGURE 12.20 Effectiveness of vacuum assistance on recovery of LNAPL.

The water that is currently being produced by the recovery system is collected and treated on site. The wastewater treatment plant reduced the chemical concentrations to desirable levels and then discharged the treated water into an existing county sanitation district industrial sewer. In this manner, all fluids produced by the recovery system were treated for disposal or reprocessed.

The LNAPL recovery program in the southwestern portion of the site was initiated in 1982. To date, the volume of LNAPL recovered is approximately 185,000 barrels, which includes about 2000 barrels from the perched pool and 183,000 barrels from the water table pool. About 28,000 barrels have been recovered to date from beneath the main portion of the refinery.

The two-pump recovery system used at the site utilizes injection wells to:

- Create a barrier to off-site migration;
- Enhance the hydraulic gradient toward free hydrocarbon recovery wells; and
- Provide for nonconsumptive water withdrawal.

The injection wells at the site have been utilized since 1982 for disposal of water generated from the hydrocarbon recovery system. The water is reinjected into the aquifer from which it was originally withdrawn so the quality of the receiving formation is not adversely affected. This type of remediation strategy also allows less strain on the wastewater handling capabilities of the facility.

The injection well system has been expanded in conjunction with the operation of the recovery system. As a result of operational constraints and locations of injection wells in service, injection rates and volumes have varied. As of January

1989, approximately 1330 acre-ft of untreated groundwater has been pumped and reinjected. Pumpage of this water has coincided with the recovery of about 183,000 barrels of LNAPL.

Making some rough approximations of recovery without two-pump systems gives an indication of the magnitude of the injection system impact. From experience at other sites with similar subsurface hydrogeologic and geologic conditions and LNAPL occurrence, and where recovery has been attempted without the creation of a cone of depression, single-pump recovery systems (typically referred to as "skimming" or "pneumatic" recovery systems) have the following recovery performance histories. Initial recovery rates from 6-in.-diameter recovery wells may be about 8 to 15 barrels/day. After about 1 to 3 months the long-term sustained recovery will drop to less than 5 barrels/day. After 1 year of operation, recovery will be about 1 barrel/day or less. The radius of influence of these wells is minimal and difficult to identify.

From six such recovery wells, the volume of LNAPL recovered in 5 years might have been:

$$
\begin{aligned}
6 \text{ wells} \times 15 \text{ barrels/day} \times 90 \text{ days} &= 8100 \text{ barrels} \\
6 \text{ wells} \times 5 \text{ barrels/day} \times 270 \text{ days} &= 8100 \text{ barrels} \\
6 \text{ wells} \times 1 \text{ barrel/day} \times 4 \text{ years} &= \underline{8760 \text{ barrels}} \\
& 24{,}960 \text{ barrels}
\end{aligned}
$$

This would be about 159,000 barrels less than that actually recovered to date. Assuming that one skimming well would recover as much as 4200 barrels in 5 years (24,960 barrels/6 wells), then 44 such recovery wells would have been required to achieve the same volume recovered.

Although this is a simplistic analysis, it points out very clearly that recovery using two-pump systems is significantly more effective than by skimming alone. This conclusion is supported by data recorded for the two respective recovery systems. The average recovery rates over the first year of production range from 50 to 150 barrels/day of LNAPL and from 350 to 3000 barrels/day of water per two-pump recovery well. In comparison, one-pump recovery wells have averaged approximately 5 barrels/day of LNAPL and 140 barrels/day of water over the first year of production. The two-pump recovery system does have disadvantages including increased operation and maintenance costs, and plugging of injection wells reflecting an increase in microbial activity. Future endeavors are anticipated to concentrate on enhanced recovery of LNAPL, continued delineation, and subsequent aquifer rehabilitation of the dissolved hydrocarbon pool.

12.6 IMPORTANCE OF LITHOFACIES CONTROL TO LNAPL OCCURRENCE AND RECOVERY STRATEGY

A 245-acre crude oil refining, processing, and storage facility located adjacent to the site discussed in Section 12.5 is bordered by other refineries and aboveground tank farms, and changed ownership around 1991. The site was initially characterized by one large LNAPL pool, and an extensive LNAPL recovery system. An LNAPL recovery system was designed and installed in 1990 and 1991, and consisted of two primary components:

- An equally spaced 48-well interceptor network oriented north–south along the hydraulically downgradient western boundary that was installed primarily for containment purposes (although the adjacent property hydraulically downgradient also had significant volumes of LNAPL, the source of which derived on site with some off-site contribution); and
- A 20-well interior recovery system.

Although LNAPL occurrence was initially depicted as one large pool, the change in ownership initiated a reevaluation of the LNAPL system and the overall aquifer restoration program and strategy by another consultant. A review of all boring logs, well construction details, and facility history through historical aerial photographs and site records was performed. As of 1994, about 161,389 barrels of total fluids, of which about 1614 barrels were LNAPL, have been recovered. In a typical 3-month period, about 8994 barrels of total fluids, of which about 90 barrels were LNAPL, were recovered. In regard to overall efficiency, this equates to an average rate of 1% LNAPL to 99% groundwater being recovered, indicating, essentially, high costs for design, installation, and monitoring, with very poor efficiency and effectiveness.

An important parameter in understanding LNAPL distribution and development of a recovery strategy in unconsolidated deposits is understanding the sedimentary architecture, as discussed in Chapter 3. Initial development of lithofacies maps is fundamental to this understanding.

The site is underlain by a sequence of unconsolidated, stratified sedimentary deposits. The water table occurs at depths of 52 to 71 ft below the ground surface, with occasional perched zones at depths of about 52 to 64 ft below the ground surface. Review of boring logs and well construction details and performance allowed for reevaluation of the subsurface environment. The soils were subsequently characterized based on relative permeability and depositional environment (Figures 12.21 and 12.22) as follows:

- Relatively low permeability alluvial soils (silt, clay, and sandy clay), which corresponds with perching layers or aquitards. These soils were interpreted to be deposited in a low-energy, suspended load, fluvial-deltaic environment outside of channel zones.
- Relatively high permeability, coarse-grained, and well-sorted sandy soils with occasional shells. These soils were interpreted to be deposited in a fluvial tidal channel axis or deltaic reworked distributary mouth bar deposits.
- Relatively moderate permeability, poorly sorted silty sand.

LNAPL occurrence was determined to reflect several factors, including, but not limited to, lithofacies control over LNAPL migration and volume, timing, rate, and

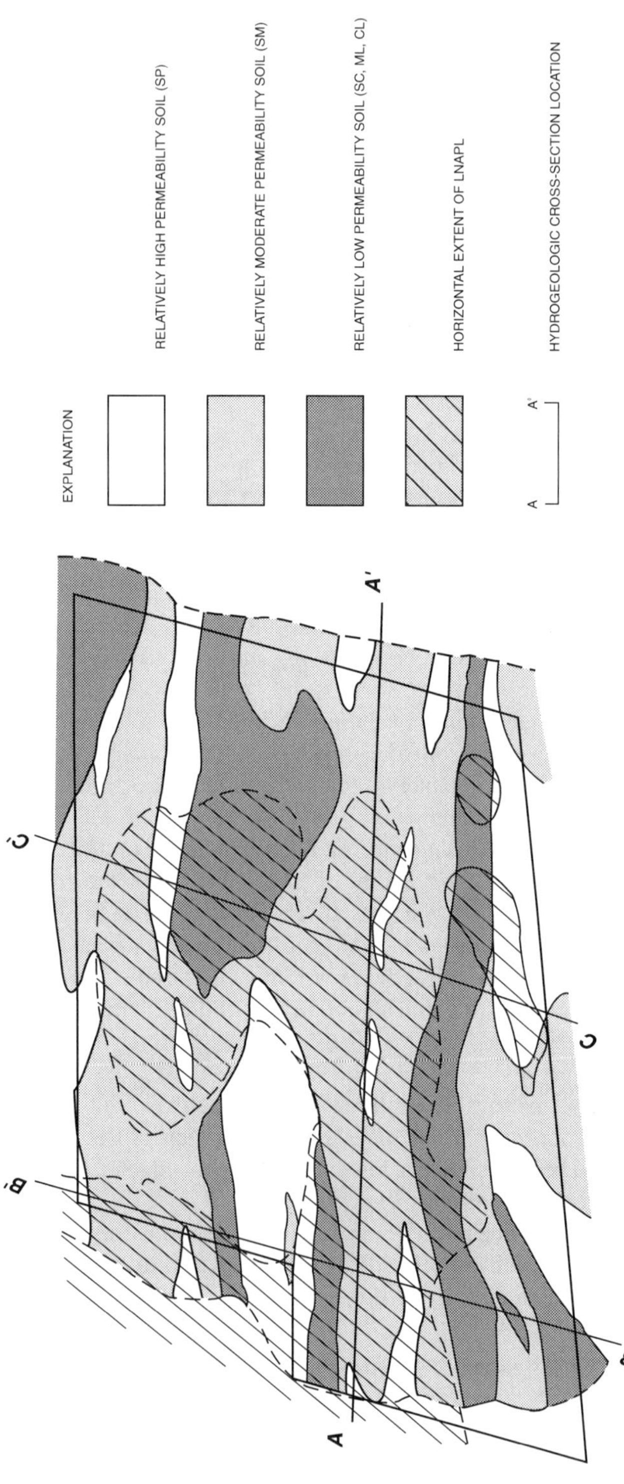

FIGURE 12.21 Horizontal view showing lithofacies control over LNAPL presence at the water table beneath a refinery site.

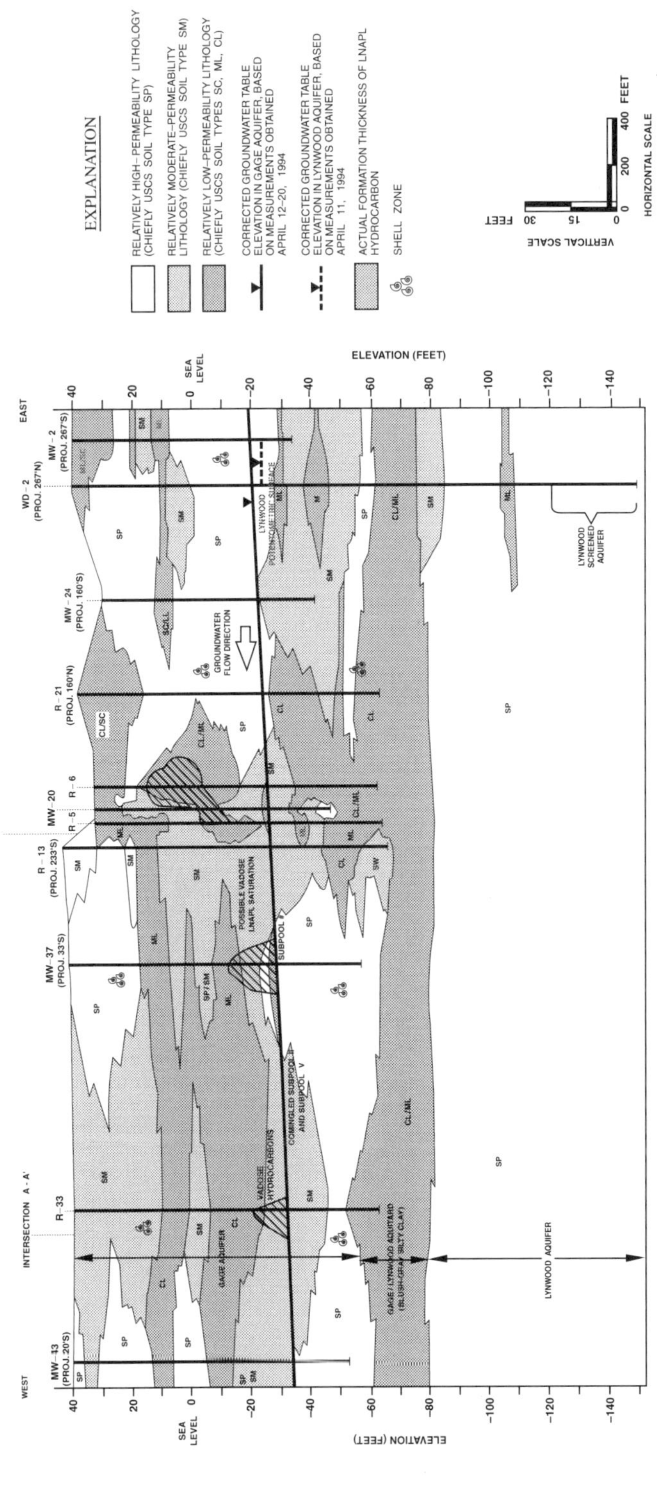

FIGURE 12.22 Hydrogeologic cross section showing lithofacies and LNAPL presence beneath a refinery site.

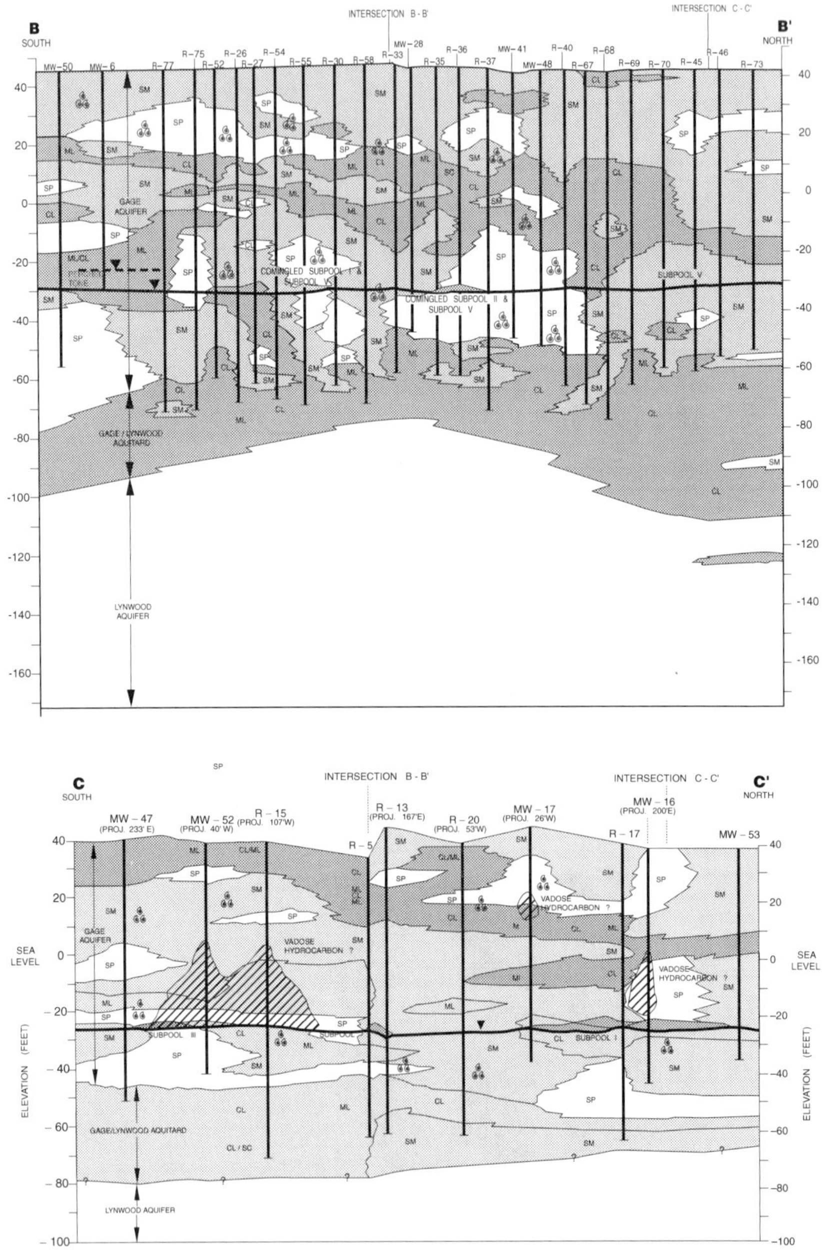

FIGURE 12.22 *(continued)*

location of release. Evidence of lithofacies contribution, in addition to forensic fingerprinting of select LNAPL samples, resulted in the demarcation of five distinct LNAPL subpools, some of which have commingled (Table 12.3). Some of these subpools were well defined by a linear zone of relatively low permeability facies, notably, between subpools II and III. More importantly, such maps provided a better understanding of why the system performed poorly, and allowed reevaluation and development of a more practical approach to LNAPL recovery and aquifer restoration, by significantly reducing the volume of LNAPL present by reconfiguration of the lateral extent and thickness of LNAPL, modifying which wells would serve as recovery wells vs. monitoring wells, and improving the ratio of total fluids to LNAPL recovered. This straightforward exercise also allowed the opportunity to reassess operational parameters and short- and long-term objectives.

12.7 REGIONAL LONG-TERM STRATEGY FOR LNAPL RECOVERY

California has a rich history of oil and gas exploration and development, dating back to 1876, the first year of commercial production. Comprising 56 counties, 30 counties produce oil and gas, 18 produce chiefly oil, while the remaining 12 produce predominantly natural gas. Hydrocarbon occurrence is predominantly oil south of 37° north latitude, with the major production in southern California within four sedimentary basins: the Los Angeles Basin, Ventura Basin, San Joaquin Basin, and Santa Maria Basin (Figure 12.23). Before 1900, the Los Angeles coastal plain comprised large Spanish land grants used primarily for agricultural purposes. With the discovery of oil, more than ten giant oil fields (ultimate recovery potential of 100 million barrels or more) have been found in the Los Angeles Basin and the Los Angeles coastal plain.

As a result, numerous refineries and associated aboveground bulk liquid storage tank farms and terminals were constructed during the early 1920s in close proximity to both petroleum production areas and the nearby shipping facilities of the Los Angeles Harbor. The majority of these facilities continued to expand their operations and areal extent through the late 1940s, while many others were initially isolated or located in moderate to heavily industrialized areas. However, with the encroachment of urban development, many are now in close proximity to densely populated light-manufacturing, commercial, and residential-zoned areas.

At present, 16 major refineries and 33 aboveground bulk liquid tank farms are situated on the Los Angeles coastal plain. The locations of these refineries and tank farms, and associated major pipeline corridors, are shown in Figure 12.24.

12.7.1 Regulatory Framework

Several refineries, tank farms, and other petroleum-handling facilities nationwide are included on the EPA 1996 National Priorities List, or are regulated under RCRA. Such facilities present several potential subsurface environmental concerns, reflecting approximately 80 years of continued operation. These concerns include:

TABLE 12.3
Summary of Reevaluated LNAPL Subpool Characteristics

Subpool Name	Subpool Location	LNAPL Product Type	API Gravity	Maximum LNAPL Thickness (ft)		Est. LNAPL Subpool Area (acres)
				Apparent	Actual	
I	North central part of the refinery	Naphtha and kerosene mixture	49.5–58.7	12.01	2.92	45.0
II	Roughly centered between the northern and southern site boundaries, upgradient limit just east of the central portion of the site, and extending downgradient to the western boundary	Kerosene, gasoline, and light gas oil mixture	35.9–58.3	13.50	2.58	40.9
III	Adjacent to the middle portion of the southern boundary of the site	Naphtha, kerosene, and light gas–oil mixture	37.7–55.3	14.06	3.40	7.2
IV	Adjacent to the eastern portion of the southern boundary of the site	Light gas–oil	31.3–32.8	13.51	1.76	1.3
V	Along the western boundary of the site, in the areas of the ARCO southwest tank farm and former Shell Ethyl Property	Light gas–oil	21.1–45.2	23.59	2.45	27.5

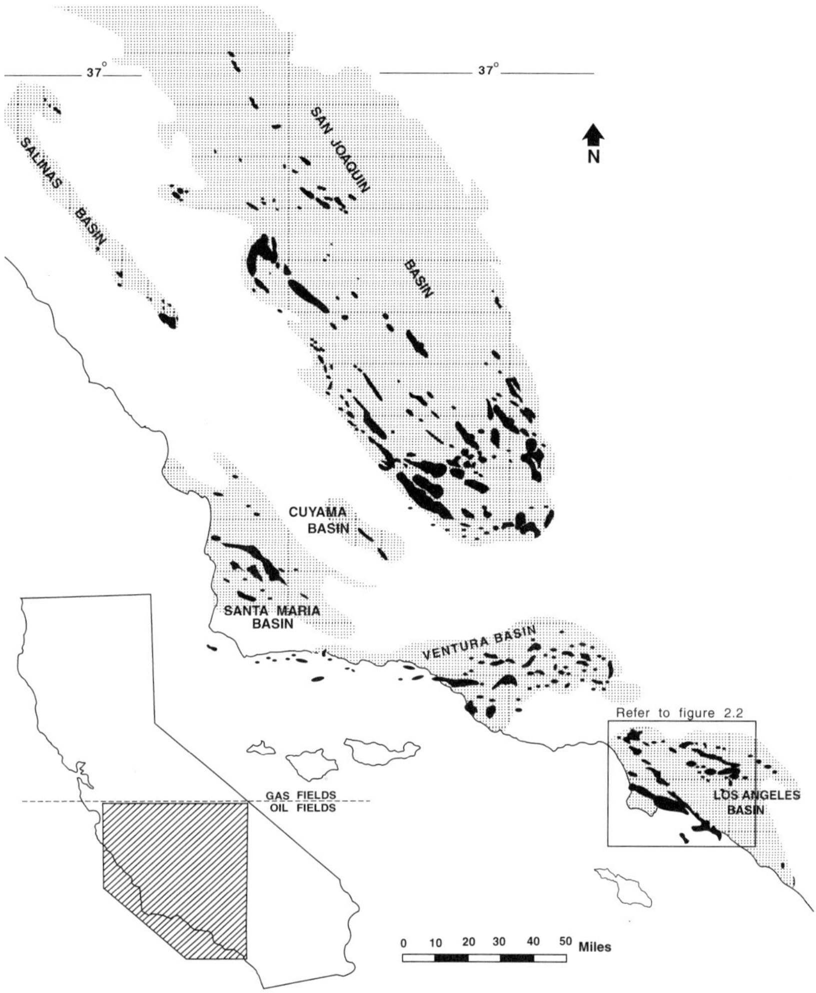

FIGURE 12.23 Major sedimentary basins and oil fields in southern California.

- Hydrocarbon-affected soil which could potentially be characterized as a hazardous material;
- Accumulation of hydrocarbon vapors posing a potential fire or explosion hazard;
- Dissolved hydrocarbons in groundwater which may adversely affect beneficial-use water-bearing zones or drinking water wells; and
- LNAPL hydrocarbon pools occurring as perched zones and generally overlying the water table which serves as a continued source of both soil and groundwater contamination.

In early 1985, oil droplets were evident on beach sands near high-priced beachfront real estate located just west of the Chevron refinery in El Segundo. The

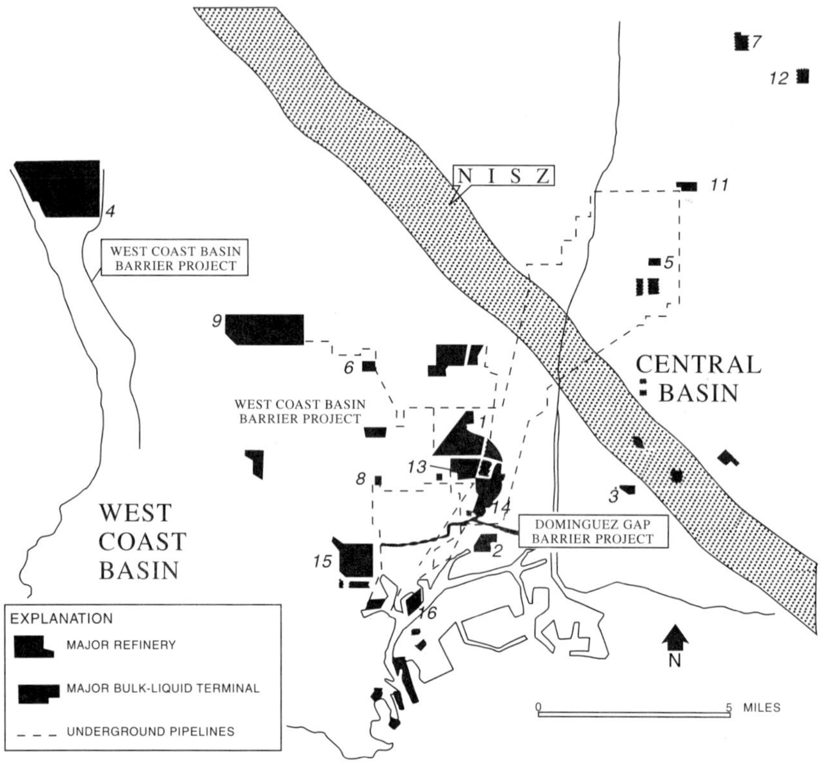

FIGURE 12.24 Location of major refineries, terminals, and pipeline corridors. (After Testa, 1992.)

immiscible hydrocarbons evidently leaked from the nearby refinery, migrated downward to the shallow water table, and then moved downward with the regional groundwater flow direction toward the ocean and beachfront properties. Elevated concentrations of hydrocarbon vapors presenting a potential explosion hazard were also detected at the bottom of a construction pit under excavation near the refinery and several adjacent homes. A significant pollution problem was recognized by the environmental regulatory community. At that time, an estimated (although exaggerated) 6,000,000 barrels of LNAPL hydrocarbon product were thought to exist beneath the majority of the 1000-acre, 80-year-old refinery.

The highly visible presence of petroleum and potential hazard to public health, safety, and welfare subsequently prompted a minimum of 16 oil refineries and tank farms to be designated as health hazards by the California Department of Health Services. This designation reflected both potential and documented occurrences of leaked hydrocarbon product derived from such facilities during 70 years of operations, which migrated through subsurface soils and accumulated as LNAPL hydrocarbon pools overlying the water table. Although several of these refineries were listed as hazardous waste sites, with assessment and remediation required under RCRA, the majority of such facilities fell under the jurisdiction of the Los Angeles

Region California Regional Water Quality Control Board (CRWQCB) Order No. 85-17, which was adopted in February 1985, reflecting the potential regional adverse impact on overall groundwater quality.

A total of 15 refineries were included under CRWQCB Order 85-17. By December 1985, this order was expanded to include aboveground bulk liquid storage facilities. Order 85-17 was the first regulatory mandate nationwide to address large-scale regional subsurface environmental impacts by the petroleum-refining industry. This order required, in part, the assessment of the subsurface presence of hydrocarbons and other associated pollutants which could affect subsurface soils and groundwater beneath and adjacent to such facilities. Specifically, the following items required addressing:

- Characterization of subsurface geologic and hydrogeologic conditions;
- Delineation of LNAPL pools including volume and chemical characterization;
- Implementation of LNAPL recovery;
- Overall aquifer restoration of dissolved hydrocarbons and associated contaminants; and
- Eventual remediation of residual hydrocarbons in soils.

At the time this order was issued, approximately six refineries had already commenced LNAPL hydrocarbon recovery programs, although such programs were limited in scope. In these cases, the large volumes of LNAPL present and the initial ease of hydrocarbon recovery proved economically favorable, since the recovered LNAPL product could be easily recycled and sold.

12.7.2 Hydrogeologic Setting

The Los Angeles coastal plain extends approximately 50 miles in a northwest–southeast direction, and it is roughly 15 to 20 miles in width between the Pacific Ocean and the base of the Puente Hills and Santa Ana Mountains. Encompassing approximately 775 square miles, the coastal plain is characterized by low relief and gentle surface gradients seaward. The low relief is interrupted by the Newport–Inglewood structural zone (NISZ), characterized by a northwesterly trending line of gentle topography extending roughly 40 miles in length. The NISZ is the major structural feature in the area and consists of echelon faults, anticlines, and domes, and is marked by a series of low hills and coastal mesas that are broken by six topographic gaps or low areas.

The Los Angeles coastal plain is divided into four principal groundwater basins. The Santa Monica Basin and West Coast Basin are located southwest of the NISZ; the Central Basin and the Hollywood Basin are situated northeast of the NISZ. The NISZ separates the West Coast Basin from the Central Basin. Hydrostratigraphic units underlying these portions of the West Coast and Central Basins where major refineries and bulk liquid terminals occur are shown in Figure 12.25. Salient hydrostratigraphic units beneath the area encompassing site No. 1, 13, and 14 (Figure 12.26) in the West Coast Basin are, in descending stratigraphic position, the Gaspur,

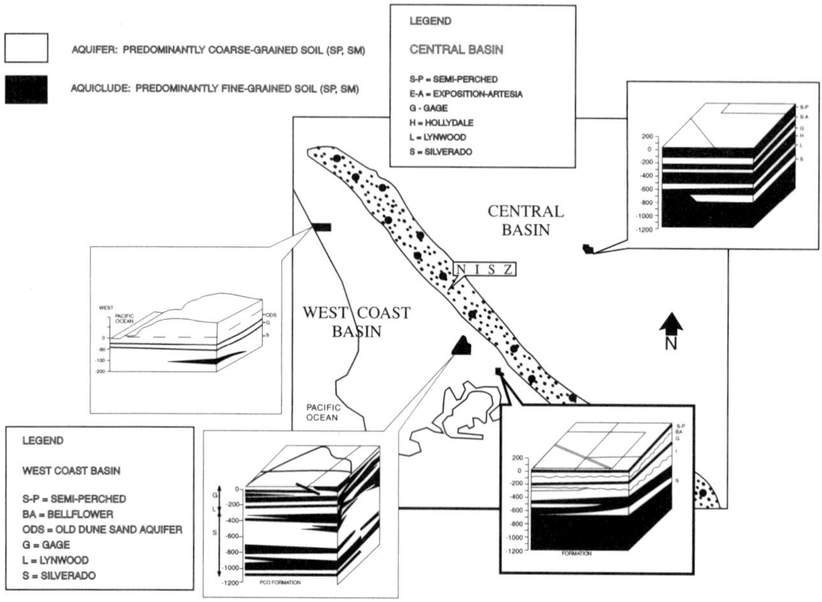

FIGURE 12.25 Subsurface hydrogeologic setting beneath major petroleum-handling facilities situated within the Central and West Coast Basins, southern California. (After Testa, 1992.)

Gage ("200-foot sand"), Lynwood ("400-foot sand"), and Silverado aquifers. The Silverado aquifer, which occurs within the Lower Pleistocene San Pedro Formation, is of very good quality and is the primary source of beneficial-use groundwater from the West Coast and Central Basins. Beds of relatively low permeability soils, which act as aquicludes or confining layers, separate these aquifers in some but not all places. In particular, angular unconformities developed along the flanks of folds within the NISZ form contact zones between younger aquifers and one or more older saturated zones.

Pertinent to the occurrence of LNAPL is the Gage aquifer, which is encountered under water table conditions. In this portion of the coastal plain, the hydrogeologic regime is influenced by numerous factors, all of which play a role in the remediation strategy. These factors include regulated pumpage from the Silverado aquifer as well as continued artificial recharge into the Gaspur, Gage, and Lynwood aquifers through injection wells associated with the Dominguez Gap Barrier Project and seawater intrusion. The NISZ also has a significant, multifaceted impact on groundwater occurrence, quality, and usage.

The area encompassed by site No. 4 (see Figure 12.25) is situated near the western margin of the West Coast Basin. This portion of the coastal plain is underlain by sand dune deposits of Holocene age, and marine and continental deposits of Pleistocene age. In descending stratigraphic order, the key hydrostratigraphic units include the Old Dune Sand aquifer, the Manhattan Beach aquiclude, the Gage aquifer, the El Segundo aquiclude, and the Silverado aquifer (see Figure 12.25). Pertinent to the occurrence of LNAPL hydrocarbon is the Old Dune Sand aquifer, which is

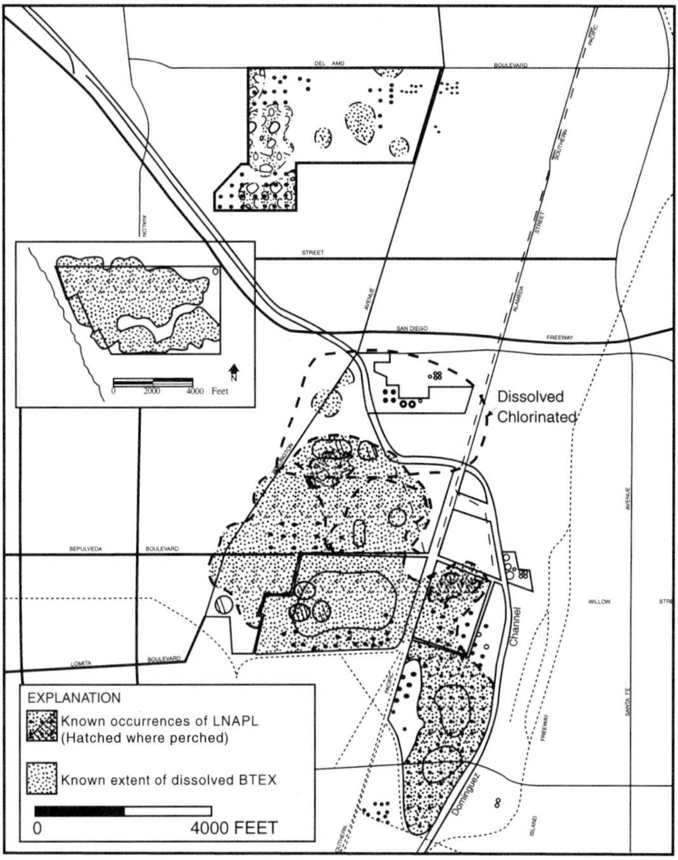

FIGURE 12.26 Spatial distribution of LNAPL beneath portions of the West Coast Basin. (After Testa, 1992.)

encountered under water table conditions. This portion of the West Coast Basin is also influenced by artificial recharge as part of the West Coast Barrier project.

Site No. 3, situated in the southern portion of the West Coast Basin, is underlain by, in descending stratigraphic position, the Semi-perched, Gage, Lynwood, and Silverado aquifers (see Figure 12.25). Most pertinent to LNAPL hydrocarbon occurrence is the Semi-perched aquifer, which is encountered under water table conditions.

Site No. 5, 7, 11, and 12 are situated in the Central Basin northeast of the NISZ (see Figure 12.25). This area is generally underlain by shallow perched and semi-perched aquifers, and nine distinct regional aquifers. In descending stratigraphic position, these aquifers are the Semi-perched, Gaspur, Exposition-Artesia, Gardena, Gage, Hollydale, Jefferson, Lynwood, and Silverado (see Figure 12.25). The Exposition-Artesia and deeper aquifers are the primary water-bearing zones tapped for municipal and industrial uses. The shallow aquifers, primarily the Semi-perched, but also to a large extent the Gaspur, in general are polluted with brines and industrial wastes, rendering them unsuitable for domestic use throughout the region. LNAPL in the Central Basin occurs chiefly within shallow, perched zones of limited lateral extent and within the Semi-perched aquifer, which is encountered under water table conditions. The deeper aquifers are generally of good water quality, although some saltwater intrusion and localized occurrences of contaminated groundwater have been reported within the Exposition-Artesia aquifer.

12.7.3 LNAPL Occurrence

The ubiquitous occurrence, areal extent, and estimated volume of major LNAPL hydrocarbon pools beneath these facilities situated on the Los Angeles coastal plain are shown in Figure 12.25. For purposes of this discussion, a pool is defined as an areally continuous accumulation of LNAPL. Two or more pools that have distinct differences in their respective physical and chemical properties are referred to as coalesced pools. Individual accumulations of relatively uniform product are referred to as subpools, since it is inferred that they have coalesced to form areally continuous occurrences. The occurrence of several pools and subpools at a particular site reflects releases from multiple sources at various times. The combined areal extent of these LNAPL pools is on the order of 1500 acres. The estimated cumulative minimum volume is on the order of approximately 1.5 million barrels; an estimated cumulative maximum volume is on the order of 7.5 million barrels. The discrepancy in LNAPL hydrocarbon volume reflects varying methodologies involved in the estimates.

The largest known subsurface accumulation of LNAPL hydrocarbon in the coastal plain area occurs beneath an active refinery (refer to site No. 4, Figure 12.25). Groundwater beneath this site occurs under unconfined water table conditions within the Old Dune Sand aquifer. The LNAPL hydrocarbon pool is elongated toward the northwest and encompasses approximately 690 acres in areal extent. Apparent LNAPL thickness in this pool ranges up to approximately 12 ft. The LNAPL pool comprises a variety of refined petroleum product types, including light oil, diesel fuel, gasoline, jet fuel, and reformat.

One of the more complex LNAPL-impacted areas is in the vicinity of sites No. 1, 13, and 14 (see Figure 12.25). Several major refineries, terminals, and pipeline corridors are located in this area. Both groundwater and LNAPL beneath this area occur within the Gage aquifer under both perched and water table conditions, at depths of 40 to 50 and 30 to 60 ft below the ground surface, respectively. Some facilities are underlain by several pools, which in turn represent two or more coalescent pools of diverse product types. The release of bulk liquid hydrocarbon product in this area has occurred over time via several pathways, including surface spillage, breached pipelines, corroded storage tank bottoms, and failed reservoir bottoms. Many of these occurrences have resulted from very slow releases over long periods of time.

Source identification of dissolved hydrocarbons in groundwater and refined LNAPL is difficult due to several factors. These factors include numerous microbiological, chemical, and physical processes, the complex historical industrial development of the area, numerous property ownership transfers, and the close proximity of several crude and petroleum-handling facilities including clusterings of refineries, bulk storage tank farms, underground pipelines in industrialized areas, and numerous underground storage tanks associated with gasoline service stations in less-industrialized areas. Several methodologies have been used to identify not only the crude and/or refined product type, but also the brand, grade, and, in some instances, the source. These methods have included routine determination of API gravity, development of distillation curves, and analysis for trace metals (notably, organic lead and sulfur). More-sophisticated methods have included gas chromatography, statistical comparisons of the distribution of paraffinic or *n*-alkane compounds of specific molecular weight, and determination of isotopic ratios of carbon and hydrogen ($^{13}C/^{12}C$ and $^{2}H/^{1}H$ or D/H, respectively) for lighter gasoline-range fractions, and $^{15}N/^{14}N$ and $^{34}S/^{32}S$ ratios for the heavier petroleum fractions.

12.7.4 LNAPL Hydrocarbon Recovery

Most major petroleum-handling facilities have implemented aquifer restoration programs, with some facilities further ahead in their respective programs relative to others. Effective recovery of LNAPL hydrocarbon can be accomplished by several means. Where perched or water table conditions are deep (greater than 30 ft below the ground surface), the approaches to LNAPL hydrocarbon recovery have conventionally included both single- and two-submersible-pump-type systems. Under shallow perched water table conditions, as commonly experienced at port facilities or shallow perched conditions, passive skimmer-type systems have also been utilized. In low-yielding formations with minimal LNAPL hydrocarbon presence, simply bailing the LNAPL periodically has been performed. More conventional one- and two-pump submersible systems have been utilized within the water table aquifer.

Overall efficiency and effectiveness of the several LNAPL hydrocarbon recovery programs in operation in the Los Angeles coastal plain are limited by numerous factors. The most important is the inability to handle and treat coproduced waters which, depending upon the size of the facility and the scale of the recovery program, can potentially exceed 1000 gpm. This reflects antiquated wastewater treatment systems (or lack thereof) at the facilities, in conjunction with existing systems functioning at capacity, cutbacks in the volume of hydrocarbon-affected groundwater that the Los Angeles County Sanitation District is willing to accept, and increased regulatory pressure against injection without treatment.

The overall efficiency and effectiveness of LNAPL hydrocarbon recovery programs have also been impacted by other factors. These factors include limitations associated with LNAPL recovery from low-yielding formations, inability to gain access to optimal recovery and off-site locations, coproduced water-handling constraints, and economic constraints.

12.7.5 Regional Long-Term Remediation Strategy

The large extent of petroleum-affected groundwater, the complexities of the regional hydrogeologic setting, and the numerous factors involved in developing short- and long-term remediation strategies have resulted in an avoidance of a site-specific approach to formulating regional objectives in the Los Angeles area groundwater basins and prioritizing these objectives. The primary objective is the protection of the Silverado aquifer and other regional beneficial-use aquifers.

Regulatory emphasis has recently focused on the following issues:

- Complete delineation of the areal extent of LNAPL, and the horizontal and vertical extent of its respective dissolved constituents;
- Enhance efficiency and effectiveness of existing LNAPL recovery programs;
- Development of analytical and numerical groundwater models to (1) predict the fate and transport of LNAPL and its dissolved constituents; (2) provide more reliable LNAPL volume determinations; and (3) enhance design for optimal groundwater cleanup strategies;
- Identify the extent and source(s) of DNAPLs and other organic compounds (e.g., chlorinated solvents, pesticides, etc.);
- Develop a source elimination program to detect leakage from aboveground tanks and underground piping in the early stages of LNAPL release; and
- Review current technologies and develop a soil cleanup strategy consistent with the depth and quantity of contaminants present.

In response to the issues outlined above, horizontal delineation of LNAPL is near completion, with off-site delineation required at a few localities. The horizontal and notably the vertical delineation of dissolved hydrocarbon constituents have yet to be addressed for most of the affected sites. Despite the large volumes of LNAPL released over time, the lateral extent of dissolved hydrocarbon constituents, notably benzene, toluene, ethylbenzene, and total xylenes (BTEX), have been limited to distances of about 200 ft or less hydraulically downgradient of most of the larger LNAPL pools at those sites where plume delineation has been addressed. The vertical extent of these constituents has not been fully ascertained. Delineation programs to address the vertical extent of dissolved hydrocarbon constituents in groundwater are currently being implemented at some sites. Enhancing the efficiency and effectiveness of existing LNAPL recovery programs is more difficult in nature, notably for the larger LNAPL recovery programs. The primary factor limiting most recovery programs is the inability of the facilities to handle the generated coproduced water.

Certainly, if protection of the Silverado aquifer is the primary objective, and the water table aquifers (i.e., Gage, Old Dune Sand, or Semi-perched aquifers) are essentially a write-off, then the intervening aquifer(s), although not used for water supply, may serve the purpose of "guardian" aquifers.

Two of the larger LNAPL hydrocarbon occurrences, site No. 1 and 4 (see Figure 12.23), formerly reinjected coproduced groundwater into generally the same hydrostratigraphic zone from which it is withdrawn; site No. 1 reinjected without treatment into the Gage aquifer, whereas site No. 4 reinjected into the Old Dune Sand aquifer. Because of the presence of dissolved hydrocarbons, notably benzene, in the coproduced water that is typically returned to the aquifer during LNAPL recovery operations, immediate application of the EPA toxicity characteristic rule may result in classification of the reinjected water as disposal of a hazardous waste. This, in turn, would terminate use of UIC Class V wells (which many of these operations currently use). These wells would then be automatically reclassified as UIC Class IV wells, which would be prohibited under RCRA. An extension was given by the EPA to allow oil companies time to comply with this rule so that LNAPL recovery efforts involving the return of benzene-affected groundwater via reinjection wells will not be readily interrupted. Although this relieves certain facilities from immediately complying with this ruling, many of these LNAPL hydrocarbon recovery programs will continue for tens of years; thus, despite the extension, the effects of future regulatory requirements and restrictions are uncertain.

Only rudimentary modeling has been performed to date to predict the fate and transport of dissolved hydrocarbons in groundwater over time. In addition, much uncertainty exists concerning the actual LNAPL volume present. Other organic compounds, notably chlorinated hydrocarbons, are of significant concern due to their high mobility in groundwater systems. However, these constituents are not typically derived from petroleum-handling facilities and are inferred to be derived from old landfill cells and chemical-related industry-type sources nearby. In any case, assessment of the occurrence, and lateral and vertical distribution of chlorinated hydrocarbons in the area, has been limited.

Petroleum-affected soil and groundwater primarily reflect historic releases although present contributions also exist at some sites. Source elimination programs are thus currently being developed on a site-specific basis. The objectives of this program are to assure that the majority of releases are historical in nature, and to identify current releases. This program in general will include:

- Periodic piping testing and replacement, as appropriate;
- Eventual conversion of underground piping to aboveground piping;
- Periodic inspection of surface and/or aboveground storage tanks; tanks will be periodically taken out of service for inspection and repaired as needed; double bottoms are also being installed, as appropriate; and
- Testing, removal, and replacement, if appropriate, of underground storage tanks.

Cleanup strategies for hydrocarbon-affected soil will most likely be the last issue to be mandated from a regulatory perspective and certainly the most difficult technically to address. This difficulty reflects the large, deep-seated volumes of residual hydrocarbon present, and the current lack of efficient, cost-effective methodologies for *in situ* remediation of residual hydrocarbons in low-permeability, fine-grained soils.

In summary, within the NISZ, structures such as folds and faults are critical with respect to the effectiveness of the zone to act as a barrier to the inland movement of salt water. A nearly continuous set of faults is aligned along the general crest of the NISZ, notably within the central reach from the Dominguez Gap to the Santa Ana Gap. The position, character, and continuity of these faults are fundamental to the discussion of groundwater occurrence, regime, quality, and usage. In addition, delineation and definition of aquifer interrelationships with a high degree of confidence are essential. The multifaceted impact of the NISZ is another aspect of the level of understanding required prior to addressing certain regional groundwater issues. A primary issue is which aquifers are potentially capable of being of beneficial use vs.

those that have undergone historic degradation. Those faults that do act as barriers with respect to groundwater flow, may, in fact, be one of several factors used in assigning a part of one aquifer to beneficial-use status as opposed to another. A second issue, based on beneficial-use status, is the level of aquifer rehabilitation and restoration to be required in association with the numerous LNAPL recovery and aquifer remediation programs being conducted within the Los Angeles coastal plain.

Overall understanding of the regional hydrogeologic conditions, including proper delineation of the relationship between the various aquifers, is essential to implementing both short- and long-term regional aquifer restoration and rehabilitation programs. The Silverado aquifer is recognized at present as the primary source for municipal supplies in the area. The NISZ, where many of the oil fields beneath the coastal plain are situated, has a significant multifaceted impact on groundwater occurrence, quality, usage, and remediation strategy. Present groundwater management programs include freshwater injection as part of the seawater intrusion barrier project, which tends to raise the water levels (or pressure head) in the major water-bearing zones, and regulated pumpage which is limited by court order, which tends to lower the water levels within the Silverado aquifer (and Lynwood aquifer). Future land-use and site-specific concerns such as wastewater handling will all play a role in the approach to remediation; thus, remediation strategies need to be based on a case-by-case basis as well as on the regional situation.

REFERENCES

Fryberger, J. S. and Shepard, D. C., 1987, *Reinjection of Water at Hydrocarbon Recovery Sites: International Symposium on Class V Injection Well Technology*: U.S. Environmental Protection Agency and Underground Injection Practices Council, 19 pp.

Henry, E. C. and Hayes, D., 1987, Practical Approach to Hydrocarbon Recovery at Marine Terminals: In *Proceedings of the National Water Well Association and American Petroleum Institute Conference on Petroleum Hydrocarbons and Organic Chemicals in Groundwater: Prevention, Detection and Restoration*, November, pp. 75–90.

Henry, E. C., Testa, S. M., Hayes, D., and Hodder, E. H., 1988, Advantages of Suction Lift Hydrocarbon Recovery Systems: Application at Three Hydrogeologic Environments in California: In *Proceedings of the National Water Well Association FOCUS Conference on Southwestern Groundwater Issues*, March, pp. 487–502.

Testa, S. M., 1992, Groundwater Remediation at Petroleum-Handling Facilities, Los Angeles Coastal Plain: In *Engineering Geology Practice in Southern California* (edited by B. W. Pipkin and R. J. Proctor), Association of Engineering Geologists, pp. 67–79.

Testa, S. M. and Paczkowski, M., 1989, Volume Determination and Recoverability of Free Hydrocarbon: *Ground Water Monitoring Review*, Winter, pp. 120–128.

Testa, S. M., Henry, E. C., and Hayes, D., 1988, Impact of the Newport–Inglewood Structural Zone on Hydrogeologic Mitigation Efforts — Los Angeles Basin, California: In *Proceedings of the National Water Well Association FOCUS Conference on Southwestern Groundwater Issues*, pp. 181–203.

Testa, S. M., Baker, D., and Avery, P., 1989, Field Studies in Occurrence, Recoverability, and Mitigation Strategy for Free Phase Liquid Hydrocarbon: In *Environmental Concerns in the Petroleum Industry* (edited by S. M. Testa), Pacific Section of the American Association of Petroleum Geologists Symposium Volume, pp. 57–81.

13 Site Closure

"All hydrocarbons naturally degrade — the only question is how much of our financial resources need be expended to enhance the process."

13.1 INTRODUCTION

It is common knowledge that all hydrocarbons naturally degrade in the subsurface environment, albeit at various rates under a variety of subsurface conditions. This simple fact is well documented in the literature, and has been previously mentioned in this book. Although biodegradable, petroleum hydrocarbons are the focus of cleanup efforts at thousands of sites throughout the United States, with considerable national effort being spent to accelerate cleanup of compounds which eventually degrade. Natural attenuation, as we commonly refer to this naturally occurring process, has been rapidly growing in momentum as an alternative to the high cost and limitations of engineered solutions in efforts to meet regulatory cleanup standards for groundwater. Our understanding of relative permeability, developed out of the petroleum industry during the 1930s, clearly demonstrated that not all petroleum hydrocarbons in the subsurface can be or will be mobilized, and that some appreciable amount of residual hydrocarbons will remain in the subsurface, regardless how innovative we try to be, and that such presence will likely exceed regulatory limits in many cases. This understanding is exemplified even more by the limitations of conventional pump-and-treat aquifer restoration strategies (discussed in Chapter 9), and the limitations posed by the presence of hydrocarbon-impacted low-permeability soils.

Natural attenuation is a remedial strategy not to be confused with "No Further Action." Perceived as an effective "bridge" between active remediation and no further action, natural attenuation provides a mechanism to achieve site closure at many sites where elevated levels of petroleum hydrocarbons and other organic compounds persist. Presented in this final chapter is discussion of biodegradation including aerobic and anaerobic reactions, fermentation and methane formation, intermediate and alternate reaction products, and biodegradation rates. This is followed by discussion of natural biodegradation, enhanced biorestoration, and conventional field procedures. From this basis, natural attenuation is further defined in the context as part of a remedial strategy to achieve site closure. As part of this discussion, review of the site-specific parameters to consider for evaluating the appropriateness of natural attenuation as a remedial alternative for soil and groundwater impacted by petroleum hydrocarbon (and other organic compounds), is presented. Several case histories are also presented where the use of a natural attenuation site closure strategy has been successfully implemented.

13.2 BIOLOGICAL DEGRADATION

Research into the fate of hazardous contaminants has included extensive studies of microbial activity in the subsurface. Work at contaminated aquifer sites has demonstrated the existence of a wide variety of subsurface microorganisms, which are metabolically active and often nutritionally diverse. Modern methods of detection have confirmed that it is not uncommon to have uniform population densities of 10^6 to 10^8 cells per gram of dry soil in uncontaminated, permeable shallow aquifers. The presence of large numbers of microorganisms is dependent upon a number of factors, including the surface area available for attachment, oxygen availability, and nutrients to support their growth. Unconsolidated sediments generally have large populations, while subsurface conditions with confining layers or limited permeability will typically have fewer organisms.

Many subsurface microorganisms are capable of degrading a variety of contaminants. Typical components of petroleum products (i.e., benzene, toluene, xylene, naphthalene, and simple aliphatic compounds) have been found to be degraded aerobically in unconfined aquifer settings. When petroleum hydrocarbons and other organics are introduced into previously uncontaminated soils, a period of acclimation may be required before the indigenous organisms are able to begin the degradation process. This period may extend for a few days to weeks or months, depending upon the variety of organisms present, availability of oxygen, nutrients, concentration of contaminant, and presence of compounds toxic to microbes.

Both aerobic and anaerobic activity has been reported; however, the majority of completed restorations have been under aerobic conditions. Successful continuance of aerobic biodegradation, even in the presence of adequate oxygen, may be limited by insufficient availability of essential nutrients, such as nitrogen and phosphorus, or adverse pH level. Concentrations of a particular contaminant must be within certain site-specific limits. If the concentration is too low, not enough carbon is available to sustain metabolic activity; conversely, if the concentration is too high, it is often toxic to the microorganisms. A proper balance of organism types, oxygen, nutrient availability, temperature, appropriate soil conditions, and degradable contaminant can result in a rapid and efficient aquifer restoration.

In the diverse, complex environment of the subsurface, microbes play a key role in the recycling of chemical building blocks where they interact with water (including dissolved ions), organics, gases, and minerals. The central position that microbes occupy in the balance of these end members is presented in Figure 13.1.

A bacterial classification system that is based on the nutrient source that provides energy for growth and reproduction is presented in Figure 13.2. Phototrophs derive their energy from sunlight and live where light is usually available. They are not active in the subsurface and are not major restorers of contaminated sites. Chemotrophic bacteria derive their energy from chemical reactions. Lithotrophic bacteria rely on inorganic chemical reactions. These fairly rare bacteria live in extreme environments, and are not considered important to remediation. Organotrophic bacteria derive their energy from organic oxidation, in one form or another. Heterotrophic bacteria use organic matter as a carbon source, whereas autotrophic bacteria use carbon sources such as CO_2 or CH_4.

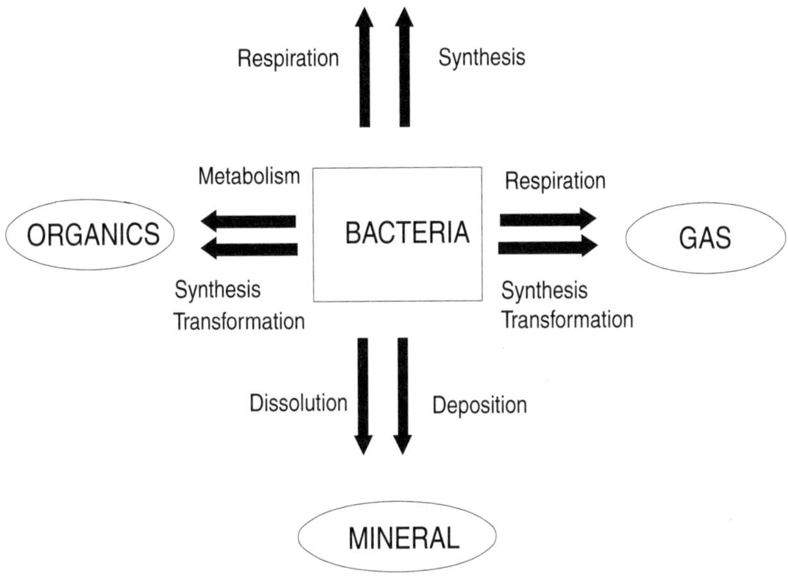

FIGURE 13.1 Microbial interactions in an aquifer system.

Heterotrophic and autotrophic bacteria are important participants in the restoration industry. Both types are indigenous to almost every site. The subsurface environment includes many thousands of species of microbes, which act in harmony to support each other. Waste products from one group become nutrients for another. When free oxygen is depleted, anaerobic activity increases. Thus, it is often convenient to consider microbiological activity as a series of processes resulting from bacterially mediated oxidation–reduction reactions.

Oxidation is defined as the reduction of electron state by addition of oxygen or removal of electrons. Thermodynamic balance requires balance. For every oxidation, there must be a corresponding reduction. Thus, for oxidation to occur, there must be a compound capable of receiving the transferred electrons. Electron acceptor compounds can include oxygen, sulfate, Fe^{3+}, phosphate, nitrate, CO_2, and certain organics.

For biodegradation to occur, everything that bacteria require for growth and reproduction must be available in the microenvironment in the immediate vicinity of the bacterium. The soil–aquifer system must provide water, attachment medium, a source of carbon, gas exchange, electron acceptor compounds, and nutrients. If any of the required items is not available, bacterial functions will be reduced or cease.

13.2.1 Aerobic Reactions

Where oxygen is readily available and all of the necessary components are present, existing bacteria reduce oxygen by the addition of protons (H^+ ions) to form water, carbon dioxide, and energy. Aerobic reactions are relatively rapid, often in terms of hours. The following general equation describes this activity:

Bacterial Classification

Source	Source Process	Description
Energy Source	Phototrophic	Uses sunlight or other light source
	Chemotrophic	Chemical reactions provide energy
	Lithotrophic	Inorganic chemical reactions
	Organotropic	Organic compounds
Carbon Source	Heterotrophic	Uses organic compounds (e.g., hydrocarbons) as carbon source
	Autotrophic	Uses inorganic compounds (e.g., carbon dioxide) as carbon source
Terminal Electron Acceptor (TEA)	Aerobic	Oxygen is TEA
	Anaerobic	A compound other than oxygen is TEA
	Facultative	Can function in the presence or absence of oxygen

Reaction	Typical Redox Potential (Eh Volt)	Electron Acceptor	End product
Oxygen Reduction	+0.80 → +0.20	O_2	H_2O, CO_2
Nitrate Reduction/ Nitrate Respiration	+0.40 → -0.20	NO_3^-	NO_2^-, N_2
Iron Reduction	+0.05 → -0.60	Fe_2O_3	Fe, O
Sulfate Reduction	-0.10 → -0.75	SO_4^-	HS^-
Methogenesis	-0.20 → -0.80	CO_2	CH_4

FIGURE 13.2 Bacterial classification based on energy source.

$$O_2 + \{CH_2O\} \rightarrow CO_2 + H_2O + \text{energy}$$

where $\{CH_2O\}$ is a general term for carbon–oxygen compounds. The limiting factor of aerobic activity is often the relatively low solubility of oxygen. At typical groundwater temperatures, oxygen is soluble only in the 6 to 10 ppm range. When dissolved oxygen reaches 0.5 ppm, most aerobic bacteria cease to function. Four parts (by weight) of oxygen are required to degrade one part hydrocarbon.

Additional oxygen can be supplied by natural processes (such as infiltrating rainwater) or by enhanced procedures including air sparging, addition of hydrogen peroxide (usable up to 200 ppm oxygen), or use of a slow-release oxygenating agent such as magnesium or calcium peroxide.

13.2.2 Anaerobic Reactions

Where oxygen is not available in free (or dissolved form), many anaerobic bacteria are capable of metabolism, by using alternate forms of electron acceptor compounds. An example of degradation is:

$$\{CH_2O\} + CaSO_4 \rightarrow CaCO_3 + H_2S + CO_2 + \text{energy}$$

Although biologically mitigated anaerobic reactions produce less energy and are significantly slower than aerobic reactions, they have certain advantages:

- Anaerobic processes circumvent the need for the addition of poorly soluble oxygen or oxygen-releasing compounds;
- Nitrate and sulfate are readily water soluble and cost-effective electron acceptors;
- Anaerobic processes are less likely to produce pore-clogging biomass; and
- The reactions require less process control.

Anaerobic bacterial reactions are much slower and produce only a fraction of the energy produced by aerobic activity. Alternate electron acceptors are generally consumed in a stepwise process:

$$O^{2-} > NO^{3-} > Fe^{3+} > SO_4^{2-}$$

13.2.3 Fermentation and Methane Formation

In settings that do not contain oxygen or alternative electron acceptors, methane production is favored. Methane production is the final step in anaerobic decomposition of organic matter.

Carbon in CH_4 can be derived from CO_2 or fermentation of organic matter. The process of methane generation is a complex multistep process, by several species of bacteria, which results in the following general reaction:

$$2\{CH_2O\} \rightarrow CH_4 + CO_2 + \text{energy}$$

Fermentation–methane reactions produce only approximately 20% of the energy produced by aerobic reactions.

13.2.4 Intermediate and Alternate Reaction Products

In the real world, very few microbiologically mitigated reactions follow the simplified reaction sequences discussed above (Figures 13.3 and 13.4). Generally, several microbial species interact to degrade the organic compounds in the soil or groundwater. During this process, intermediate products are formed, which may be acted upon further by other microbes. Sometimes solid matter is precipitated, gases are formed (which escape), or more soluble products result.

13.2.5 Biodegradation Rates

When an organic compound in liquid form is released into the subsurface, it migrates according to the processes as previously discussed. Within a short period of time (hours, days, weeks) the indigenous bacteria begin to adjust to the presence of organic compound, and attempt to use this new substrate (food) as carbon or an energy source. Assuming that the material is not initially deleterious to all of the bacteria

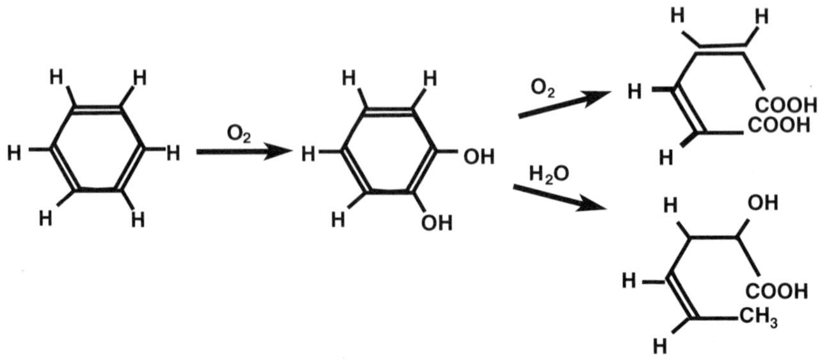

FIGURE 13.3 Possible degradation steps for an aromatic hydrocarbon.

species, the bacteria begin to utilize this material. The rate of bacterial colony growth where the nutrient, electron acceptors, and water supply is limited to that available at the time of a release (i.e., a closed system) is illustrated in Figure 13.5.

Initially, the cells require a period of time, the lag phase, to adjust to the altered environment. No cell multiplication occurs until the individual cells have almost doubled in size. After the adaptation is complete, the cells begin to reproduce exponentially. With each period of cell division, the population doubles. Some cells

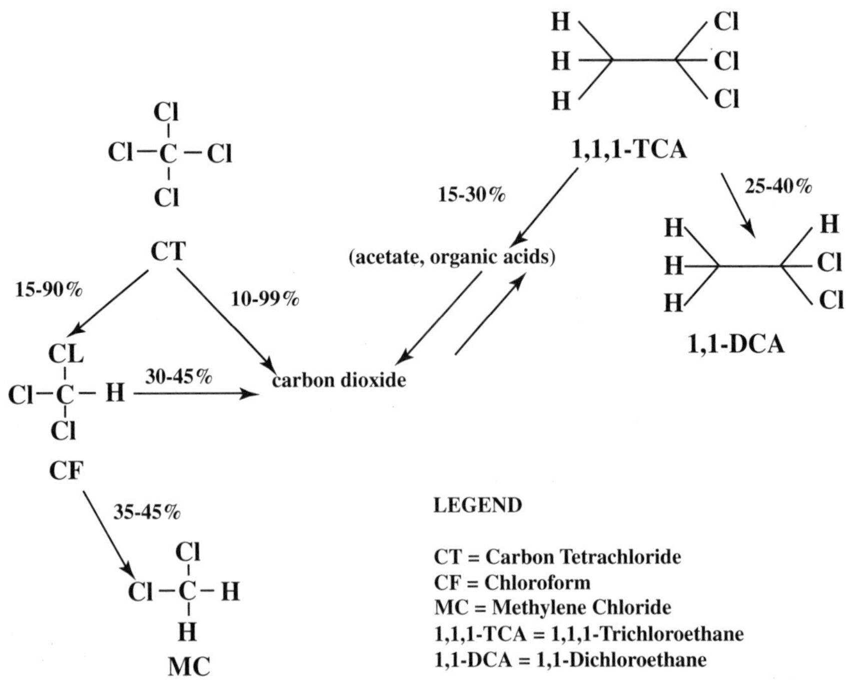

FIGURE 13.4 Possible degradation steps for a chlorinated hydrocarbon.

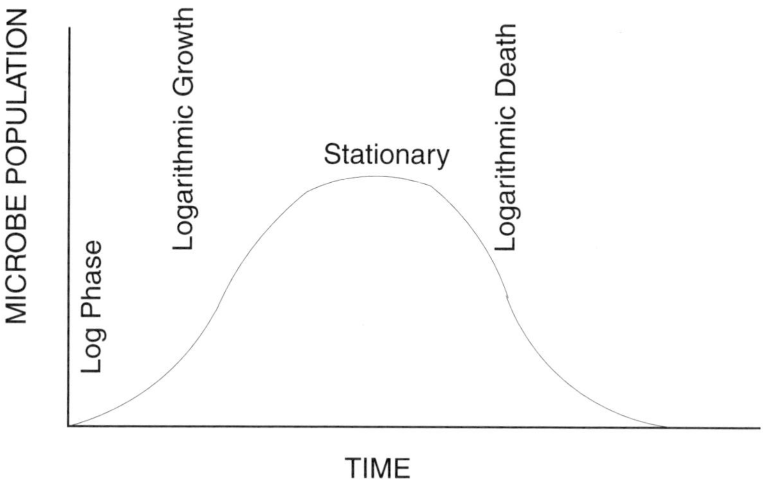

FIGURE 13.5 Characteristic bell-shaped bacterial growth curve.

can reproduce every 10 min, while others require a much longer period of time. The rate of growth depends upon the type of bacteria, availability of nutrients and water, and proper temperature.

The rapid growth phase is called the exponential growth phase. A cell with a generation time of 20 min could theoretically produce 2.2×10^{43} cells per day. Approximately 1 trillion cells weigh 1 g. If this rate of multiplication continued for one complete day, the cell mass would weigh more than the Earth. Obviously, this cannot occur. Growth is controlled by the food supply and the ability to dispose of waste products. As the growth slows and eventually stabilizes, the stationary phase is reached, where no net reproductive growth occurs.

Without renewal of the food supply, or adequate waste removal, the colony will enter the logarithmic death phase. The number of cells will decline as rapidly as it initially increased. Ultimately, the entire culture expires and the cycle is complete.

If an open system with renewal of substrate, nutrients, water, and electron acceptors can be supplied, the growth rate of the bacteria population is able to continue for an extended period of time until the remediation is complete. The Monod equation describes the type of bacterial growth that can be expected in an open system:

$$\mu = \frac{\mu_{max} * s}{K_s + S} \tag{13.1}$$

where μ = specific bacterial growth at a point in time, μ_{max} = maximum bacterial growth, S = available substrate at a point in time, and K_s = half velocity constant (the concentration of substrate at $0.5\mu_{max}$).

A typical bacterial growth according to the Monod prediction is depicted in Figure 13.6. As the concentration of substrate increases, the rate of microbial growth

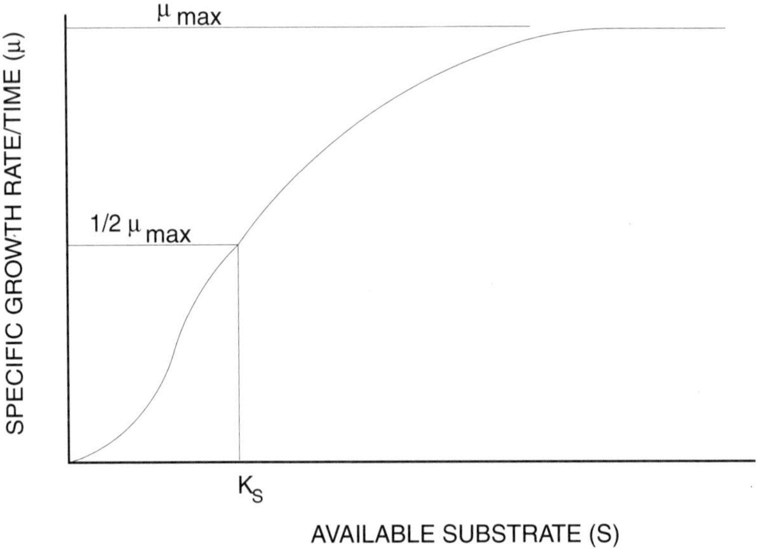

FIGURE 13.6 Bacterial growth in an open system.

increases. Once μ_{max} is achieved, the addition of more substrate does not result in a corresponding increase in specific bacterial growth. Other limitations, such as availability of space or electron acceptor transfer rate, limit population growth.

When monitoring the degradation of a set of chemicals that are being degraded by a community of bacteria, prediction of the rate of restoration can be a difficult task. In almost all of the dynamic (open) bacteria-mitigated degradation systems, the rate of chemical degradation is directly related to the concentration of the chemicals present. The power rate law describes the biodegradation of an organic chemical that is proportional to the concentration of the organic chemical:

$$-dC/dt = kC^n \tag{13.2}$$

where C = concentration of the organic chemical, k = biodegradation rate constant, and n = the order of the reaction (1, 2, etc.).

Many common organic chemicals are degraded at a first-order rate, which assumes that the rate of biodegradation depends only on the concentration of the contaminant and the rate coefficient (k). The rate coefficient can be calculated by the equation:

$$k = (2.303/t)\log(a/x) \tag{13.3}$$

where k = rate constant, 1/time, t = time between analyses, a = initial concentration (at first analysis), and x = concentration at second analysis.

The half-life of that chemical (under the operative conditions at that specific site) can be calculated. Half-life is the length of time required for the concentration of a chemical to be reduced to one half of the initial concentration. For example, if 3 mg/l of benzene dissolved in water has a half-life of 0.8 years, the concentration at that time will be 1.5 mg/l. After an additional 0.8 years, the concentration will be 0.75 mg/l. The half-life can be calculated as:

$$t_{1/2} = 0.693/k \tag{13.4}$$

where $t_{1/2}$ = half-life and k = rate constant.

Considerable care must be exercised in the selection of the rate constant for each constituent to avoid significantly over- or underpredicting actual decay rates. It is usually most effective to evaluate the results of several analytical events, several weeks apart. Half-lives for common petroleum hydrocarbons and organic compounds encountered in groundwater are presented in Table 13.1.

TABLE 13.1
Half-Lives of Common Petroleum Hydrocarbons and Organics in Soil and Groundwater

	Biodegradation Half-Life[a,b] (h at 25°C)	
Compound	**Soil**	**Groundwater**
Acetone	24–168	48–336
Anthracene	1,200–11,040	2,400–22,080
Benzene	120–384	240–17,280
1,2-Dichlorobenzene	672–4,320	1,344–8,640
1,1-Dichloroethane	768–3,696	1,344–8,640
1,2-Dichloroethane	2,400–4,320	2,400–8,640
1,1-Dichloroethylene	672–4,320	1,344–3,168
1,2-Dichloroethylene	672–4,320	1,344–69,000
DDT	17,520–1.4×10^5	384–2.7×10^5
Ethylbenzene	72–240	144–5,472
Methyl ethyl ketone	24–168	48–336
Methyl *t*-butyl ether	672–4,320	1,344–8,640
Naphthalene	398–1,152	24–6,192
1,1,1,2-Tetrachloroethane	16–1,604	16–1,604
1,1,2,2-Tetrachloroethane	10.7–1,056	10.7–1,056
Tetrachloroethylene	4,320–8,640	8,640–17,280
Tetraethyl lead	168–672	336–1,344
1,2,4-Trichlorobenzene	672–4,320	1,344–8,640
1,1,2-Trichloroethane	3,263–8,760	3,263–17,520
Trichloroethylene	4,320–8,640	7,704–39,672
Toluene	96–528	168–672
Vinyl chloride	672–4,320	1,344–69,000
Xylenes	168–672	336–8,640

[a] Data obtained from Howard et al. (1991).

[b] Half-lives reflect unacclimated aerobic conditions.

TABLE 13.2
Composite Soil Analytical Results

Compound	Initial Concentration (ppm)	Final Concentration (ppm)	Days Elasped	Half-Life (days)
TPH as Gasoline	7485	950	545	184
Benzene	50	4.5	668	192
Toluene	210	26	668	221

As an example of the use of half-lives, a buried pipeline carrying natural gas condensate situated between a natural gas–processing facility and a refinery ruptured under a small intermittent stream. This stream flows in a narrow, deep valley toward a farm pond several hundred feet downstream. The pond serves as an important watering hole for a cattle ranch. Following repair of the pipeline, it was important to determine how long the streambed would continue to contribute contaminants to the pond. Soil samples were periodically retrieved from select locations along the streambed at several time intervals, and analyzed for TPH-gasoline, benzene, and toluene, in addition to general water quality indicators (i.e., ion balance, conductivity, dissolved oxygen, etc.) as summarized in Table 13.2.

As natural biological attenuation occurs, the rate of degradation of the contaminants in the soil can be estimated by calculating their half-lives. At the end of each half-life period, the concentration of the contaminant will be reduced by one half. Beginning with a concentration starting time "0," the sequence then is $^1/_2$, $^1/_4$, $^1/_8$, … for each succeeding half-life time period assuming all processes continue uninterrupted.

The half-life for TPH as gasoline was determined as follows:

$$
\begin{aligned}
K &= \left(\frac{2.303}{t}\right)\log\left(\frac{a}{x}\right) \\
&= \left(\frac{2.303}{545}\right)\log\left(\frac{7485}{950}\right) = 0.004 \\
&= t_{1/2} = 0.693 / K \\
&= t_{1/2} = 0.693 / 0.004 = 184 \text{ days}
\end{aligned}
\tag{13.5}
$$

where K = biodegradation constant, t = time between samples in days, a = initial concentration, x = final concentration, and $t_{1/2}$ = half-life in days. The calculated half-lives, as summarized in Table 13.2 for 184, 192, and 221 days, indicated that biodegradation was occurring. Further testing after 1020 days confirmed the effectiveness of natural biological and related processes. Although these rates are rather slow, further testing of groundwater and soil noted significant concentrations of sulfides and methane interpreted to reflect a reducing or anaerobic environment.

13.3 NATURAL BIODEGRADATION

In some undisturbed subsurface systems, an equilibrium is established. Bacteria have acclimated to food sources, water availability, and electron acceptor types. The number and variety of microbial cells are balanced in this system. If the system is aerobic, the microbial activity continues at the rate of oxygen resupply. If the system is anaerobic, the rate of activity cannot exceed the accessibility of alternate electron acceptors. Generally, the subsurface (lower than the plant root zone) is relatively deficient in available carbon and electron acceptors. Under these normal semi-equilibrium conditions, a soil or aquifer system can consume organic materials within a reasonable range. When a chemical release is introduced into a well-established soil system, the system must change to react to this new energy source. The bacterial balance readjusts, in an effort to acclimate to the new carbon source.

In an initially aerobic system, the demand for electron acceptors (oxygen) may exceed the replaceable supply. If the concentration of contaminant is sufficiently high, it may be toxic to the bacteria. The result is often that aerobic activity only continues on the fringe of the plume where the contaminent concentration is relatively low and oxygen is more available. For example, biological consumption of benzene requires approximately three to four parts oxygen for every part hydrocarbon. If groundwater has a dissolved oxygen content of 4 ppm, the microorganisms can only degrade 1 ppm benzene before the oxygen is consumed. If oxygen recharge is not immediate, the system alters to alternative electron acceptors or the microbial activity may cease. Since the water solubility of most gasoline components is higher than oxygen solubility, further degradation cannot continue until the dissolved chemicals migrate (or disperse) to come into contact with oxygenated water.

The zones of degradation around a typical hydrocarbon plume are illustrated in Figure 13.7. The residual hydrocarbon mass is retained in place by capillary forces, and migrates by volatilization (in the unsaturated zone) and dissolution (in the saturated zone). Note the narrow aerobic margin where oxygen has not yet been depleted. Clean groundwater is present where abundant oxygen and nutrients (nitrate and phosphates) are available. The anaerobic core experiences little, if any, remedial activity. In this type of unaided natural system, the plume continues to expand until the rate of degradation at the fringe equals the rate of release from the residual core.

In a few well-documented instances, natural bioremediation has been remarkably successful. At these sites, sources of oxygen and nutrients have been available. A rise in the water table has brought contaminants to microbes and declines have enabled oxygen migration to the subsurface. The site of a former pumping station was underlain by a shallow alluvial water table aquifer, with seasonal fluctuations of 4 to 5 ft. At the time of demolition, the aquifer under this site was contaminated with LNAPL as gasoline and dissolved constituents such as BTEX. After a pumping recovery system had removed the recoverable LNAPL, the question of dissolved product removal was debated. During the next 2 years of monitoring, the concentration of dissolved hydrocarbons continually declined more rapidly than the calculated dispersion.

It was determined that rainwater infiltration from a fertilized wheat field overlying the aquifer provided sufficient excess nitrogen and phosphorus to support

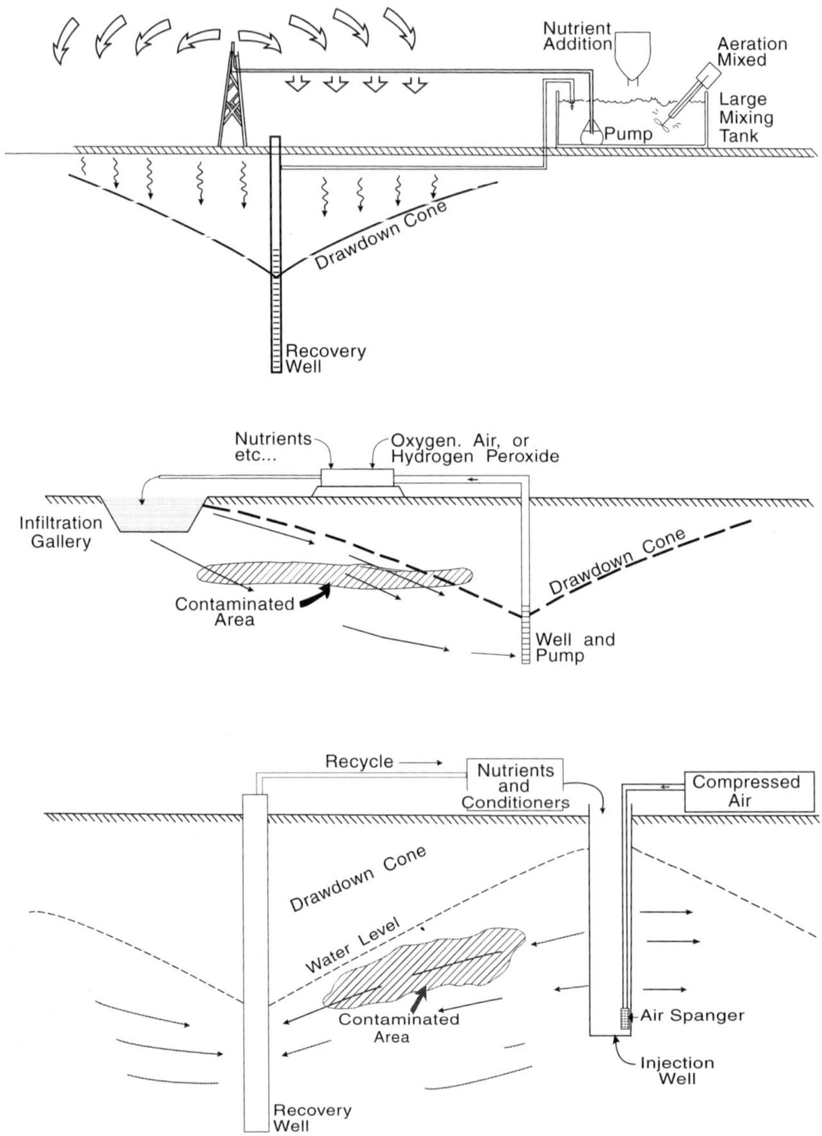

FIGURE 13.7 Schematic cross sections of enhanced biorestoration of unconfined aquifers.

the bioactivity. Rising and falling of the water table brought the food source (gasoline) to the biomass that was suspended above the water table. In addition, oxygen was readily available due to the high permeability and the shallowness of the water table (undoubtedly, some volatilization of the gasoline also occurred). This example illustrates that natural biological systems can adapt to chemical releases and remediate themselves, if assisted and not overloaded with new sources of contamination.

13.4 ENHANCED BIORESTORATION

Enhanced biorestoration is a means by which naturally occurring processes are deliberately manipulated to increase or enhance the rate of cleanup. Biological activity in the subsurface is controlled by the availability of one or more of the necessary metabolic requirements such as an electron acceptor or nutrient. Although electron acceptors are most often the limiting factor, inadequate availability of nitrogen, phosphorus, or micronutrients (such as potassium, copper, or even vitamins) can restrict optimum restoration. When the proper balance of these factors is established, the rate of chemical degradation is maximized.

Under most circumstances microorganisms are attached to solid soil particles and await the arrival of water, nutrients, and electron acceptors. When the biomass is above the water table, the dependence is upon migration of nutrients and diffusion of oxygen downward (or upward via capillary action). Bacterial colonies that develop below the saturated zone are dependent upon liquid phases for the delivery of necessary growth media.

The controls for the rate of biological activity can be summarized as follows:

- The stoichiometry of the metabolic process;
- The concentration of the required nutrients in the mobile phases;
- The advective flow of the mobile phases or the steepness of concentration gradients within the phases;
- The opportunity for colonization in the subsurface by metabolically capable organisms;
- The toxicity of wastes or a co-occurring material.

Enhanced biorestoration is thus the process of providing all the materials necessary for optimum degradation of the contaminant of concern. This process can include direct injection of air or nutrients, circulation of groundwater to distribute materials, addition of cometabolites (such as sugar) where additional sources of carbon and attached oxygen are needed, or injection of special inhibitors to limit precipitation of pore-plugging iron oxides.

Trial and error is often the typical procedure that is used to implement enhanced biorestoration. In simple cases of gasoline cleanups, these may be appropriate; however, when the chemicals involved are recalcitrant (difficult to degrade), toxic, or present in a complex geologic environment (i.e., low-permeability soil, lateral or vertical heterogeneities, etc.), enhanced biorestoration can be difficult, and risk assessment–type analyses may be more suited for a particular site.

13.5 FIELD PROCEDURES

During the past several years, the number of successful bioremediation projects has steadily increased. Remediation professionals have experimented with many techniques, some more effective than others. The following paragraphs discuss some of the more efficient procedures used to introduce nutrients and oxygen (or other electron acceptors).

A geologic and hydrogeologic investigation of the site is an important first step to evaluate the physical setting for water availability, pore sizes, direction and rate of groundwater flow, and other factors that are needed to assess the "best" enhancement procedure. LNAPL recovery is not noted to remove as much free-phase product as possible. Concurrently, a laboratory analysis is made to determine if the indigenous microflora can degrade the contaminant. Further testing of site materials is completed to evaluate the oxygen quantities needed, the proper balance of nutrients, optimum pH, and the presence of toxicants, and to detect other controlling factors. The primary decision is whether to treat the contaminated water above- or belowground. If the decision is made to treat the water aboveground, it is pumped from wells or sumps, treated in aboveground facilities (similar to a wastewater plant) and either returned to flush the aquifer through injection wells (infiltration galleries) or discharged to another disposal outlet. If the water and soil are in the subsurface, plans must be made to provide oxygen and nutrients in sufficient quantities to be effective.

A typical scenario of operation is to place recovery wells strategically at locations that will provide hydraulic containment of the contaminated area while pumping. Injection wells (or infiltration galleries) are located at the edge of the capture zone. Nutrients and oxygen supplied through the injection wells (infiltration galleries) migrate toward the recovery wells. Additional wells for oxygen feeding are often installed throughout the site.

Oxygen is usually the most difficult nutrient to supply. Water in the shallow, uncontaminated subsurface is normally found to contain between 3 and 4 ppm dissolved oxygen (DO). Since water can contain 8 to 10 ppm DO, this value can vary significantly. When contaminants are present, the oxygen is quickly consumed and often is reduced to < 1/2 ppm where aerobic biological activity ceases. Increased oxygen can be provided by air sparging, pure oxygen injection, addition of hydrogen peroxide, or introduction of metal peroxides (calcium or magnesium).

Air sparging is usually accomplished by forcing air through a short section of well screen or a porous ceramic block. As the air enters the aquifer, it passes through the soil pores and forms very small bubbles, which increase the transfer of oxygen by diffusion. When properly installed below the aquifer interface, the air is delivered under pressure, which forces it to disperse. Up to 10 ppm DO can be provided.

Hydrogen peroxide (H_2O_2) has been found to be a good source of oxygen. It can deliver much higher concentrations than the microflora can tolerate. Often an "inhibitor" is added to control the release of monatomic oxygen to a level of 200 ppm near the injection point and near 100 ppm in the zone of bioactivity. At concentrations >100 ppm the H_2O_2 may degas and form bubbles that plug the pore spaces. Iron in the soil can catalyze the decomposition of the H_2O_2. Additional phosphates may be necessary to precipitate the iron to limit its adverse reactions.

Commercial-grade pure oxygen has also been used at remediation sites, with concentrations of 40 to 50 ppm in the water being achieved. However, the use of pure oxygen has several disadvantages. Pure oxygen is relatively expensive, may bubble out of solution before microbes can use it, and may be an explosion hazard if not handled properly. Ozone (O_3) has also been used successfully at remedial projects. Where adequate industrial supplies of this oxidant are commercially available, it provides a good alternative to hydrogen peroxide.

Metal peroxides (both calcium and magnesium) have recently been found to provide a steady supply of oxygen. The peroxide compounds are available in finely ground powder form. Usually they are blended with inert compounds and supplied in fabric tubes that are sized for installation in wells. Once in contact with water, the oxygen releases from the powder and is transported by either diffusion or the groundwater. When properly blended, the rate of oxygen release can be controlled to meet the demand of the microbes. Sometimes the finely ground powder is injected directly into the aquifer through push tubes without the necessity of wells.

In the unsaturated zone, soil venting is an effective and inexpensive procedure to provide air. When the bulk of the product is held above the water table, supplies of air can be provided by the use of vacuum wells located in this zone. The flow of air is drawn either from the subsurface to the well (Figure 13.8a) or through vent wells (Figure 13.8b). The alternative benefit of this approach is that volatile portions of the product are removed by the lower-pressure operation. Soil vapor venting is particularly well suited to less permeable silt and clay soils.

Biological treatment of organic chemical contamination in the subsurface is an effective restoration procedure for many types of organic contamination (especially light petroleum products). Successful use of this technique relies upon optimization of a large number of factors including biological species, soil structure, contaminant concentration, contaminant chemical structure, nutrient balance , subsurface hydraulic conditions, economic considerations, and regulatory interaction. Because of the

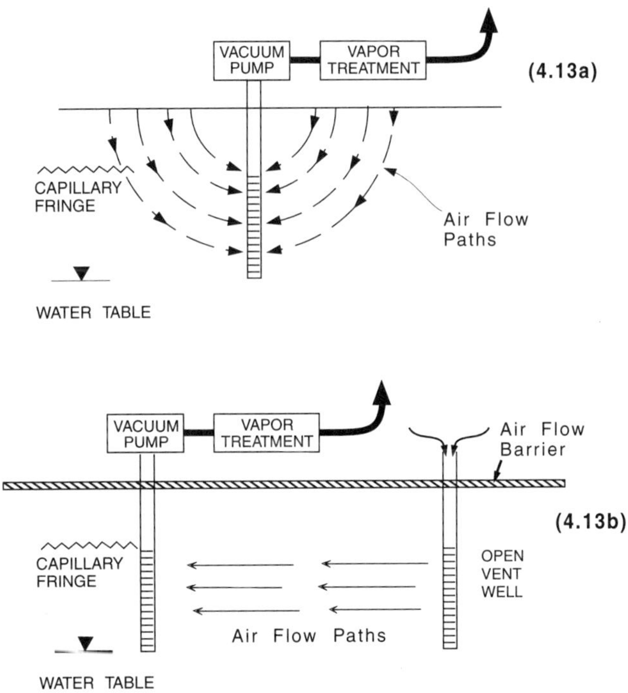

FIGURE 13.8 Schematic showing forced air soil venting.

complexity of this type of restoration, most effective project teams are composed of a variety of technical specialists, each practicing his or her profession.

13.6 NATURAL ATTENUATION AS A REMEDIAL STRATEGY

Natural attenuation is defined as unenhanced physical, chemical, and biological processes that act to limit the migration and reduce the concentration of contaminants in the subsurface. Aerobic bioremediation is the most important process in regard to petroleum hydrocarbons because this process is capable of destroying a large percentage of hydrocarbon contaminant mass under certain conditions. Destruction occurs as a result of bacteria oxidizing reduced materials such as hydrocarbons to obtain energy. Their metabolism removes electrons from the hydrocarbon donor via a number of enzyme-catalyzed steps along respiratory or electron transport chains to the final electron receptor, typically oxygen. The metabolized hydrocarbon ends up as a new cell mass with the by-products carbon dioxide, water, and the growth of new microorganisms.

Other significant natural attenuation processes include volatilization, dispersion, and adsorption. Volatilization, for example, can significantly reduce hydrocarbon-contaminant mass in soil. Light-end volatilized hydrocarbons are typically degraded in the vadose zone or slowly released to the atmosphere. Dispersion is the primary mechanism in groundwater for transporting soluble contaminants away from their source areas where little degradation occurs, to where they can be readily degraded, typically at the fringes of a plume where oxygen levels in groundwater are not depleted. Alternatively, adsorption can limit migration. Organic-rich soils such as peat may be effective adsorbers of petroleum, implying containment, which can be an effective way to manage low-risk sites. These secondary processes are not viewed as significantly important because they do not result in contaminant destruction, but rather in mass transfer. However, these secondary processes should still be considered as playing an important role in natural attenuation.

It was not too long ago when most state cleanup regulations did not address natural attenuation, and, in fact, the use of natural attenuation as a remedial alternative ran counter to the "anti-degradation" policies of many states. There was, and still is in some cases, a regulatory and public perception that the generator must use engineering and technology to clean up a particular site, that natural attenuation is no more than a loophole used to avoid the high cost of remediation. In the mid-1990s, agencies responsible for establishing cleanup levels for soils and groundwater at impacted sites began moving toward a position that, at least at some sites, less than pristine conditions will prevail where cleanup to regulatory levels (i.e., background levels, maximum contaminant levels, or more stringent levels based on cancer risks, etc.) was not economically or technologically feasible. In California, for example, these areas were referred to as "Non-attainment Areas" or Ground Water Quality Limited Zones. In such areas, soil and groundwater may not need to meet promulgated regulatory cleanup levels. The demonstration required that a generally relied-upon condition be reached where "appropriate management can limit future environmental and health risks," or, in other words, the hydrogeologic environment was well known, no significant plume migration persisted, the source was either removed or isolated, asymptotic levels were reached, other remedial alternatives were fully evaluated, etc. Such demonstrations ignored natural attenuation.

One of the primary reasons for this perception is that many view natural attenuation as equivalent to no further action. The difference between the use of "natural attenuation" and "no further action" as a remedial strategy is that natural attenuation needs to be demonstrated, often implying that additional site characterization and development of a groundwater monitoring phase for an acceptable period of time are necessary. By demonstration, it is meant that there is evidence that contaminant reduction has been achieved, and that intrinsic bioremediation is likely responsible for that reduction. Natural attenuation if properly demonstrated increases the overall protection of the environment by either containment or destruction of contaminants. No further action, on the other hand, implies no additional investigation is required regardless of whether the contaminants of concern are degrading or migrating.

Natural attenuation should not be perceived as a permanent remedy or as a means to achieve certain cleanup levels, but rather as (1) an interim measure until future technologies are developed, (2) a managerial tool for reducing site risks, and (3) a bridge from active engineering (i.e., pump-and-treat, vapor extraction, etc.) to no further action.

No further action may be preferable to natural attenuation in certain instances. Very low risk situations may be better served by no further action since it eliminates the need for continued monitoring and further documentation. Sites with low levels of contaminants or nondiscernible plumes may be better candidates for no further action. Furthermore, very minor releases of hydrocarbons to the subsurface may not be sufficient to support bioremediation. Alternatively, sites with elevated levels of contaminants in nonpotable aquifers may be better addressed through conduct of a risk assessment.

13.7 EVALUATION OF PARAMETERS

There exist several methods or lines of evidence to demonstrate whether or not natural attenuation is occurring. Because sites can vary dramatically in their complexity and amount of effort required, the level of documentation that can be reasonably obtained will vary. What is important is that because many impacted sites will be or are difficult to clean up completely to the satisfaction of all parties, the use of natural attenuation will likely be at minimum considered at some stage of the project or remedial action. It is thus prudent to generate evidence and documentation to assess the suitability of natural attenuation as part of the site subsurface characterization process.

The fundamental parameters for evaluation of natural attenuation can be divided into four general groups (Table 13.3):

- Hydrogeologic factors
- Chemical characteristics
- Biological characteristics and
- Circumstantial factors

TABLE 13.3
Primary Parameters for Evaluation of Natural Attenuation as a Remedial Strategy

Parameter Group	Parameter for Evaluation
Hydrogeologic	Gradient
	Permeability
	Recharge
	Moisture content/field capacity
	Depth to impacted (contaminated) area
	Groundwater depth
	Dissolved oxygen
	Soil gas
	Extent of contamination/plume stability
Chemical	Hydrocarbon type
	Chromatographic evidence
	Hydrocarbon concentration
	Soil pH
	Nitrogen and phosphorus
Biological	Microscopic examination
	Plate counts
	Total heterotrophs
	Petroleum degraders
	Total organic carbon
Circumstantial	Time required for cleanup
	Age of release

13.7.1 Hydrogeologic Factors

Hydrogeologic factors for consideration include aquifer type, hydrogeologic gradient, permeability, recharge capability, depth to groundwater, moisture content/field capacity, dissolved oxygen (DO), depth to contamination, extent of contamination, and plume stability.

Aquifer Type: Aquifers or water-bearing zones characterized by relatively high quantities of naturally occurring carbon can favorably influence biodegradation of highly chlorinated solvents, whereas high quantities of anthropogenic carbon will influence biodegradation of highly chlorinated solvents, but at a faster rate. Low concentrations of carbon, but high concentrations of DO, may influence the biodegradation of vinyl chloride, but not favorably influence biodegradation of the highly chlorinated solvents, which require anaerobic conditions.

Groundwater Gradients: Groundwater gradients should be consistent seasonally, with moderate steepness such that a steady flow of electron acceptors is supplied to the plume, without being too steep to cause migration of a plume beyond the ability of microbes to contain it.

Permeability: The rate of microbial ability to metabolize hydrocarbons is limited primarily by the availability of electron acceptors and nutrient supply. In general, uniform soil zones of moderate to high permeability are more favorable for natural attenuation because of their ability to transmit fluids. Deposits that tend to channelize groundwater flow may be undesirable. Hydraulic conductivities $>10^{-9}$ are considered acceptable.

Recharge: Strong recharge of meteoric water to subsurface water-bearing zones provides an annual source of oxygen-enriched water and, in fertilized areas, also provides nutrients for microbial growth. In addition, for releases to the vadose zone, the downward infiltration of wetting fronts displaces oxygen-depleted and CO_2-rich soil gas with fresh, oxygenated gas. However, excessive ponding of water or heavy precipitation-type climates can cause water sealing to occur. This stops diffusion of oxygen and may cause microbial activity to become curtailed.

Moisture Content/Field Capacity: Within the vadose zone, moisture content is important since microbial growth is limited by excessively wet or dry soil. Moisture content, expressed as a percentage of the field (or holding) capacity, indicates the ratio of moisture to air in the soil. The recommended range for optimal growth is between 40 and 70%.

Depth to Contamination: In general, the shallower the release in the vadose zone, the more rapid the diffusion of soil gas, and the greater the indigenous microbial density.

Dissolved Oxygen: For groundwater plumes, the presence of DO is critical for maintaining aerobic conditions. A DO concentration of at least 1 to 2 mg/l is considered a minimum value to sustain a microbial population. Anaerobic conditions, if present, may cause growth of bacteria capable of degrading hydrocarbons using alternate electron acceptors such as iron (Fe^{3+}) and nitrate NO_3^+). DO should be a standard field parameter and be measured as part of any groundwater program where remediation may potentially be required. DO needs to be measured in a closed cell, and should not be measured on groundwater samples retrieved with bailers. Submersible low-flow or bladder-type pumps are recommended for this purpose.

Soil Gas: The minmum O_2 concentration that can support aerobic metabolism in unsaturated soil is approximately 1%. O_2 diffuses into soil because of pressure gradients, and CO_2 moves out of soil because of diffusivity gradients. Excess water restricts the movement of O_2 into and through the soil. A minimum air-filled pore volume of 10% is considered adequate for aeration. Soil gas surveys using a mobile geoprobe unit have become a valuable tool to demonstrate a zone of enhanced microbial metabolism in the subsurface.

Extent of Contamination/Plume Stability: Defining the extent of subsurface contamination, both in the vadose zone and in groundwater, is a fundamental objective of any investigation and lays the foundation for the natural attenuation alternative. An adequate groundwater monitoring well network is necessary to determine adequately the lateral and vertical extent of hydrocarbon-impacted soil and groundwater. If the release is relatively recent, it becomes important to demonstrate that natural attenuation is limiting plume migration. Older releases are anticipated to have generally stabilized. For groundwater, asymptotic concentration limits should be achieved. Regardless, it is critical to demonstrate plume stability through the use of soil borings, soil gas data, monitoring wells, among other techniques.

13.7.2 Chemical Characteristics

Chemical characteristics of importance to natural attenuation processes include petroleum hydrocarbon or organic compound type, concentration, pH, and nitrogen and phosphorus content.

Hydrocarbon Type: The light hydrocarbons generally degrade more readily than heavier hydrocarbons. A significant percentage of hydrocarbons found in light- to medium-range distillates (i.e., gasoline, jet fuels, diesel, etc.) are all amenable to biodegradation. The monoaromatic hydrocarbons (i.e., BTEX) are the most soluble and degrade the easiest, followed by the straight-chain alkanes. Compounds that are resistant to degradation include the isoprenoids (branched-chain alkanes) such as pristane and phytane hydrocarbons, and asphaltenes. For diesel-range hydrocarbons, a useful indication of degradation is the C_{17}/pritane ratio. Biodegradability of gasoline additives such as MTBE is considered poor and characterized by much lower natural attenuation potential relative to dissolved BTEX. This is evidenced by its persistence in groundwater, and the extensive size of MTBE-dissolved plumes reported in the literature.

Chlorinated solvents behave differently relative to fuel components in the subsurface, and may require both aerobic and anaerobic conditions at varying times during the natural attenuation process for complete biodegradation to carbon dioxide and chloride ions to occur. Hydrocarbon compounds associated with highly chlorinated (or oxidized) solvents such as TCE and PCE have a limited number of hydrogen atoms (or electrons) and thus do not serve as electron donors but rather electron acceptors. Biodegradation of the highly chlorinated solvents can be accomplished through reductive chlorination, as electron donors, or through cometabolism. With reductive chlorination, DO is the preferred electron acceptor from a thermodynamic perspective. Thus, oxygen if present will undergo reduction reactions prior to the less thermodynamically favorable chlorinated compounds. Highly chlorinated solvents thus biodegrade more favorably under anaerobic conditions, and do not typically proceed to biodegrade under aerobic conditions. Under anaerobic conditions, biodegradation is electron donor and carbon source limited. Other sources for electron donors such as naturally occurring organic carbon or anthropogenic carbon must be present, or biodegradation will be limited.

Vinyl chloride is the least-oxidized chlorinated aliphatic hydrocarbon, and may serve as an electron donor. A vinyl chloride molecule consists of more hydrogen atoms relative to chloride atoms (3 to 1); thus, reductive dechlorination is not favorable to biodegradation. However, under aerobic conditions, vinyl chloride can serve as an electron donor with oxygen as an electron acceptor.

Cometabolism refers to the degradation of the chlorinated solvent as a by-product of the degradation of other substrates by microorganisms, and does not benefit the microorganism. As the degree of dechlorination decreases, the cometabolism rates increase. Thus, less oxidized or chlorinated solvents such as chlorinated ethenes (excluding PCE) biodegrade more favorably under aerobic conditions.

Chromatographic Evidence: Sample chromatograms often contain evidence that natural attenuation has occurred. If chromatograms of fresh petroleum that was released can be obtained, then a solid comparison can be made between fresh and weathered samples relatively soon after being released in many circumstances.

Hydrocarbon Concentration: The concentration of hydrocarbons in both soil and groundwater is important to consider. High concentrations of BTEX (>10 mg/kg) and other volatile hydrocarbons have solvent-type properties that are toxic to membranes. High total petroleum hydrocarbon (TPH) concentrations require more oxygen and nutrients for degradation. TPH concentrations over 10,000 mg/kg can also be detrimental by inhibiting water and air flow by obstruction of soil pores, and may be biologically unavailable. For these reasons, active remediation of release source areas by excavation, LNAPL recovery, or vapor extraction is always recommended when practical. Conversely, low concentrations of hydrocarbons may not be enough to stimulate microbial growth.

Soil pH: Soil pH should be in the range of 6 to 8, to maintain cell turgidity and promote enzymatic reactions. Soil buffers, such as carbonate minerals, can be valuable in neutralizing acidic groundwaters as a result of high CO_2 concentrations because of microbiological activity.

Nitrogen and Phosphorus

Soil with carbon/nitrogen/phosphoros ratios of 100:10:1 are recommended for optimal bacterial growth. However, suboptimal ratios are not thought to be an impediment to intrinsic bioremediation as oxygen is typically the factor limiting microbial growth in the subsurface. Low concentrations of these nutrients are typically recycled and made available to new microbial colonies through growth, death, and decay of older colonies.

13.7.3 Biological Characteristics

Biological characteristics include microscopic and chromatographic examinations, plate counts, total heterotrophs, petroleum degrading, total hydrogen degrading, bacteria microscopic examination, and TOC.

Microscopic Examination: Microscopic examination involves a slurry derived from fresh soil that is examined under high magnification for bacterial type, cell health, protozoans, and approximate number of bacteria.

Plate Counts: Plate counts involve indirectly determining the number of organisms in soil (direct counting is difficult and tedious). Plate count techniques are selective and designed to detect microorganisms with particular growth forms. Soil suspensions undergo serial dilutions and each dilution is placed in a single substrate (agar). After incubation, each colony is counted, and each colony is associated with a single viable microbial unit capable of propagation, otherwise termed a colony-forming unit (CFU). Plate counts tend to underestimate the actual number of bacteria typically by an order of magnitude. The most common tests are for total heterotrophs and petroleum degraders.

Total Heterotrophs: This tests for all bacteria capable of using organic carbon as an energy source. The number obtained from this test is compared with background samples from similar soil types of similar depths, and compared with the number of hydrocarbon-degrading bacteria within the same sample.

Petroleum Degraders: This test cultures bacteria that are able to grow using petroleum as the sole carbon source. A popular variation of this test is called the

Sheen Screen method, which relies on the ability of microorganisms to emulsify oil. It is important that the petroleum type used as the carbon source resemble the petroleum encountered at the site. If the concentrations of the CFUs are a significant percentage of the total heterotrophs, and significantly elevated above background levels, then the indigenous microbial population has adapted to the release of hydrocarbons. CFUs greater than 1×10^{-5} are considered capable of supporting significant biodegradation.

Total Organic Carbon: Formations with a significant TOC content greater than the petroleum hydrocarbon content, as with peat-rich soil, can compete with the hydrocarbon-degrading microbes for oxygen and nutrients. A representative sample should thus be analyzed for TOC if organic matter is observed in soil samples or borings.

13.7.4 Circumstantial Factors

Circumstantial factors include the time required for cleanup and the age of release. The time required for natural attenuation to reduce contaminant levels effectively is difficult to predict, although certain models are available. Time should not, however, be a controlling factor. Sites, for example, that are situated in industrial areas, airfields, etc. are not likely to change land use in the foreseeable future and thus have the advantage of time. If it can be demonstrated that contaminant levels are declining, then absolute time for cleanup should not be critical providing the site is being monitored. However, natural attenuation may not be appropriate for sites that must be remediated in accordance with best-available technology to cleanup standards within compressed time frames at sensitive sites.

The relative age of the release can play an important role in the use of natural attenuation. Microbial populations need to adapt to the presence of petroleum hydrocarbons before they can metabolize the hydrocarbons and reproduce in concentrations above background levels. With older releases, if natural attenuation is occurring, associated groundwater plumes will, one hopes, have stabilized and have begun to shrink, and maximum seasonal water table fluctuations will have occurred and distributed LNAPL over a greater area. This maximum distribution results in less obstruction to groundwater flow through the zone of contamination, dissolves out the more toxic compounds, and brings electron acceptors (oxygen) to the microbes. It is thus much easier to gather evidence for natural attenuation in older releases than in more recent ones. Natural attenuation is more difficult to document in recent releases or when the release has not stabilized.

13.8 CASE HISTORIES

13.8.1 Natural Attenuation of Diesel-Range Hydrocarbons in Soil

The Washrack/Treatment Area (WTA) Superfund site at McChord Air Force Base is located south of Tacoma in western Washington State. The portion of the site selected for remedial action in the Record of Decision (ROD) is the location of a release of hydrocarbons to a shallow aquifer. LNAPL was discovered over approximately 8 acres of the site. The source was assumed to be excess fuels from an engine-testing house disposed of via infiltration trenches about 20 to 30 years ago.

The site is situated on a glacial outwash plain, and surface deposits consist of fill and undisturbed gravel and sand outwash containing minor silt and peat. Underlying the outwash at a depth of 20 or more feet is glacial till. Groundwater occurs under water table conditions within the highly permeable outwash deposits at a depth of 8 to 12 ft across the site.

Recovery of LNAPL via trenches was specified in the ROD but found to be unsuccessful due to its discontinuous and relatively thin apparent thickness of less than 1 in. A backhoe test pit program revealed the hydrocarbons to be distributed over a distinct 3-ft-thick smear zone. Observation of the hydrocarbons in the smear zone indicated significant weathering may have occurred as evidenced by a silver-gray staining of petroleum on the sand and gravel.

The fuel types that were released were never documented. Chromatography characterized the free-product as consisting of approximately 25% gasoline-range hydrocarbons (C_6 to C_{12}) and 75% light diesel-range hydrocarbons. Jet fuel was likely the source for the volatile hydrocarbon fraction. Samples indicated insignificant levels of BTEX in either the affected soil or the groundwater. It was suspected that BTEX concentrations were originally present at much higher concentrations (because of the JP-4 component), but have been essentially eliminated through volatilization, solubilization, and intrinsic bioremediation. Data from soil samples collected at 1-ft intervals across the smear zone indicated that the upper 2 ft contain $^1/_{10}$ the total hydrocarbons of the lower 1 ft. The boundary between the upper and lower smear zones was collated to the average seasonal low water table. In other words, soil in the upper smear zone was seasonally unsaturated and soil in the lower smear zone was seasonally saturated.

The dominantly unsaturated nature of the upper smear zone was also distinct in that it had a substantially lower percentage (10%) of total volatile (gasoline-range) vs. total recoverable (diesel-range) hydrocarbons. In contrast, the lower smear zone contained 25% total volatile hydrocarbons. The difference was thought to be due to the dominantly saturated nature of the lower smear zone, which limited the amount of volatilization that could occur. However, the dramatic difference in TPH concentration was thought to be largely attributable to intrinsic bioremediation.

A case to manage the release and avoid active remediation was made by documenting that natural attenuation had occurred and was likely to continue in the future. Data were compiled to support two conclusions: (1) that natural attenuation was a still active, ongoing process expected to continue in the future and (2) that hydrocarbon levels had decreased since the release. To establish that natural attenuation was occurring, as suspected based on visual observations previously discussed, soil samples were retrieved for microcosm testing (i.e., total heterotrophs and hydrocarbon degraders), nutrient and pH, microscopic examination, chromatographic examination, and total-organics evaluation. Microscopic examination indicated that the upper smear zone contained healthy-looking cells. Protozoans were observed in the sample, indicating a well-developed, mature microbial population. Petroleum degraders comprised 20% of the heterotrophs and were present at approx-

TABLE 13.4
Site Conditions Favorable for Natural Attenuation of Diesel-Range Hydrocarbons in Soil, McChord Air Force Base, Washington

Parameter for Evaluation	Site-Specific Conditions
Soil permeability	Unconsolidated outwash gravels and sands; $K = 2$ cm/s
Recharge	Annual rainfall +45 in.; annual water table fluctuation typically 2 to 3 ft
Groundwater depth	8 to 12 ft
Soil moisture content	Upper smear zone — 18% of field capacity
Soil pH	Neutral at 6.8
Soil gas O_2 and CO_2	Not measured
Hydrocarbon type	Jet fuel and light diesel mixture
Hydrocarbon concentration	Upper smear zone average 1000 mg/kg TPH[a]
	Lower smear zone average 10,000 mg/kg
Total heterotrophs	Upper smear zone 10^6 CFU/g soil
	Lower smear zone 10^5 CFU/g soil
	Background 10^2 CFU/g soil
Total hydrocarbons degraders[b]	Upper smear zone 10^5 CFU/g soil
	Lower smear zone 10^5 CFU/g soil
	Background 10^2 CFU/g soil
Microscopic examination of bacteria	Upper smear zone: bacteria intact, healthy appearance, many protozoans
	Lower smear zone: bacteria grainy, unhealthy appearance, no protozoans
Chromatographic interpretation	Loss of volatiles and degradation of *n*-alkanes in upper smear zone; lower smear zone relatively less degraded
Percent volatiles[c]	Upper smear zone: 10%
	Lower smear zone: 25%

[a] TPH measured as sum of total volatile and extractable hydrocarbons.
[b] Determined using Sheen Screen.
[c] Percent volatiles defined as (total volatile hydrocarbons/total extractable hydrocarbons) × 100.

imately 10^5 CFU/g. In contrast, background samples contained 10^5 CFU/g total heterotrophs of which fewer than 1 in 1000 were petroleum degraders. The carbon–nitrogen ratio was 20:1, which is considered favorable for the relatively slow growth of bacteria via intrinsic bioremediation. The chromatographs indicated a loss of volatiles and degradation of the semivolatile alkane portion of the diesel-range hydrocarbons in the upper smear zone as compared with the relatively unweathered lower smear zone. The specific site conditions favorable for supporting natural attenuation are summarized in Table 13.4.

It was decided that natural attenuation with long-term monitoring was appropriate for the site based on the low risk the affected soil presented. An explanation of significant differences (ESD) was prepared by the regulators to revise the remedy selected in the ROD. In the ESD, active remediation of the upper and lower smear zone was rejected by the regulators for several reasons:

- Natural attenuation had reduced overall hydrocarbon concentrations and would be expected to continue, although slowly, in the future.
- In-place hydrocarbon-affected soil in the lower smear zone did not contain a regulatory driver compound such as benzene that could be expected to be a continued source, adversely affecting groundwater.
- The cost for active remediation of the 8-acre site was judged impractical and not warranted considering the limited risk present.

13.8.2 Natural Attenuation of Asymptotic Gasoline-Range Hydrocarbons in Groundwater

The site is a carwash facility located in the city of Escondido, in southern California. A sequence of investigative and remedial activities has been performed since 1987 that established the presence, nature, and extent of subsurface hydrocarbons derived from former USTs. Since conducting these activities, the former USTs have been removed and the hydrocarbon-affected soil surrounding the USTs has been excavated. Additionally, all potentially recoverable LNAPL (i.e., gasoline) has been removed, leaving only residual hydrocarbon saturation (i.e., smear zone) in place at the water table. Subsequent pump-and-treat activities have reduced dissolved BTEX concentrations to asymptotic levels (Figure 13.9).

Deposits encountered beneath the site are unconsolidated, stratified, laterally discontinuous, Quaternary age alluvial sediments consisting predominantly of fine-

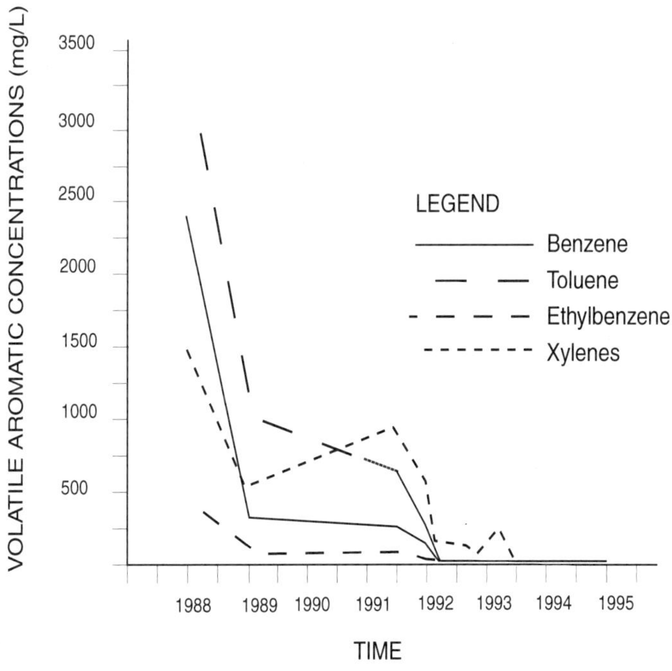

FIGURE 13.9 Graph showing asymptotic conditions.

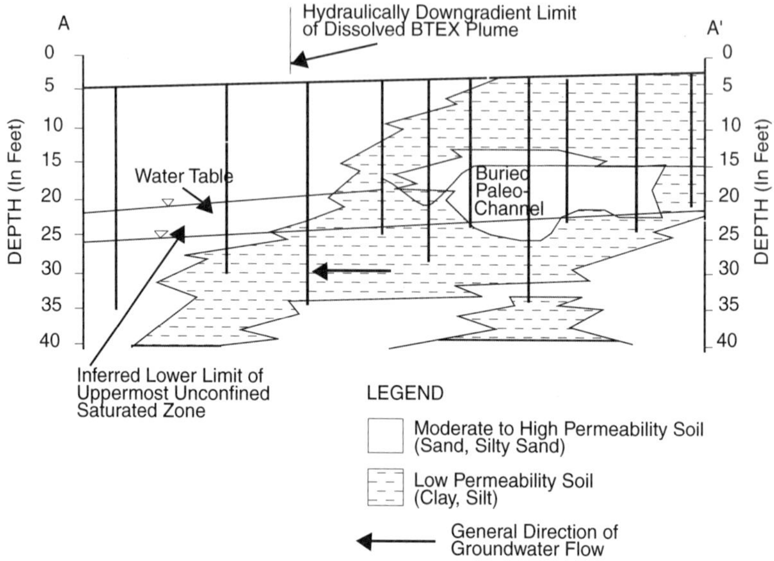

FIGURE 13.10 Hydrogeologic cross section showing stratigraphic control of distribution of LNAPL.

grained soil (i.e., clay and silt) interbedded with discrete, fine-to-coarse-grained silty-sand lenses that occupy small, discrete sand lenses or larger unfilled alluvial channels (Figure 13.10). A buried sandy alluvial channel exists directly beneath the site and coincides with the shallow saturated zone and capillary fringe impacted by the old release. This sandy lithsome was also reflected by anisotrophy in the configuration of the capture zone observed during conduct of pumping tests.

Groundwater occurs under unconfined (semiperched), heterogeneous, anisotrophic conditions at depths of about 12 to 15 ft below the ground surface. Groundwater flow is generally toward the northwest as gradients varying from 0.008 to 0.02 ft/ft. Asymptotic benzene concentrations range between 1.2 and 22 μg/l in monitoring and extraction wells, whereas toluene, ethylbenzene, and total xylenes concentrations have been reduced below their respective maximum contaminant level (MCL).

The lateral extent of the dissolved hydrocarbon plume has continued to decrease, although no significant decrease in dissolved benzene has occurred since January 1992, nor is anticipated to occur in the future as a result of continuation of conventional pump-and-treat (Figure 13.11). Aerobic biodegradation has also been documented to exist since 1992, and is ongoing as indicated by the inverse relationship between dissolved BTEX and DO. The dissolved plume is also situated hydraulically upgradient from a relatively larger LNAPL pool and dissolved hydrocarbon plume; separation of plumes has historically been maintained and has continued to be maintained.

To reach the ultimate objective of no further action, a remedial approach of natural attenuation was proposed. This approach, summarized in Table 13.5, is based upon the following criteria:

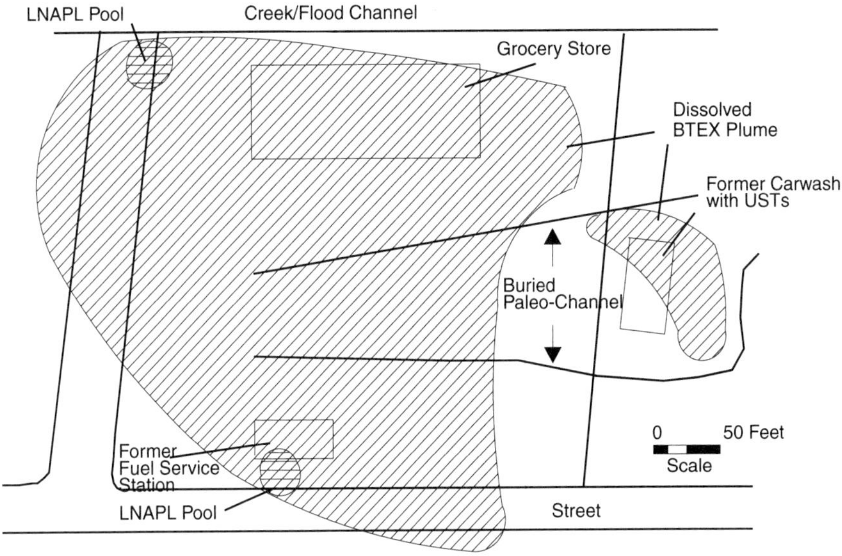

FIGURE 13.11 General areal extent of LNAPL and dissolved BTEX plume.

- Source recovery operations have been completed (i.e., hydrocarbon-affected soil and LNAPL have been removed).
- The dissolved BTEX plume is stable and will not migrate off site to potentially sensitive receptors.
- The adverse affected aquifer is a low-risk aquifer containing non-beneficial-use groundwater.
- Asymptotic concentrations of dissolved BTEX in groundwater have been attained and anticipated to decrease further with time, although not to below MCL for benzene.

TABLE 13.5
Site Conditions Favorable for Natural Attenuation of Gasoline Constituents in Groundwater

Parameter for Evaluation	Site-Specific Conditions
Source recovery	Former USTs and hydrocarbon-impacted soil removed
Groundwater depth	Shallow water table conditions at 12 to 15 ft below the ground surface
Gradient	Seasonally consistent
Extent of contamination	Extent of plume reduced via pump-and-treat
Plume stability	Plume is stable; asymptotic conditions reached
Dissolved oxygen	Aerobic biodegradation occurring as evidenced by inverse relationship between dissolved BTEX and DO
Continued groundwater monitoring	Adequate groundwater monitoring system in place
Water-bearing zone quality	Non-beneficial-use groundwater

- Asymptotic conditions in the treatment system influent are evident through detected low BTEX concentrations.
- Aerobic biodegradation of the BTEX plume is ongoing as indicated by the inverse relationship between dissolved BTEX and DO.
- Bioattenuation of the BTEX plume is anticipated to occur following system shutdown assuming a stoichiometric ratio of 1 ppm BTEX to 3 ppm DO.
- The site maintains an adequate groundwater-monitoring system capable of detecting any potential rebound of dissolved BTEX concentrations should they occur following system shutdown.

13.8.3 Natural Attenuation of Elevated Gasoline-Range Hydrocarbons in Groundwater

The sites are two utility maintenance yards that lie adjacent to one another. In both cases the former USTs were excavated and removed, and the hydrocarbon-affected soil surrounding the former USTs was excavated and reused/recycled via incorporation as an ingredient in the production of cold-mix asphalt.

Deposits encountered beneath the site are unconsolidated, stratified, laterally discontinuous, Quaternary age alluvial and shallow-lake sediments consisting predominantly of fine-grained soil (i.e., clay and silty), with occasional coarse-grained soil (i.e., sand and gravelly sand) from the ground surface to about 8.0 ft, and 15 to 20 ft below the ground surface. Groundwater is encountered under water table conditions at shallow depths of about 5 to 9 ft below the ground surface, and generally flows toward the northwest at a gradient of about 0.004 ft/ft.

Two small, isolated, localized, delineated, and naturally contained LNAPL pools of limited lateral extent and minimal actual thickness exist in the areas of the former USTs, although the associated dissolved BTEX plumes have commingled (Figure 13.12). BTEX concentrations range up to 14,000, 15,300, 1300, and 10,700 μg/l, respectively, in close proximity to the LNAPL pools.

Due to the characteristic presence of significantly elevated dissolved solids, groundwater is nonusable. A feasibility study was performed to demonstrate that limited natural attenuation of the dissolved BTEX plume was occurring, and could be enhanced. The shallow water-bearing zone was found to be deficient in nitrogen and phosphorus (nondetectable to 24 mg/l and 0.11 to 0.40 mg/l, respectively), resulting in low (10^2 CFU/ml or less) to moderate (10^3 to 10^4 CFU/ml) levels of heterotrophic- and gasoline-degrading bacteria. However, baseline temperature (25 to 28°C) and pH (6.80 to 7.44) levels were favorable. DO levels were generally aerobic (i.e., 1.0 to 3.6 mg/l), and could be supplemented with the addition of oxygen-releasing compounds (i.e., MgO_2), and possibly macronutrients (i.e., nitrogen and phosphorus).

Natural attenuation was favorably received and approved as an appropriate remedial alternative for these two sites. Periodic gauging and monitoring of key parameters will continue for a minimum period of 2 to 3 years to demonstrate continued stability and/or decline in dissolved BTEX prior to site closure.

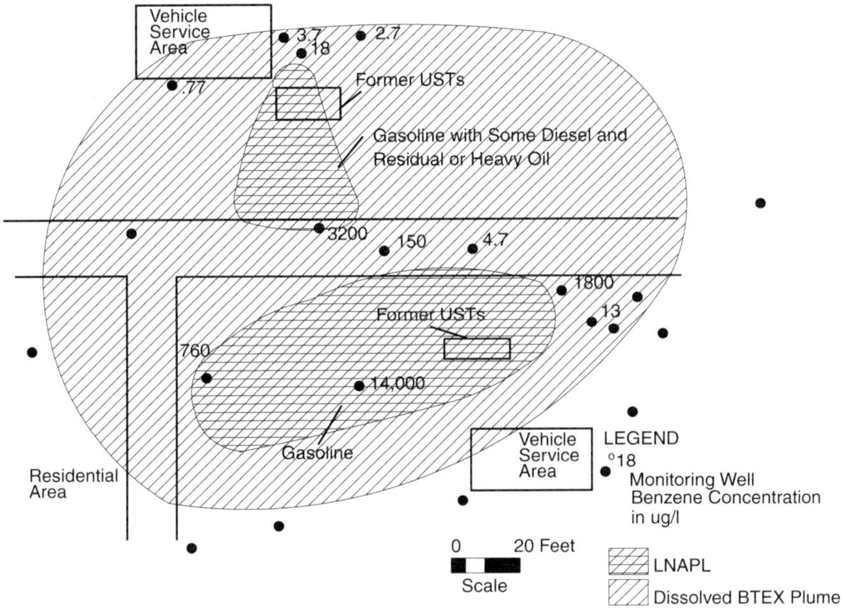

FIGURE 13.12 Areal extent of LNAPL and commingled dissolved hydrocarbon plumes at adjacent sites.

REFERENCES

Alleman, B. C. and Leeson, A. (editors), 1999, *Natural Attenuation of Chlorinated Solvents, Petroleum Hydrocarbons, and Other Organic Compounds*: Battelle Press, Columbus, OH, 402 pp.

American Petroleum Institute, 1987, *Field Study of Enhanced Subsurface Biodegradation of Hydrocarbons Using Hydrogen Peroxide as an Oxygen Source*: American Petroleum Institute, Publication No. 4448, Washington, D.C., 76 pp.

Andrews, T., Zervas, D., and Greenburg, R. S., 1997, Oxidizing Agent Can Finish Cleanup Where Other Systems Taper Off: *Soil and Groundwater Cleanup*, July, pp. 39–42.

Barker, J. F. and Patrick, G. C., 1985, Natural Attenuation of Aromatic Hydrocarbons in a Shallow Sand Aquifer: *Groundwater Monitoring Review*, Vol. 7, No. 1, pp. 64–71.

Barker, J. F. and Wilson, J. T., 1997, Natural Biological Attenuation of Aromatic Hydrocarbons under Anaerobic Conditions: In *Subsurface Remediation* (edited by C. H. Ward, J. A. Cherry, and M. R. Scalf), Ann Arbor Press, Chelsea, MI, pp. 289–300.

Borden, R. C., Daniel, R. A., LeBrun, L. E., and Davis, C. W., 1997, Intrinsic Biodegradation of MTBE and BTEX in a Gasoline-Contaminated Aquifer: *Water Resources Research*, Vol. 33, No. 5, pp. 1105–1115.

Brady, P. V., Brady, M. V., and Borns, D.J., 1998, *Natural Attenuation, CERCLA, RBCA's, and the Future of Environmental Remediation*: CRC Press/Lewis Publishers, Boca Raton, FL, 245 pp.

Brown, E. J. and Braddock, J. F., 1990, Sheen Screen, a Miniaturized Most Probable Number Method for Enumeration of Oil-Degrading Micro-Organisms: *Applied and Environmental Microbiology*, December, pp. 3895–3896.

Christensen, L. B. and Larson, T. M., 1993, Method for Determining the Age of Diesel Oil Spills in the Soil: *Groundwater Monitoring and Remediation*, Fall, Vol. 13, No. 4, pp. 142–149.

Christensen, T. H., Lyngkilde, J., Nielsen, P. H., Albrechtsen, H. J., and Heron, G., 1997, Natural Biological Attenuation: Integrative Transition Zones: In *Subsurface Remediation* (edited by C. H. Ward, J. A. Cherry, and M. R. Scalf), Ann Arbor Press, Chelsea, MI, pp. 329–341.

Davis, G. B., Johnston, C. D., Patterson, B. M., Bucer, C. and Dennett, M., 1988, Estimation of Biodegradation Rates Using Respiration Tests during In-Situ Bioremediation of Weathered Diesel NAPL: *Ground Water Monitoring & Remediation*, Spring, pp. 122–132.

Davis, J. W., Klier, N. J., and Carpenter, C. L., 1994, Natural Biological Attenuation of Benzene in Ground Water beneath a Manufacturing Facility: *Ground Water*, Vol. 32, No. 2, pp. 215–226.

Koenigsberg, S., 1997, Enhancing Bioremediation: *Environmental Protection*, February, pp. 19–22.

Lyman, W. J., Reidy, P. J., and Levy, B., 1992, *Mobility and Degradation of Organic Contaminants in Subsurface Environments*: C. K. Smoley, Inc., Chelsea, MI, 395 pp.

Madson, E. L., 1991, Determining in Situ Biodegradation: *Environmental Science and Technology*, Vol. 25, No. 10, pp. 1663–1673.

McAllister, P. M. and Chiang, C. Y., 1994, A Practical Approach to Evaluating Natural Attenuation of Contaminants in Ground Water: *Ground Water Monitoring and Remediation*, Spring, Vol. 14, No. 2, pp. 161–173.

National Research Council, 1993, *In-Situ Bioremediation — When Does It Work?* National Academy Press, Washington, D.C.

Odencrantz, J. E., 1998, Implications of MTBE for Intrinsic Remediation of Underground Fuel Tank Sites: *Remediation*, Vol. 8, No. 3, pp. 7–16.

Odermatt, J. R., 1997, Setting Rational Limits on Natural Attenuation: *Soil and Groundwater Cleanup*, February–March, pp. 30–33.

Ollila, P. W., 1996, Evaluating Natural Attenuation with Spreadsheet Analytical Fate and Transport Models: *Ground Water Monitoring & Remediation*, Fall, Vol. 16, No. 4, pp. 69–75.

Rifai, H. S., 1997, Natural Aerobic Biological Attenuation: In *Subsurface Remediation* (edited by C. H. Ward, J. A. Cherry, and M. R. Scalf), Ann Arbor Press, Chelsea, MI, pp. 411–427.

Salanitro, J. P., 1993, The Role of Bioattenuation in the Management of Aromatic Hydrocarbon Plumes in Aquifers: *Ground Water Monitoring and Remediation*, Fall, 1993, Vol. 13, No. 4, pp. 150–161.

Semprini, L., 1997, In Situ Transformation of Halogenated Aliphatic Compounds under Anaerobic Conditions: In *Subsurface Remediation* (edited by C. H. Ward, J. A. Cherry, and M. R. Scalf), Ann Arbor Press, Chelsea, MI, pp. 429–450.

Shaffner, R., Jr., Hawkins, E., and Wieck, J., 1996, A Look at Degradation of CAH's: *Soil and Groundwater Cleanup*, May, pp. 20–21.

Smith, C. S., 1998, Evaluating a Site for Natural Attenuation: EPA Region IV's New Guidance: *Remediation*, Vol. 8, No. 3, pp. 17–35.

Suflita, J. M. and Mormile, M. R., 1993, Anaerobic Biodegradation of Known and Potential Gasoline Oxygenates in the Terrestrial Subsurface: *Environmental Science and Technology*, Vol. 27, No. 5, p. 976.

Testa, S. M., 1994, *Geological Aspects of Hazardous Waste Management*: CRC Press/Lewis Publishers, Boca Raton, FL, 537 pp.

Testa, S. M. and Colligan T., 1994, Cleaning Up Sites Naturally: An Often Ignored Remedial Alternative: In *Proceedings of the National Water Well Association/American Petroleum Institute Conference on Petroleum Hydrocarbons and Organic Chemicals in Ground Water*, Houston, TX, 1994, pp. 289–303.

Tucker, W. A., 1992, UST Corrective Action: When Is It Over? *Groundwater Monitoring Review*, Vol. 12, No. 2, pp. 5–8.

U.S. Environmental Protection Agency, 1998, *EPA Seminars: Monitored Natural Attenuation for Ground Water*: Office of Research and Development, EPA/625/K-98/001.

Washburn, S. and Edelmann, K. G., 1999, Development of Risk-Based Remediation Strategies: *Practice Periodical of Hazardous, Toxic and Radioactive Waste Management*, April, pp. 77–82.

Weinstein, W., 1996, Mixing Up a Microbial Cocktail: *Environmental Protection*, May, pp. 24–25.

Wickramanayake, G. B. and Hinchee, R. E. (editors), 1998, *Natural Attenuation — Chlorinated and Recalcitrant Compounds*: Battelle Press, Columbus, OH, 379 pp.

Wiedemeier, T. H. and Pound, M. J., 1998, Natural Attenuation can be an Option for Chlorinated Solvents: *Soil and Groundwater Cleanup*: May, pp. 34–36.

Wiedemeier, T. H., Swanson, M. A., Wilson, J. T., Kampbell, D. H., Miller, R. N., and Hansen, J. E., 1996, Approximation of Biodegradation Rate Constants for Monoaromatic Hydrocarbons (BTEX) in Ground Water: *Ground Water Monitoring & Remediation*, Summer, Vol. 16, No. 3, pp. 186–194.

Wilson, J. T., 1997, Enhanced Biological Electron Acceptor H_2O_2: In *Subsurface Remediation* (edited by C. H. Ward, J. A. Cherry, and M. R. Scalf), Ann Arbor Press, Chelsea, MI, pp. 451–461.

Wisconsin Department of Natural Resources, 1994, *Naturally Occurring Biodegradation as a Remedial Action Option for Soil Contamination — Interim Guidance (Revised)*: Wisconsin Department of Natural Resources, Madison, WI, August, 16 pp.

Yeh, C. K. and Novak, J. T., 1994, Anaerobic Biodegradation of Gasoline Oxygenates in Soil: *Water Environment Research*, Vol. 66, No. 5, pp. 744–752.

APPENDIX A
API Gravity and Corresponding Weights and Pressure at 60°F

API Gravity	Weight lb/gal	lb/bbl	lb/ft³	psi/ft	ft/psi
10	8.328	349.8	62.37	0.4331	2.309
15	8.044	337.9	60.24	0.4183	2.391
20	7.778	326.7	58.25	0.4045	2.472
25	7.529	316.2	56.39	0.3916	2.554
26	7.481	314.2	56.03	0.3891	2.570
27	7.434	312.2	55.68	0.3866	2.587
28	7.387	310.3	55.33	0.3842	2.603
29	7.341	308.3	54.99	0.3818	2.619
30	7.296	306.4	54.65	0.3795	2.635
31	7.251	304.5	54.31	0.3771	2.652
32	7.206	302.7	53.97	0.3748	2.668
33	7.163	300.9	53.65	0.3726	2.684
34	7.119	298.8	53.33	0.3703	2.701
35	7.076	297.2	53.00	0.3860	2.717
36	7.034	295.4	52.69	0.3650	2.733
37	6.993	293.7	52.38	0.3637	2.750
38	6.951	291.9	52.07	0.3616	2.765
39	6.910	290.2	51.76	0.3594	2.782
40	6.870	288.6	51.46	0.3574	2.798
41	6.830	286.9	51.16	0.3553	2.815
42	6.790	285.2	50.86	0.3532	2.831
43	6.752	283.6	50.58	0.3512	2.847
44	6.713	282.0	50.29	0.3492	2.864
45	6.675	280.4	50.00	0.3472	2.880
46	6.637	378.8	49.72	0.3453	2.896
47	6.600	277.3	49.44	0.3433	2.913
48	6.563	275.7	49.17	0.3414	2.929
49	6.526	274.2	48.89	0.3395	2.945
50	6.490	272.7	48.62	0.3376	2.962
51	6.455	271.2	48.36	0.3358	2.978
52	6.420	269.7	48.09	0.3340	2.994
53	6.385	268.2	47.83	0.3321	3.011
54	6.350	266.7	47.58	0.3304	3.029
55	6.316	265.4	47.32	0.3286	3.043
56	6.283	264.0	47.07	0.3269	3.059
57	6.249	262.6	46.82	0.3251	3.076
58	6.216	261.2	46.57	0.3234	3.092
59	6.184	259.8	46.33	0.3217	3.108
60	6.151	258.3	46.09	0.3200	3.125
61	6.119	257.1	45.85	0.3184	3.141

APPENDIX A *(continued)*
API Gravity and Corresponding Weights and Pressure at 60°F

API Gravity	Weight			psi/ft	ft/psi
	lb/gal	lb/bbl	lb/ft³		
62	6.087	255.8	45.61	0.3167	3.158
63	6.056	254.5	45.37	0.3151	3.174
64	6.025	253.2	45.14	0.3135	3.190
65	5.994	251.9	44.91	0.3119	3.206
70	5.845	245.6	43.80	0.3041	3.288
75	5.703	239.7	42.74	0.2968	3.369
80	5.568	234.0	41.73	0.2897	3.452
85	5.440	228.6	40.76	0.2831	3.533
90	5.316	223.4	39.84	0.2767	3.614
95	5.199	218.5	38.96	0.2706	3.695
100	5.086	213.8	38.12	0.2647	3.778

APPENDIX B
Specific Gravity Corresponding to API Gravity

API	Decimal Parts of a Degree									
Gravity	0.0	0.1	0.2	0.3	0.4	0.5	0.6	0.7	0.8	0.9
10	1.0000	0.9993	0.9986	0.9979	0.9972	0.9965	0.9958	0.9951	0.9944	0.9937
11	0.9930	0.9923	0.9916	0.9909	0.9902	0.9895	0.9888	0.9881	0.9874	0.9868
12	0.9861	0.9854	0.9847	0.9840	0.9833	0.9826	0.9820	0.9813	0.9806	0.9799
13	0.9792	0.9786	0.9779	0.9772	0.9765	0.9759	0.9752	0.9745	0.9738	0.9732
14	0.9725	0.9718	0.9712	0.9705	0.9698	0.9692	0.9685	0.9679	0.9672	0.9665
15	0.9659	0.9652	0.9646	0.9639	0.9632	0.9626	0.9619	0.9613	0.9606	0.9600
16	0.9593	0.9587	0.9580	0.9574	0.9567	0.9561	0.9554	0.9548	0.9541	0.9535
17	0.9529	0.9522	0.9516	0.9509	0.9503	0.9497	0.9490	0.9484	0.9478	0.9471
18	0.9465	0.9459	0.9452	0.9446	0.9440	0.9433	0.9427	0.9421	0.9415	0.9408
19	0.9402	0.9396	0.9390	0.9383	0.9377	0.9371	0.9365	0.9358	0.9352	0.9346
20	0.9340	0.9334	0.9328	0.9321	0.9315	0.9309	0.9303	0.9297	0.9291	0.9285
21	0.9279	0.9273	0.9267	0.9260	0.9254	0.9248	0.9242	0.9236	0.9230	0.9224
22	0.9218	0.9212	0.9206	0.9200	0.9194	0.9188	0.9182	0.9176	0.9171	0.9165
23	0.9159	0.9153	0.9147	0.9141	0.9135	0.9129	0.9123	0.9117	0.9111	0.9106
24	0.9100	0.9094	0.9088	0.9082	0.9076	0.9071	0.9065	0.9059	0.9053	0.9047
25	0.9042	0.9036	0.9030	0.9024	0.9018	0.9013	0.9007	0.9001	0.8996	0.8990
26	0.8984	0.8978	0.8973	0.8967	0.8961	0.8956	0.8950	0.8944	0.8939	0.8933
27	0.8927	0.8922	0.8916	0.8911	0.8905	0.8899	0.8894	0.8888	0.8883	0.8877
28	0.8871	0.8866	0.8860	0.8855	0.8849	0.8844	0.8838	0.8833	0.8827	0.8822
29	0.8816	0.8811	0.8805	0.8800	0.8794	0.8789	0.8783	0.8778	0.8772	0.8767
30	0.8762	0.8756	0.8751	0.8745	0.8740	0.8735	0.8729	0.8724	0.8718	0.8713
31	0.8708	0.8702	0.8697	0.8692	0.8686	0.8681	0.8676	0.8670	0.8665	0.8660
32	0.8654	0.8649	0.8644	0.8639	0.8633	0.8628	0.8623	0.8618	0.8612	0.8607
33	0.8602	0.8597	0.8591	0.8586	0.8581	0.8576	0.8571	0.8565	0.8560	0.8555
34	0.8550	0.8545	0.8540	0.8534	0.8529	0.8524	0.8519	0.8514	0.8509	0.8504
35	0.8498	0.8493	0.8488	0.8483	0.8478	0.8473	0.8468	0.8463	0.8458	0.8453
36	0.8448	0.8443	0.8438	0.8433	0.8428	0.8423	0.8418	0.8413	0.8408	0.8403
37	0.8398	0.8393	0.8388	0.8383	0.8378	0.8373	0.8368	0.8363	0.8358	0.8353
38	0.8348	0.8343	0.8338	0.8333	0.8328	0.8324	0.8319	0.8314	0.8309	0.8304
39	0.8299	0.8294	0.8289	0.8285	0.8280	0.8275	0.8270	0.8265	0.8260	0.8256
40	0.8251	0.8246	0.8241	0.8236	0.8232	0.8227	0.8222	0.8217	0.8212	0.8208
41	0.8203	0.8198	0.8193	0.8189	0.8184	0.8179	0.8174	0.8170	0.8165	0.8160
42	0.8156	0.8151	0.8146	0.8142	0.8137	0.8132	0.8128	0.8123	0.8118	0.8114
43	0.8109	0.8104	0.8100	0.8095	0.8090	0.8086	0.8081	0.8076	0.8072	0.8067
44	0.8063	0.8058	0.8054	0.8049	0.8044	0.8040	0.8035	0.8031	0.8026	0.8022
45	0.8017	0.8012	0.8008	0.8003	0.7999	0.7994	0.7990	0.7985	0.7981	0.7976
46	0.7972	0.7967	0.7963	0.7958	0.7954	0.7949	0.7945	0.7941	0.7936	0.7932
47	0.7927	0.7923	0.7918	0.7914	0.7909	0.7905	0.7901	0.7896	0.7892	0.7887
48	0.7883	0.7879	0.7874	0.7870	0.7865	0.7861	0.7857	0.7852	0.7848	0.7844
49	0.7839	0.7835	0.7831	0.7826	0.7822	0.7818	0.7813	0.7809	0.7805	0.7800
50	0.7796	0.7792	0.7788	0.7783	0.7779	0.7775	0.7770	0.7766	0.7762	0.7758
51	0.7753	0.7749	0.7745	0.7741	0.7736	0.7732	0.7728	0.7727	0.7720	0.7715
52	0.7711	0.7707	0.7703	0.7699	0.7694	0.7690	0.7686	0.7682	0.7678	0.7674

APPENDIX B *(continued)*
Specific Gravity Corresponding to API Gravity

API Gravity	Decimal Parts of a Degree									
	0.0	0.1	0.2	0.3	0.4	0.5	0.6	0.7	0.8	0.9
53	0.7669	0.7665	0.7661	0.7657	0.7653	0.7649	0.7645	0.7640	0.7636	0.7632
54	0.7628	0.7624	0.7620	0.7616	0.7612	0.7608	0.7603	0.7599	0.7595	0.7591
55	0.7587	0.7583	0.7579	0.7575	0.7271	0.7567	0.7563	0.7559	0.7555	0.7551
56	0.7547	0.7543	0.7539	0.7535	0.7531	0.7527	0.7523	0.7519	0.7515	0.7511
57	0.7507	0.7503	0.7499	0.7495	0.7491	0.7487	0.7483	0.7479	0.7475	0.7471
58	0.7467	0.7463	0.7459	0.7455	0.7451	0.7447	0.7443	0.7440	0.7436	0.7432
59	0.7428	0.7424	0.7420	0.7416	0.7412	0.7408	0.7405	0.7401	0.7397	0.7393
60	0.7389	0.7385	0.7381	0.7377	0.7374	0.7370	0.7366	0.7362	0.7358	0.7351
61	0.7351	0.7347	0.7343	0.7339	0.7335	0.7332	0.7328	0.7324	0.7320	0.7316
62	0.7313	0.7309	0.7305	0.7301	0.7298	0.7294	0.7290	0.7286	0.7283	0.7279
63	0.7275	0.7271	0.7268	0.7264	0.7260	0.7256	0.7253	0.7249	0.7245	0.7242
64	0.7238	0.7234	0.7230	0.7227	0.7223	0.7219	0.7216	0.7212	0.7208	0.7205
65	0.7201	0.7197	0.7194	0.7190	0.7186	0.7183	0.7179	0.7175	0.7172	0.7168
66	0.7165	0.7161	0.7157	0.7154	0.7150	0.7146	0.7143	0.7139	0.7136	0.7132
67	0.7128	0.7125	0.7121	0.7118	0.7114	0.7111	0.7107	0.7103	0.7100	0.7096
68	0.7093	0.7089	0.7086	0.7082	0.7079	0.7075	0.7071	0.7068	0.7064	0.7061
69	0.7057	0.7054	0.7050	0.7047	0.7043	0.7040	0.7036	0.7033	0.7029	0.7026
70	0.7022	0.7019	0.7015	0.7012	0.7008	0.7005	0.7001	0.6998	0.6995	0.6991

Note: The specific gravity corresponding to an API gravity of 32.7, for example, will be found on line "32" in column "0.7."

APPENDIX C
Viscosity and Specific Gravity of Common Petroleum Products

Liquid	Specific Gravity	40°F	60°F	70°F	80°F	100°F	120°F	130°F	140°F	160°F
Crankcase Oils — Automobile Lubricating Oils										
SAE 10	.88–.935	1,500–2,400	600–900	—	300–400	170–220	110–130	—	75–90	60–65
SAE 20	.88–.935	2,400–9,000	900–3,000	—	400–1,100	220–550	130–280	—	90–170	65–110
SAE 30	.88–.935	9,000–14,000	3,000–4,400	—	1,100–1,800	550–800	280–400	—	170–240	110–150
SAE 40	.88–.935	14,000–19,000	4,400–6,000	—	1,800–2,400	800–1,100	400–550	—	240–320	150–200
SAE 50	.88–.935	19,000–45,000	6,000–10,000	—	2,400–4,000	1,100–1,800	550–850	—	320–480	200–280
SAE 60	.88–.935	45,000–60,000	10,000–17,000	—	4,000–6,000	1,800–2,500	850–1,200	—	480–580	280–380
SAE 70	.88–.935	60,000–120,000	17,000–45,000	—	6,000–10,000	2,500–4,000	1,200–1,800	—	580–900	380–500
Transmission Oils — Automobile Transmission Gear Lubricants										
SAE 90	.88–.935	14,000	5,500	—	2,200	1,100	650	—	380	240
SAE 140	.88–.935	35,000	12,000	—	5,000	2,200	1,200	—	650	400
SAE 250	.88–.935	160,000	50,000	—	18,000	7,000	3,300	—	1,700	1,000
Tars										
Coke oven–tar	1.12+	—	—	3,000–8,000	—	650–1,400	—	—	—	—
Gas house–tar	1.16–1.3	—	—	15,000–300,000	—	2,000–20,000	—	—	—	—
Crude Oils										
TX, OK	.81–.916	—	—	100–700	—	34–210	—	—	—	—
WY, MT	.86–.88	—	—	100–1,100	—	46–320	—	—	—	—
CA	.78–.92	—	—	100–4,500	—	34–700	—	—	—	—
PA	.8–.85	—	—	100–200	—	38–86	—	—	—	—

APPENDIX C *(continued)*
Viscosity and Specific Gravity of Common Petroleum Products

Liquid	Specific Gravity	40°F	60°F	70°F	80°F	100°F	120°F	130°F	140°F	160°F
Crude Oils *(continued)*										
Water	1.0	31.5	31.5	—	31.5	31.5	31.5	—	31.5	31.5
Gasoline	.68–.74	30	30	—	30	30	30	—	30	30
Jet Fuel	.74–.85	35	35	—	35	35	35	—	35	35
Kerosene	.78–.82	42	38	—	34	33	31	—	30	30
Fuel Oil and Diesel Oil										
No. 1 fuel oil	.82–.95	40	38	—	35	33	31	—	30	30
No. 2 fuel oil	.82–.95	70	50	—	45	40	—	—	—	—
No. 3 fuel oil	.82–.95	90	68	—	53	45	40	—	—	—
No. 5A fuel oil	.82–.95	1,000	400	—	200	100	75	—	60	40
No. 5B fuel oil	.82–.95	1,300	600	—	490	400	330	—	290	240
No. 6 fuel oil	.82–.95	—	70,000	—	20,000	90,000	1,900	—	900	500
No. 2D diesel fuel oil	.82–.95	100	68	—	53	45	40	—	36	35
No. 3D diesel fuel oil	.82–.95	200	120	—	80	60	50	—	44	40
No. 4D diesel fuel oil	.82–.95	1,600	600	—	280	140	90	—	68	54
No. 5D diesel fuel oil	.82–.95	15,000	5,000	—	2,000	900	400	—	260	160
Navy No. 1 fuel oil	.989	4,000	1,100	—	600	380	200	—	170	90
Navy No. 2 fuel oil	1.0	—	24,000	—	8,700	3,500	1,500	—	900	480
Gas	.887	180	90	—	60	50	45	—	—	—
Insulating	—	350	150	—	90	65	50	—	45	40

Lard	.912–.925	1,100	600	—	380	287	180	—	140	90
Linseed	.925–.939	1,500	500	—	250	143	110	—	85	70
Raw Menhadden	.933	1,500	500	—	250	140	110	—	80	70
Neats foot	.917	—	1,000	—	430	230	160	—	100	80
Olive	.912–.918	1,500	550	—	320	200	150	—	100	80
Palm	.924	1,700	700	—	380	221	160	—	120	90
Peanut	.920	1,200	500	—	300	195	150	—	100	80
Quenching	—	2,400	900	—	450	250	180	—	130	90
Rape seed	.919	2,400	900	—	450	250	180	—	130	90
Rosin	.980	28,000	7,800	—	3,200	1,500	900	—	500	300
Rosin (wood)	1.09	Extremely viscous								
Sesame	.923	1,100	500	—	290	184	130	—	90	60
Soya bean	.927–.98	1,200	475	—	270	165	120	—	80	70
Sperm	.883	360	250	—	170	110	90	—	70	60
Turbine (light)	.91	500	350	—	230	150	—	—	—	—
Turbine (heavy)	.91	3,000	1,400	—	700	330	200	—	150	100
Whale	.925	900	450	—	275	170	140	—	100	80

Source: A.F.L. Industries Catalog, West Chicago, IL.

APPENDIX D
Viscosity Conversion Table

SSU Seconds Saybolt Universal	SSF Seconds Saybolt Furel	Kinematic Viscosity Centistokes (cSt)	Seconds Redwood (standard)
31	—	1.00	29
35	—	2.56	32.1
40	—	4.30	36.2
50	—	7.40	44.3
60	—	10.20	52.3
70	12.95	12.83	60.9
80	13.70	15.35	69.2
90	14.44	17.80	77.6
100	15.24	20.20	85.6
150	19.30	31.80	128
200	23.5	43.10	170
250	28.0	54.30	212
300	32.5	65.40	254
400	41.9	87.60	338
500	51.6	110.0	423
600	61.4	132	508
700	71.1	154	592
800	81.0	176	677
900	91.0	198	762
1,000	100.7	220	896
1,500	150	330	1,270
2,000	200	440	1,690
2,500	250	550	2,120
3,000	300	660	2,540
4,000	400	880	3,380
5,000	500	1,100	4,230
6,000	600	1,320	5,080
7,000	700	1,540	5,920
8,000	800	1,760	6,770
9,000	900	1,980	7,620
10,000	1,000	2,200	8,460
15,000	1,500	3,300	13,700
20,000	2,000	4,400	18,400

Notes: Kinematic viscosity (stokes) = absolute viscosity (poises)/specific gravity; 1 centistoke = stoke/100; 1 centipoise = poise/100; 1 stoke = 100 centistokes; 1 poise = 100 centipoises. The term *centipoises* is referred to commonly as a measure of kinematic viscosity. Convert centipoises to centistokes by dividing by the specific gravity of the solution at the operating temperature. Plotting viscosity: If viscosity is known at any two temperatures, the viscosity at other temperatures can be obtained by plotting the viscosity against temperature in °F on log paper. The points lie in a straight line.

Source: A.F.L. Industries Catalog, West Chicago, IL.

Index

A

B

C

D

E

F

G

H

I

J

K

L

M

N

O

P

Q

R

S

T

U

V

W

Z